Photonic Signal Processing

Techniques and Applications

OPTICAL SCIENCE AND ENGINEERING

Founding Editor
Brian J. Thompson
University of Rochester
Rochester, New York

1. Electron and Ion Microscopy and Microanalysis: Principles and Applications, *Lawrence E. Murr*
2. Acousto-Optic Signal Processing: Theory and Implementation, *edited by Norman J. Berg and John N. Lee*
3. Electro-Optic and Acousto-Optic Scanning and Deflection, *Milton Gottlieb, Clive L. M. Ireland, and John Martin Ley*
4. Single-Mode Fiber Optics: Principles and Applications, *Luc B. Jeunhomme*
5. Pulse Code Formats for Fiber Optical Data Communication: Basic Principles and Applications, *David J. Morris*
6. Optical Materials: An Introduction to Selection and Application, *Solomon Musikant*
7. Infrared Methods for Gaseous Measurements: Theory and Practice, *edited by Joda Wormhoudt*
8. Laser Beam Scanning: Opto-Mechanical Devices, Systems, and Data Storage Optics, *edited by Gerald F. Marshall*
9. Opto-Mechanical Systems Design, *Paul R. Yoder, Jr.*
10. Optical Fiber Splices and Connectors: Theory and Methods, *Calvin M. Miller with Stephen C. Mettler and Ian A. White*
11. Laser Spectroscopy and Its Applications, *edited by Leon J. Radziemski, Richard W. Solarz, and Jeffrey A. Paisner*
12. Infrared Optoelectronics: Devices and Applications, *William Nunley and J. Scott Bechtel*
13. Integrated Optical Circuits and Components: Design and Applications, *edited by Lynn D. Hutcheson*
14. Handbook of Molecular Lasers, *edited by Peter K. Cheo*
15. Handbook of Optical Fibers and Cables, *Hiroshi Murata*
16. Acousto-Optics, *Adrian Korpel*
17. Procedures in Applied Optics, *John Strong*
18. Handbook of Solid-State Lasers, *edited by Peter K. Cheo*
19. Optical Computing: Digital and Symbolic, *edited by Raymond Arrathoon*
20. Laser Applications in Physical Chemistry, *edited by D. K. Evans*
21. Laser-Induced Plasmas and Applications, *edited by Leon J. Radziemski and David A. Cremers*

22. Infrared Technology Fundamentals, *Irving J. Spiro and Monroe Schlessinger*
23. Single-Mode Fiber Optics: Principles and Applications, Second Edition, Revised and Expanded, *Luc B. Jeunhomme*
24. Image Analysis Applications, *edited by Rangachar Kasturi and Mohan M. Trivedi*
25. Photoconductivity: Art, Science, and Technology, *N. V. Joshi*
26. Principles of Optical Circuit Engineering, *Mark A. Mentzer*
27. Lens Design, *Milton Laikin*
28. Optical Components, Systems, and Measurement Techniques, *Rajpal S. Sirohi and M. P. Kothiyal*
29. Electron and Ion Microscopy and Microanalysis: Principles and Applications, Second Edition, Revised and Expanded, *Lawrence E. Murr*
30. Handbook of Infrared Optical Materials, *edited by Paul Klocek*
31. Optical Scanning, *edited by Gerald F. Marshall*
32. Polymers for Lightwave and Integrated Optics: Technology and Applications, *edited by Lawrence A. Hornak*
33. Electro-Optical Displays, *edited by Mohammad A. Karim*
34. Mathematical Morphology in Image Processing, *edited by Edward R. Dougherty*
35. Opto-Mechanical Systems Design: Second Edition, Revised and Expanded, *Paul R. Yoder, Jr.*
36. Polarized Light: Fundamentals and Applications, *Edward Collett*
37. Rare Earth Doped Fiber Lasers and Amplifiers, *edited by Michel J. F. Digonnet*
38. Speckle Metrology, *edited by Rajpal S. Sirohi*
39. Organic Photoreceptors for Imaging Systems, *Paul M. Borsenberger and David S. Weiss*
40. Photonic Switching and Interconnects, *edited by Abdellatif Marrakchi*
41. Design and Fabrication of Acousto-Optic Devices, *edited by Akis P. Goutzoulis and Dennis R. Pape*
42. Digital Image Processing Methods, *edited by Edward R. Dougherty*
43. Visual Science and Engineering: Models and Applications, *edited by D. H. Kelly*
44. Handbook of Lens Design, *Daniel Malacara and Zacarias Malacara*
45. Photonic Devices and Systems, *edited by Robert G. Hunsberger*
46. Infrared Technology Fundamentals: Second Edition, Revised and Expanded, *edited by Monroe Schlessinger*
47. Spatial Light Modulator Technology: Materials, Devices, and Applications, *edited by Uzi Efron*
48. Lens Design: Second Edition, Revised and Expanded, *Milton Laikin*
49. Thin Films for Optical Systems, *edited by Francoise R. Flory*
50. Tunable Laser Applications, *edited by F. J. Duarte*
51. Acousto-Optic Signal Processing: Theory and Implementation, Second Edition, *edited by Norman J. Berg and John M. Pellegrino*

52. Handbook of Nonlinear Optics, *Richard L. Sutherland*
53. Handbook of Optical Fibers and Cables: Second Edition, *Hiroshi Murata*
54. Optical Storage and Retrieval: Memory, Neural Networks, and Fractals, *edited by Francis T. S. Yu and Suganda Jutamulia*
55. Devices for Optoelectronics, *Wallace B. Leigh*
56. Practical Design and Production of Optical Thin Films, *Ronald R. Willey*
57. Acousto-Optics: Second Edition, *Adrian Korpel*
58. Diffraction Gratings and Applications, *Erwin G. Loewen and Evgeny Popov*
59. Organic Photoreceptors for Xerography, *Paul M. Borsenberger and David S. Weiss*
60. Characterization Techniques and Tabulations for Organic Nonlinear Optical Materials, *edited by Mark G. Kuzyk and Carl W. Dirk*
61. Interferogram Analysis for Optical Testing, *Daniel Malacara, Manuel Servin, and Zacarias Malacara*
62. Computational Modeling of Vision: The Role of Combination, *William R. Uttal, Ramakrishna Kakarala, Spiram Dayanand, Thomas Shepherd, Jagadeesh Kalki, Charles F. Lunskis, Jr., and Ning Liu*
63. Microoptics Technology: Fabrication and Applications of Lens Arrays and Devices, *Nicholas Borrelli*
64. Visual Information Representation, Communication, and Image Processing, *edited by Chang Wen Chen and Ya-Qin Zhang*
65. Optical Methods of Measurement, *Rajpal S. Sirohi and F. S. Chau*
66. Integrated Optical Circuits and Components: Design and Applications, *edited by Edmond J. Murphy*
67. Adaptive Optics Engineering Handbook, *edited by Robert K. Tyson*
68. Entropy and Information Optics, *Francis T. S. Yu*
69. Computational Methods for Electromagnetic and Optical Systems, *John M. Jarem and Partha P. Banerjee*
70. Laser Beam Shaping, *Fred M. Dickey and Scott C. Holswade*
71. Rare-Earth-Doped Fiber Lasers and Amplifiers: Second Edition, Revised and Expanded, *edited by Michel J. F. Digonnet*
72. Lens Design: Third Edition, Revised and Expanded, *Milton Laikin*
73. Handbook of Optical Engineering, *edited by Daniel Malacara and Brian J. Thompson*
74. Handbook of Imaging Materials: Second Edition, Revised and Expanded, *edited by Arthur S. Diamond and David S. Weiss*
75. Handbook of Image Quality: Characterization and Prediction, *Brian W. Keelan*
76. Fiber Optic Sensors, *edited by Francis T. S. Yu and Shizhuo Yin*
77. Optical Switching/Networking and Computing for Multimedia Systems, *edited by Mohsen Guizani and Abdella Battou*
78. Image Recognition and Classification: Algorithms, Systems, and Applications, *edited by Bahram Javidi*

79. Practical Design and Production of Optical Thin Films: Second Edition, Revised and Expanded, *Ronald R. Willey*
80. Ultrafast Lasers: Technology and Applications, *edited by Martin E. Fermann, Almantas Galvanauskas, and Gregg Sucha*
81. Light Propagation in Periodic Media: Differential Theory and Design, *Michel Nevière and Evgeny Popov*
82. Handbook of Nonlinear Optics, Second Edition, Revised and Expanded, *Richard L. Sutherland*
83. Polarized Light: Second Edition, Revised and Expanded, *Dennis Goldstein*
84. Optical Remote Sensing: Science and Technology, *Walter Egan*
85. Handbook of Optical Design: Second Edition, *Daniel Malacara and Zacarias Malacara*
86. Nonlinear Optics: Theory, Numerical Modeling, and Applications, *Partha P. Banerjee*
87. Semiconductor and Metal Nanocrystals: Synthesis and Electronic and Optical Properties, edited by *Victor I. Klimov*
88. High-Performance Backbone Network Technology, *edited by Naoaki Yamanaka*
89. Semiconductor Laser Fundamentals, *Toshiaki Suhara*
90. Handbook of Optical and Laser Scanning, *edited by Gerald F. Marshall*
91. Organic Light-Emitting Diodes: Principles, Characteristics, and Processes, *Jan Kalinowski*
92. Micro-Optomechatronics, *Hiroshi Hosaka, Yoshitada Katagiri, Terunao Hirota, and Kiyoshi Itao*
93. Microoptics Technology: Second Edition, *Nicholas F. Borrelli*
94. Organic Electroluminescence, *edited by Zakya Kafafi*
95. Engineering Thin Films and Nanostructures with Ion Beams, *Emile Knystautas*
96. Interferogram Analysis for Optical Testing, Second Edition, *Daniel Malacara, Manuel Sercin, and Zacarias Malacara*
97. Laser Remote Sensing, *edited by Takashi Fujii and Tetsuo Fukuchi*
98. Passive Micro-Optical Alignment Methods, *edited by Robert A. Boudreau and Sharon M. Boudreau*
99. Organic Photovoltaics: Mechanism, Materials, and Devices, *edited by Sam-Shajing Sun and Niyazi Serdar Saracftci*
100. Handbook of Optical Interconnects, *edited by Shigeru Kawai*
101. GMPLS Technologies: Broadband Backbone Networks and Systems, *Naoaki Yamanaka, Kohei Shiomoto, and Eiji Oki*
102. Laser Beam Shaping Applications, *edited by Fred M. Dickey, Scott C. Holswade and David L. Shealy*
103. Electromagnetic Theory and Applications for Photonic Crystals, *Kiyotoshi Yasumoto*
104. Physics of Optoelectronics, *Michael A. Parker*
105. Opto-Mechanical Systems Design: Third Edition, *Paul R. Yoder, Jr.*
106. Color Desktop Printer Technology, *edited by Mitchell Rosen and Noboru Ohta*
107. Laser Safety Management, *Ken Barat*

108. Optics in Magnetic Multilayers and Nanostructures, *Štefan Višňovský*
109. Optical Inspection of Microsystems, *edited by Wolfgang Osten*
110. Applied Microphotonics, *edited by Wes R. Jamroz, Roman Kruzelecky, and Emile I. Haddad*
111. Organic Light-Emitting Materials and Devices, *edited by Zhigang Li and Hong Meng*
112. Silicon Nanoelectronics, *edited by Shunri Oda and David Ferry*
113. Image Sensors and Signal Processor for Digital Still Cameras, *Junichi Nakamura*
114. Encyclopedic Handbook of Integrated Circuits, *edited by Kenichi Iga and Yasuo Kokubun*
115. Quantum Communications and Cryptography, *edited by Alexander V. Sergienko*
116. Optical Code Division Multiple Access: Fundamentals and Applications, *edited by Paul R. Prucnal*
117. Polymer Fiber Optics: Materials, Physics, and Applications, *Mark G. Kuzyk*
118. Smart Biosensor Technology, *edited by George K. Knopf and Amarjeet S. Bassi*
119. Solid-State Lasers and Applications, *edited by Alphan Sennaroglu*
120. Optical Waveguides: From Theory to Applied Technologies, *edited by Maria L. Calvo and Vasudevan Lakshiminarayanan*
121. Gas Lasers, *edited by Masamori Endo and Robert F. Walker*
122. Lens Design, Fourth Edition, *Milton Laikin*
123. Photonics: Principles and Practices, *Abdul Al-Azzawi*
124. Microwave Photonics, *edited by Chi H. Lee*
125. Physical Properties and Data of Optical Materials, *Moriaki Wakaki, Keiei Kudo, and Takehisa Shibuya*
126. Microlithography: Science and Technology, Second Edition, *edited by Kazuaki Suzuki and Bruce W. Smith*
127. Coarse Wavelength Division Multiplexing: Technologies and Applications, *edited by Hans Joerg Thiele and Marcus Nebeling*
128. Organic Field-Effect Transistors, *Zhenan Bao and Jason Locklin*
129. Smart CMOS Image Sensors and Applications, *Jun Ohta*
130. Photonic Signal Processing: Techniques and Applications, *Le Nguyen Binh*

Photonic Signal Processing

Techniques and Applications

Le Nguyen Binh

CRC Press is an imprint of the
Taylor & Francis Group, an **informa** business

First published in paperback 2024

Published 2008
CRC Press
2385 NW Executive Center Drive, Suite 320, Boca Raton FL 33431

and by CRC Press
4 Park Square, Milton Park, Abingdon, Oxon, OX14 4RN

CRC Press is an imprint of Taylor & Francis Group, LLC

Publisher's Note
The publisher has gone to great lengths to ensure the quality of this reprint but points out that some imperfections in the original copies may be apparent.

ISBN 13: 978-0-8493-3762-8 (hbk)
ISBN 13: 978-1-138-74684-8 (pbk)
ISBN 13: 978-1-315-22119-9 (ebk)

DOI: 10.1201/9781420019520

To the memory of my father and to my mother

To Phuong and Lam

Contents

Chapter 2

Chapter 3

Chapter 4

Chapter 6

Preface

Photonic signal processing (PSP) is very attractive because it has the potential to overcome the electronic limits for processing ultra-wideband signals. Furthermore, PSP provides signal conditioning that can be integrated in-line with fiber optic systems. Several techniques have been proposed and reported for the implementation of the photonic counterparts of conventional electronic signal processing systems. Signal processing in the photonic domain offers significant improvement of signal quality.

This book was written to address the emerging techniques of processing and manipulating of signals propagating in an optical domain. This means the pulses or signal envelopes are complex or modulating the optical carriers. Naturally, the applications of such processing techniques in photonics are essential to illustrate their usefulness. The change of the transmission cable from coaxial and metallic waveguide to flexible optically transparent glass fiber has allowed the processing of ultra-high-speed signals in the microwave and millimeter-wave domain to the photonic domain in which the delay line can be implemented in the fiber lines, which are very lightweight and space efficient.

Chapter 1 gives a brief historical perspective of PSP and an introduction to a number of photonic components essential for photonic processing systems, including, but not exclusively, optical amplification devices, optical fibers, and optical modulators. Chapter 2 discusses the representation of photonic circuits using signal-flow graph techniques, which have been employed in electrical circuits since the 1960s; however, they have been adapted for photonic domains in which the transmittance of a photonic subsystem determines the optical transfer function of a photonic subsystem. The coherent and incoherent aspects of photonic circuits are important in terms of whether the field or the power of the lightwaves should be used as well as because the length of the photonic processors must be less than that of the coherence length.

Photonic signal processors such as differentiators and integrators are described in Chapter 3. Their applications in the generation of solitons and in optically amplified fiber transmission systems are described in Chapter 4. Chapter 5 illustrates the compensation of dispersion using photonic processors. Chapter 6 explains the design of optical filters using photonic processing techniques.

Many individuals have contributed, either directly or indirectly, to this book. Thanks are due to Associate Professor John Ngo of Nanyang Technological University (NTU), Singapore, for the work that he conducted during his time as a Ph.D. candidate at Monash University; Associate Professor Shum Ping of the School of Electrical and Electronic Engineering of NTU; Dr. Wenn Jing Lai; and Steven Luk. I appreciate the help of my CRC Press/Taylor & Francis editor, Taisuke Soda, and the encouragement of his colleague, Theresa Delforn. Last but not least, I thank my wife Phuong and my son Lam for their support and for putting up with my busy writing schedule of the book and my daily teaching and research commitments at the university, which certainly took away a large amount of time that we could have spent together.

Le Nguyen Binh
Melbourne, Australia

Author

Le Nguyen Binh, Ph.D., received a B.E.(Hons.) and a Ph.D. in electronic engineering and integrated photonics in 1975 and 1980, respectively, both from the University of Western Australia, Nedlands, Western Australia. He has worked extensively in the fields of optical communication systems and networks and photonic signal processing in academic and industrial environments, including the advanced technology research laboratories of Siemens and Nortel Networks. He is currently director of the Center for Telecommunications and Information Engineering and a reader in the Department of Electrical and Computer Systems Engineering of Monash University. He has made outstanding contributions in the fields of integrated optics, optical communication systems and networks, and photonic signal processing, and has published more than 150 papers in leading journals and refereed conferences. He holds numerous patents related to photonics and optical communications.

1 Principal Photonic Devices for Processing

This opening chapter provides the fundamentals of some principal photonic devices that are essential for processing of ultrafast electrical and electronic signals in the photonic domain, including optical communications and microwave photonics. It describes the basic structures, the benefits, and the needs of photonic signal processing for the advancement of such systems. The device structures, fabrication, and characteristics of operating functions of the optical amplifiers, optical filters, and optical modulators are examined. The transmittance of photonic components can be represented, in Chapter 2, in the form of signal flow and blocks of photonic signal subsystems.

1.1 OPTICAL FIBER COMMUNICATIONS

The proposed dielectric waveguides for guiding lightwaves by Kao and Hockham in 1966 [1] have revolutionized the transmission of broadband signals and ultrahigh capacity over ultra-long global telecommunication systems and networks. In the 1970s, the reduction of the fiber losses over the visible and infrared spectral regions was extensively investigated. The demonstration of the pure and low loss silica-based glass fiber was achieved in the early 1970s. Since then, the reduction of the fiber dispersion with the theoretical development and manufacturing of weakly guiding single-mode fibers were extensively conducted. The research, development, and commercialization of optical fiber communication systems were then progressed with practical demonstrations of higher and higher bit rates and longer and longer transmission distance. Significantly, the installation of optical fiber systems was completed in 1978 [2]. Since then, fiber systems have been installed throughout the world and interconnecting all continents of the globe with terrestrial and undersea systems.

The primary reason for such exciting research and development is that the frequency of the lightwave is in the order of a few hundred terahertz and the low loss windows of glass fiber are sufficiently wide so that several tens of terabits per second capacity of information can be achieved over ultra-long distance, whereas the carrier frequencies of the microwave and millimeter-wave are in the tens of gigahertz.

Indeed in the 1980s, the transmission speed and distance were limited because of the ability of regeneration of information signals in the optical or photonic domain. Data signals were received and recovered in the electronic domain; then the lightwave sources for retransmission were modulated. The distance between these regenerators was limited to 40 km for installed fiber transmission systems. Furthermore, the dispersion-limited distance was longer than that of the attenuation-limited distance, and no compensation of dispersion was required.

The repeaterless distance could then be extended to another 20 km with the use of coherent detection techniques in the mid-1980s; heterodyne and homodyne techniques were extensively investigated. However, the complexity and demands on the extreme narrow line width of the lightwave source at the transmitter and the local oscillator for mixing at the receiver restrict the applications of coherent processing.

This attenuation was eventually overcome with the invention of the optical amplifiers in 1987 [3,4] using Nd or Er doping in silica fiber. Optical gain of 20–30 dB can be easily obtained. Hence, the only major issue was the dispersion of lightwave signals in long-haul transmission. This leads to extensive search for the management methods for compensation of dispersion. The simplest method can be the use of dispersion compensating fibers inserted in each transmission span, hence the phase reversal of the lightwave and compensation in the photonic domain. This is a form of photonic signal processing.

This leads to the development of high and ultrahigh bit rate transmission system. The bit rate has reached 10, 40, and 80 Gb/s per wavelength channels in the late 1990s. Presently, the transmission systems of several wavelength channels each carrying 40 Gb/s are practically proven and installed in a number of routes around the world.

Recently, 10 Gb/s Ethernet has been standardized and the 100 Gb/s rate is expected to be implemented in the future. Furthermore, soliton systems are also attracting much attention with bit rate reaching 160 or 320 Gb/s with an ultrashort pulse generating from mode-locked (ML) lasers [5]. At this ultrafast speed, the processing of signals in the electronic domain is no longer possible and thus signal processors in the photonic domain are expected to play a major role in these systems.

1.2 PHOTONIC SIGNAL PROCESSORS

1.2.1 PHOTONIC SIGNAL PROCESSING

The term signal processing means extracting or modifying information from signals at the receiver end [6]. This requires the modification of the frequency, phase, and amplitude of the received signals [7–30]. This function is essential in all communication systems as the signals arrived at the receiver end usually contain both information and disturbances, such as distortion and noises that must be filtered and eliminated. In addition, if the information data are transmitted at ultrahigh speed then it may be very useful to down convert the signals to a lower frequency region so that the manipulation and detection of the signals can be then recovered in the electronic domain. This down conversion process is normally termed as photonic mixing. The noise removal process is filtering. Other photonic processing processes can be frequency discrimination, phase comparison, signal correlation, photonic differentiation and integration, analog-to-digital optical conversion, etc. [29,31] which are emerging as potential processors for future ultrafast photonic systems. Implementing these photonic processors for processing signals at speed above 10 GHz would offer significant advantages and overcome the bottlenecks caused by conventional electronic signal processors. Furthermore, the all in-line fiber-optic and integrated optic structures of photonic signal processors are inherently compatible with optical fiber transmission and microwave fiber-optic systems, hence ease of connectivity with built-in signal conditioning. Photonic signal processing functions can also be implemented in parallel as photons do not interact in a linear optical medium and hence, lightwaves can be implemented in parallel in the same optical pipeline.

Unlike the traditional approach, the idea of photonic signal processing is to process the signals while they are still in the photonic domain. Therefore, these processors can be located in-line within the fiber-optic transmission or photonic systems. In a simple photonic system, there would be an optical transmitter consisting of a laser source, which is modulated either directly or externally by an optical modulator of electro-optic (EO), acousto-optic, or magneto-optic effects, a transmission/guiding optical medium, optical devices, such as amplifiers, phase comparators using interferometric devices.

The first proposal of the processing of signals in the photonic domain was coined by Wilner and van der Heuvel in 1976 [7] indicating that the low loss and broadband of the transmittance of the single-mode optical fibers would be the most favorable condition for processing of broadband signals in the photonic domain. Since then, several topical developments in the field, including integrated optic components, subsystems, and transmission systems have been reported and contributed to optical communications [32–42]. Almost all traditional functions in electronic signal processing have been realized in the photonic domain.

1.2.2 Some Processor Components

Ever since the proposal and invention of photonic signal processor was developed, the structures of these processors have continuously been evolved to the invention of new photonic devices, such as optical amplifying devices, photonic filters of either fiber resonance circuits or fiber Bragg gratings (FBG). These components provide a strong influence on the implementation of photonic signal processors. It is thus important to understand the operation principles and characteristics of these photonic devices. We briefly describe here the operation principles and characteristics, their circuit representation in the form of the photonic transmittance transfer functions would then be given in the next chapter.

1.2.2.1 Optical Amplifiers

An optical amplifier is an important component in incoherent fiber-optic signal processors because it can compensate for optical losses as well as providing design flexibility resulting in potential applications. Optical amplifiers can provide signal amplification directly in the optical domain. The operational principles, characteristics, and performances of the optical amplifiers are described. The gain of an optical amplifier is generated by the processes of stimulated scattering induced by nonlinear (NL) scattering in an optical fiber, or stimulated emission caused by a population inversion in a lasing medium. The former process is utilized by stimulated Raman scattering fiber amplifiers and stimulated Brillouin scattering and parametric fiber amplifiers; the later types are based on NL effects in fibers. The latter process is employed by semiconductor laser amplifiers (SLAs) or rare earth-doped fiber amplifiers. Furthermore, integrated optic waveguides employing silica on silicon planar technology can be doped with erbium (Er) ions to for compact optical amplifiers [43–71]. In this section, we give a brief description of the fiber Er-doped amplifiers, which is mostly widely used in current long-haul optically amplified transmission systems and processing networks.

Other than compensating for the transmission loss, boosting optical power, and increasing the receiver sensitivity, the EDFA is attractive due to its high gain (30 dB typically gain), low noise, large optical bandwidth, and polarization insensitivity characteristics. One example is the incorporating of the EDFA in a recirculating loop using a directional coupler and its compensating for the coupler splitting power, hence generating a lightwave oscillator. If a pulse is coupled to this oscillator then a periodic train of signals can then be generated from each round of circulation of the lightwave signals.

Figure 1.1 shows the schematic diagram of an EDFA incorporating the pump lasers, the gain medium silica-based doped with Er. Figure 1.2 shows the relevant energy level diagram of Er ions in a silica host. Although the Er ions can be pumped at various lengths to obtain population inversion in a 1550 nm transition occurring between $^4I_{13/2}$ and $^4I_{15/2}$, only two longest pump wavelengths at 980 and 1480 nm shown in Figure 1.2 are of practical interest. For these pump wavelengths, the gain efficiency with respect to the pump power is at maximum and the available semiconductor pump sources at these wavelengths are available.

1.2.2.2 Pumping Characteristics

Depending on the selected pump wavelength, Er-doped fiber can be approximately described as a three- or two-level laser system. Figure 1.3 shows the absorption and

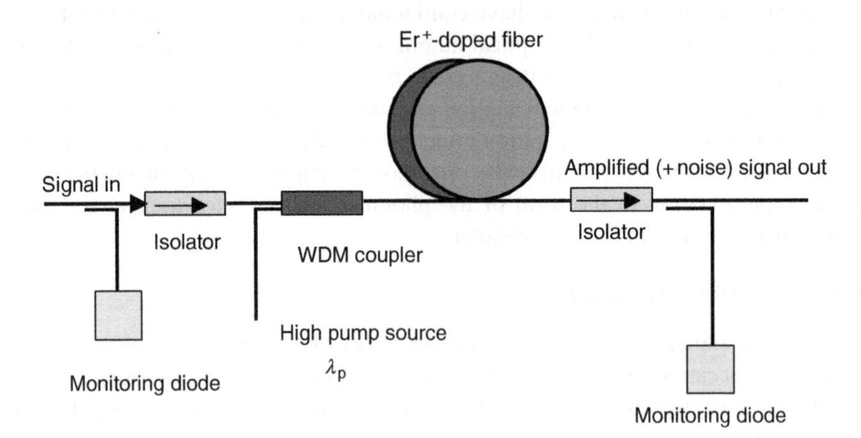

FIGURE 1.1 Schematic diagram of a forward-pumped erbium (Er)-doped fiber amplifier. WDM multiplexer combines signal and pump wavelengths into the erbium-doped fiber. Monitoring diodes are used for feedback control of the amplification factor through adjusting the pump power level.

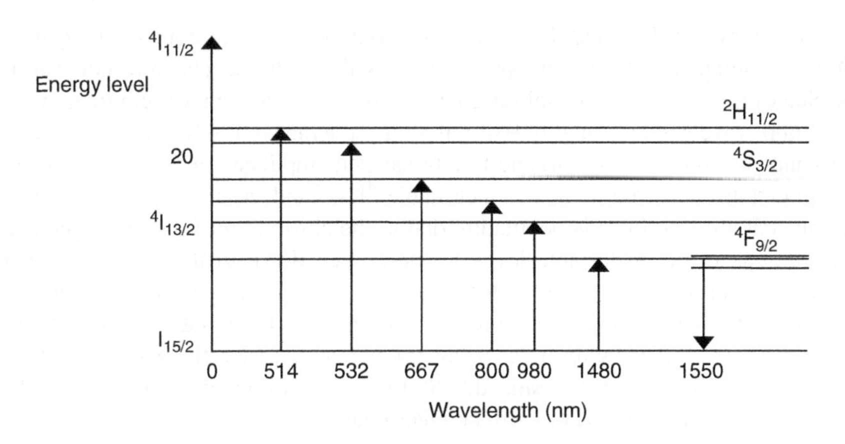

FIGURE 1.2 Energy level diagram of erbium-doped silica glass.

emission of the Er ion with respect to wavelength. The fiber host is GeO_2-doped silica fiber. To theoretically analyze, optimize the design and characteristics of Er-doped fiber amplifiers (EDFA). These levels involved in this laser system experience the Stark effect resulting in splitting of the energy levels. Since the energy levels are split with an uneven internal subpopulation, distribution makes it possible to pump Er^+:glass directly in an $^4I_{13/2}$ level and to achieve overall population

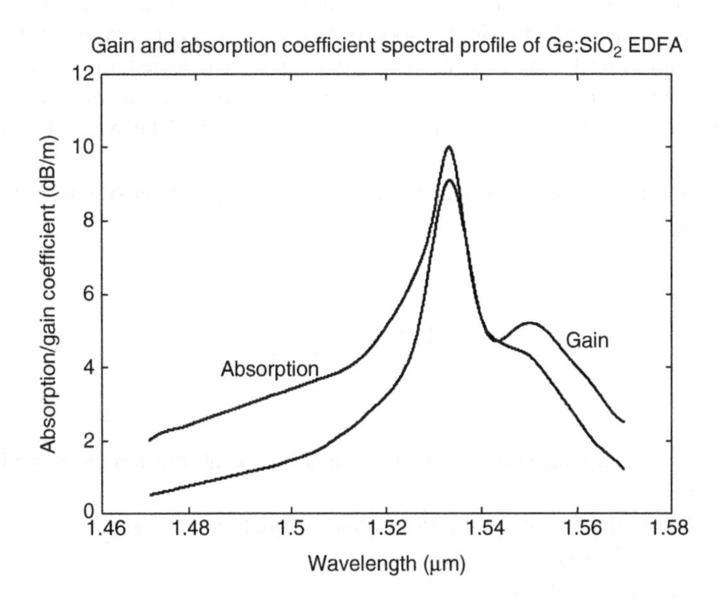

FIGURE 1.3 Absorption and gain (emission) spectra of a typical Er-doped alumina-germano-silicate fiber. Note the gain is greater than the absorption in the wavelength range 1510–1560 nm where the fiber loss is also at minimum.

inversion between $^4I_{13/2}$ and $^4I_{15/2}$ levels. This pumping scheme, corresponding to the 1480 nm pump wavelength, would not be possible if the levels were not split by the Stark effect. This energy splitting level also contributes to a certain tolerance to the pump wavelengths (roughly 980 nm with a tolerance of $+10$ and -10 nm and $+15$ and -15 nm at 1480 nm) and to a broadened amplification line width. Optical amplification occurs between the metastable $^4I_{13/2}$ and ground $^4I_{15/2}$ levels from the inversion of the Er ions population due to the absorption of the pump photons. The Er ions in the metastable level are de-excited down to the ground level by stimulated emission (providing optical gain) and spontaneous emission (resulting in noises). Pumping efficiency of about 0.8 dB/m was been obtained at 820 nm. Although it is smaller by an order of magnitude compared with 980 and 1480 nm pumping, 30 dB gain can be realized with 40–50 mW of pumping power, level has already possible by using GaAs semiconductor laser.

1.2.2.3 Gain Characteristics

Figure 1.3 shows the absorption and emission spectrum exhibited by Er ions in a typical Al_2O:GeO_2 silica fiber over the 1450–1600 nm spectral range. The gain spectrum is dependent on the presence of the dopants, alumina or Germania within the core region. The plots clearly illustrate the possibility to achieve gain between 1530 and 1560 nm, where the loss of single-mode silica-based optical fibers is at minimum.

The amplification of input signal to the Er-doped fiber combines the two effects, the confinement of the pump beam to the core-doped regions and the absorption of energy of the pump beam. Depending on the pump wavelength the single-mode Er-doped fiber can support a single-mode or few-order-mode operating region. That means a nonuniformly pumped or inverted over its whole length. This situation requires a specific model, which is based on a standard atomic rate equation. A thorough and extensive description of this modeling of the EDFA can be found in Ref. [3] [35].

The gain provided at the wavelength λ by an Er-doped fiber of length L can be described by

$$G_{\mathrm{dB}}(\lambda) = \sigma_{\mathrm{e}}(\lambda)N_2 - \sigma_{\mathrm{a}}(\lambda)N_1 \tag{1.1}$$

$$\text{with} \quad N_i = \frac{1}{N_{\mathrm{t}}}\int_0^L N_i(z)\,\mathrm{d}z \quad \text{for } i = 1,2$$

where

$\sigma_{\mathrm{e}}(\lambda)$ and $\sigma_{\mathrm{a}}(\lambda)$ are the stimulated emission and ionic absorption cross sections, respectively
N_1 and N_2 are the excited and ground state population densities
N_{t} is the total density of Er^+ ions
L the length of the doped fiber

Now let us assume that the pump beam propagates in the same direction as the signal, referred as forward pumping or a copropagating scheme. Such a pumping and signal input scheme is shown in Figure 1.1. The input and pump beams are multiplexed in a WDM coupler to propagate in the same direction. The monitoring diodes are used to control the pump power at an appropriate level for the signal input power to ensure that the signal is not amplified in the saturation region.

If the pump power is too small with respect to the length L of the doped fiber, signal photons are amplified in the first portion of the positive gain, then reabsorbed in the second fiber portion of the negative gain. This is because of the decay of the pump beam along the length, leading to noninverted populations in the second section of the gain medium.

Increasing the pump power will guarantee a total population inversion over the whole length of the doped amplifying fiber and a higher gain. Once that inverted population is achieved by large pump power and regardless of the input signal power, there is no benefit to increase the pump power. This means that the length of the doped fiber must be optimized for a maximum gain with respect to the pump power and operational input power.

For large signal input power P_s^{in}, i.e., the saturation regime, the maximum output power that can be extracted from the amplifier is maximum for a 1480 nm pumping wavelength, compared to a 980 nm pumping wavelength, since the photon energy of the former is 1.34×10^{-19} J and the later is 2.03×10^{-19} J. These figures must be compared to the photon energy of the signal at 1550 nm of 1.28×10^{-19} J. This follows the principle of energy conservation expressed in terms of photon flux ϕ, the number of photons per second.

Let ϕ_p^{in} and ϕ_s^{in}, ϕ_s^{out} the fluxes of the pump photons and signal photons at the input and output of the doped fiber, respectively, we then have, by definition

$$\phi_p^{in} = \frac{P_p^{in}}{h\nu_p} \tag{1.2}$$

$$\phi_s^{in} = \frac{P_s^{in}}{h\nu_s} \tag{1.3}$$

$$\phi_s^{out} = \frac{P_s^{out}}{h\nu_s} \tag{1.4}$$

where h is Planck's constant. The fluxes of photons must fulfill according to the conservation of energy

$$\frac{P_s^{out}}{h\nu_s} \leq \frac{P_p^{in}}{h\nu_p} + \frac{P_s^{in}}{h\nu_s} \tag{1.5a}$$

or

$$P_s^{out} \leq P_s^{in} + P_p^{in}\frac{\lambda_p}{\lambda_s} \tag{1.5b}$$

Equation 1.5b shows that the maximum output power can be extracted by selecting the wavelength ratio between the pump and signal beams. For a given pump power, the larger the pump wavelength the higher the signal power output. This is the reason for the use of the 1480 nm pump beam together with the reason that the fiber would operate in the single-mode region. However, at this wavelength there is a need to filter the pump signal at the output of the fiber to avoid the mixing of the pump and signal power beam at the receiver.

The power spectral density of the forward amplified spontaneous emission (ASE) can be approximated by

$$S_{\mathrm{ASE}} = 2n_{\mathrm{sp}}h\nu(G_{\mathrm{EDF}} - 1)\Delta\nu \tag{1.6}$$

where

G_{EDF} is the small signal optical gain of the Er-doped fiber

$\Delta\nu$ is the effective bandwidth (in Hz) in which S_{ASE} is expressed, the spontaneous emission bandwidth

The spontaneous emission factors related to the inversion of Er ions population by

$$n_{\mathrm{sp}}(\lambda_s, \lambda_p) = \frac{\eta_s N_2}{\eta_s N_2 - N_1} \tag{1.7}$$

where $\eta_s = \sigma_e(\lambda_s)/\sigma_a(\lambda_s)$. In the limit of the high pumping power, when the spontaneous emission factor n_{sp} reaches its minimum value we have

$$n_{\mathrm{sp}}^{\mathrm{min}}(\lambda_s, \lambda_p) = \frac{1}{1 - \left[\dfrac{\sigma_e(\lambda_p)\sigma_a(\lambda_s)}{\sigma_a(\lambda_p)\sigma_e(\lambda_s)}\right]} \tag{1.8}$$

The spontaneous emission factor of Equation 1.8 is achieved if the emission is nonradiative, i.e., $\sigma_e(\lambda_p) = 0$. This corresponds to the three-level laser system with $\lambda_p = 980$ nm where the decay is nonradiative from the level $^4I_{11/2}$. When 1480 nm wavelength laser pumping is used, $\sigma_e(\lambda_p) \neq 0$ the spontaneous emission factor n_{sp} is greater than 1 and we expect a degradation of the signal-to-noise ratio (SNR). This degradation of the SNR is commonly referred as the EDF noise figure (NF) and its lower limit is 3 dB, the so-called quantum limit. In practice n_{sp} is always greater than 1 (even at 980 nm) and the input coupling loss C_1 from the transmission fiber to the amplifying fiber has to be taken into account for actual EDFA modules. Therefore, the NF of EDFA is expressed as $2n_{\mathrm{sp}}/C_1$.

$$\mathrm{NF}_{\mathrm{EDFA}} = 2n_{\mathrm{sp}}/C_1 \tag{1.9}$$

Amplified spontaneous emission is generated over a continuum of wavelengths spanning the entire spectrum of the fiber gain spectrum. Because of the nonuniform spectral response of the n_{sp} and G_{EDF} over the amplification bandwidth, the spectral distribution of S_{ASE} is not flat. This ASE is then the source of optical noises, leading to different noise term of the electrical signal after the optoelectronic conversion in the photodetector. The quantity of the ASE depends strongly on the gain value required for signal amplification.

When the input power is increased, the optical amplifier is progressively reaching saturation: the gain and subsequently the power spectral density are decreased and vice versa for the noise. This noise behavior has a crucial impact on the design of optical-amplified fiber systems. In the case of long concatenated optical amplifiers, the ASE noises generated at each EDFA section are accumulated and further amplified at these succeeding amplifiers. If in-line optical amplifiers are operated with constant output total power, the signal power will gradually decrease along the chain to the worst effect of the increase in optical noises power. In addition, the spectrum of the ASE noise at the end of spectrum leads to self-filtering effects and thus limits the useful optical bandwidth of the signal. This is critical for designing the wavelength-division multiplexed (WDM) optical communication systems.

Typical properties of the Er-doped fiber are the Er ions doped in confined region of 2–4 μm core diameter as shown in Figure 1.4. The Er ions concentration is a few hundreds ppm, leading to a typical attenuation of about 10 dB/m at 1532 nm when the doped fiber is in an unpumped mode of operation. The mode field diameter of the doped fiber is about 4 μm that is much smaller than that of standard optical fibers (for optical communication systems). Thus, a splicing loss of about 2 dB is expected from these two mismatched fibers. This splicing loss can be reduced if the standard fiber is tapered; the loss can be as low as 0.2 dB. Optical isolator may be added to the two ends of the doped fibers in order to prevent the amplification medium from oscillating and behaving as a laser.

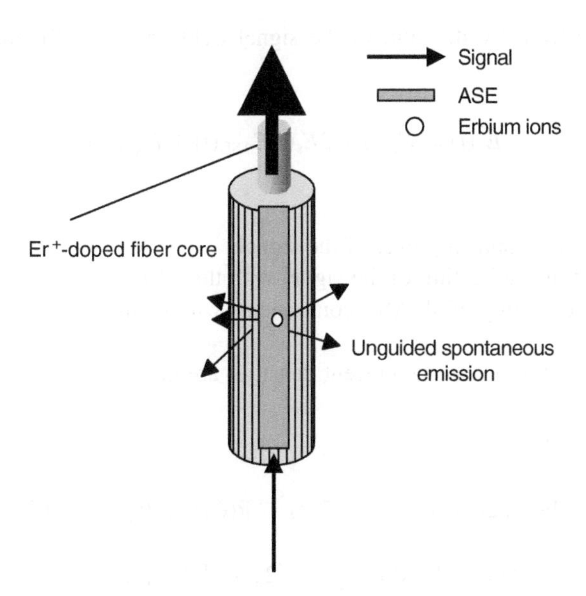

FIGURE 1.4 Generation of an amplified spontaneous emission (ASE) with an optical amplifier.

1.2.3 Noise Considerations of EDFAs and Impact on System Performance

1.2.3.1 Noise Considerations

Optical amplification is obtained in the expense of the optical noise added to output signal of the systems. The optical ASE noise is spreading over the entire spectrum much wider than the signal itself. This noise density is integrated over the noise spectrum of the photodetector and thus it has a major effect on the receiver noise characteristics. To avoid such an integrated effect, an optical filter is usually placed in front of an optical filter whose bandwidth is sharp and covering that of the signals. Whether optical amplifiers are used as an in-line amplifiers, power amplifiers, or preamplifiers, the noises fallen into the signal bandwidth will severely limit the maximum repeaterless distance. The predominant noise term generated by optical amplifiers, namely the signal-ASE beat noise, is directly related to the EDFA gain and NF.

An optoelectronic amplifier consists of a photodiode, an electronic preamplifier, an AGC control circuit, a low pass filter, and a decision circuit. The received signals are sampled by a clock recovery circuitry. The photodiode converts the incoming optical power into a photocurrent $i_s(t)$ according to the square law:

$$i_s(t) = G\Re\langle E_s(t)\rangle \tag{1.10}$$

where

G is the average avalanche gain factor
$\Re$ is the photodiode responsivity
$E_s(t)$ is the electromagnetic field of the incoming optical waves

This field is effectively the sum of the signal field and the ASE field. Thus, we have

$$E^2(t) = E_s^2(t) + 2E_s(t)E_{ASE}(t) + E_{ASE}^2(t) \tag{1.11}$$

where

$E_s^2(t)$ is the mean optical power of the signal
$2E_s(t)E_{ASE}(t)$ is the beating of the signal with the ASE
$E_{ASE}^2(t)$ is the beating of all ASE components with themselves

The mean noise power spectral current densities are given by

- Shot noises

$$\text{For signal: } \mathrm{d}(i_{s\text{-}s}^2) = [2qG^{2+x}\Re(GP_s + P_{sp} + I_d)]\mathrm{d}f \tag{1.12a}$$

$$\text{For ASE: } \mathrm{d}(i_{s\text{-ASE}}^2) = [2qG^{2+x}\Re S_{ASE}\Delta\nu]\mathrm{d}f \tag{1.12b}$$

- Beat noises

$$\text{For signal-ASE: } \mathrm{d}(i^2_{\text{b-ASE}}) = [2qG^{2+x}\Re^2 S_{\text{ASE}} P_{\text{s}}]\mathrm{d}f \tag{1.12c}$$

$$\text{For ASE-ASE: } \mathrm{d}(i^2_{\text{SE-ASEA}}) = [4\Re^2 G^{2+x} S^2_{\text{ASE}} \Delta\nu]\mathrm{d}f \tag{1.12d}$$

where
q is the electronic charge
G is the avalanche gain of the photodiode
x is the multiplication factor of the photodiode
P_s and S_{ASE} are the detected signal power and noise spectral density, respectively
$\Delta\nu$ is the optical bandwidth of the optical filter placed in front of the photodetector
df is the differential frequency interval when measuring the noise power spectrum

The electronic noises of preamplifier S_{eq} to give the total equivalent noise current at the input of the electronic amplifier are denoted as $\mathrm{d}(i^2_{\text{Neq}})$; this is included in the signal noise current or can be treated separately. The total noise power can then be found by integrating it over the system bandwidth.

The unit of S_{ASE} is usually expressed in dBm/(0.1 nm) because the optical filter and useable optical bandwidth are in order of much less than 1.0 nm and the optical power noise is measured in microwatts. The main purpose of the optical filter is to reject the unwanted ASE power on either side of the signal wavelength and to reduce the ASE-ASE beating noises. In practical systems, the noise terms of the remnant pump power or the relative intensity noise of the laser transmitter are negligible.

The shot noises of the signals are depending on whether a 0 or a 1 is received. The total currents generated at the optical receiver with optical amplifiers are thus given by integrating the noise density over the electrical bandwidth of the systems (normally about 0.7 of the bit rate for a PCM modulation hierarchy). These noise currents (square) are given by

$$i^2_{\text{SN}}(\text{“1”}) = \left[\sum \mathrm{d}(i^2_{\text{N}}(\text{“1”})\right] B_{\text{e}} \tag{1.13a}$$

$$i^2_{\text{SN}}(\text{“1”}) = \left[\sum \mathrm{d}(i^2_{\text{N}}(\text{“0”})\right] B_{\text{e}} \tag{1.13b}$$

where B_e is the 3 dB bandwidth of the electrical low pass filter placed after the electronic preamplifier, or effectively the bandwidth of the electronic preamplifier. We can thus integrate the total spectral densities over the electrical bandwidth. Once the total noises generated by the ASE and the total quantum shot noises and electronic noises equivalent current are known, the bit error rate (BER) can be found with ease. The BER is thus given by

$$\text{BER} = Q(\delta) \tag{1.14a}$$

where δ is given by

$$\delta = \frac{i_1 - i_0}{i_{\text{SN}}(\text{``1''}) + i_{\text{SN}}(\text{``0''})} \tag{1.14b}$$

The average photo currents corresponding to the "low" or mark and "high" or space levels are detected by the photodetector. The signal generated currents are related to the mean signal power as

$$i_1 = \Re G P_{\text{s}}^{\text{``1''}} = \Re G \frac{2T_{\text{ex}}}{1 + T_{\text{ex}}} P_{\text{s}} \tag{1.15a}$$

$$i_0 = \Re G P_{\text{s}}^{\text{``0''}} = \Re G \frac{2}{1 + T_{\text{ex}}} P_{\text{s}} \tag{1.15b}$$

where T_{ex} is an extension ratio between "1" and "0" photocurrents. It is also assumed that the equal probability between "0" and "1" digital levels for optimum threshold level of the decision circuit. Thus, once the signal power P_{s} can be found the required optical power at the input of the optical amplified electronic receiver can be converted into the receiver sensitivity for a certain bit rate and BER.

The δ factor can also be expressed as a function of the SNR which is usually expressed in dB/(0.1 nm) as

$$\text{SNR} = 10 \log \frac{P_{\text{s}}}{S_{\text{ASE}} \Delta \nu} \tag{1.16}$$

where $\Delta\nu = 0.1$ nm, i.e., 1.25×10^{10} at 1550 nm (normally the passband of the optical filter is placed immediately after the EDFA to reduce ASE noise). The optical SNR corresponds to the ratio between the signal power and the ASE power at the signal wavelength. To guarantee negligible degradation caused by the EDFA noise during data stream amplification, SNR must be greater than 20 dB/(0.1 nm) at the input of the photodetector at 2.5 Gb/s.

Typical DFB semiconductor transmitters deliver power in the range of 0–3 dBm. An EDFA can be used as a power amplifier within the transmitter to boost the launched power and to extend the fiber span of the systems. In practical systems, the input power is normally about 0 dBm and the EDFA is deeply saturated. The generated ASE power is low, resulting in a large SNR and in a nonsignificant amount of ASE-signal beating at the photodiode level. In the optical receiver terminal, the optical fiber amplifier can be used to optimize the receiver sensitivity, hence the fiber length of a repeaterless span.

To perform as an efficient optical amplifier, the gain of EDFA must be sufficiently large to make all the noise term much smaller than that of the ASE-beat term. In this case, we have the signal-dependent electronic noises (i_{Neq}) no longer the dominating noise of the total noise. Thus from Equation 1.13b, we have

$$\delta_{\max} = \frac{(2\Re G P_{\text{is}}^{\text{“1”}} G_{\text{EDF}})^2}{2G^2\Re^2 S_{\text{ASE}} B_{\text{o}} G_{\text{EDF}} P_{\text{is}}^{\text{“1”}} B_{\text{e}}} \tag{1.17a}$$

where

P_{is} is the input optical signal power at the input of the EDF amplifier

G_{EDF} is the optical power gain

or

$$\delta_{\max} = \frac{P_{\text{is}}^{\text{“1”}} G_{\text{EDF}}}{2q\ S_{\text{ASE}} B_{\text{o}} B_{\text{e}}} \tag{1.17b}$$

We note that as soon as the amplifier gain G_{EDF} in dB reaches 20 dB, the δ-factor reaches the maximum value; thus, it is not beneficial to increase the optical gain higher than 20 dB.

Therefore, the new receiver sensitivity for a preoptical amplifier receiver can be found by converting the P_{is} to dBm for a value of δ corresponding to a certain BER (e.g., $BER = 10^{-9}$ for $\delta = 6$).

1.2.3.2 Fiber Bragg Gratings

With the discovery of the photosensitivity of silica fiber, especially the Ge-doped silica, by Hill et al. in 1978 [72], a new type of fiber-based component has been created by writing patterns so that there are regions in which the refractive index can be increased or decreased. Hence, if an interference pattern is imposed on the fiber, a grating can be generated. Hence, lightwaves can be reflected if the period of the grating matches that of the lightwave. Thus, the phase matching condition would lead to the filtering effects on the reflected spectrum. The center of the passband happens exactly when the phase matching is at the Bragg wavelength λ_B.

$$\lambda_B = 2n_{\text{eff}}\Lambda \tag{1.18}$$

where

Λ is the period of the grating

n_{eff} is the effective refractive index of the guided mode, i.e., the ratio of the propagation of the guided lightwave along the Z-direction and that of the wave vector

The grating reflection coefficient can be determined during the fabrication of the grating, usually depending on the exposure time of the silica fiber under the UV illumination.

The wavelength-dependent reflection property makes FBG a very attractive sampling element used for the construction of discrete time signal photonic processors. In the case of grating-based photonic signal processors, the sampling time can be controlled via the spacing of the grating, usually by pyroelectric effects or by stressing the FBG. Thus, the center wavelength can be selective and thus a tunable wavelength processing system can be generated or demultiplexing of different wavelength lightwave channels in a WDM system [73,74]. Chirped fiber grating can also be employed in which the period is continuously varied. This allows the simultaneous reflection of different wavelength lightwaves but different delay time, hence dispersion compensation of modulated lightwave channels. Arrays of FBGs can be formed in cascade or parallel to generate discrete photonic signal processors.

The reflectance transfer function of a uniform FBG of length L is given by [75]

$$\rho = \frac{-\kappa \sinh\left(\sqrt{\kappa^2 - \hat{\sigma}^2 L}\right)}{\sigma^2 \sinh\left(\sqrt{\kappa^2 - \hat{\sigma}^2 L}\right) + j\sqrt{\kappa^2 - \hat{\sigma}^2} \cosh\left(\sqrt{\kappa^2 - \hat{\sigma}^2 L}\right)} \tag{1.19}$$

where κ is the AC coupling coefficient given by

$$\kappa = \frac{\pi}{\lambda} \nu \delta n_{\mathrm{eff}} \tag{1.20}$$

where δn_{eff} is the average change of the effective index of the grating over the whole length of the FBG. $\hat{\sigma}$ is the DC self-coupling coefficient and is related to a detuning factor ($= \delta + \sigma$) given by

$$\delta = 2\pi n_{\mathrm{eff}} \left(\frac{1}{\lambda} - \frac{1}{\lambda_{\mathrm{B}}}\right) \tag{1.21}$$

$$\sigma = \frac{2\pi}{\lambda} \delta n_{\mathrm{eff}} \tag{1.22}$$

Under the operating condition of incoherence processor, the reflection in power is given as the modulus (magnitude) of the transmittance transfer function which can be written as

$$r = \frac{\sinh^2\left(\sqrt{\kappa^2 - \hat{\sigma}^2 L}\right)}{\cosh^2\left(\sqrt{\kappa^2 - \hat{\sigma}^2 L}\right) - \dfrac{\hat{\sigma}^2}{\kappa^2}} \tag{1.23}$$

The phase change of the reflection transfer function can be found as

$$\theta = \tan^{-1}\left[\frac{\mathrm{Im}(\rho)}{\mathrm{Re}(\rho)}\right] \tag{1.24}$$

The maximum reflectance is thus given as

$$r = \tanh^{-1}\kappa L \tag{1.25}$$

This shows that the grating reflectivity can be increased monotonically as a function of the length of the grating. The dimensionless product κL is the measure of the grating strength.

1.3 OPTICAL MODULATORS

1.3.1 INTRODUCTORY REMARKS

The engineering of optical transmission systems has been pushed to the maximum limit. We have witnessed the transmission bit rate increases from 2.5 Gb/s in the mid-1980s to 40 G and even 160 Gb/s in the 1990s and the early twenty-first century [37,38,76–99]. These transmission rates are beyond the capability of direct modulation of the laser sources. External modulation technique is the method, which would modulate the lightwaves generated from the laser sources whose line width must also be very narrow, usually around 100 MHz.

Most of advanced optical modulators are operating based on two principal physical effects: the linear electro-optic effect (EOE) and the electro-absorption effect (EAE). Both effects depend on the applied electric field, which makes the modulators as voltage-controlled devices. The applied field changes the refractive index or the absorption rate via the EOE and the EAE, respectively. This is used in changing the phase of the optical waves along the propagation path of an interferometer and hence the constructive or destructive output.

Lightwaves can have various characteristics, which can be modulated for carrying information, including the intensity, the frequency, the phase, and the polarization. Among these, the intensity modulation is very popular in the normal ON–OFF keying (OOK) and until recently the phase modulation has been considered for efficient bandwidth. Optical sources of different carrier frequencies, particularly their stability plays a very important role in WDM and DWDM optical fiber communications. The spacing between the centers of the wavelengths of operation is naturally very critical in system and network performance. Thus, it is very crucial for both simulation and hardware implementation to ensure that the optical carriers generated represent the true physical picture of the practical optical sources.

Unlike direct modulation of the laser diodes, external modulation modulates the lightwaves continuously generated by the laser; the lithium niobate ($LiNbO_3$) modulators can be designed to be chirp free, adjustable chirp, negative or positive chirp which can be employed to compensate for the fiber dispersion. In analog systems, the linearization of the external modulator can provide very low modulation distortion.

This section introduces the principles of modulation of the lightwaves using EO, electro-absorption (EA), and interferometric effects in $LiNbO_3$ uniaxial or polymer materials systems. The waveguides can be either diffused or rib structures. Some numerical designs of optical waveguides are given. The modulation phenomena are described for the EA and Mach–Zehnder interferometric modulators (MZIMs).

1.3.2 Lithium Niobate Optical Modulators

The maturity of $LiNbO_3$ technology has warranted the realization of devices that provide high speed switching and modulation. Among the more mature and widely used devices that provide broadband modulation are the modulators, which are based on interferometric and directional coupler structure [76]. Our interest in this section is the design and applications of the *Y*-branch interferometric modulators.

There exist two versions of interferometric type modulators. One is the Mach–Zehnder type structure that employs a 2×2 directional couplers as splitter and combiner, whereas the other one is the 3 dB (i.e., symmetric) *Y*-branch splitting and combining interferometer. The *Y*-branch interferometer is the most popular structure for optical modulation devices when only one input and one output optical ports are required. In this section, we outline the relevant $LiNbO_3$ technology and the operational principles of such modulators.

1.3.2.1 Optical-Diffused Channel Waveguides

1.3.2.1.1 Property of $LiNbO_3$

Optical waveguide devices based upon $LiNbO_3$ substrates are presently in an advanced state and employed as the external modulators in most photonic transmitters in advanced optical communication systems. $LiNbO_3$ has several important characteristics that make it attractive for waveguide devices. It is relatively easily processed and its material properties have been studied and documented extensively [78]. $LiNbO_3$ is a dielectric crystal which belongs to the rhombohedral space group, 3 mm. It possesses a high Curie temperature point, large EO and acousto-optic effects, large nonlinear optical coefficients, and high birefringence. The $LiNbO_3$ crystal is defined in three dimensions with three principal axes namely the *X*, *Y*, and *Z* (or *C*-) axes. The axes form the basis of specifying the particular orientation or cutting of the wafers. An *X*-cut crystal implies that the direction normal to the flat surface of the wafer is parallel to the *X*-principal axis, whereas a *Z*-cut crystal indicates that the direction normal to the flat surface of the wafer is parallel to the *Z* or *C*-principal axis.

Owing to its birefringence or doubly refracting properties, $LiNbO_3$ as an uniaxial birefringent type has two refractive indices, known, respectively, as the ordinary and extraordinary refractive indices, which can be represented by an ellipsoid in the three-dimensional (3-D) space. The light polarized in the *Z*-axis would see the extraordinary index, whereas for light polarized in the *X*- or *Y*-crystal axis, the ordinary index would be effective. The refractive index seen by polarization in any direction other than a principal axis would depend on the direction of polarization with respect to the principal axes.

Figure 1.5 shows the index ellipsoid of the $LiNbO_3$ uniaxial crystal. Alternatively in the plane of the wave propagation, we have the refractive index contours for a positive uniaxial material, such as $LiNbO_3$ crystal as shown in Figure 1.6. The diagrams are drawn relative to the *C*-axis as the principal axis. Lightwaves traveling in the device are not propagating along a principal axis, but in some arbitrary direction. In $LiNbO_3$ optical modulators, we usually use the crystals with the cuts in either *X*- or *Y*-axes as illustrated in Figures 1.6b and c, respectively. These

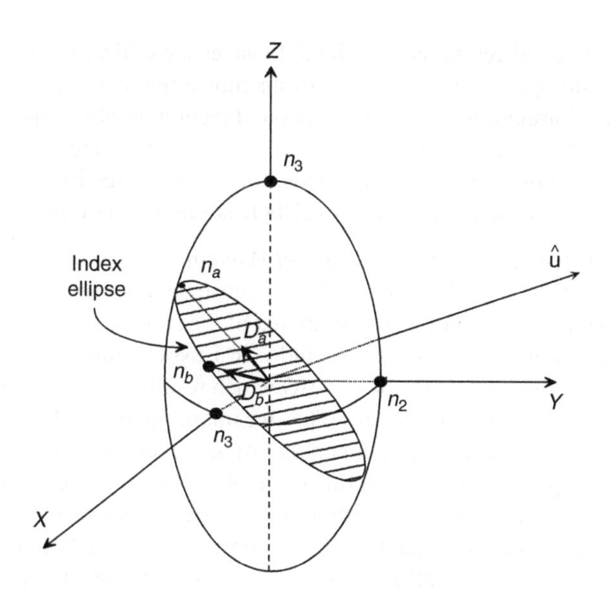

FIGURE 1.5 Index ellipsoid with ordinary index $n_o(n_2)$ and $n_e(n_3)$ the extraordinary index. n_a and n_b are refractive indices seen by lightwaves polarized other than the principal crystal axis.

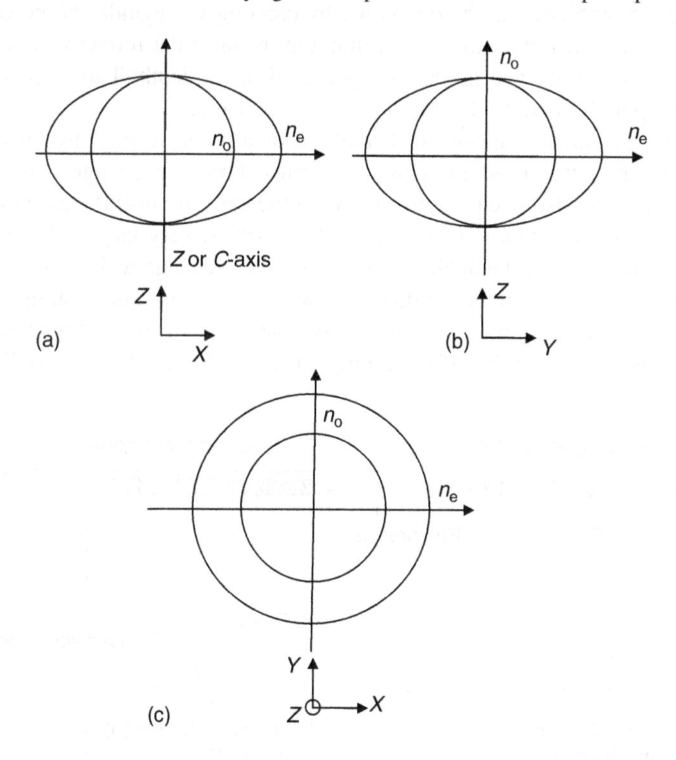

FIGURE 1.6 Refractive index contours for $LiNbO_3$ crystal with *Z*- or *C*-denoted as the principal axis. (a) Lightwaves propagation in the *Z*-axis polarized along *Y*-cut $LiNbO_3$, (b) prop- direction *Z*-axis and *X*-cut crystal, (c) *Y*-prop and *Y*-cut crystal.

effective refractive indices as seen by the lightwaves are critical when one uses it as an EO or acousto-optic modulator so as to maximize the tensor coefficients of the relevant effect. Furthermore, the propagation direction is also dependent on the guidance of such lightwaves in optical wave guides, which are fabricated by in-, out-diffusion or proton exchange methods. In the next section, the fabrication techniques of the optical waveguides in $LiNbO_3$ are briefly described.

1.3.2.1.2 Fabrication of Optical Channel Waveguide

The formation of waveguide in the $LiNbO_3$ substrate involves the creation of a region of higher refractive index relative to that of the substrates. Several techniques [81] have been employed to form waveguides in $LiNbO_3$. Initially, the waveguide was formed by thermal out-diffusion of Li_2O which results in an increased refractive index for the extraordinary index, n_e. This technique involves the placing of the $LiNbO_3$ substrate in a furnace at a temperature of in the region of 980°C–1075°C for 5–10 h [81,82] that causes the out-diffusion of Li_2O from the substrate, hence creating a layer of region of higher refractive index at the surface of the substrate. In addition to being limited to guiding light in only one polarization, the achievable index change is small and therefore provides waveguide modes whose confinement is relatively weak. Furthermore, channel waveguides cannot be formed conveniently except by etching the substrate to form ridge waveguides.

The above problems can be overcome by creating waveguides based on the in-diffusion of a thin metallic layer, e.g., titanium, to raise the refractive index of the substrate on the surface and graded along the diffusion depth. This technique is by far the most popular and established method since 1976.

Optical channel waveguides in $LiNbO_3$ can also be formed by an exchange process similar to that used for glass substrates. Proton exchange using benzoid and other acids has also been employed. As in the case of out-diffused waveguides, only the extraordinary index n_e is increased. However, very large index difference in only one direction can be achieved using proton exchange technique.

The fabrication of titanium-diffused waveguides is quite straightforward. Although there are slight variations, a typical step of the fabrication of the channel waveguides is by liftoff patterning shown in Figure 1.7. The polished and

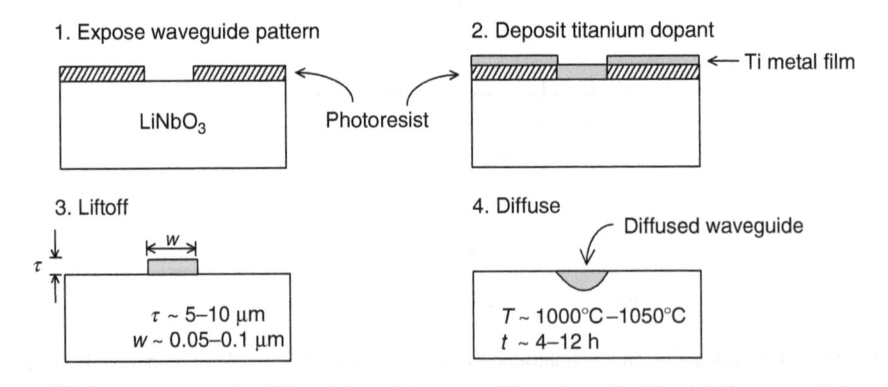

FIGURE 1.7 Steps of fabrication of Ti-diffused optical waveguide in $LiNbO_3$.

cleaned crystal is spun coated with a thin layer of photoresist. A mask with the desired waveguide pattern is placed in contact with the crystal and then exposed to UV light. Upon developing to remove the exposed photoresist, a window corresponding to the waveguide pattern is left in the photoresist. Titanium (Ti) is then deposited over the entire crystal by rf-sputtering, by e-beam deposition or by using a resistively heated evaporator. The crystal is then placed in a photoresist solvent, which removes the photoresist and the unwanted titanium, leaving the desired strip of titanium. Instead of liftoff, the patterning can also be achieved by depositing Ti over the entire surface and selectively removing it outside the desired waveguide region.

The Ti-patterned substrate is then placed in a diffusion furnace under oxygen flow for diffusion at temperatures that range from 980°C to 1050°C for typical diffusion times of 5–10 h. The lower limit (temperature) results in an overly long diffusion time, while the upper limit is set by the desire to remain below the Curie temperature (~1125°C) to avoid the need to repole the crystal after diffusion. Various ambient conditions have been used for the diffusion. Historically, the heating and diffusion cycles were performed in an atmosphere of flowing argon bubbled through a water column. A process of cool down after diffusion is performed in flowing oxygen to allow reoxidation of the crystal to compensate for oxygen loss during diffusion. The water vapor treatment was initially employed to reduce the photorefractive effect. Later, it was realized that this process also reduces Li_2O out-diffusion that can cause unwanted planar guiding for the extraordinary polarization. Typically, 80% relative humidity in the flow gas is sufficient to eliminate unwanted guiding at diffusion temperature as high as 1050°C.

Diffusion process is controlled so that a single-mode waveguide with a low coupling loss, i.e., matched spot size with that of the optical fiber, can be achieved [85,86]. Various fabrication parameters and conditions can be varied independently to control important waveguide parameters, such as waveguide width, effective depth, and peak index change. The waveguide depth depends upon the diffusion time, the diffusion temperature, and to a lesser extent on the Ti strip width. The waveguide width is photolithographically defined although increased somewhat by lateral diffusion. Most importantly, the peak waveguide-substrate index change depends, for fixed diffusion temperature and time, upon the Ti and density.

1.3.2.1.3 Guided Modes of Ti:LiNbO$_3$ Channel Waveguide

The mode distribution of Ti:$LiNbO_3$ channel waveguide, a graded index waveguide, plays a significant role in the design of the optical modulators and switches. Efficient design and fabrication of such optical devices require good knowledge of the modal characteristics of the relevant channel waveguide. The key features of the fabrication of Ti:$LiNbO_3$ waveguide can be understood, and hence the modulation efficiency for Mach–Zehnder optical modulator.

Over the past decades, much work has been done in fabricating low loss, minimum mode size Ti-diffused channel waveguide [30,34]. From these references, the diffusion process involved in the fabrication of $LiNbO_3$ waveguide and its relevant diffusion profile can be used for the design of diffused channel optical waveguides.

1.3.2.1.4 Refractive Index Profile of Ti:$LiNbO_3$ Waveguide

When the Ti metal is diffused, Ti-ion distribution spreads more widely than the initial strip width. The profiles can be described by the sum of an error function, while the Ti-ion distributions perpendicular to the substrate surface can be approximated by a Gaussian function [86]. This, of course, is true only if the diffusion time is long enough to diffuse all the Ti metal into the substrate. We consider this case as having the finite dopant source. However, if the total diffusion time is shorter than needed to exhaust the Ti source, the lateral diffusion profile would take up the sum of the complementary error function while the depth index profile is given by the complementary function [83]. This case is considered to have had an infinite doping source [12]. In our study, we would assume that there is sufficient time for the source to be fully diffused because in most practical waveguide, it is undesirable to have Ti residue deposited on the surface of the waveguide because this will increase the propagation loss. This increase in propagation loss is a result of stronger interaction with the $LiNbO_3$ surface (and thus an increased scattering loss) as the modes become more weakly guided.

In general, the two-dimensional (2-D) refractive index distribution of a weakly guiding channel waveguide

$$n(x, y) = n_b + \Delta n(x, y) = n_b + \Delta n_0 f(x) g(y) \tag{1.26}$$

where

n_b is the refractive index of the bulk (substrate)

$\Delta n(x, y)$ is the variation of the refractive index in the guiding region in the vertical and lateral directions

$\Delta n(x, y)$ in our diffusion model is essentially a separable function where $f(x)$ and $g(y)$ are the functions that describe the lateral and perpendicular diffusion profile and Δn_0 is known as the surface index change after the diffusion time. The surface index change is defined as the change of refractive index on the substrate just below the center of the Ti strip. In other words, it is the refractive index when both $f(x)$ and $g(y)$ assume unity value. The variation of the refractive index can be represented as [12]

$$\begin{aligned}\Delta n(x, y) &= \frac{dn}{dc}\tau \int_{-w/2}^{w/2} \frac{2}{d_y\sqrt{\pi}} \exp\left[-\left(\frac{y}{d_y}\right)^2\right] \frac{1}{d_x\sqrt{\pi}} \exp\left[-\left(\frac{x-u}{d_x}\right)^2\right] du \\ &= \Delta n_0 f(x) g(y)\end{aligned} \tag{1.27}$$

where

$$f(x) = \frac{1}{2} \frac{\left\{\mathrm{erf}\left[x + \left(\frac{w}{2}\right)/d_x\right] - \mathrm{erf}\left[x - \left(\frac{w}{2}\right)/d_x\right]\right\}}{\mathrm{erf}\left(\frac{w}{2d_x}\right)} \tag{1.28}$$

$$g(x) = \exp\left[-\left(\frac{y}{d_y}\right)^2\right] \tag{1.29}$$

and

$$\Delta n_0 = \frac{\mathrm{d}n}{\mathrm{d}c}\frac{2}{\sqrt{\pi}}\frac{\tau}{d_y}\mathrm{erf}\left(\frac{w}{2d_x}\right) \tag{1.30}$$

with

$$d_x = 2\sqrt{D_x t}$$

and

$$d_y = 2\sqrt{D_y t} \tag{1.31}$$

where

- t is the total diffusion time
- c is the Ti concentration
- (d_x and d_y) and (D_x and D_y) are the diffusion lengths and constants in the x and y directions, respectively
- τ and w are the initial Ti strip thickness and width, respectively
- $\mathrm{d}n/\mathrm{d}c$ is the change of index per unit change in Ti metal concentration

Figures 1.8 and 1.9 show the diffusion profile across and into the $LiNbO_3$ substrate. The diffusion conditions are the same for both cases. Figure 1.10 shows the variation of the surface index with respect to the initial Ti strip width. The change of surface index would reach a saturated value when increasing the width. The change of surface index can be given by

$$\Delta n_0 = \frac{\mathrm{d}n}{\mathrm{d}c}\frac{2}{\sqrt{\pi}}\frac{\tau}{d_y} \tag{1.32}$$

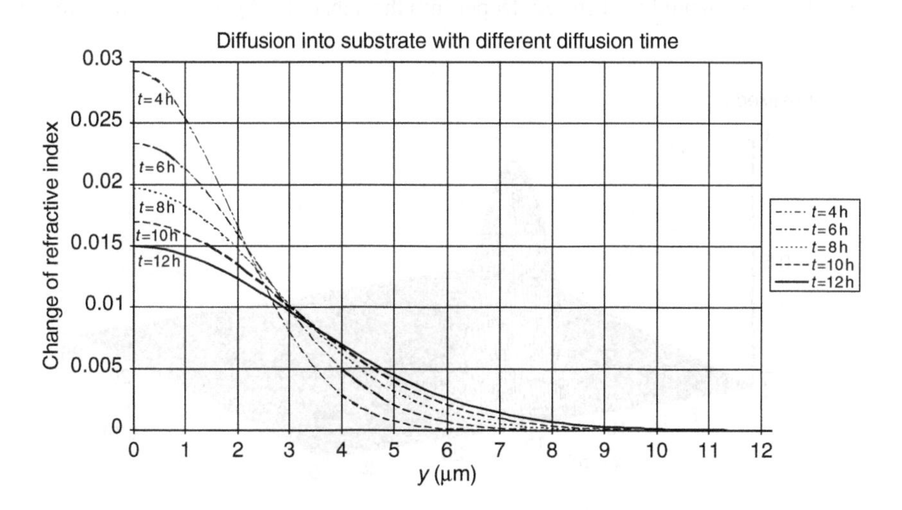

FIGURE 1.8 Depth index variation with increasing diffusion time.

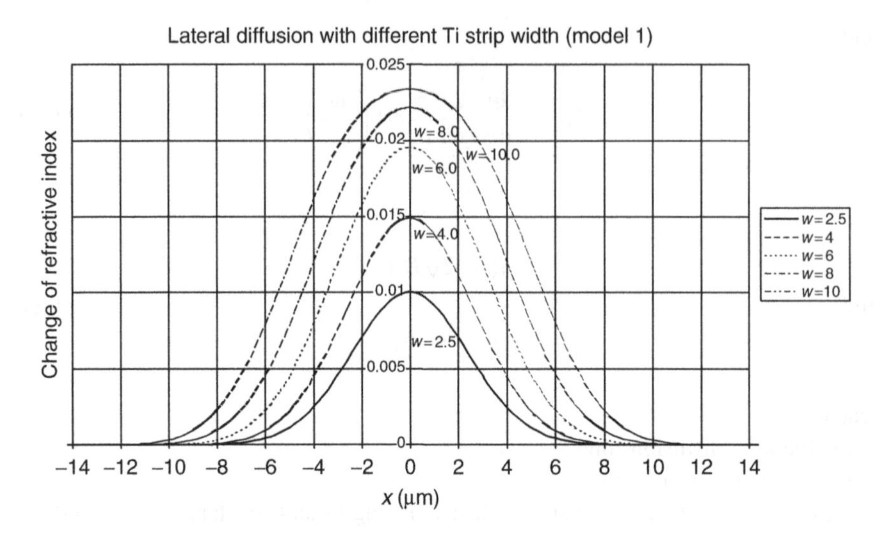

FIGURE 1.9 Lateral diffusion variation with increasing Ti strip width.

Any increase in the surface index must be from a higher thickness of the Ti strip, or a decrease in diffusion depth, d_y, which involves either an increase or decrease in diffusion temperature. According to the work of Fukuma and Noda [83], the diffusion length is very close to one another in both lateral and depth directions (isotropic diffusion) at 1025°C for *Z*-cut $LiNbO_3$ crystal. An increase in temperature greater than that would result in a higher diffusion constant in the depth direction and lower value for lateral diffusion and vice versa for diffusion temperature lower than 1025°C. The diffusion length can also be changed by controlling the diffusion time. Essentially, longer diffusion time would means a lower surface index change as most of the Ti source would be diffused deeper into the substrate. Again, we could predict

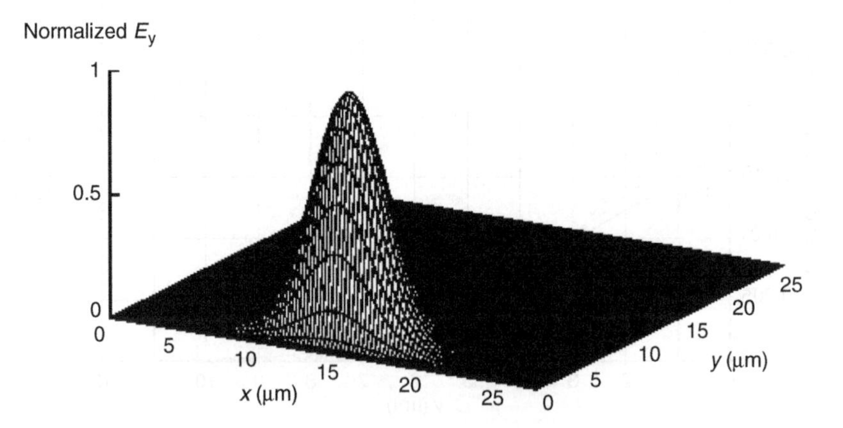

FIGURE 1.10 Typical (transverse magnetic) TM modal field of a diffused channel waveguide.

that a higher change of surface index since not all the Ti metal is exhausted. Typical fabrication condition and parameters are assumed to be $T = 1025°C$, $\tau = 1100$ Å, $dn/dc = 0.625$, d_x and d_y are both equal to 3.2 μm. We can observe that by controlling the width of the initial Ti strip width, one can vary the change of refractive index and the relative size of the channel waveguide, thus enabling us to control the number of mode and the mode spot size that can be supported by the diffused channel waveguide.

In general, a narrow initial Ti width would give a near cutoff mode for the refractive index change would be too small. The optical mode would be weakly confined, thus giving a larger mode size. As we increase the Ti width, the refractive index difference would be higher and the waveguide mode would be better confined and has a smaller mode size. This is important to minimize the coupling loss between the single-mode fiber and the modulator waveguide. However, the mode size would increase with further increase in Ti width due to a larger physical size of the waveguide. The change of surface index can also be controlled by varying the thickness of the Ti strip. As implied in Equation 1.32, the surface index change is proportional to the Ti strip thickness, τ. Penetrable Ti thickness at 1000°C–1050°C for 6 h would be around 50–80 nm [30]. If the Ti strip is too thin, the refractive index change approaches cutoff conditions. All these characteristics and the distribution of the guided modes are illustrated in the next section.

1.3.2.1.5 Mode Distribution: Numerical and Experimental

The parameters of the diffusion profile can be used for simulation and design of the modal characteristics of Ti:$LiNbO_3$ waveguide. Experimental work reported by Suchoski and Ramaswamy [89] in the fabrication of minimum mode size low loss Ti:$LiNbO_3$ channel waveguide can be confirmed with simulated waveguide modes. We analyze the *Z*-cut *Y*-propagating material as shown in Figure 1.11 and likewise in Figure 1.12 for different crystal orientations. For this particular crystal cut, the relevant optical field would be (transverse magnetic) TM polarized, which corresponds to the polarization along the extraordinary index axis of the crystal. Hence, the change of refractive index of the extraordinary index, n_e, is expected. The TM-polarized mode width and depth, which is defined as $1/e$ intensity full width and full depth, are measured for Ti:$LiNbO_3$ waveguides fabricated under the condition where $T = 1025°C$ for 6 h. The sample waveguides have Ti thickness ranging from 500 to 1100 Å, and Ti strip widths ranging from 2.5 to 10 μm. We thus can see that the mode size increases as the Ti strip width is decreased from 4 to 2.5 μm. This increase is more pronounced especially with the thinner Ti films, hence smaller Δn, because the waveguides become closer to cutoff. The TM mode depth and width decrease as the Ti thickness is increased from 500 to 800 Å. However, for 4 μm strip widths, the mode size does not decrease further for Ti films thicker than 800 Å. This is an indication that it is not possible to diffuse any more Ti into the substrate for Ti thickness of more than 800 Å for 6 h diffusion time. It can be focused on Ti thickness that ranges between 700 and 800 Å because it is the thickness that gives minimum mode sizes, which is ideal for the design of optical modulator for maximizing the overlap integral between the guided mode and the applied modulating field. To achieve this, we must first estimate the suitable diffusion parameter to be used in the

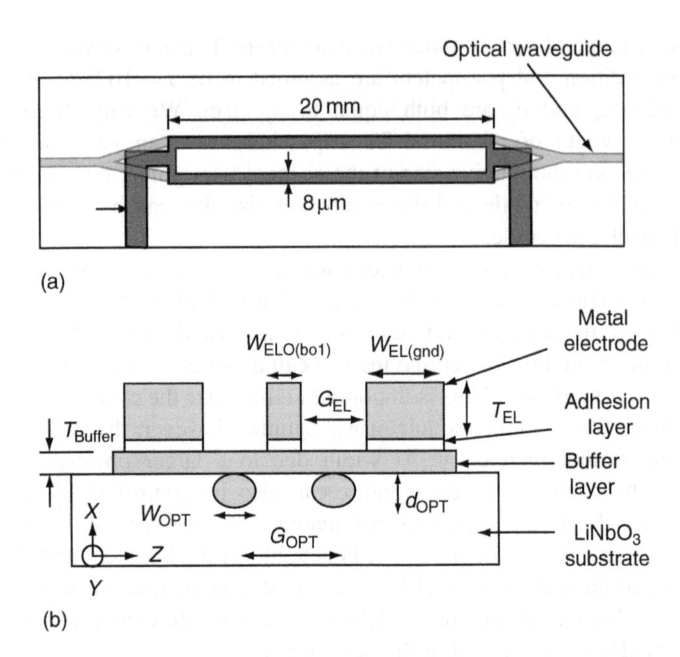

FIGURE 1.11 A schematic diagram of a dual-drive Mach–Zehnder intensity modulator: (a) planar view and (b) cross-sectional view. The dimensions shown are typical values only.

simulation. Various values of $\mathrm{d}n/\mathrm{d}c$ have been reported [24]. Measurements reported by Minakata et al. [81] show the change of extraordinary index n_e per Ti concentration as

$$\frac{\mathrm{d}n_\mathrm{e}}{\mathrm{d}c} = 0.625 \tag{1.33}$$

The nominal values for diffusion constants, D_x and D_y, obtained from the work of Fukuma and Noda [83] were both measured to be 1.2×10^{-4} $\mu\mathrm{m}^2/\mathrm{s}$ at the nominated temperature which is 1025°C giving both diffusion length of d_x and d_y the value of 3.2 μm. With these nominal parameters, the optical channel waveguides with a Ti thickness, τ of 700 Å can be simulated. The following figures are the result of our simulation compared to the experimental one and some illustrations of the TM mode profile. Figure 1.13 shows that the simulated results appear to have overestimated both Γ_x and Γ_y. Such discrepancy is anticipated fabrication of diffused waveguide is subjected to many changes. Various reports [76–88] have shown that even though the nominal diffusion condition can be very much the same, the measured diffusion parameters can still differ greatly from one another due to possible differences in stoichiometry between different crystals and measurement techniques. Therefore, there would certainly be some uncertainties that lie in fabrication parameters and also the application of the refractive index model. Such uncertainty can be compensated by adjusting the value of $\mathrm{d}n/\mathrm{d}c$ and also D_x and D_y. We find that by adjusting the following diffusion parameters where $\mathrm{d}n/\mathrm{d}c = 0.8$, $D_x = 1.4 \times 10^{-4}$ $\mu\mathrm{m/s}^2$ and

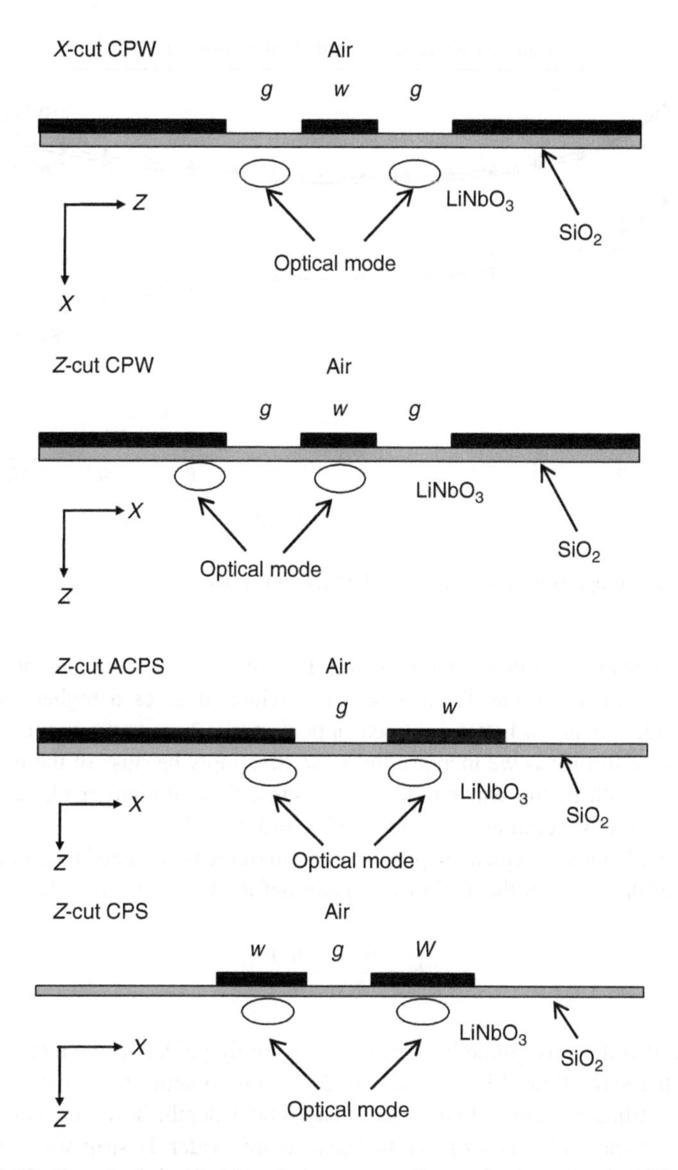

FIGURE 1.12 Commonly used electrode structure and crystal orientation in $LiNbO_3$.

$D_y = 1.1 \times 10^{-4}$ $\mu m/s^2$, the properties of the estimated modes correspond well within design limit with the experimental waveguides for the case where the waveguide is well guided as shown in Figure 1.13. The plots with circular graph markers are extracted from Ref. [30] whereas the plots with square graph markers are simulated results. The simulated and measured result matches to within 5%. The results show that the mode width, Γ_x corresponds well to the experimental result with differences of less than 3%. The mode depth, Γ_y, however, matches only to within 8%. Effectively, the larger initial mode spot size is because of the lower refractive

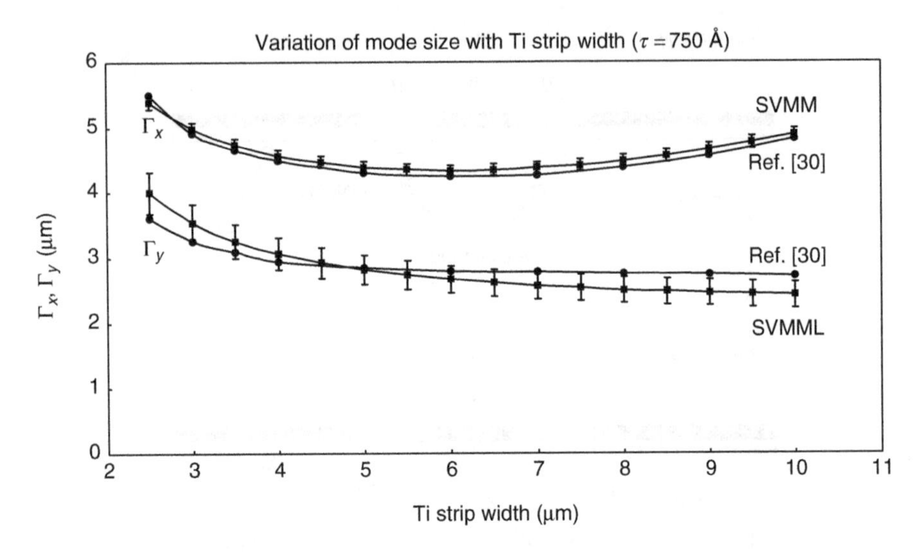

FIGURE 1.13 Comparison of simulated and experimental mode sizes for $\tau = 700$ Å.

index change resulted from a much narrower Ti width, thus causing the optical mode to be less confined. As the Ti strip becomes wider, it gives a higher change of refractive index, hence a better confined optical mode. The mode width, however, would increase further as we increase the Ti width simply because of the increase in the physical width of the waveguide. At the same time, the larger physical width would enable the waveguide to support higher-order modes.

Figure 1.14 shows a typical dispersion relation of the normalized mode index b as a function of the width of the Ti thin-film layer before the diffusion, defined as [37]

$$b = \frac{(n_{\text{eff}}^2 - n_{\text{s}}^2)}{2\Delta n \cdot n_{\text{s}}} \tag{1.34}$$

It is shown that the waveguide becomes more strongly guided as we increase the Ti width. At the same time, higher-order mode begins to settle in as the strip width becomes significantly larger than 6 μm. The modal depth, however, continues to decrease with the wider Ti strip width because any wider Ti strip width does not affect the diffusion depth. It only increases the surface index, thus giving smaller modal depth. The surface index, however, only reaches a maximum value as we increase Ti width w. This is shown in Figure 1.17. Therefore, as the Ti strip width is increased to a certain point, the modal depth ceases to decrease further at any apparent rate, as shown by both the experimental and simulated results. At this point, lateral diffusion dominates. It is worth to be reminded that the limiting case of increasing width in Ti width is a planar waveguide. Similarly, with the same diffusion parameter, the experiment where Ti thickness is increased to 800 Å can be compared. Fouchet et al. [85] had shown the relation between refractive index change $\Delta n_{\text{e,o}}(Z)$ and Ti concentration $C(Z)$ in the mathematical form of

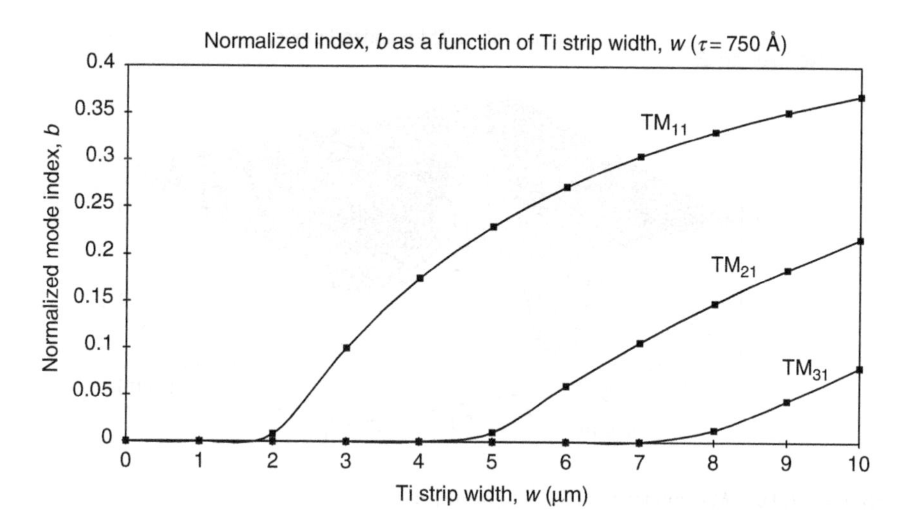

FIGURE 1.14 Normalized mode index, b as a function of Ti strip width, w.

$$\Delta n_{e,o}(Z) = A_{e,o}(C_o, \lambda)\ [C(Z)]^{\alpha_{e,o}} \tag{1.35}$$

The expression shows that the proportionality coefficient $A_{e,o}$ depends not only on the wavelength λ but also on the diffusion parameters which are characterized by C_o, the Ti surface concentration.

The mode intensity distribution of a higher-order mode is illustrated in Figure 1.15 and its field in Figure 1.16. The diffusion profile of Ti ions in the substrate can be seen as

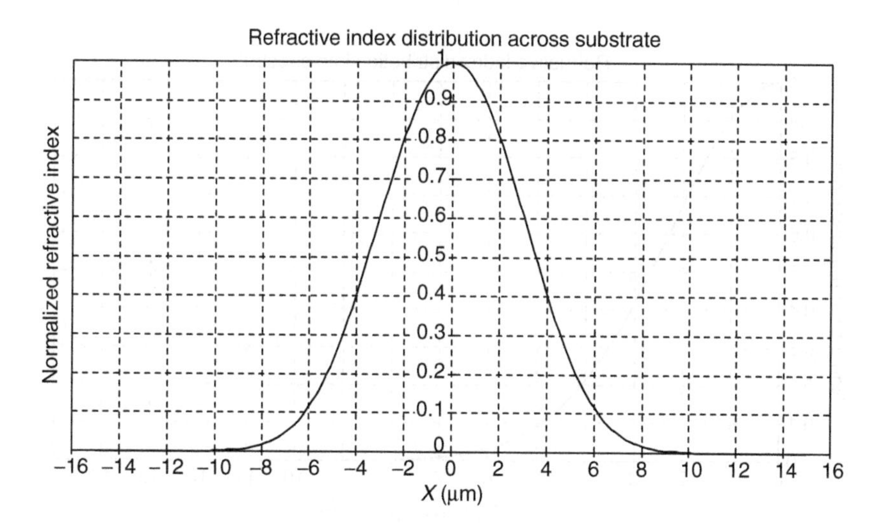

FIGURE 1.15 Lateral diffusion profile of Ti ions.

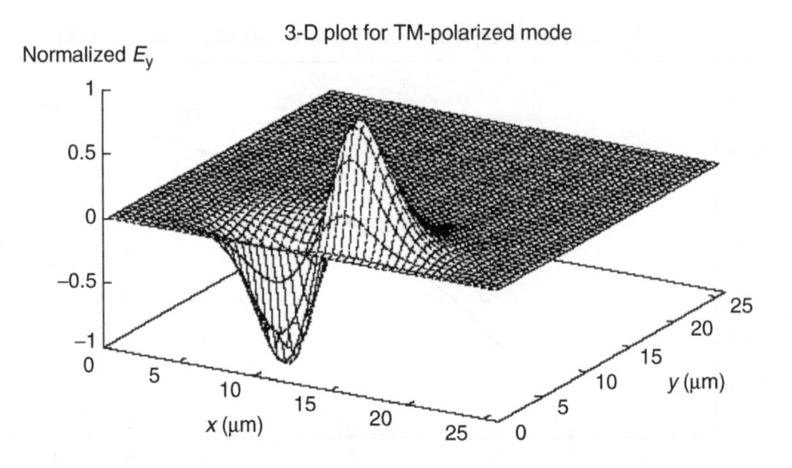

FIGURE 1.16 TM_{21} mode: $\tau = 750$ Å, $w = 10$ μm.

shown in Figure 1.17. The surface concentration of Ti is shown in Figure 1.18 and the variation of the mode field distribution with respect to the diffusion time is shown in Figure 1.19.

1.3.2.2 Linear Electro-Optic Effect

The linear electro-optic (Pockels) effect, which is the basis for active wave guide device control, provides a change in refractive index proportional to the applied electric field. The way in which this index change results in optical switching, intensity modulation, filter tuning, etc., and depends upon the device configurations. A voltage V applied to the electrodes placed over or alongside the waveguide,

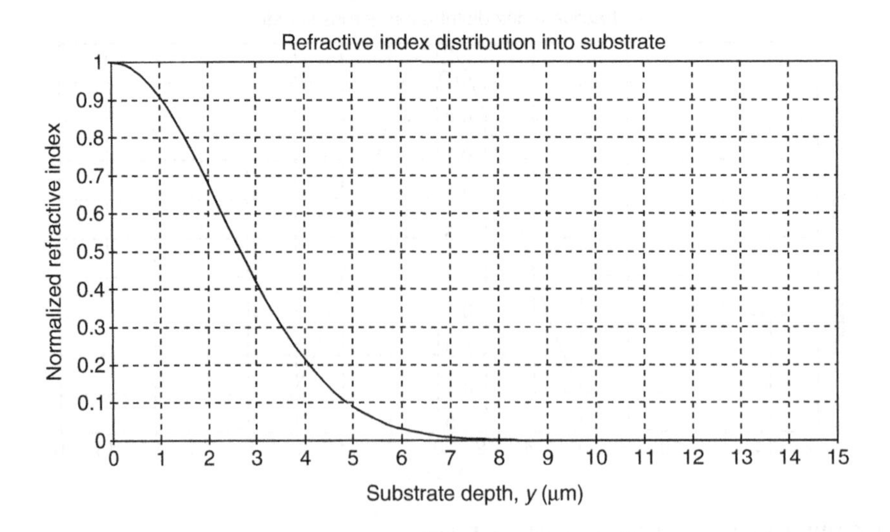

FIGURE 1.17 Depth of the Ti diffusion profile.

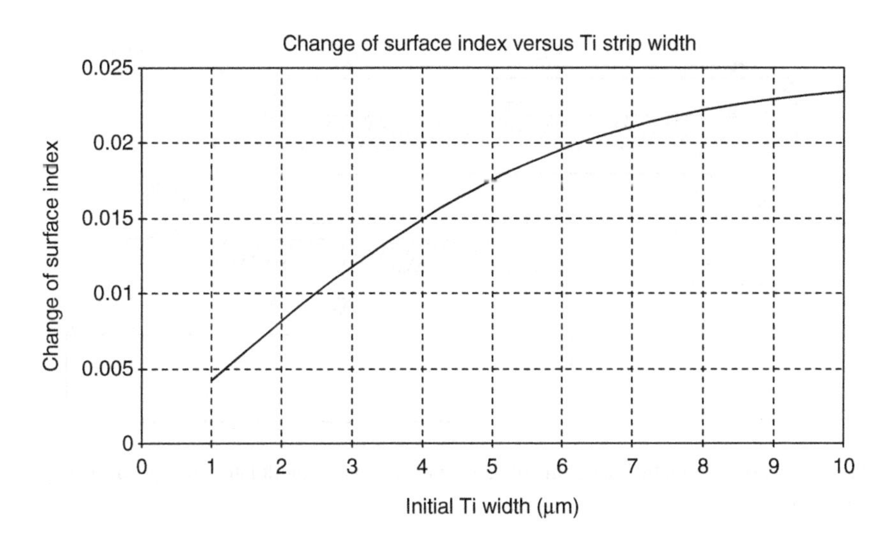

FIGURE 1.18 Change of surface index as a function of initial Ti width.

as shown in Figure 1.20, created an internal electric field of approximate magnitude $|E| \approx V/G$, G being the width of the electrode gap [76,87,88].

The linear change in the coefficients of the index ellipsoid due to an applied electric field (E_j) along the principal crystal axis is [76]

$$\Delta\left(\frac{1}{n_i^2}\right) = \sum_{j=1}^{3} r_{ij}E_j \quad \text{or} \quad (\Delta n)_i = -\frac{n^3}{2}\sum_{j=1}^{3} r_{ij}E_j \tag{1.36}$$

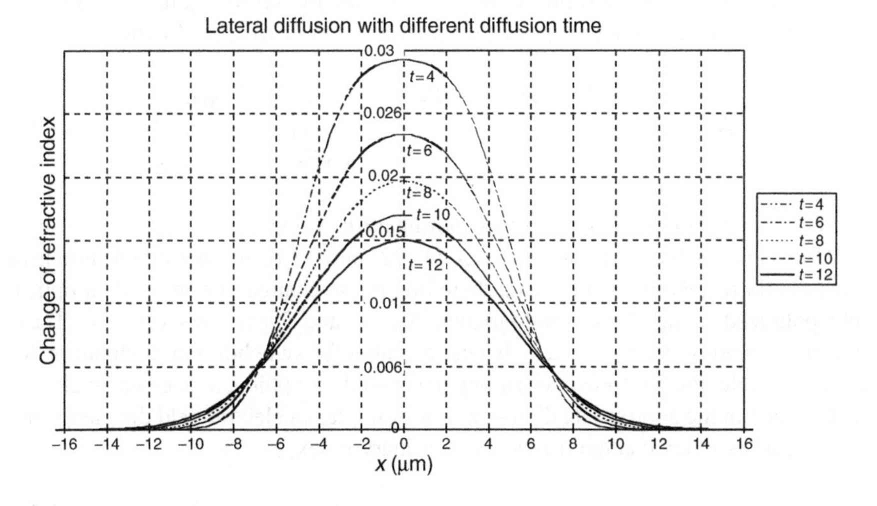

FIGURE 1.19 Lateral diffusion variation of Ti with increasing diffusion time.

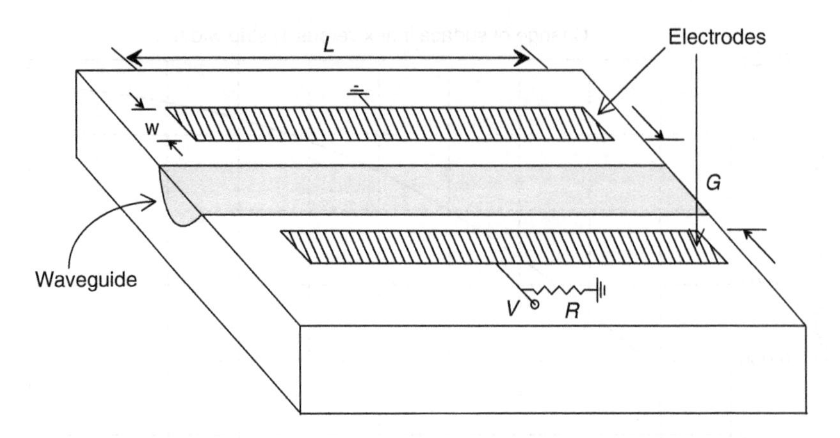

FIGURE 1.20 Schematic diagram of an optical phase modulator using electro-optic effects (EOE).

where $i = 1, 2, \ldots, 6$ and r_{ij} is the 6×3 EO tensor matrix for crystals of class 3 m. Due to the symmetry of the tensor components, the 6×6 tensor matrix can be reduced to 6×3 which takes the form

$$\begin{bmatrix} 0 & -r_{22} & r_{13} \\ 0 & r_{22} & r_{13} \\ 0 & 0 & r_{33} \\ 0 & r_{51} & 0 \\ r_{51} & 0 & 0 \\ -r_{22} & 0 & 0 \end{bmatrix} \tag{1.37}$$

It is noted that in this tensor system $r_{61} = -r_{22} = 3.4$ pm/V, $r_{42} = r_{51} = 28$ pm/V, $r_{12} = r_{22}$, and $r_{23} = r_{13} = 8.6$ pm/V. By inserting the EO tensor r_{ij}, the six values of Δn_i can be written as the elements of a symmetric 3×3 matrix. For $LiNbO_3$, we have

$$\Delta n_{ij} = -\frac{n^3}{2}\begin{pmatrix} -r_{22}E_y + r_{13}E_z & -r_{22}E_x & r_{51}E_x \\ -r_{22} & r_{22}E_y + r_{13}E_z & r_{51}E_y \\ r_{51}E_x & r_{51}E_y & r_{33}E_x \end{pmatrix} \tag{1.38}$$

where n is either the ordinary n_o or extraordinary n_e value.

Utilization of the diagonal elements 11, 22, and 33 of the perturbed refractive index matrix results in an index and, therefore, phase change, for an incident optical field polarized along the crystallographic X-, Y-, and Z-axis, respectively. These diagonal elements effect an index change, essential for switches and modulators, for the optical field aligned (polarized) along the crystallographic j, axis given an electric field applied in the appropriate direction. For example, an electric field directed along E_z (E_3) causes index change (to the extraordinary index, $j = 3$)

$$\Delta n_{33} = -\left(n^3/2\right) r_{33} E_z \tag{1.39}$$

The electrode orientation relative to the waveguide needed to generate E_z depends, of course, upon the orientation of the crystal used. The orientation is frequently specified by the cut, the direction perpendicular to the flat surface on which the waveguide is fabricated.

The off-diagonal elements of the tensor matrix (Equation 1.39), on the other hand, represent the electro-optically induced conversion or mixing between orthogonal polarization components. For example,

$$\Delta n_{13} = -\left(n^3/2\right) r_{51} E_x \tag{1.40}$$

This represents a rotation of the index ellipsoid that causes a coupling proportional to the r_{51} coefficient between the otherwise orthogonal E_{o1} and E_{o3} optical fields due to an electric field applied in the x-direction (E_x). Utilization of off-diagonal EO elements is necessary to induce polarization change in Ti:$LiNbO_3$ waveguides. It is essential for polarization control devices.

The devices with Z-, X-, or Y-cut $LiNbO_3$ are shown in Figure 1.21. When the electrodes are placed on either side of the waveguide, the horizontal electric field $E_{||}$ is applied while the vertical electrical field $E_\perp$ is employed with one electrode placed directly over the waveguide. In the latter case, an insulating buffer layer is required to eliminate loss to the TM-like polarization, as discussed earlier. In either case, the crystal orientation should typically be chosen to use the largest EO coefficient, r_{33} ($=30.9 \times 10^{-12}$ m/V). The contour plot of a TM guided mode can be seen in Figure 1.22 and 1.23. Its 3-D distribution is shown in Figure 1.24.

The total phase shift over the interaction length L is then [76]

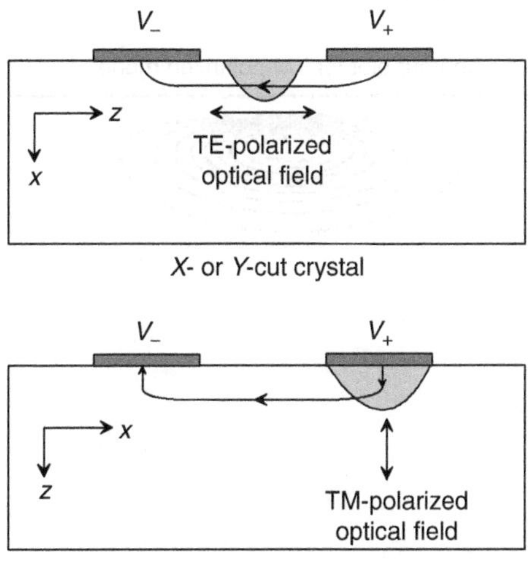

FIGURE 1.21 Position of waveguide and polarization of electrical and optical field for different crystal cut to manipulate the largest electro-optic (EO) coefficient of $LiNbO_3$, r_{33}.

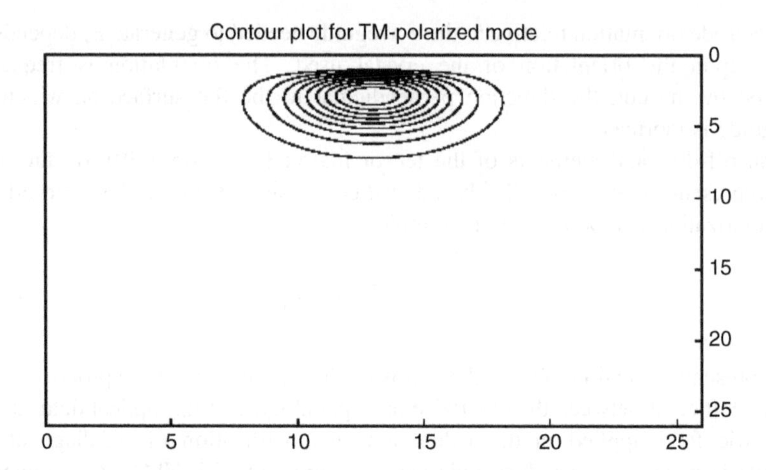

FIGURE 1.22 Contour plot of the (transverse magnetic) TM modal field of a diffused channel waveguide. The unit of the horizontal axis is micro-meters and the vertical axis is arbitrary units.

$$\Delta\beta L = -\pi^3\Gamma\frac{V}{g}\frac{L}{\lambda} \tag{1.41}$$

where Γ is the overlap integral defined as

$$\Gamma = -\frac{\pi n_e^3 r_{33} L}{\lambda}\iint |E_o(x,y)|^2 E_m(x,y)\mathrm{d}x\,\mathrm{d}y \tag{1.42}$$

where

$E_o(x, y)$ is the normalized optical field polarized along the z-crystal axis
E_m is the applied microwave

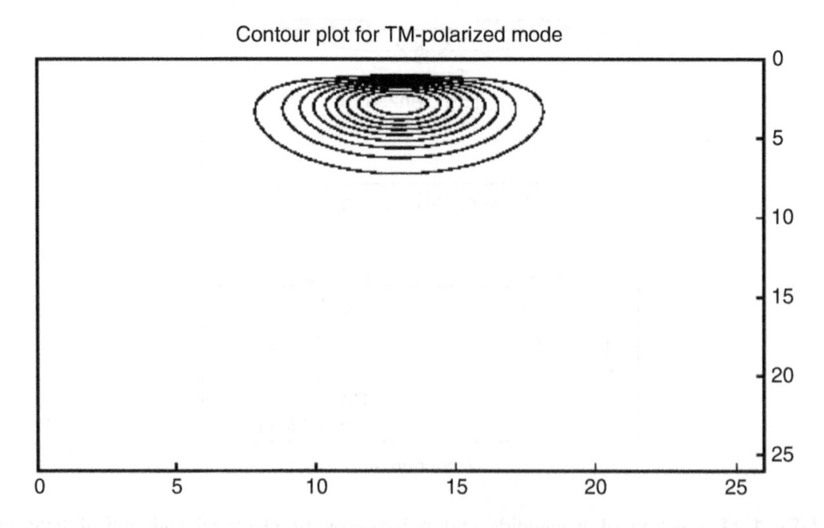

FIGURE 1.23 Vertical mode profile of a diffused channel waveguide. The unit of the horizontal axis is micro-meters and the vertical axis is arbitrary units.

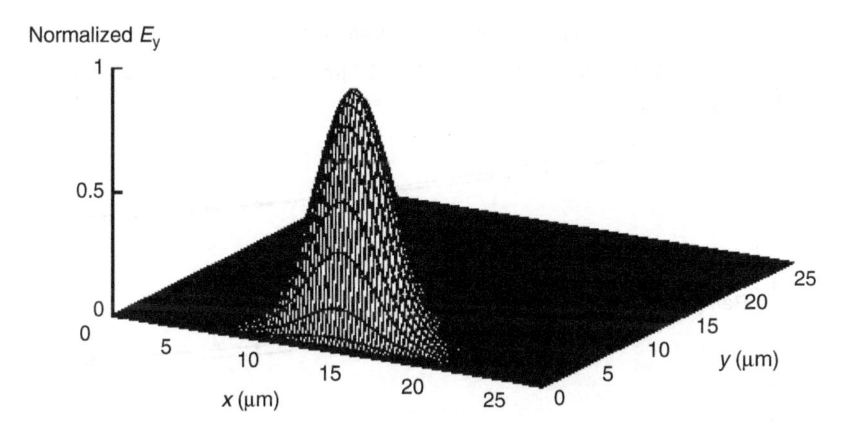

FIGURE 1.24 Horizontal mode profile of a diffused channel waveguide.

For design purpose, $E_o(x, y)$ can be defined as

$$|E_o(x, y)|^2 = \frac{4y^2}{w_x w_y^3 \pi} \exp\left[-\left(\frac{x-p}{w_x}\right)^2\right] \exp\left[-\left(\frac{y}{w_y}\right)^2\right] \tag{1.43}$$

where the widths of 1/e intensity are 2 w_x and 1.376 w_y, respectively and p is the peak position of the optical field in the lateral direction. The experimental results [10] showed that the quasi-TM mode had, in 1/e intensity of 3.9 μm in the lateral direction and a width of 2.8 μm in the depth direction, been achieved by the diffusion of 4 μm width and 0.08 μm thick Ti strip, carried out for 6 h at 1025°C. The above expression can be used to model the overlap integral for simulation purposes with w_x and w_y both being ~2 μm as shown in Figure 1.25. In fact, the mode modeling of the waveguide would enable us to predict the mode sizes; thus, the values of w_x and w_y of a waveguide are fabricated under a particular diffusion condition as shown in Figure 1.23.

The voltage for a π phase shift, see Equation 1.41, denoted as V_π, is defined as

$$V_\pi = \frac{\lambda g}{n_e^3 r_{33} L \Gamma} \tag{1.44}$$

In the design of a switch or modulator, it is advantageous to obtain a low drive voltage, wide modulation bandwidth, and low optical insertion loss. It can be shown that the achievable bandwidth also scales as $1/L$. Therefore, for a given EO material and wavelength, one tries to minimize the voltage length product, $V \times L$, by optimizing the lateral geometric parameter G/Γ. It can also be shown that the modulation bandwidth and, more importantly, the insertion loss also depend upon the lateral geometric parameters of the traveling wave electrode, thus resulting in an interdependence that may require performance trade-offs.

In the summary for X-cut Y-propagating crystal, a horizontal electric field of the lightwaves is parallel to the surface of the crystal and hence along the Z-direction, the

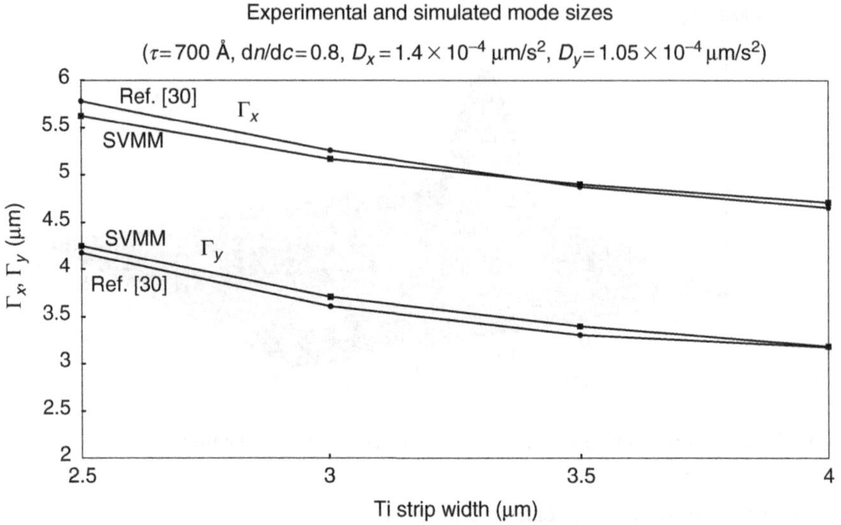

FIGURE 1.25 Overlap integral of the guided mode in the x and y directions with the driving electric signals as calculated using the approximation method semi-vectorial mode modeling (SVMM).

transverse electric (TE)-polarized optical guided mode would be modulated. For X-cut crystals, an electric field vertical to the substrate is modulated. For optical waveguide along the Y-axis of the crystal the TM optical field would be modulated. The TM mode is polarized along the vertical axis, the Z-axis. For both cases, the electric field of the traveling RF waves and the guided lightwaves is aligned along the Z-axis, thence r_{33} is used which is the most effective EO coefficient.

The modulation of the guided lightwaves is also strongly dependent on the velocity matching between them. The dielectric constant of $LiNbO_3$ varies drastically from the microwave to optical region due to large ionic contribution to its electronic bonding structure. For Z-cut substrate, the relative dielectric constants for the parallel and perpendicular to the crystal plane are $\varepsilon_{ry} = 43$ and $\varepsilon_{rz} = 28$, respectively. Thus for a transmission line along the Y-direction, the effective relative dielectric constant under quasi-static approximation can be approximated by

$$\varepsilon_{\text{eff}} = \sqrt{\varepsilon_{ry}\varepsilon_{rz}} \tag{1.45}$$

which is about 35. The microwave refractive index depends on the structure of the traveling wave electrodes whether it is coplanar waveguide (CPW) or asymmetric CPW. For the common CPW, the effective microwave index can be estimated by

$$n_{\mu} = [(\varepsilon_{\text{eff}} + 1)/2]^{1/2} \tag{1.46}$$

and thus much larger (about four times) than that of the optical refractive index (square root of the dielectric constant) of about 2.17. As a result, the RF microwave is traveling much slower than that of the lightwaves. Thence, the matching of the velocities between the RF and optical waves is very critical to extend the bandwidth of the optical modulators.

The most popular method is using a SiO_2 buffer layer deposited on the surface of the substrate and the electrode is fabricated on top of this buffer layer of thickness of about 1 μm to reduce the effective microwave index of the traveling wave, and hence increase the traveling wave velocity.

1.3.3 Electro-Absorption Modulators

1.3.3.1 Electro-Absorption Effects

The EA effects arise when the EA coefficient of the material in a single-mode optical waveguide results in the modulation of the intensity of the lightwaves passing through. Group III–V semiconductors, such as InGaAs, InAlAs, InAsP, InGaP, and InGaAlAs grown on GaP substrate, are commonly used as the EA-guided media for optical modulation.

The EA effects have two distinct types in bulk (the Franz–Keldysh effect [FKE]) and quantum well-confined (Stark effect) structures which can be used to produce the electro-refractive Mach–Zehnder modulator (MZM) with appropriate detuning energy. The band diagram of the FKE in bulk medium is shown in Figure 1.26a and b under the influence of an applied electric field. E_c and E_v represent the conduction and valence band, respectively.

In practice, the quantum well-confined structures are commonly used with the design and integration of optical waveguide structures. An EA waveguide usually consists of p-i-n semiconductor layers sandwiched with the p- and n-type layers, hence layers of high and low refractive indices. This is similar to a rib waveguide structure, which provides the confinement of the lightwaves. The lateral confinement of the propagating lightwaves is normally achieved by deep mesa etch, the ridge waveguide structure. The deep-etched ridge waveguide can then be planarized using polyimide or regrown semi-insulating InP.

When an electric field is applied across a quantum well (QW), the band diagram is tilted; hence, the electron and hole confinements are changed and the energy gap between the valence and conduction bands is reduced as illustrated in Figure 1.26. As a result, the exciton absorption peaks are shifted to longer wavelength region, the redshift. Furthermore, the exciton line width is also broadened due to the influence of the applied electric field. This causes a significant increase of the optical absorption at the edge of the spectrum in the longer wavelength region.

In a semiconductor QW, both electrons and holes are tightly confined in a narrow well so that the electron and hole energies are quantized to form discrete sub-bands energy levels. In the design of optical modulator for advanced optical communication system, it is desirable that (1) the optical waveguide is single mode and (2) the guided mode is well confined to enable high efficiency in modulation voltage and hence high ON–OFF distinct ratio and ultra-broadband. Furthermore, the total insertion loss is also very critical. Typically, EA propagation loss is about 15–20 dB/cm while that of Ti:$LiNbO_3$ is only just 0.22 dB/cm. Three sources contribute to this high propagation loss: the first is the residual absorption loss in the active layer; the second is the

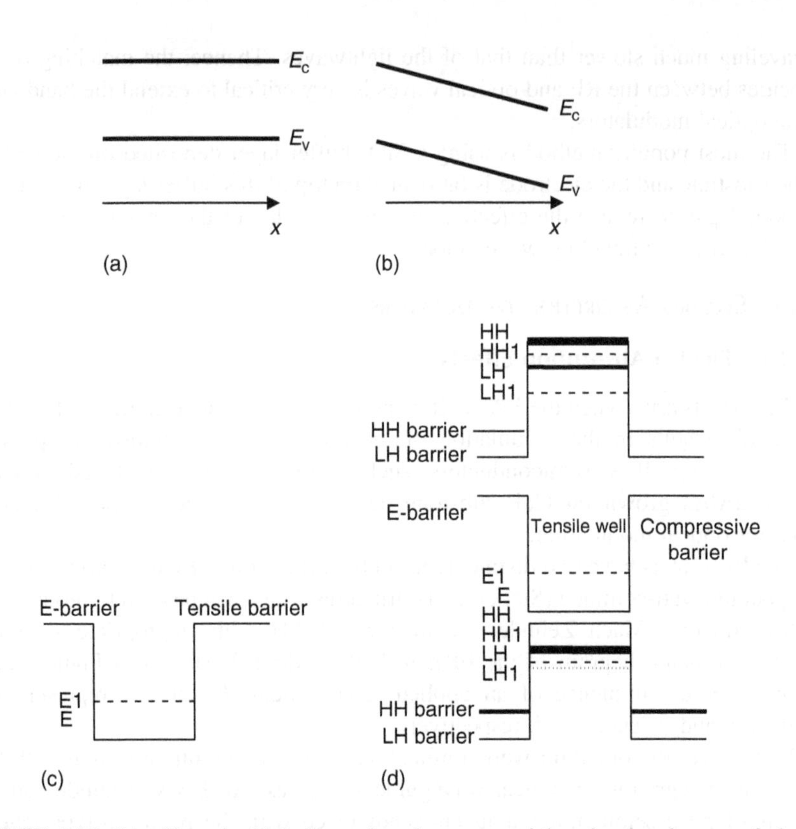

FIGURE 1.26 EA effects in (a) bulk semiconductor material under the absence of electric field, (b) under applied field and in quantum well (QW) structure, (c) compressive quantum well (QW) and tensile barrier, and (d) tensile QW and compressive barrier.

interband absorption loss induced by free carriers in the highly doped layers; and the third is the scattering loss caused by the roughness of the side walls of the ridge waveguides and the interfaces between waveguide regions.

The multiquantum well (MQW) structures can have high electron barrier and low hole barrier by introducing compressive strain at the well layer and tensile strain at the barrier layer as shown in Figures 1.26c and d. Furthermore, the control of the high and low barriers by compression and tensile would allow the design and implementation of the refractive index profile of the ridge waveguide and hence the design of polarization insensitive QW EA modulators. Consequently, the typical width of the QW is about 80–100 Å to achieve high electron barrier and relatively low hole barrier to increase the escape of the heavy holes out of the well [11]. A schematic diagram of the EA modulator is shown in Figure 1.27 using the group III–V compound semiconductor layers structured in a QW. The quantum-confined Stark effect ensures the confinement of the electron and holes. As a result, they overlap and interact strongly and form a bond similar to a hydrogen atom. This is called, as mentioned above, the exciton, whose spectra have a strong absorption and localized in the vicinity of the wavelength corresponding to the bandgap of the QW. When an electric field is applied to the

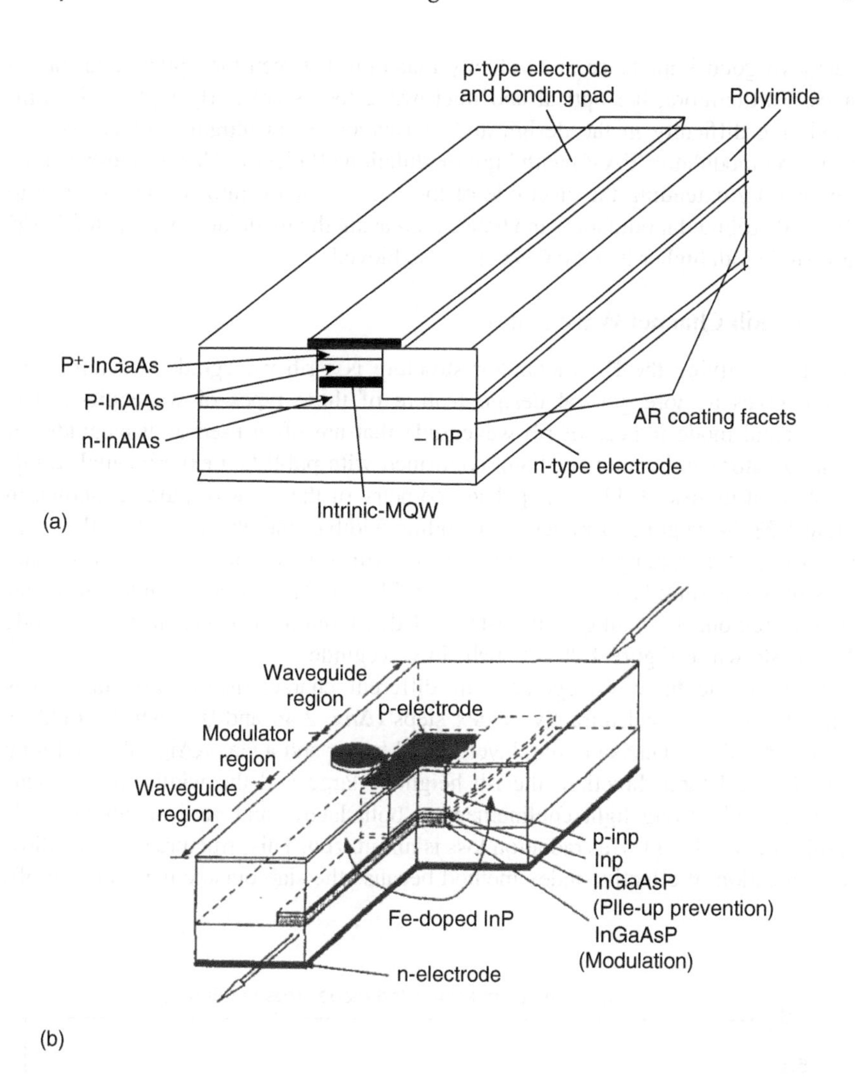

FIGURE 1.27 Schematic of a multiquantum well (MQW) electro-absorption modulator (EAM): (a) generic structure and (b) a high-speed modulator.

electrode, electrons and holes are forced to the opposite ends of the QW and are physically separated. Thence, the spatial overlap or interaction of these carriers is reduced and the excitonic absorption is decreased and the spectra broadened. This thus makes possible the modulation of the lightwaves passing through the optical waveguides formed by the QW structure whose refractive index must be formed so that guided modes can exist. This type of waveguide is usually called the rib waveguide. The typical dimensions of the modulator are waveguide width of 1–3 μm, depth of 30–50 μm, and length of 50–300 μm. Ohmic contacts are formed across the rib waveguide and a low dielectric constant polyimide is used to reduce the device capacitance hence higher bandwidth. For a typical EA waveguide, it is very difficult to achieve good impedance and velocity matching between the optical and microwaves. Furthermore, the optical and microwave losses are fairly high. This compounds the difficulty in the design and fabrication of an ultrahigh speed electro-absorption modulator (EAM) with high modulation efficiency. This difficulty can be overcome by extending the electrode of the EA modulator into several sections to achieve distributed modulation and hence increasing the modulator bandwidth. EAM with bandwidth higher than 50 GHz can be achieved.

1.3.3.2 Rib Channel Waveguides

Having identifying the EA modulator structure is a rib waveguide structure. This section gives a more general design feature of these types of waveguides. The fundamental mode indices of rib waveguide that are often used as waveguide for EA modulator can be simulated and confirmed with published experimental results of polarized modes [3–11,72–75]. The geometry of the rib waveguide is shown in Figure 1.28. Waveguide parameters, including width of the rib w, height of the rib h, thickness of the guiding layer underneath the rib d, index of the substrate n_s, and index of the guiding layer n_g, are listed in Table 1.1. The refractive index of the air cladding region, n_c is unity. The 3-D field distribution of a typical guided mode TM_{31} is shown in Figure 1.29 for such rib waveguide.

Each of the three waveguides has different characteristics. Structure 1 has relatively large vertical refractive index steps ($\Delta n = 2.44$ and 0.1) which could, e. g., correspond to a GaAs guiding layer bound by air and a $Ga_{0.75}Al_{0.25}As$ confining layer. In the lateral direction, the rib height is large and the width narrow. This structure, with strong light confinement in both lateral and vertical directions, is useful for curved guides, as radiation loss is minimized. This structure does not allow the application of effective index method because the slab outside the rib is cut off.

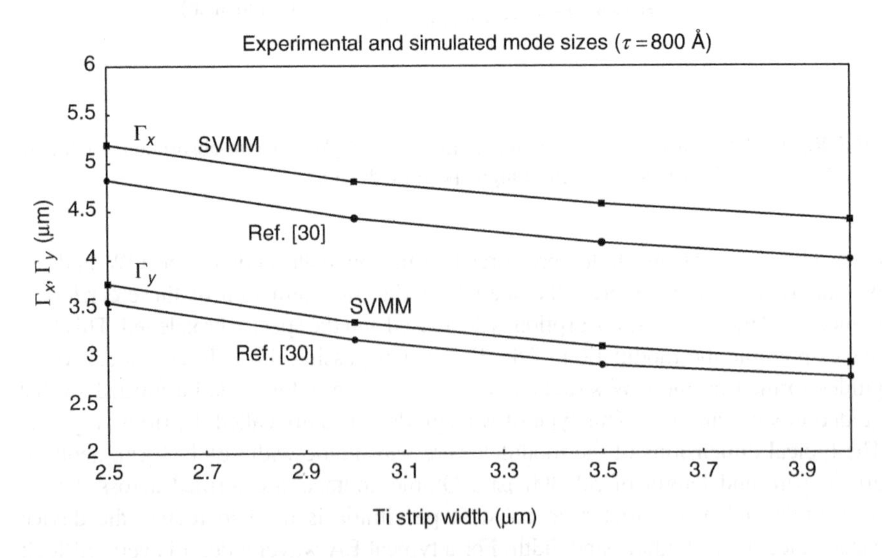

FIGURE 1.28 Comparison of simulated and experimental mode sizes for $\tau = 800$ Å.

TABLE 1.1
Typical Parameters of a Three Rib Waveguide at 1550 nm Wavelength

Rib Guide	n_g	n_s	d (μm)	h (μm)	w (μm)
Structure 1	3.44	3.34	0.2	1.1	2
Structure 2	3.44	3.36	0.9	0.1	3
Structure 3	3.44	3.435	3.5	2.5	4

Structure 2 shows a weakly guiding feature. In this case, the rib height is much less, allowing the mode to extend laterally. This is particularly useful for directional coupler structures, as strong coupling between adjacent guides leads to short coupling lengths. The guiding layer thickness is made small to give a thin mode shape in the vertical direction, and thus low voltage operation. Essentially, this structure is strongly confined vertically and weakly confined horizontally. Such features enable the application of effective index method because the small etch step and large width to height ratio are the conditions of validity of this approximate method.

Structure 3 gives a good matching of its spot size and that of the single-mode fiber. Insertion loss is a crucial parameter for most waveguide devices, and is determined by propagation loss and losses due to mode mismatch. Fresnel reflection loss is also important, but can be reduced to insignificant levels by using $\lambda/4$ antireflection coatings. Mode profiles of a circularly symmetric optical fiber and a waveguide will, in general, be different, due to the differing refractive indices of the semiconductor and the fiber, and also the differing shapes of the modes. The effects of both these factors may be alleviated by the use of appropriate waveguide designs. In Structure 3, the guiding layer is relatively thick, and the stripe width and height are adjusted to give a more symmetric mode shape. In this structure, the slab mode is near cutoff. Again, it is also to be pointed out that because the rib height is nearly twice the slab thickness and the rib width is less than the rib height, the accuracy of the effective index method is expected to be poor. Figures 1.30 through 1.35 show

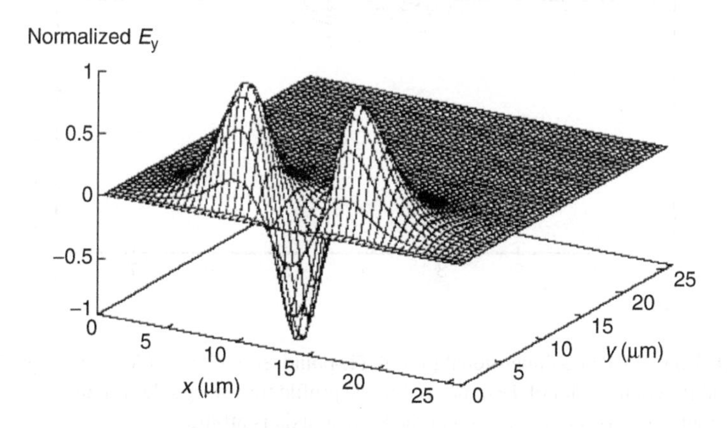

FIGURE 1.29 TM_{31} mode: $\tau = 750$ Å, $w = 10$ μm.

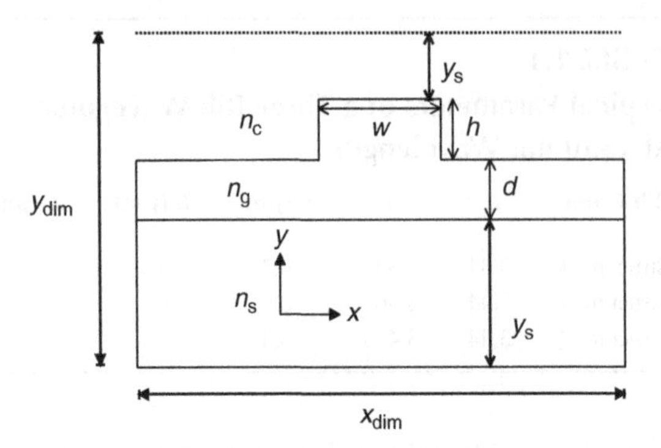

FIGURE 1.30 Typical structure of rib waveguide.

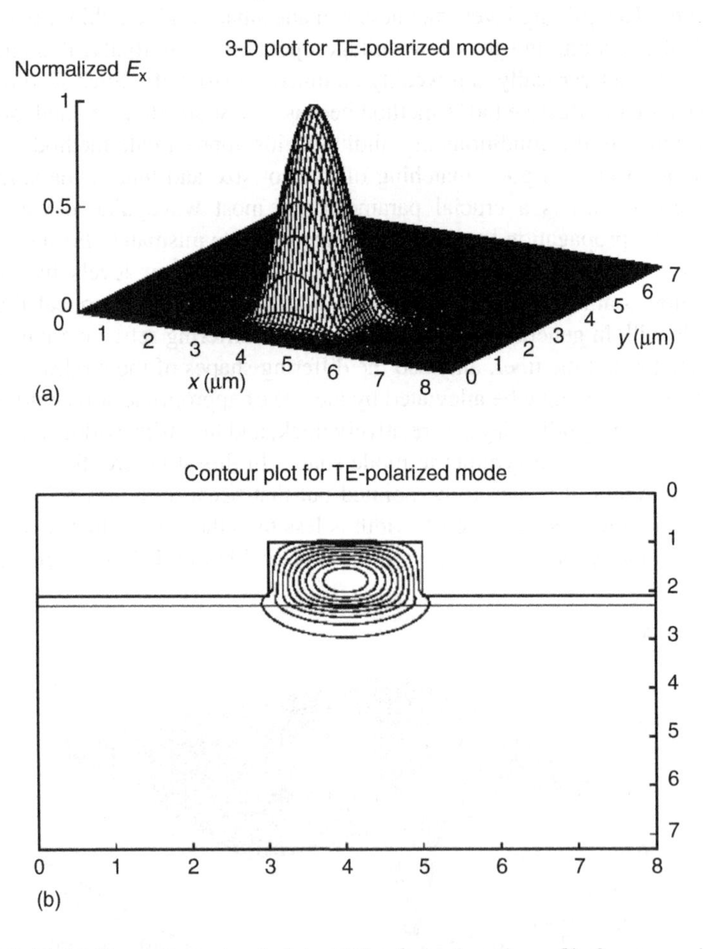

FIGURE 1.31 (a) Three-dimensional plot of TE-polarized mode profile for waveguide structure 1 and (b) contour plot of TE-polarized mode profile for waveguide structure 1. The unit of the horizontal axis is micro-meters and the vertical axis is arbitrary units.

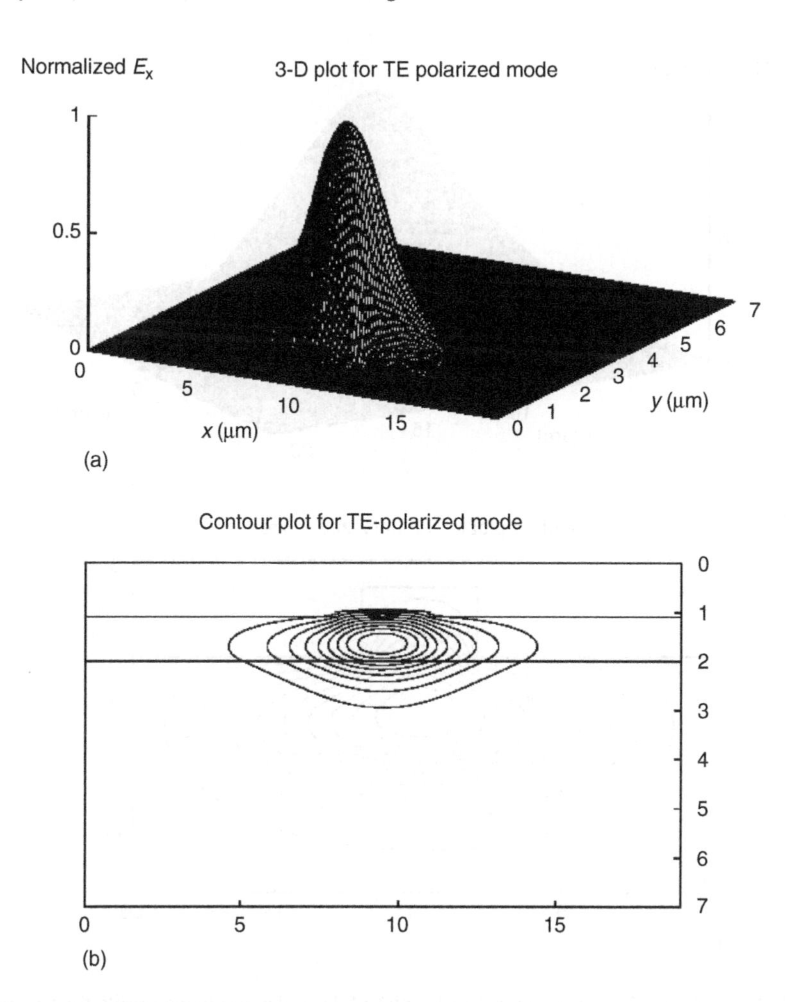

FIGURE 1.32 (a) Three-dimensional (3-D) plot of TE-polarized mode profile for waveguide structure 2 and (b) contour plot of TE-polarized mode profile for waveguide structure 2.

the contour plots and 3-D simulated plot of the TE-polarized mode of the three waveguide structures using finite difference method with grid sizes h_x and h_y are 0.1.

1.3.3.2.1 Higher-Order Modes

Higher-order eigenmodes of rib waveguides can be estimated. Table 1.2 gives typical values of the parameters of the waveguide structure. Figures 1.36 through 1.39 show the fundamental mode and the leading asymmetric mode of the guided TE-polarized optical field.

The leading asymmetric mode of Figure 1.38 can be obtained with an initial eigenvalue that is close to the eigenvalue of the leading asymmetric mode. The eigenvalue of the fundamental mode (see Figure 1.36) can be calculated as 347.78889. Other eigenmodes can also be estimated without difficulty by using the finite difference model. We note that there are only a limited number of eigenmodes

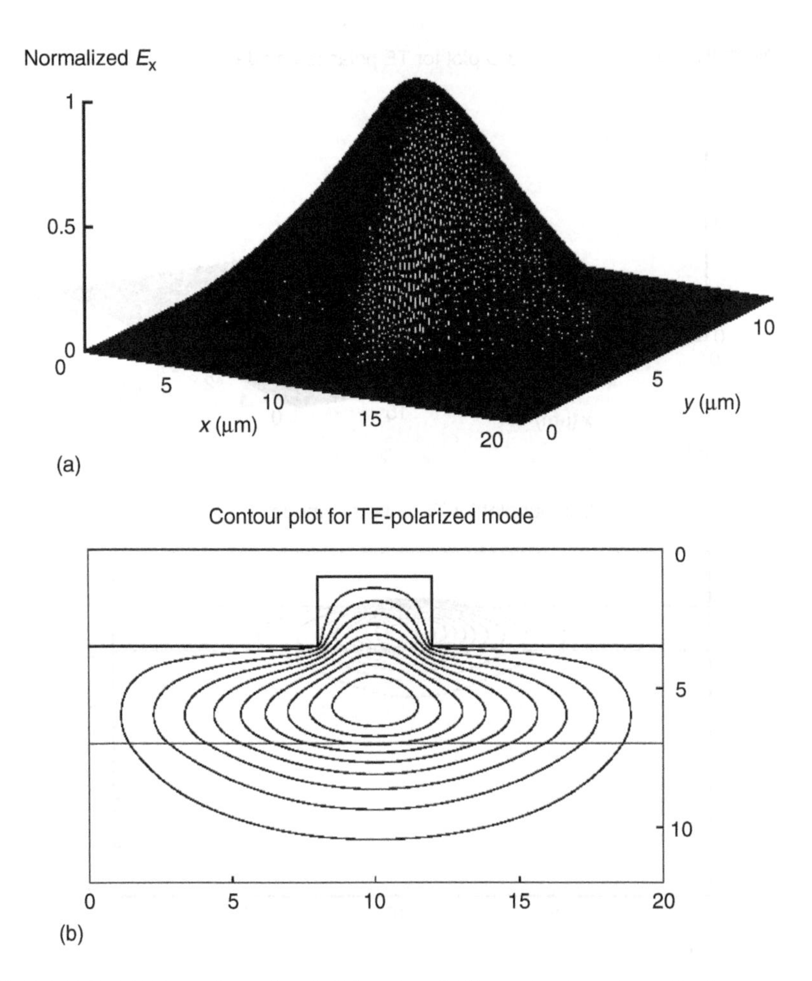

FIGURE 1.33 (a) Three-dimensional (3-D) plot of TE-polarized mode profile for waveguide structure 3 and (b) contour plot of TE-polarized mode profile for waveguide structure 3.

possibly supported by a certain waveguide structure. These modes can be illustrated in Figures 1.40 and 1.41.

1.3.4 Operational Principles and Transfer Characteristics

1.3.4.1 Electro-Optic Mach–Zehnder Interferometric Modulator

1.3.4.1.1 Operating Principles

The operational principle of such device is simple. The waveguide has a single accessible input and output port. It has been an especially popular structure for ON/OFF modulation because of its simplicity. The *Y* branches at the input and output port, with an appropriate branching angle, typically less than 1°, serve as a 3 dB splitter and combiner. Figure 1.6 shows a schematic diagram with a typical configuration of the MZIMs. The asymmetric coplanar electrode structure is

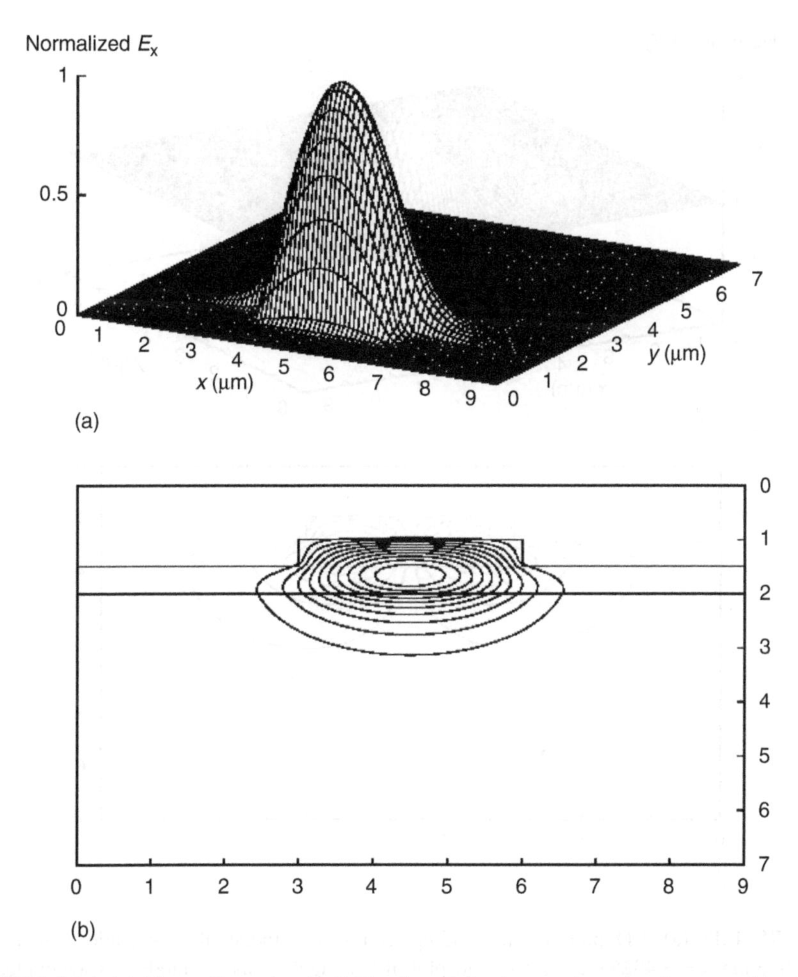

FIGURE 1.34 (a) Three-dimensional (3-D) plot of fundamental mode of waveguide. The calculated effective index = 3.4133105. (b) Contour plot of the fundamental mode of the waveguide.

employed in this configuration. The input light waves are split into equal paths. Each propagates over an individual arm of the interferometer. The arms are to be sufficiently separated to prohibit evanescent coupling between them, usually less than 1° angle of arc at the *Y*-junction for diffused optical waveguides. The lengths of the optical paths of the two arms are usually equal. Therefore, if no phase shift is introduced between the propagating light waves, the two split lightwaves combine in phase, constructively at the output *Y*-branch 3 dB combiner and continue to propagate undiminished in the output waveguide. However, if an EO phase difference of π is introduced between the two components, the combined mode distribution would become a double-mode-like; such mode is supported by the single-mode waveguide and hence would radiate into the substrate and the transmitted lightwave is at a minimum state.

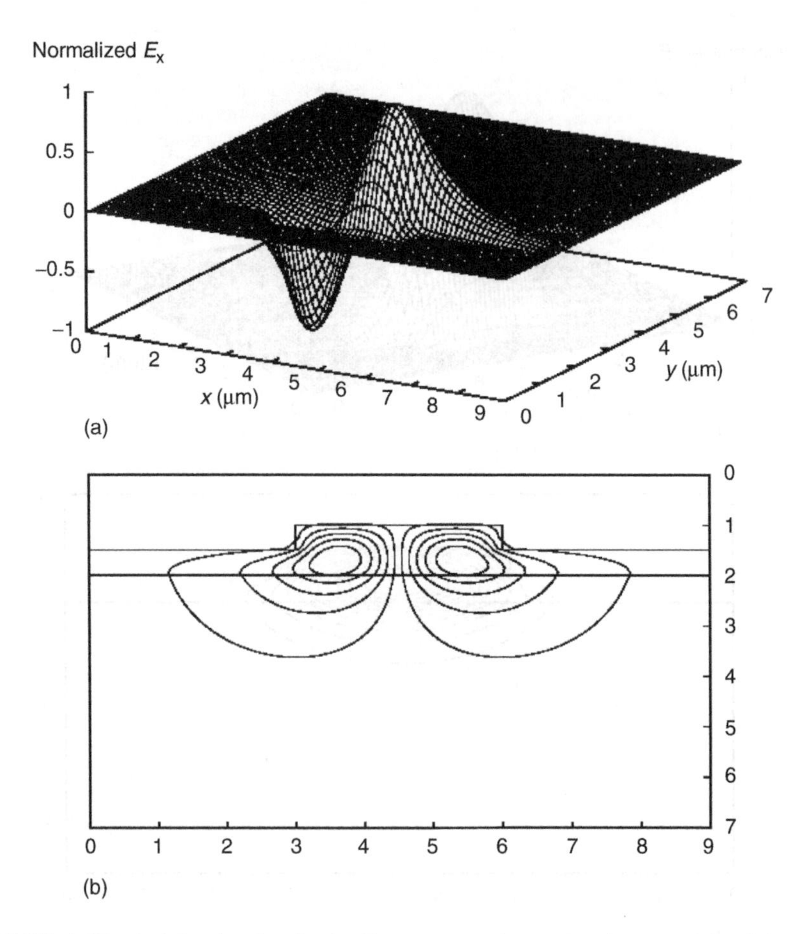

FIGURE 1.35 (a) 3-D plot for the leading asymmetric mode of waveguide. Calculated effective index = 3.4025302. (b) Contour plot of leading asymmetric mode of waveguide.

Similar to the phase modulator discussed above, the phase shift can be introduced by mounting traveling wave electrodes on the waveguide to provide the electric field that can perturb the refractive index electro-optically. Hence, the voltage V_π would be the voltage that causes a phase shift of π hence switching the modulator off. The expression for V_π is the same as the one of the phase modulators in Equation 1.44 except that the overlap integral, Γ, would be larger

TABLE 1.2
Parameters of a Rib Waveguide at 1550 nm Wavelength

Guide	n_g	n_s	d (μm)	h (μm)	w (μm)
	3.44	3.40	0.5	0.5	3

TABLE 1.3
Modulators and Modulation Techniques for Long-Haul Optical Transmission

Modulation–Amplitude (Amplitude Shift Keying [ASK] or ON–OFF Keying [OOK])	Phase Modulation (Phase Shift Keying [PSK] or Differential Phase shift Keying [DPSK])	$LiNbO_3$ Mach–Zehnder Interferometric Modulator (MZIM)	Electro-Absorption Modulator (EAM)	Electro-Optic (EO) Polymer MZIM
Non-return-to-zero (NRZ)	n.a.	Yes	Yes	Yes
Return-to-zero (RZ)	n.a.	Yes	Yes	Yes
Carrier-suppressed return-to-zero (CS-RZ)	n.a.	Yes	No	Yes
Duo-binary	n.a.	Yes	No	Yes
Multilevel	n.a.	Yes	Yes	Yes
n.a.	Binary PSK	Yes	No	Yes
n.a.	DPSK	Yes	No	Yes
n.a.	Differential quadrature PSK (DQPSK)	Yes	No	Yes
n.a.	Multilevel M-ary PSK	Yes	No	Yes

than the one of the phase modulators with a push–pull phase shift that introduces a phase shift of equal and opposite magnitude in each arm. Such push–pull phase shift is achievable by using the appropriate traveling electrode structure. Table 1.3 tabulates the driving conditions and voltage level required for different advanced modulation formats for optical modulators of $LiNbO^3$ and semiconductor types which are commonly used in photonic signal processor.

1.3.4.1.2 Design Consideration

There are four major design criteria for high quality traveling wave electrodes. They are (1) good microwave-optical velocity match; (2) low microwave electrode losses; (3) a low V_π; and (4) an electrode characteristic impedance near 50 Ω. With $LiNbO_3$, it is difficult to concurrently satisfy all these requirements, and hence trade-offs may be necessary. A good microwave-optical velocity match involves having the effective indices of the optical mode and the guided microwave mode to be as close as possible so that the achievable bandwidth is higher. The general expression for the achievable small signal bandwidth [76] is given by

$$\text{Bandwidth} \approx 1.4c/[\pi L(n_m - n_o)] \tag{1.47}$$

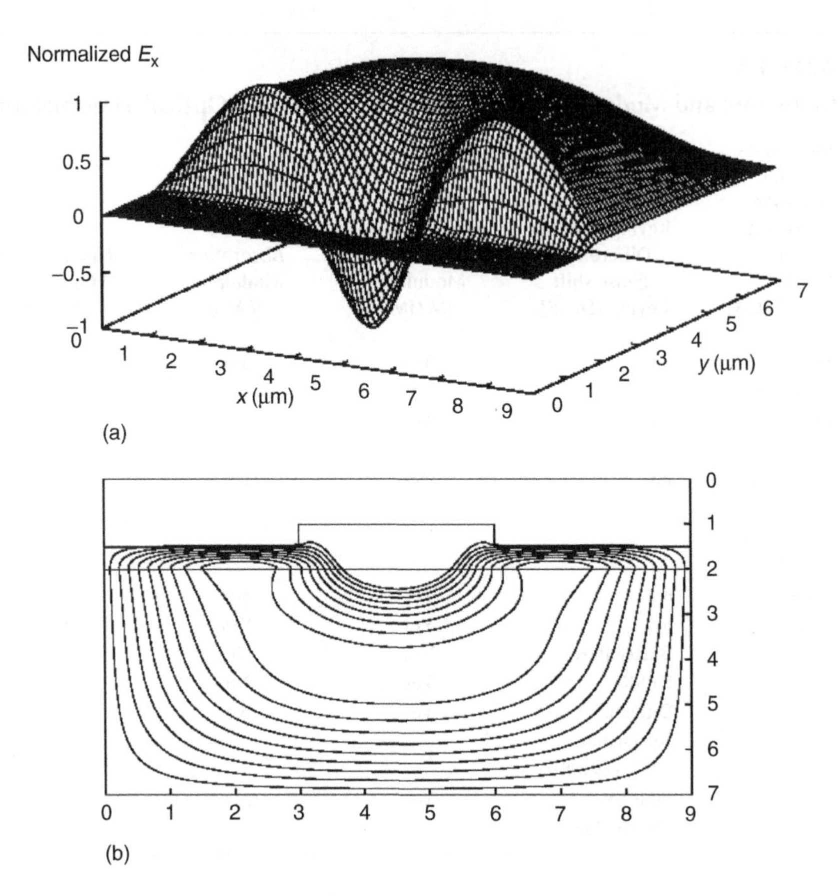

FIGURE 1.36 (a) 3-D plot of the third-order mode which is not supported by the waveguide structure. Calculated effective index = 3.3980958, normalized index = −0.0473143. (b) Contour plot of radiated mode.

where n_o and n_e are the effective microwave index and the effective index of the optical mode. As we can see in this expression, the achievable bandwidth is significantly larger if the difference between n_m and n_o is small. For $LiNbO_3$, the effective index n_o of the guided optical mode is ≈2.146 at 1.3 μm and 2.138 at 1.5 μm. For Z-cut $LiNbO_3$, the effective index n_m of the guided microwave mode, derived by using a static approximation, is given as

$$n_m = \left[(\varepsilon_x \varepsilon_z)^{1/2} + \frac{1}{2}\right]^{1/2} \tag{1.48}$$

where ε_x and ε_z are the relative dielectric constants, which take up the value of 43 and 28, respectively. The above equation gives n_m the value of 4.225, which is nearly twice the value of n_o. Thus, in $LiNbO_3$, compared to the static value, the

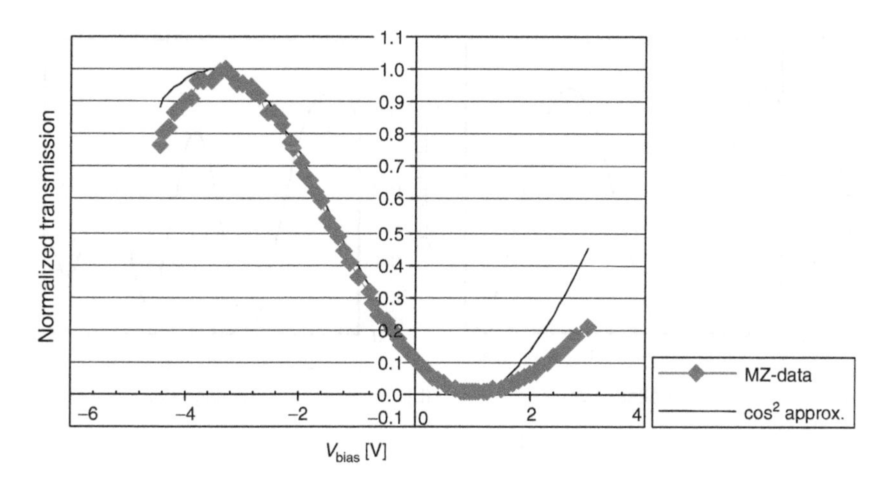

FIGURE 1.37 Measured transmission function and theoretical $\cos^2$ fit for $LiNbO_3$ MZIM.

microwave mode should be sped up by a factor of nearly 2 to obtain velocity match and broadband performance. This can be accomplished by employing thick traveling wave electrodes in conjunction with intervening SiO_2 buffer layer to provide for partial propagation of the microwave mode in lower dielectric constant medium.

Similar to the phase modulator mentioned earlier, a low V_π is achieved by appropriate design of the electrode width to gap ratio, w/g and the fabrication of optical waveguide with relatively low mode size so that the electro-optical overlap integral can be maximized. While electrode width to gap ratio affects the overlap integral Γ, it also affects the characteristic impedance Z of the traveling electrodes. Furthermore, the physical dimensions of the waveguide will also affect the microwave electrode losses with the copper losses being the major loss factor.

The design of thick traveling electrodes with intervening SiO_2 buffer layer is not a trivial one. It needs to be designed in such a way that the design criteria outlined above are satisfied as closely as possible. It is therefore important to understand the characteristics of different traveling wave electrode structures namely CPW, coplanar strip (CPS), and asymmetric coplanar strips (ACPS) and the effect of their geometrical parameters, such as interelectrode gap g, electrode width w, electrode thickness t, and the buffer layer thickness t_b on the design consideration. Owing to the fact that the electrode thickness and SiO_2 buffer layer thickness are involved, analytical technique such as conformal mapping is no longer be applicable. That is why numerical technique becomes advantageous. In the next section, we provide a general description of how the traveling wave electrodes are mounted onto the diffused waveguide.

1.3.4.1.3 Traveling Wave Electrodes

After the formation of the diffused waveguide structure, a SiO_2 thin-film layer may be deposited to enhance the matching between the lightwaves and the microwave or

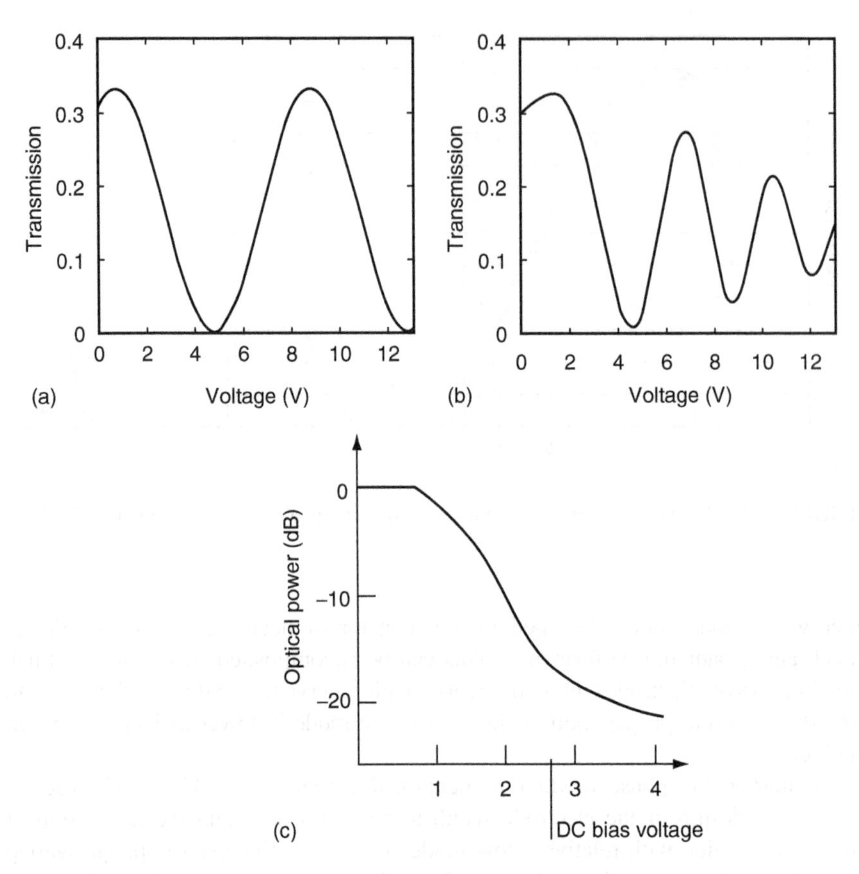

FIGURE 1.38 Transmission function of (a) EO $LiNbO_3$ modulator, (b) electro-absorption modulator (EAM), and (c) one cycle of EAM. *Note*: The complete depletion of the $LiNbO_3$ and nonzero minimum of the EAM.

millimeter wave. The traveling electrode and earth plane are then deposited onto the crystal surface and aligned on top of or at an appropriate position over the waveguide structure such that an optimum field is established for the largest EOE. If an electrode is to be placed on top of the waveguide, an intermediate (typically dielectric) buffer layer is needed to reduce current induced heating losses to the TM-polarized (optical mode polarized perpendicular to the crystal plane). SiO_2 is frequently employed, typically deposited by atmospheric chemical vapor deposition (CVD). For long wavelength ($\lambda = 1.55$ μm) operation, a 0.2 μm thick CVD SiO_2 layer eliminates measurable loading loss. Various kinds of dielectric materials, such as Al_2O_3, Si_3N_4, and indium tin oxide (ITO), have also been tested as buffer layer [13]. Among them, ITO is one of the most interesting materials, which behaves as a dielectric material, capable of being a buffer layer at optical frequencies and as a conductive material, capable of being an electrode at microwave frequencies. Its application, being often used to improve the stability against the DC drift, is directed

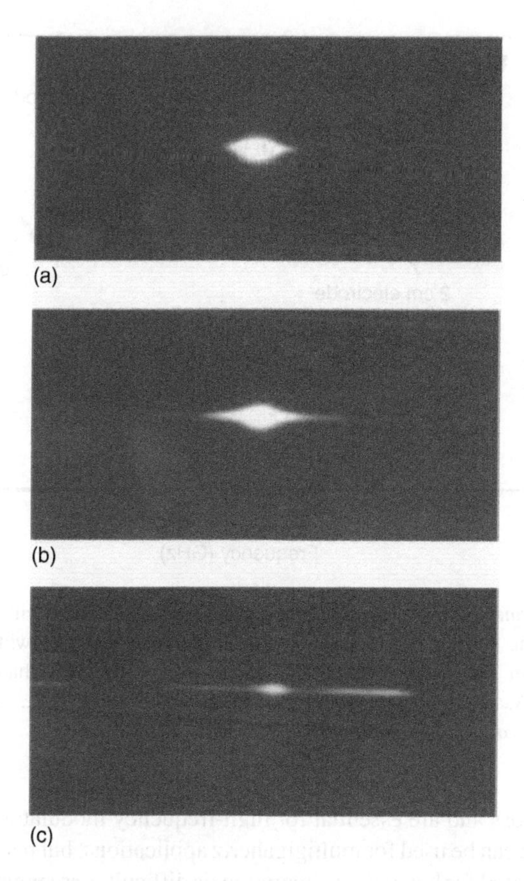

FIGURE 1.39 Guided mode profile exhibiting the effect of waveguide bleaching. (a) Guided mode intensity distribution of a polymer optical before the bleaching, (b) and (c) (transverse electric) TE- and (transverse magnetic) TM-polarized mode intensity distribution, respectively, after the polymer structures exposed to intense light source. (From Oh, M.-C., Zhang, H., Zhang, C., Erlig, H., Chang, Y., Tsap, B., Chang, D., Szep, A., Steier, W.H., Fetterman, H. R., and Dalton, L.R., *IEEE J. Selected Top. Quantum Electron.*, 7, 826–835, 2001. With permission.)

to eliminating the dielectric buffer layer, but without suffering the optical losses caused by metals. Another means to minimize the DC drift, which prevents application of the DC bias voltage required for some device applications, is to etch away the buffer layer in the electrode gap region [76,93,94].

Low loss superconducting electrodes as a transmission line for traveling wave signals together with a shielding plane have also been employed to demonstrate 26.5 GHz 3 dB bandwidth MZIMs at 4.2 K. The choice of electrode material depends upon the application. For relatively low speed (modulation frequencies below 1 GHz), an evaporated aluminum thin-film electrode (~0.2 μm thick), with a flash of Ti for adhesion, and then thick gold-plated layer is sufficient. Very narrow gap electrodes (~1 μm) can be defined by liftoff technique of the lithography. Thicker films are used to

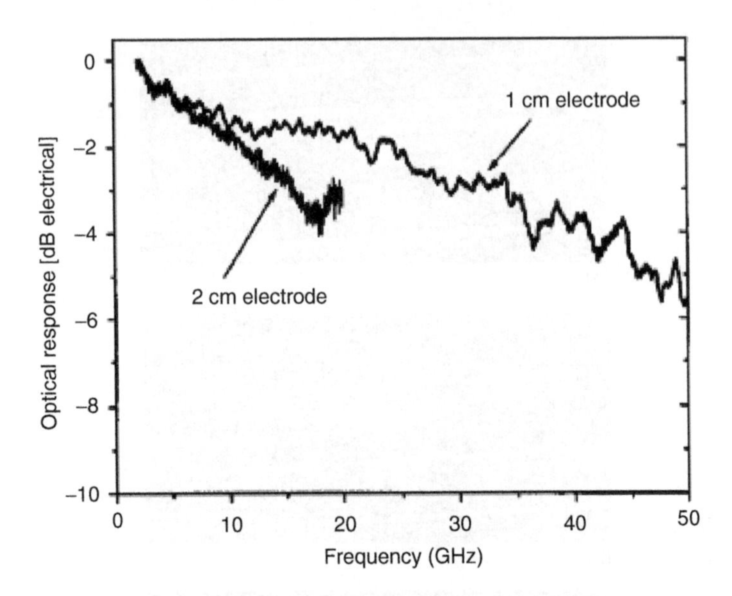

FIGURE 1.40 Frequency response of the polymer MZ modulator with 10 and 20 mm long electrodes. The shorter device exhibits the 3-dB electrical bandwidth over 30 GHz from 2 GHz. (From Oh, M.-C., Zhang, H., Zhang, C., Erlig, H., Chang, Y., Tsap, B., Chang, D., Szep, A., Steier, W.H., Fetterman, H.R., and Dalton, L.R., *IEEE J. Selected Top. Quantum Electron.*, 7, 826–835, 2001. With permission.)

reduce electrical loss and are essential for high-frequency modulators. Indeed, about 1–2 μm thick gold can be used for multigigahertz applications, but patterning the small gap, which is essential for low voltage operation, is difficult. Larger electrode gap (~30 μm) can be chemically etched.

The preferred electrode material for very high speed devices is gold, which can be electro-plated several micrometers thick with an electrode gap as small as 5 μm. The electrode fabrication steps are shown in Figure 1.42. A thin gold seed layer is

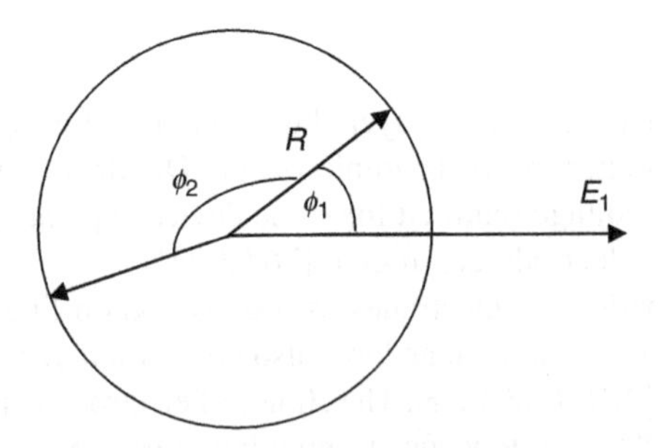

FIGURE 1.41 Phasor representation of the two paths of the MZIM including the two phases of the two paths of the interferometer with respect to the reference phase of the input optical field.

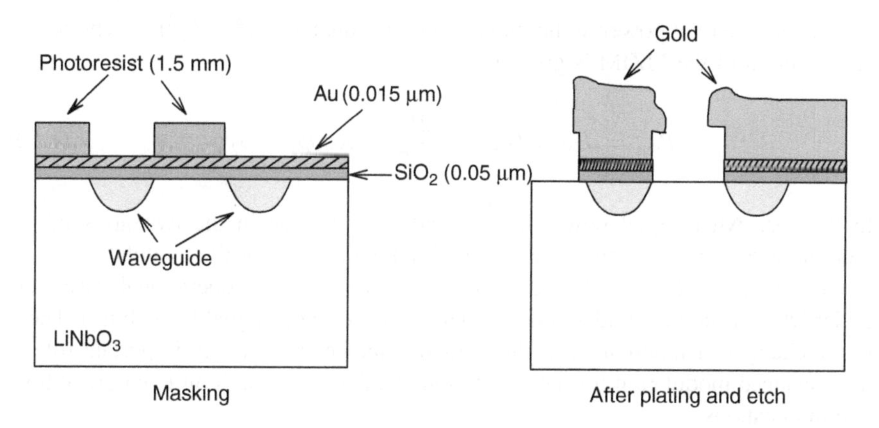

FIGURE 1.42 Gold electrodes of the traveling wave electrode under real electroplating.

deposited over the entire crystal; photoresist is spun and the electrode pattern is aligned over the waveguides opening a window in the resist. The crystal is then placed in a plating bath and gold plated to a thickness of 2–3 μm. Recently, there were reports of traveling wave modulators with achievable gold thickness as high as 15–20 μm [93]. Afterwards, the gold seed layer is removed in the gap region. Gold plated in this fashion has conductivity comparable to bulk conductivity. Good techniques of electrode fabrication are essential for high-performance Ti:$LiNbO_3$ modulators.

1.3.5 Modulation Characteristics and Transfer Function

1.3.5.1 Transfer Function

1.3.5.1.1 MZIM

Modulation efficiency is the most important property of all types of optical modulators. For analog systems, the linearity of the transfer function is critical that is the incremental change determining the mall signal and distortion of the output signals. While in digital systems the ON–OFF extinction ratio and the required voltage swing, modulation efficiency is determined by the transfer function of the modulator which, in turn, determines the design of the modulators, the waveguide confinement of the modes, and the traveling wave electrodes.

For a voltage-controlled MZIM the transfer function is the output power of the lightwaves versus the applied voltage. This transfer function determines not only the modulation efficiency but also the linear and NL chirp regions of the operation. The optical filed amplitude at the output of the MZIM is given by

$$E_o = \frac{1}{\sqrt{2}}(E_1 e^{j\phi_1} + E_2 e^{j\phi_2}) \quad (1.49)$$

where E_1 and E_2 are the field amplitudes of the optical paths 1 and 2 of the interferometer and ϕ_1 and ϕ_2 represent the phase changes imposed on the lightwaves propagating through the two paths. Assuming no loss at the splitter and combiner

Y-junction, the total power at the input would be equal to $(E_1^2 + E_2^2)^{1/2}$. The power at the output of the MZIM is given by

$$P_{\text{out}} = |E_o|^2 = 0.5\left[1 + \frac{2E_1E_2}{P_{\text{in}}}\cos(\phi_1 - \phi_2)\right] \tag{1.50}$$

In the case when equal power is split and loss is same in the two arms of the modulator $E_1 = E_2$, we can obtain a simplified expression for the modulator.

The phase difference $\phi_1 - \phi_2$ can be assigned for two operational cases: (1) under DC bias and (2) under small signal (linear) or large signal (switching). These phase changes or modulation of the refractive indices play a very important role in the advanced modulation formats of the long haul and ultrahigh bit rate optical fiber communications.

1.3.5.1.1.1 Single Drive MZIM

When only one arm of the MZIM is modulated by the RF traveling wave electrode, the phase difference becomes

$$\Delta\phi = \phi_1 - \phi_2 + \Phi = \Gamma\Delta n\frac{2\pi}{\lambda}L + \Phi \tag{1.51}$$

where

Δn is the change of the refractive index of the path through which the EOE occurs

Γ is the confinement factor of the optical field and the electric traveling wave field as defined in the previous section

λ is the operating wavelength

L is the modulation length or the length of the traveling wave electrode

Φ is the DC phase due to the bias voltage of the MZIM which can be expressed by

$$\Phi = \pi\frac{V_B}{V_\pi} \tag{1.52}$$

where

V_B is the DC bias voltage

V_π is the voltage required to generate a total phase change at DC of π radians that is at the minimum transmission point of the interferometer

Thus, the transmittance of the MZIM can be defined as the ratio between the output and the input power. This can be given as by using Equations 1.50 through 1.52.

$$T_{\text{MZIM}} = 0.5\left\{1 + \frac{2E_1E_2}{P_{\text{in}}}\cos\left[\pi\frac{v(t)}{V_\pi}\right] + \Phi\right\}$$

or (1.53)

$$T_{\text{MZIM}} = 0.5\left\{1 + r\cos\left[\pi\frac{v(t)}{V_\pi}\right] + \Phi\right\}$$

where

$v(t)$ is the time-dependent signals applied to the electrode superimposing on the DC bias voltage

$r = E_1E_2/P_{in}$ and $r = 1$ when the input optical field is equally split into the two branches of the interferometer

Figure 1.38 shows the typical theoretical curve and measured transfer curve of the MZIM. It is noted that there is a shift of the maximum transmission point. This can happen normally due to the drift of the DC voltage due to static charge deposited on the surface of the crystal. This effect is dealt with in the next sections.

1.3.5.1.1.2 Dual-Drive MZIM

In a dual-drive MZIM, there are two hot electrodes which are positioned over the two paths of the interferometers. Thus, there are two different voltages V_π. However, the design of MZIM requires these voltages equal.

If the two paths of a dual drive have independent drive voltages of V_1 and V_2, the transmission transfer function is given by

$$E_o(\text{MZIM}) = \frac{E_i}{2}\left[\exp\left(j\pi\frac{V_1}{V_\pi} + \Phi_1\right) + \exp\left(j\pi\frac{V_1}{V_\pi} + \Phi_1\right)\right] \tag{1.54}$$

where we have assumed that the V_π is the same for the two hot electrodes, Φ_1 and Φ_2 denote the DC phases excited by the bias voltages applied to the electrodes. We can represent Equation 1.54, in phasor form as shown in Figure 1.41. The output electric field is the vectorial difference between the two rotating vectors in the circle having a radius of R (with a maximum value of $P_{in}/2$). These phase differences can be designed for modulation formats, e.g., phase shift keying, DPSK, DQPSK, or M-ary PSK. However, these dual-drive modulators suffer a significant set back when driven at its ultimate high frequency regions. We can see that, from Equation 1.54, if the amplitudes of the two paths of the interferometer are not identical, e.g., different losses or attenuation coefficients, then the phases and amplitudes of the output field would be severely disturbed. This difficulty becomes very serious when the dual-drive MZIM is driven at it ultimate bandwidth (above 40 GHz). The traveling wave electrode normally would suffer different scattering and microwave losses of the two electrodes. In practice, single-drive MZIM would be more preferable than the dual-drive modulator unless it is really essential, especially when driven at the ultrahigh speed.

1.3.5.1.2 EA Modulators

The transfer function of an EA modulator is given by

$$T_{EA}(V) = t_0\exp[-\Gamma_{EA}\alpha(V)L] \tag{1.55}$$

where

t_0 is the transmission loss at zero applied voltage

Γ_{EA} is the overlap factor of the EA traveling wave electric field and the optical mode distribution

$\alpha(V)$ is the variation of the absorption coefficient due to the applied voltage
L is the modulation length, typically the electrode length

We could observe that the transfer function, when normalized with the loss at zero applied voltage, is an oscillating function whose amplitude is an exponential decaying fashion with nonzero distinction ratio. The EA is also accompanied by the generation of the photocurrent. Thus, under the scenario when the input optical power is high, the generated photocurrent may affect the performance of the modulator, including the photonic absorption and the modulator bandwidth.

1.3.5.2 Extinction Ratio for Large Signal Operation

Large signal operation of an optical modulator is important for digital optical fiber transmission, including several modulation formats NRZ, RZ OOK, CSRZ OOK, DPSK, DQPSK, etc., in which the voltage signals can swing over 0–2 V_{π} to achieve the required phase or amplitude. Thus, a high ON/OFF ratio is very crucial for the operation of optical modulators employed in optical fiber communication systems. For MZIMs, the ON or OFF state is achieved when the phase difference is at $m\pi$ (with $m = 0, 2, 4, \ldots, n$) or $(2n + 1)\pi$ (with $n = 0, 1, 2$). The extinction ON–OFF ratio (OOR) is defined as the ratio of the optical power outputs at the ON and OFF states which can be derived from Equations 1.54 and 1.55) as

$$\mathrm{OOR} = \frac{P_o(\max)}{P_o(\min)} = \frac{(E_1 + E_2)^2}{(E_1 - E_2)^2} \tag{1.56}$$

This OOR would be considered as the ratio between the bright and dark fringes in the spatial interference system. If an OOR of 20 dB (or 1% intensity OFF ratio as compared with the ON state) is required, then the power difference between E_1 and E_2 would be about 2 dB. Thus, the optical waveguide propagation in the two branches of the interferometer is critical to obtain the maximum extension ratio.

The switching between ON and OFF states needs a voltage V_{π}; hence, a small V_{π} is preferred to minimize the complexity of microwave components and amplifiers to drive the modulator, especially when operating for 40 Gb/s transmission system. At this rate, not only the electronic amplifier is critical but also the electrical paths that require phase delay matching of the electrical length and the reflection and scattering issues. The V_{π} is strongly proportional to g/L, i.e., proportional with the gap between the hot electrode and the ground plane and inversely proportional to the effective length of the electrode. Hence, a low V_{π} would lead to larger stray capacitance and hence lower modulation bandwidth.

For EAM to achieve small voltage swing and high OOR, MQW is preferred due to strong extension exciton effect and mode confinement in the QW structure. From Equation 1.56, the OOR of an EA modulator decays exponentially with Γ, α, and L which are dependent on the width of the rib waveguide (or equivalently to the electrode gap g of the MZIM). This width cannot be altered significantly as it must satisfy the single-mode guide eigen-solution conditions.

1.3.5.3 Small Signal Operation

Unlike the digital large signal operation, small or linear operation of the modulator is required when analog transmission is required such as in the case of analog fiber-optic link, photonic signal processing, etc. In these cases, the linearity of the transfer function plays a very important role. In such an RF link, the link gain is defined as [97]

$$G = \frac{P_{\text{RF-o}}}{P_{\text{RF-i}}} \tag{1.57}$$

that is the ratio between the total input RF power to that of the optical transmitter and the received RF electronic power at the output of the link. Thus for an ideal 50 Ω RF traveling wave electrode and a photodetector with a responsivity $\Re$, we have

$$G = \frac{P_{\text{RF-o}}}{P_{\text{RF-i}}} = \frac{v_{\text{i}}^2/R_{\text{i}}}{(P_{\text{o}}\Re)^2 R_{\text{o}}} \tag{1.58}$$

where P_{o} is the output power at the end of the RF link that is directly related to the output power at the output of the optical modulator and the exponential attenuation factor of the length L_f of the optical fiber by

$$P_{\text{o}} = P_{\text{o}} - Txe^{-\alpha Lf}$$

Assuming that the photodetector responsivity is close to unity, the ratio $P_{\text{o}}/v_{\text{i}}$ of Equation 1.58 is the transfer ratio of the optical modulator at a particular bias point of its transfer curve. We can now consider the gain of the RF link for the two cases of MZIM and EAM as (1) for MZIM, the maximum gain would occur at the phase quadrature point $V_{\pi/2}$ with the gain of $\pi/(2V_{\pi})$ and (2) for EAM, the maximum slope gain happens at around the midpoint to 0.7 of the normalized transfer curve.

1.3.5.4 DC Bias Stability and Linearization

The bias stability has been developed and proven over several months using automatic biasing circuitry. However, when the bias voltage is changed suddenly, the drift of the bias point is normally observed. This is due to the distribution of charges under the variation of the applied voltage. Temperature bias drift of the optical field at the output of $LiNbO_3$ MZIM is one of the inevitable problems affecting the long-term stability of these photonic components. In addition, there is the DC drift phenomena due to the charges deposited on the surface of the crystal or SiO_2 buffer layer. The thermal drift is a reversible process but if combined with the DC drift they would accelerate the permanent damages to these devices. Because the principal origin of the thermal drift is the pyroelectric charges generated in the $LiNbO_3$ substrates, the thermal drift of an *X*-cut $LiNbO_3$ MZIM is mostly suppressed, in contrast to that of *Z*-cut types.

Linearization of the transfer function of the MZIM is important. This characteristic is well defined as it follows a $\cos^2$ profile. Two modulators can be used in tandem to cancel the higher-order harmonics by controlling the bias and amplitude and phase of the electrical signals modulating the modulators [96,98].

1.3.6 Chirp in Modulators

1.3.6.1 General Aspects

Generally, the intensity modulation is derived from the optical phase modulation that would normally lead to variation of the phase of the optical sinusoidal waves, hence the chirping effects [98]. This effect is very important for digital optical communication systems either by impairment or as a predistortion to compensate for the linear fiber dispersion effects.

The envelope of the lightwaves at the output of an optical modulator is usually complex, i.e., it contains both the magnitude and phase components. The temporal complex envelop of the optical field of a Gaussian shape pulse at the input of the optical fiber or at the output of the modulator before transmission ($z=0$) can be expressed by

$$A(0,t) = |A(0,t)|\exp\left[-\frac{(1+jC)}{2}\frac{t^2}{T_0^2}\right] \tag{1.59}$$

where

$j=(-1)^{1/2}$

T_0 is the full-width half mark (FWHM) of the intensity of the Gaussian pulse

C is the chirp parameter

The chirp is assumed to be linear in this case. Equation 1.59 indicates the envelope with an instantaneous frequency increases linearly with the carrier frequency along the raising and falling pulse edges. If $C < 0$, the chirp is up-chirp; otherwise if $C > 0$, then we have down-chirp. Alternatively, we refer $C > 0$ as positive chirp and $C < 0$ as negative chirp. Thus, the carrier frequency change is related to the phase as given by

$$\delta\omega(t) = -\frac{\delta\phi}{\delta t} = \frac{C}{T_0}t \tag{1.60}$$

with ϕ is the phase of the pulse envelop $A(0,t)$. For the transmission of the Gaussian pulse in the linear regime of a single optical fiber with a dispersion parameter β_2, the FWHM of a chirp pulse at a distance z is given by [32]

$$\frac{T_z}{T_0} = \left[\left(1+\frac{C\beta_2 z}{T_0}\right)^2 + \left(\frac{\beta_2 z}{T_0}\right)^2\right] \tag{1.61}$$

We can thus observe that the pulse width can be shortened over some distance if the chirp factor C is negative and optimum at $C = -2$. This means that the optical modulator can be prechirped to compensate for an initial distance of the transmission span. Indeed, this is normally done using 10 Gb/s optical transmission systems.

A more generic expression of the field at the output of an optical modulator with the real and imaginary parts (loss term) of the refractive index, $n_m(t) = n_R(t) + jn_I(t)$ can be written as

$$A(0,t) = |A(0,t)|\exp[-j\phi(t)] = |A(0,t)|\exp\{-jk_0[n_R(t) + jn_I(t)]\}L \qquad (1.62)$$

where k_0 is the wave number in vacuum. Alternatively, the definition of the chirp factor can be adopted from Henry's alpha parameter α, which is related to the change of the intensity I of the pulse, defined as [94]

$$\alpha_H = -2I\frac{\delta\phi}{\delta I} \qquad (1.63)$$

This definition is similar to the line width enhancement factor defined in the case of direct modulation. Thence the carrier frequency variation is given by

$$\delta\omega = -\frac{\delta\phi}{\delta t} = \frac{\alpha_H}{2I}\frac{dI}{dt} \qquad (1.64)$$

This frequency change takes the positive sign for the raising edge (chirp up) and negative sign for falling edge (chirp down).

Using the analytical expression of the chirp of an external modulator, the transmission bandwidth of an SMF of length L can be obtained. The output field $A(z,t)$ of the guided lightwaves after propagation length z of the fiber can be evaluated in the frequency domain using the Fourier transform around the central optical frequency of the wavelength channel as

$$A_0(z,t) = \frac{1}{2\pi}\int_{-\infty}^{+\infty} A(0,\omega)\exp\left\{j\left[\omega t - \beta L - \beta_1(\omega - \omega_0) - \frac{\beta_2}{2}(\omega - \omega_0)^2\right]\right\}d\omega \qquad (1.65)$$

$A(0,\omega)$ is the Fourier transform or spectrum of the optical field at the output of the modulator. Assuming that the pulse follows a Gaussian profile, the 3 dB bandwidth of the fiber including the chirp factor is then given by estimating the pulse width and ignoring the pure delay due to β_1, from Equation 1.65 [32]:

$$B_R = \frac{1}{2\left[2(\alpha_H^2 + 1)^{1/2} + 2\alpha_H\right]^{1/2}|\beta_2 L|} \qquad (1.66)$$

This is consistent with the chirp parameter C given in Equation 1.61 and the maximum transmission bit rate happens with the chirp factor of -2 which is because of the pulse compression caused by the blueshift chirp.

1.3.6.2 Modulation Chirp

In the modulation of the lightwaves, we have the field at the output of the modulator and is thus given by

$$A_o(V) = A_i(V)\left[(\delta\phi/\delta\phi)\,(\delta A_i/\delta V)^{-1}\right] \quad (1.67)$$

Case 1: MZIM

In general, the MZIM with dual-drive electrodes can be expressed with its field at the output by using Equation 1.56 and the refractive index changes due the EOE created by the modulation voltages applied to the two arms of the interferometers as

$$\begin{aligned} A(0,t) &= \exp[-j\phi(t)] \\ &= 1(2)^{-1/2}\{A_1 \exp[-jk_0\Delta n_1(V)L + j\phi_0] + A_2 \exp[-jk_0\Delta n_2(t)L]\} \end{aligned} \quad (1.68)$$

where

Δn_1 and Δn_2 are the refractive index changes in arms 1 and 2 of the interferometer, respectively
L is the effective length of the two electrodes, assuming that they are identical
A_1 and A_2 are the amplitudes of the optical fields after the Y-branch splitter

In case that the splitter is a 3 dB type, then the field is split equally between the two arms and hence the field of Equation 1.68 can be written as

$$\begin{aligned} A(0,t) &= \exp[-j\phi(t)] \\ &= A_i 2(2)^{-1/2}(\cos[k_0\Delta n_1(V)L + (\phi_0/2)][\exp\{-j[k_0\Delta n_1(t)L + \phi_0]\}]) \end{aligned} \quad (1.69)$$

Thus for an applied signal voltage $v(t)$, we have

$$A(0,t) = A_i\frac{2}{\sqrt{2}}\left(\cos\left[\frac{\pi}{2}\frac{v(t)}{V_\pi} + (\phi_0/2)\right]\exp\left\{-j\left[\frac{\pi}{2}\frac{v(t)}{V_\pi} + (\phi_0/2)\right]\right\}\right) \quad (1.70)$$

The chirp factor α_H can then be obtained from this equation as

$$\alpha_H = K\tan\left(\frac{\pi}{2}\frac{v(t)}{V_\pi} + \frac{\phi_0}{2}\right) \quad (1.71)$$

with K as an arbitrary constant. We can now observe that when $\phi_0 = 0$, that is biasing the MZIM at the maximum transmission point then at the zero level (OFF-position) of the modulation voltage, the chirp approaches infinity; thus, it is unacceptable for system application. The MZIM would thus be normally biased at the phase quadrature point or at the minimum transmission point, i.e., V_π to avoid the difficulty in the uncontrollable chirp.

For push–pull operation of a dual-drive modulator, the voltages applied to the two traveling wave electrodes are set in complete symmetric and opposite in sign and then we have the changes of the refractive indices of the two paths in opposite sign. Hence even at zero DC bias, we still have

$$A(0,t) = \frac{1}{\sqrt{2}} A_i \left\{ \cos\left[\frac{\pi}{2}\frac{v(t)}{V_\pi}\right] \right\} \quad (1.72)$$

Thus in this case, there are no chirp effects at all voltage levels imposed on the output optical field of the MZIM provided that the traveling wave RF field would be in synchronization with the guided lightwaves. Furthermore, it requires that any electrical paths applied to these electrodes are electrically matched. Otherwise, when operating the MZIM at ultrahigh speed in the region higher than 40 GHz, this mismatch would create the phase difference and hence chirp effects. The fabrication of the electrodes, especially the electroplating of the brick wall-like structures, would no doubt not be identical and would vary from one device to the other. This is the principal reason why it is preferable that equipment manufacturers select single drive MZIM for systems operating at this speed.

Case 2: EAM

The chirp factor α_H of the EAM can be estimated using the optical absorption coefficient and then the change of the refractive index as a function of the optical frequency. The typical α_H for EAM is greater than 2 at low bias. In general, the chirp factor of EAM decreases with an increase of the drive voltage and with a lower operating wavelength.

1.3.7 Electro-Optic Polymer Modulators

Polymer EO waveguide devices have also been investigated as the ultrahigh speed modulation components. The operating bandwidth of this polymeric modulator can reach higher than 100 GHz, thus allowing signal processing of millimeter-wave signals. There have been various experimental results to demonstrate the merits of EO polymer devices, which potentially offer ultrahigh-speed modulation based on an inherent velocity match of RF and optical waves, integration of electronic circuits with polymer waveguides, multilevel-stacked integrated optical circuits, and polarization controlling waveguide devices [100]. However, to be accepted in system applications, these EO polymer devices must demonstrate modest insertion loss, good long-term thermal stability, and sufficient photostability at the communication wavelengths. Typical EO polymers have higher absorption loss than $LiNbO_3$, especially at 1550 nm wavelength, due to the vibration overtones. Recent works show that the losses can be decreased by the substitution of fluorine or chlorine for the hydrogen. The thermal stability of the EO polymer materials can also be improved by employing a high backbone polymer, a side chain attached chromophore, and a cross-linked polymer matrix. However, in these polymers with an enhanced thermal stability, the efficiency of poling was reduced and the EO coefficient was sacrificed to improve the thermal stability. Recently, an EO polymer based on a dendritic structure has demonstrated both high EO coefficient and excellent poled order stability at 85°C. The photostability of EO polymer waveguides under 1300 nm wavelength illumination has been investigated. It was found that some EO polymer materials degrade because of the chemical reaction with excited oxygen. To develop an EO polymer material suitable for commercial devices, material synthesis research has been directed toward the optimization of several requirements simultaneously including reduced insertion loss, long-term thermal stability at 85°C, and long-term photostability at 10 mW input power at the communication wavelengths while maintaining a high EO coefficient [100–128].

Furthermore, there have been significant advances on all of these issues based on a guest–host polymer consisting of highly NL chromophore with phenyltetraene bridge (CLD) and amorphous polycarbonate (APC). This chromophore was mixed in a polymethylmethacrylate (PMMA) host to achieve a push–pull modulator with a driving voltage of 0.8 V at 1300 nm. In addition to this highly NL chromophore, APC was identified as a promising host material because of its high thermal stability and low loss at 1550 nm wavelength [118]. These high NL properties would be advantageous for photonic switching and other NL photonic applications.

Similar to the device structures of the EAM and $LiNbO_3$ MZIM, the polymer optical waveguides and traveling wave electrode structures are required for ultra-broadband operations. The single-mode waveguides used in polymer modulators have a buried rib structure that can be fabricated by dry etching in oxygen. Single-mode waveguides are required to provide a high extinction ratio of the modulator. EO polymers have a relatively high index of refraction compared to the passive cladding polymers. Hence, to keep the waveguide operating in the single-mode region, it has been widely accepted that the thickness of the core layer must be thin enough to suppress higher-order modes in its slab waveguide. At 1550 nm, due to the high refractive index reference between the guiding region and the substrate, it requires a typical thickness of the EO polymer layer to be less than 1.5 μm. As a result, it is difficult to maintain the required fabrication tolerance to achieve a high extinction ratio in the modulator. In the case of 3-D rib waveguides, it has been shown that the single-mode condition does not require the thin core layer. Similar structures are shown in Figures 1.30a and b.

Even if the core layer is so thick that it could support several modes in its slab waveguide, it is still possible to design the rib waveguide to confine only one mode by radiating the higher-order modes horizontally into the slab modes. Then the oversized rib waveguide will support only the single mode without additional radiation loss. This approach can be used to design large-core single-mode waveguides in EO polymer devices. The intensity of the fundamental guided mode, shown in Figure 1.39, is consistent with the plots of the mode distribution of rib waveguides shown in Figure 1.30b. The dispersion curve given in Figure 1.43 can be used for the design of rib waveguides.

The thickness of the core layer affects the EO modulation efficiency by changing the film thickness across the electrodes. It also affects the fiber coupling loss to the waveguide because of the mode-size mismatch. A 3 μm rib width and thin height can be selected to obtain sufficient confinement in the lateral direction due to the high effective index contrast. For 6–8 μm wide waveguides, a refractive index difference of 0.003 is required to produce the single-mode waveguide with good confinement.

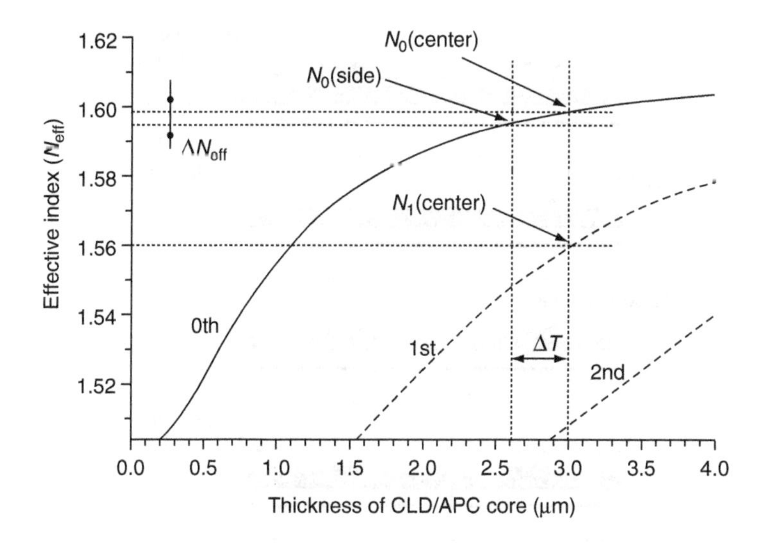

FIGURE 1.43 Dispersion curve for the design of polymer rib waveguide structure. (From Oh, M.-C., Zhang, H., Zhang, C., Erlig, H., Chang, Y., Tsap, B., Chang, D., Szep, A., Steier, W. H., Fetterman, H.R., and Dalton, L.R., *IEEE J. Selected Top. Quantum Electron.*, 7, 826–835, 2001. With permission.)

The waveguide consists of UV15 lower cladding (1.504), CLD/APC core layer (1.612), and UFC170 upper cladding (1.488). In the rib waveguide structure, the waveguide side needs to be etched by to obtain high refractive index difference so that the fundamental mode does not propagate through the rib waveguide. This type of rib waveguide can be proven to be effective for interferometric modulation. The minimum thickness of the cladding layers is determined by setting the metallic absorption loss due to the surface plasmon coupling to the upper or lower electrodes to an acceptable value. The total thickness of the waveguide consisting of lower cladding, core, and upper cladding should be as thin as possible to minimize. For the polymers used in the modulators, the minimum thickness for negligible metallic loss is approximately 8 μm for the 3 μm core waveguide. The upper and lower claddings are 2.5 μm thick. Even if the thickness of the core is reduced to less than 3 μm, the total thickness of the device cannot be decreased because the thinner core waveguide requires the thicker cladding layers. Figure 1.44 illustrates the steps of the fabrication of a polymer EO MZ modulator. Initially, the substrate is coated with a thin layer of Cr–Au; then a thin layer of polymer is spin coated and a rib waveguide structure is exposed and etched to form rib waveguide structures. A superstrate layer can also used to cover the rib waveguide so as to create a buried rib waveguide structure. This buried structure is necessary to minimize the coupling between the fiber spot size and that of the planar optical waveguide for efficient modulation. Traveling wave electrodes are deposited on top of the ribs as shown in Figure 1.20. The output intensity distribution across the polymer waveguide, shown in Figure 1.39, which can be confirmed with those modeled in Figure 1.30 but note also the TE- and TM-polarized modes of the rib waveguide. It is noted that the electric field of the

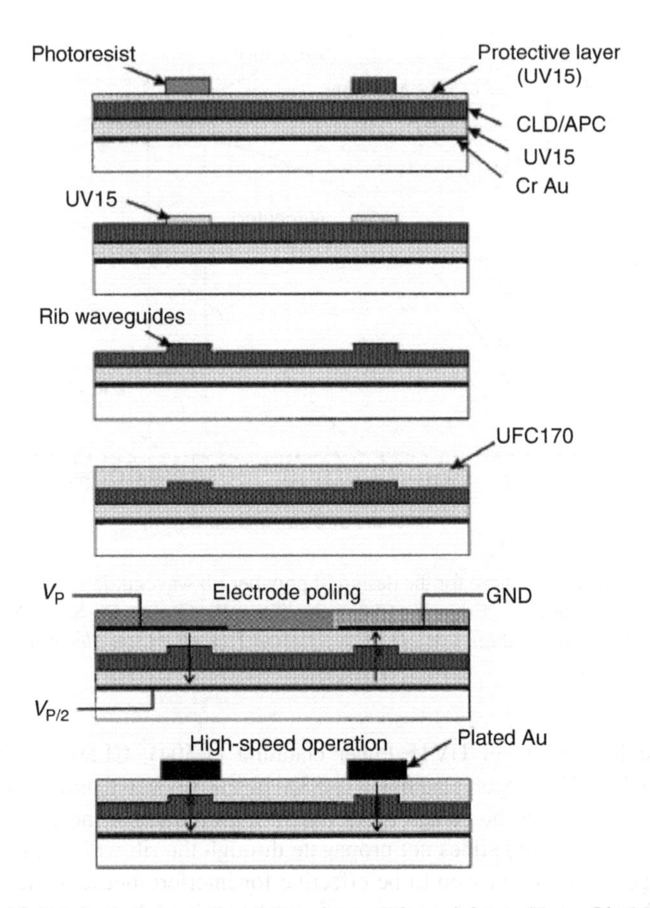

FIGURE 1.44 Fabrication steps of an MZ polymer EO modulator. (From Oh, M.-C., Zhang, H., Zhang, C., Erlig, H., Chang, Y., Tsap, B., Chang, D., Szep, A., Steier, W.H., Fetterman, H. R., and Dalton, L.R., *IEEE J. Selected Top. Quantum Electron.*, 7, 826–835, 2001. With permission.)

traveling wave is vertical due to the metallic layer deposited on the substrate. This allows much more efficient modulation voltage for the destructive interference at the output. The DC V_π as shown in the transmission curve in Figure 1.45 is much smaller than that of the $LiNbO_3$ of about 1.5 V. A near 30 GHz 3 dB bandwidth can be achieved. Much higher bandwidth modulators (up to 100 GHz) can also be fabricated in polymer material structures.

1.3.8 Modulators for Photonic Signal Processing

Over the last two decades, the MZIM have been exploited extensively in the modulation of lightwaves for applications in optical communication systems and networks as well as in photonic sensor networks. This section classifies the uses of MZIM in terms of specific applications rather than the types of the devices $LiNbO_3$ or EAM. We also note that although the linearity of the EAM is better than that of the

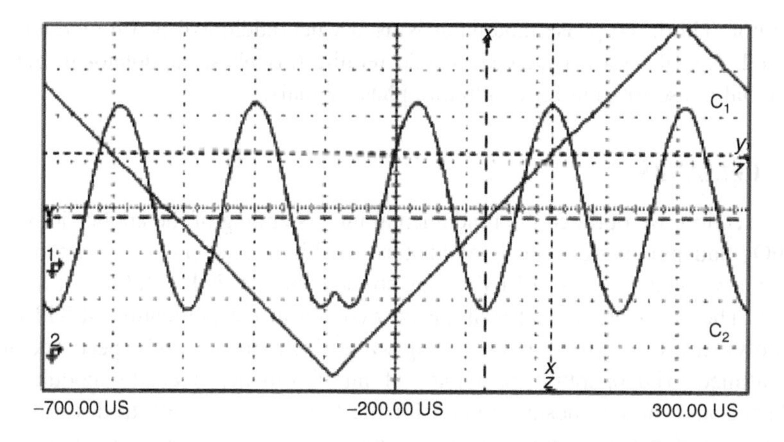

FIGURE 1.45 Transmission curve of an MZIM polymer EO modulator V_π is about 1.4 V with a ramp driven voltage input to the electrode. (From Oh, M.C., Zhang, H., Zhang, C., Erlig, H., Chang, Y., Tsap, B., Chang, D., Szep, A., Steier, W.H., Fetterman, H.R., and Dalton, L.R., *IEEE J. Selected Top. Quantum Electron.*, 7, 826–835, 2001. With permission.)

$LiNbO_3$ but suffers from its nonzero extinction. Thus, cross talk would be existed. Most applications now employ $LiNbO_3$ MZIMs [99].

The performance of long-haul terrestrial and undersea systems can be degraded by a combination of the transmission fiber NL-induced refractive index and chromatic dispersion. The nonlinearity is normally generated by the total aggregate optical power by the multiplexed optically modulated lightwave channels. The linear chromatic dispersion is the resulting phenomena due to the refractive index of the fiber materials and the guided mode effective index changes with respect to the optical frequencies of the signals on both sides of the optical carrier frequency. The ultimate aims of the modulation techniques, that is photonic signal processing of the lightwave carrier, for the transmission system applications are (1) to generate efficient signal bandwidth, (2) to suppress the carrier and hence to total aggregate power of the transmitting channels, and (3) to generate efficient formats for efficient detection and hence increase the receiver sensitivity.

In general, the modulation formats can be (1) either amplitude modulation, that is the OOK or amplitude shift keying (ASK), (2) phase modulation, e.g., phase shift keying (PSK), differential PSK (DPSK), differential quadrature PSK (DQPSK) differential DQPDSK, (3) frequency modulation, e.g., frequency shift keying (FSK), and (4) polarization modulation.

The generic arrangement of a photonic transmitter for advanced modulation formats can be illustrated as shown in Figure 1.43 in which three modulators can be used. First, the pulse carver is employed and biased at the phase quadrature or at the minimum transmission point so as to generate the periodic clock pulses. This is essential when RZ pulses are required. Otherwise for NRZ, this pulse carver modulator is to be biased at the maximum transmission point to pass the signals. Second, the data modulator is used for modulating the data sequence from a bit pattern generator. Third, the phase modulator is used when additional phase modulation is required, e.g., the format DQPSK would require this phase modulator to generate additional phase rotation for quadrature phase signals.

1.4 REMARKS

In this section, we outlined the fundamental understanding of the applications of Ti: $LiNbO_3$ and polymeric optical modulators for high-speed photonic systems. The elementary characteristics of $LiNbO_3$ are studied and the EOE exhibit by $LiNbO_3$ is shown. The key features and techniques involved in the fabrication of Ti:$LiNbO_3$ and polymeric waveguides and traveling wave electrodes for high-speed operations are outlined. The operating principles of interferometric intensity modulator are explained and the key design criteria for the traveling wave electrodes to achieve wideband performance are summarized. Optical modulators could be operating for effective ultrafast processing if (1) the guided optical modes in rib and diffused waveguide are matched and hence the mode size and effective index of the relevant waveguides for optimum interaction with the traveling wave electric field; (2) the microwave properties such as the characteristic impedance Z and microwave effective index n_m of the traveling wave electrodes can be matched with the effective index of the optical guided mode, hence the fabrication of vertical brick wall-like electrodes is critical; and (3) the effect of a more subtle geometrical factor of the electrode structures such as the trapezoidally shaped electrode structure that would normally be encountered, when very thick electrode can be corrected to improve the optical–electrical velocity match. These high-speed optical modulators are found to be very significant for processing of ultrafast signals and modulation of the phase of frequency of the lightwave carriers for processing and communication systems.

REFERENCES

1. C. Kao and G. Hockham, Dielectric-fiber dielectric waveguides for optical frequencies, *Proc. IEEE*, 133(3), 1151–1158, 1966.
2. T.C. Cannon, D.L. Pope, and D.D. Sell, Installation and performance of the Chicago lightwave transmission systems, *IEEE Trans. Commun.*, 26(7), 1056–1060, 1978.
3. E. Desurvire, *Erbium-Doped Fiber Amplifiers: Principles and Applications*, New York: John Wiley & Sons, 1994.
4. R.J. Mears, L. Reekie, I.M. Jauncy, and D.N. Payne, Low noise erbium doped fiber amplifier operating at 1.54 μm, *Electron. Lett.*, 23, 1026–1028, 1987.
5. T. Hirooka, M. Nakazawa, F. Futami, and S. Watanabe, A new adaptive equalization scheme for a 160-Gb/s transmitted signal using time-domain optical Fourier transformation, *IEEE Photon. Technol. Lett.*, 16(10), 2371–2373, 2004.
6. B. Moslehi, J.W. Goodman, M. Tur, and H.J. Shaw, Fiber-optic lattice signal processing, *Proc. IEEE*, 72, 909–930, 1984.
7. K. Wilner and A.P. van der Heuvel, Fiber-optic delay lines for microwave signal processing, *Proc. IEEE*, 64(5), 805–807, 1976.
8. K.P. Jackson, S.A. Newton, B. Moslehi, M. Tur, C.C. Cutler, J.W. Goodman, and H.J. Shaw, Optical fiber delay-line signal processing, *IEEE Trans. Microwave Theory Tech.*, MTT-33, 193–210, 1985.

9. C.C. Wang, High-frequency narrow-band single-mode fiber-optic transversal filters, *J. Lightwave Technol.*, LT-5, 77–81, 1987.
10. R.I. MacDonald, Switched optical delay-line signal processors, *J. Lightwave Technol.*, LT-5, 856–861, 1987.
11. D.M. Gookin and M.H. Berry, Finite impulse response filter with large dynamic range and high sampling rate, *Appl. Opt.*, 29, 1061–1062, 1990.
12. A. Ghosh and S. Frank, Design and performance analysis of fiber-optic infinite-impulse response filters, *Appl. Opt.*, 31, 4700–4711, 1992.
13. B. Moslehi, Fiber-optic filters employing optical amplifiers to provide design flexibility, *Electron. Lett.*, 28, 226–228, 1992.
14. B. Moslehi and J.W. Goodman, Novel amplified fiber-optic recirculating delay line processor, *J. Lightwave Technol.*, 10, 1142–1147, 1992.
15. B. Moslehi, K.K. Chau, and J.W. Goodman, Optical amplifiers and liquid-crystal shutters applied to electrically reconfigurable fiber optic signal processors, *Opt. Eng.*, 32, 974–981, 1993.
16. J. Capmany and J. Cascon, Optical programmable transversal filters using fiber amplifiers, *Electron. Lett.*, 28, 1245–1246, 1992.
17. J. Capmany and J. Cascon, Discrete time fiber-optic signal processors using optical amplifiers, *J. Lightwave Technol.*, 12, 106–117, 1994.
18. S. Sales, J. Capmany, J. Marti, and D. Pastor, Solutions to the synthesis problem of optical delay line filters, *Opt. Lett.*, 20, 2438–2440, 1995.
19. S. Sales, J. Capmany, J. Marti, and D. Pastor, Novel and significant results on the non-recirculating delay line with a fiber loop, *IEEE Photon. Technol. Lett.*, 7, 1439–1440, 1995.
20. S. Sales, D. Pastor, J. Capmany, and J. Marti, Fiber-optic delay-line filters employing fiber loops: Signal and noise analysis and experimental characterization, *J. Opt. Soc. Am. A*, 12, 2129–2135, 1995.
21. D. Pastor, S. Sales, J. Capmany, J. Marti, and J. Cascon, Amplified double coupler fiber-optic delay line filter, *IEEE Photon. Technol. Lett.*, 7, 75–77, 1995.
22. M.C. Vazquez, B. Vizoso, M. Lopez-Amo, and M.A. Muriel, Single and double amplified recirculating delay lines as fiber-optic filters, *Electron. Lett.*, 28, 1017–1019, 1992.
23. M.C. Vazquez, R. Civera, M. Lopez-Amo, and M.A. Muriel, Analysis of double-parallel amplified recirculating optical-delay lines, *Appl. Opt.*, 33, 1015–1021, 1994.
24. C. Vazquez, M. Lopez-Amo, M.A. Muriel, and J. Capmany, Performance parameters and applications of a modified amplified recirculating delay line, *Fiber Integrated Opt.*, 14, 347–358, 1995.
25. B. Vizoso, C. Vazquez, R. Civera, M. Lopez-Amo, and M.A. Muriel, Amplified fiber-optic recirculating delay lines, *J. Lightwave Technol.*, 12, 294–305, 1994.
26. B. Vizoso, I.R. Matias, M. Lopez-Amo, M.A. Muriel, and J.M. Lopez-Higuera, Design and application of double amplified recirculating ring structure for hybrid fiber buses, *Opt. Quantum Electron.*, 27, 847–857, 1995.
27. E.C. Heyde and R.A. Minasian, A solution to the synthesis problem of recirculating optical delay line filters, *IEEE Photon. Technol. Lett.*, 6, 833–835, 1994.
28. E.C. Heyde, Theoretical methodology for describing active and passive recirculating delay line systems, *Electron. Lett.*, 31, 2038–2039, 1995.
29. J. Capmany, J. Cascon, J.L. Martin, S. Sales, D. Pastor, and J. Marti, Synthesis of fiber-optic delay line filters, *J. Lightwave Technol.*, 13, 2003–2012, 1994.
30. S. Sales, J. Capmany, J. Marti, and D. Pastor, Experimental demonstration of fiber-optic delay line filters with negative coefficients, *Electron. Lett.*, 31, 1095–1096, 1995.
31. A.V. Oppenheim and R.W. Schafer, *Discrete-Time Signal Processing*, Englewood Cliffs, NJ: Prentice-Hall, 1989.

32. G.P. Agrawal, *Fiber-Optic Communication Systems*, New York: John Wiley & Sons, 1992.
33. S.K. Sheem and T.G. Giallorenzi, Single-mode fiber-optical power divider: Encapsulated etching technique, *Opt. Lett.*, 4, 29–31, 1979.
34. M.J.F. Digonnet and H.J. Shaw, Analysis of a tunable single mode optical fiber coupler, *IEEE J. Quantum Electron.*, QE-18, 746–754, 1982.
35. S. Shimada and H. Ishio (Eds.), *Optical Amplifiers and Their Applications*, New York: John Wiley & Sons, 1994, chapters 1–5.
36. J.C. Cartledge and G.S. Burley, The effect of laser chirping on lightwave system performance, *J. Lightwave Technol.*, 7, 568–573, 1989.
37. S.E. Miller, Integrated optics: An introduction, *Bell Syst. Tech. J.*, 48: 2059–2069, 1969.
38. T. Tamir (Ed.), *Integrated Optics*, New York: Springer-Verlag, 1975.
39. L.D. Hutcheson, *Integrated Optical Circuits and Components*, New York: Marcel Dekker, 1987.
40. S.E. Miller and I.P. Kaminow (Eds.), *Optical Fiber Telecommunications II*, New York: Academic Press, 1988.
41. R.G. Hunsperger, *Integrated Optics: Theory and Technology*, New York: Springer-Verlag, 3rd ed., 1991.
42. J.M. Senior, *Optical Fiber Communications: Principles and Practice*, London: Prentice Hall, 2nd ed., 1992.
43. M. Kawachi, Silica waveguides on silicon and their application to integrated-optic components, *Opt. Quantum Electron.*, 22, 391–416, 1990.
44. M. Kawachi, Recent progress in planar lightwave circuits, *Proceedings of the IOOC*, 10th International Conference on Integrated Optics and Optical Fiber Communication, Hong Kong, vol. 3, pp. 32–33, 1995.
45. C.H. Henry, Silica planar waveguides, *Proceedings of the IREE*, 19th Australian Conference on Optical Fiber Technology, Melbourne, pp. 326–328, 1994.
46. B.H. Verbeek, C.H. Henry, N.A. Olsson, K.J. Orlowsky, R.F. Kazarinov, and B.H. Johnson, Integrated four-channel Mach–Zehnder multi/demultiplexer fabricated with phosphorous doped SiO_2 waveguides on Si, *J. Lightwave Technol.*, 6, 1011–1017, 1988.
47. T. Kitoh, N. Takato, K. Jinguji, M. Yasu, and M. Kawachi, Novel broad-band optical switch using silica-based planar lightwave circuit, *IEEE Photon. Technol. Lett.*, 4, 735–737, 1992.
48. M. Okuno, K. Kato, Y. Ohmori, M. Kawachi, and T. Matsunaga, Improved 8×8 integrated optical matrix switch using silica-based planar lightwave circuits, *J. Lightwave Technol.*, 12, 1597–1606, 1994.
49. R. Nagase, A. Himeno, M. Okuno, K. Kato, K. Yukimatsu, and M. Kawachi, Silica-based 8×8 optical matrix switch module with hybrid integrated driving circuits and its system application, *J. Lightwave Technol.*, 12, 1631–1639, 1994.
50. Y. Hida, N. Takato, and K. Jinguji, Wavelength division multiplexer with passband and stopband for 1.3/1.55 μm using silica-based planar lightwave circuit, *Electron. Lett.*, 31, 1377–1379, 1995.
51. K. Okamoto, K. Takiguchi, and Y. Ohmori, 16-Channel optical add/drop multiplexer using silica-based array-waveguide gratings, *Electron. Lett.*, 31, 723–724, 1995.
52. K. Okamoto, M. Ishii, Y. Hibino, Y. Ohmori, and H. Toba, Fabrication of unequal channel spacing array-waveguide grating multiplexer modules, *Electron. Lett.*, 31, 1464–1465, 1995.
53. K. Sasayama, M. Okuno, and K. Habara, Photonic FDM multichannel selector using coherent optical transversal filter, *J. Lightwave Technol.*, 12, 664–669, 1994.

54. K. Oda, S. Suzuki, H. Takahashi, and H. Toba, An optical FDM distribution experiment using a high finesse waveguide-type double ring resonator, *IEEE Photon. Technol. Lett.*, 6, 1031–1034, 1994.
55. S. Suzuki, M. Yanagisawa, Y. Hibino, and K. Oda, High-density integrated planar lightwave circuits using SiO_2–GeO_2 waveguides with a high-refractive index difference, *J. Lightwave Technol.*, 12, 790–796, 1994.
56. S. Suzuki, K. Oda, and Y. Hibino, Integrated-optic double-ring resonators with a wide free spectral range of 100 GHz, *J. Lightwave Technol.*, 13, 1766–1771, 1995.
57. E. Pawlowski, K. Takiguchi, M. Okuno, K. Sasayama, A. Himeno, K. Okamato, and Y. Ohmori, Variable bandwidth and tunable centre frequency filter using transversal-form programmable optical filter, *Electron. Lett.*, 32, 113–114, 1996.
58. K. Okamoto, M. Ishii, Y. Hibino, and Y. Ohmori, Fabrication of variable bandwidth filters using arrayed-waveguide gratings, *Electron. Lett.*, 31, 1592–1594, 1995.
59. K. Takiguchi, K. Jinguji, K. Okamato, and Y. Ohmori, Dispersion compensation using a variable group-delay dispersion equaliser, *Electron. Lett.*, 31, 2129–2194, 1995.
60. K. Takiguchi, S. Kawanishi, H. Takara, K. Okamato, K. Jinguji, and Y. Ohmori, Higher order dispersion equaliser of dispersion shifted fiber using a lattice-form programmable optical filter, *Electron. Lett.*, 32, 755–757, 1996.
61. C.H. Henry, G.E. Blonder, and R.F. Kazarinov, Glass waveguides on silicon for hybrid optical packaging, *J. Lightwave Technol.*, 7, 1530–1539, 1989.
62. Y. Yamada, A. Takagi, I. Ogawa, M. Kawachi, and M. Kobayashi, Silica-based optical waveguide on terraced silicon substrate as hybrid integration platform, *Electron. Lett.*, 29, 444–446, 1993.
63. S. Mino, K. Yoshino, Y. Yamada, M. Yasu, and K. Moriwaki, Optoelectronic hybrid integrated laser diode module using planar lightwave circuit platform, *Electron. Lett.*, 30, 1888–1890, 1994.
64. E.E.L. Friedrich, M.G. Oberg, B. Broberg, S. Nilsson, and S. Valette, Hybrid integration of semiconductor lasers with Si-based single-mode ridge waveguides, *J. Lightwave Technol.*, 10, 336–339, 1992.
65. Y. Yamada, H. Terui, Y. Ohmori, M. Yamada, A. Himeno, and M. Kobayashi, Hybrid-integrated 4×4 optical gate matrix switch using silica-based optical waveguides and LD array chips, *J. Lightwave Technol.*, 10, 383–390, 1992.
66. G. Nykolak, M. Haner, P.C. Becker, J. Shmulovich, and Y.H. Wong, Systems evaluation of an Er^{3+}-doped planar waveguide amplifier, *IEEE Photon. Technol. Lett.*, 5, 1185–1187, 1993.
67. K. Hattori, T. Kitagawa, M. Oguma, Y. Ohmori, and M. Horiguchi, Erbium-doped silica-based waveguide amplifier integrated with a 980/1480 nm WDM coupler, *Electron. Lett.*, 30, 856–857, 1994.
68. R.N. Ghosh, J. Shmulovich, C.F. Kane, M.R.X. de Barros, G. Nykolak, A.J. Bruce, and P.C. Becker, 8-mW threshold Er^{3+}-doped planar waveguide amplifier, *IEEE Photon. Technol. Lett.*, 8, 518–520, 1996.
69. K. Takiguchi, K. Okamoto, Y. Inoue, M. Ishii, K. Moriwaki, and S. Ando, Planar lightwave circuit dispersion equalizer module with polarization insensitive properties, *Electron. Lett.*, 31, 57–58, 1995.
70. N. Takato, K. Jinguji, M. Yasu, H. Toba, and M. Kawachi, Silica-based single-mode waveguides on silicon and their application to guided-wave optical interferometers, *J. Lightwave Technol.*, 6, 1003–1010, 1988.
71. K. Oda, N. Takato, H. Toba, and K. Nosu, A wide-band guided-wave periodic multi/demultiplexer with a ring resonator for optical FDM transmission systems, *J. Lightwave Technol.*, 6, 1016–1023, 1988.

72. K.O. Hill, Y. Fuji, D.C. Johnson, and B.S. Kawasaki, Photosensitivity in optical fiber waveguides: Application in reflection filter fabrication, *Appl. Phys. Lett.*, 32(10), 647–649, 1978.
73. D.B. Hunter and R.A. Minassian, Microwave optical filters using in-fiber Bragg grating arrays, *IEEE Microwave Guided Wave Lett.*, 6(9), 103–105, 1996.
74. D.B. Hunter and R.A. Minassian, Tunable optical transversal filter based on chirped gratings, *Electron. Lett.*, 31(25), 2205–2207, 1995.
75. T. Ergodan, Fiber grating spectra, *IEEE J. Lightwave Technol.*, 15(8), 1277–1294, 1997.
76. R.C. Alferness, Guided wave devices for optical communication, *IEEE J. Quantum Electron.*, QE-17, 946–959, 1981.
77. R.C. Alferness, Waveguide electrooptic modulators, *IEEE Trans. Microwave Theory Tech.*, MTT-30(8), 1121–1137, 1982.
78. INSPEC, Properties of Lithium Niobate, Emis Data Reviews Series no. 5, INSPEC publication, pp. 131–146.
79. D.S. Smith, H.D. Riccius, and R.P. Edwin, Refractive indices of lithium niobate, *Opt. Commun.*, 17, 332–335, 1976.
80. D.F. Nelson and R.M. Mikulyak, Refractive indices of congruently melting lithium niobate, *J. Appl. Phys.*, 45, 3688–3689, 1974.
81. M. Minakata, S. Shaito, and M. Shibata, Two dimensional distribution of refractive index changes in Ti-diffused $LiNbO_3$ waveguides, *J. Appl. Phys.*, 50, 3063–3067, 1979.
82. M. Valli and A. Fioretti, Fabrication of good quality Ti:$LiNbO_3$ planar waveguides by diffusion in dry and wet O_2 atmospheres, *J. Modern Opt.*, 35, 885–890, 1985.
83. M. Fukuma and J. Noda, Optical properties of titanium-diffused $LiNbO_3$ strip waveguides and their coupling-to-a-fiber characteristics, *Appl. Opt.*, 20, 591–597, 1980.
84. W.K. Burns, P.H. Klein, E.J. West, and L.E. Plew, Ti diffusion in Ti:$LiNbO_3$ planar and channel optical waveguides, *J. Appl. Phys.*, 50, 6175–6182, 1979.
85. S. Fouchet, A. Carenco, C. Daguet, R. Guglielmi, and L. Riviere, Wavelength dispersion of Ti induced refractive index change in $LiNbO_3$ as a function of diffusion parameters, *IEEE J. Lightwave Technol.*, LT-5, 700–708, 1987.
86. J. Ctyroky, M. Hofman, J. Janta, and J. Schrofel, 3-D analysis of LiNbO3: Ti channel waveguides and directional couplers, *IEEE J. Quantum Electron.*, QE-20, 400–409, 1984.
87. M.D. Feit, J.A. Fleck, Jr., and L. McCaughan, Comparison of calculated and measured performance of diffused channel-waveguide couplers, *J. Opt. Soc. Am.*, 73, 1296–1304, 1983.
88. W.C. Chuang, W.C. Chang, W.Y. Lee, J.H. Leu, and W.S. Wang, A comparison of the performance of $LiNbO_3$ travelling wave phase modulators with various dielectric buffer layers, *J. Opt. Commun.*, 14, 142–148, 1993.
89. P.G. Suchoski and R.V. Ramaswamy, Minimum mode size low loss Ti:$LiNbO_3$ channel waveguides for efficient modulator operation at 1.3 μm, *IEEE J. Quantum Electron.*, QE-23, 1673–1679, 1987.
90. G.L. Li and P.K.L. Yu, Optical intensity modulators for digital and analog applications, *IEEE J. Lightwave Technol.*, 21, 2010–2030, 2003.
91. G.L. Li, C.K. Sun, S.A. Pappert, W.X. Chen, and P.K.L. Yu, Ultra-high speed travelling-wave electro-absorption modulator—design and analysis, *IEEE Trans. Microwave Theory Tech.*, MTT-47, 1177–1183, 1999.
92. H. Chung, W.S.C. Chang, and E.L. Adler, Modelling and optimization of travelling-wave $LiNbO_3$ interferometric modulators, *IEEE J. Quantum Electron.*, 27, 608–617, 1991.

93. K. Yoshida, Y. Kanda, and S. Kohjiro, A travelling wave-type LiNbO3 optical modulator with super-conducting electrodes, *IEEE Trans. Microwave Theory Tech.*, MTT-47, 1201–1205, 1999.
94. E.L. Wotten, K.M. Kissa, A. Yi-Yan, E.J. Murphy, D.A. Lafaw, P.F. Hallemeier, D. Maack, D.V. Attanasio, D.J. Fritz, G.J. McBrien, and D.E. Bossi, Review of lithium niobate for fiber-optic communications, *IEEE J. Selected Areas Quantum Electron.*, 6, 69–82, 2000.
95. C.H. Bulmer, W.K. Burns, and S.C. Hiser, Pyro-eletric effects in $LiNbO_3$ channel waveguide devices, *Appl. Phys. Lett.*, 48, 1036–1038, 1986.
96. U.V. Cummings and W.B. Bridges, Distortion in linearized electro-optic modulators, *IEEE Trans. Microwave Theory Tech.*, MTT-43, 2184–2197, 1995.
97. U.V. Cummings and W.B. Bridges, Bandwidth of linearized electro-optic modulators, *J. Lightwave Technol.*, 16, 1482–1490, 1998.
98. F. Koyama and K. Iga, Frequency chirping in external modulators, *J. Lightwave Technol.*, 6, 87–93, 1988.
99. L.N. Binh, Lithium niobate optical modulators: Devices and applications, *J. Cryst. Growth*, 288, 180–187, 2006.
100. C.C. Teng, Traveling-wave polymeric optical intensity modulator with more than 40 GHz of 3-dB electrical bandwidth, *Appl. Phys. Lett.*, 60(13), 1538–1540, 1992.
101. D. Chen, H.R. Fetterman, A. Chen, W.H. Steier, L.R. Dalton, W. Wang, and Y. Shi, Demonstration of 110 GHz electro-optic polymer modulators, *Appl. Phys. Lett.*, 70, 3335–3337, 1997.
102. S.T. Kowel, S. Wang, A. Thomsen, W. Chan, T.M. Leslie, and N.P. Wang, High field poling of electrooptic etalon modulators on CMOS integrated circuits, *IEEE Photon. Technol. Lett.*, 7(7), 754–756, 1995.
103. S. Kalluri, M. Ziari, A. Chen, V. Chuyanov, W.H. Steier, D. Chen, B. Jalali, H. Fetterman, and L.R. Dalton, Monolithic integration of waveguide polymer electro-optic modulators on VLSI circuitry, *IEEE Photon. Technol. Lett.*, 8, 644–646, 1996.
104. M. Hikita, Y. Shuto, M. Amano, R. Yoshimura, S. Tomaru, and H. Kozawaguchi, Optical intensity modulator in a vertically stacked coupler incorporating electro-optic polymer, *Appl. Phys. Lett.*, 63(9), 1161–1163, 1993.
105. S.M. Garner, S.-S. Lee, V. Chuyanov, A. Chen, A. Yacoubian, and W.H. Steier, Three dimensional integrated optics using polymers, *IEEE J. Quantum Electron.*, 35, 1146–1155, 1999.
106. M.-C. Oh, W.-Y. Hwang, and K. Kim, TE/TM polarization converter using twisted optic-axis waveguides in poled polymers, *Appl. Phys. Lett.*, 70(17), 2227–2229, 1997.
107. I. McCulloch and H. Yoon, Fluorinated NLO polymers with improved optical transparency in the near infrared, *J. Polym. Sci. A*, 33, 1177–1183, 1995.
108. T.C. Kowalczyk, T.Z. Kosc, K.D. Singer, A.J. Beuhler, D.A. Wargowski, P.A. Cahill, C.H. Seager, M.B. Meinhardt, and S. Ermer, Crosslinked polyimide electro-optic materials, *J. Appl. Phys.*, 78(10), 5876–5883, 1995.
109. H.-T. Man and H.N. Yoon, Long term stability of poled side-chain nonlinear optical polymer, *Appl. Phys. Lett.*, 72(5), 540–542, 1998.
110. W. Sotoyama, S. Tatsuura, and T. Yoshimura, Electro-optic side-chain polyimide system with large optical nonlinearity and high thermal stability, *Appl. Phys. Lett.*, 64(17), 2197–2199, 1994.
111. H. Ma, B. Chen, T. Sassa, L.R. Dalton, and K.-Y. Jen, Highly efficient and thermally stable nonlinear optical dendrimer for electrooptics, *J. Am. Chem. Soc.*, 123(5), 986–987, 2001.

112. M.A. Mortazavi, H.N. Yoon, and C.C. Teng, Optical power handling properties of polymeric nonlinear optical waveguides, *J. Appl. Phys.*, 74(8), 4871–4876, 1993.
113. M.A. Mortazavi, K. Song, H. Yoon, and I. McCulloch, Optical power handling of nonlinear polymers, *Polym. Prepr.*, 35, 198–199, 1994.
114. Y. Shi, W. Wang, W. Lin, D.J. Olson, and J.H. Bechtel, Double-end crosslinked electro-optic polymer modulators with high optical power handling capacity, *Appl. Phys. Lett.*, 70(11), 1342–1344, 1997.
115. C. Zhang, A.S. Ren, F. Wang, L.R. Dalton, S.-S. Lee, S.M. Garner, and W.H. Steier, Thermally stable polyene-based NLO chromophore and its polymers with very high electro-optic coefficient, *Polym. Prepr.*, 40, 49–50, 1999.
116. Y. Shi, C. Zhang, H. Zhang, J.H. Bechtel, L.R. Dalton, B.H. Robinson, and W.H. Steier, Low (sub-1-volt) halfwave voltage polymeric electrooptic modulators achieved by controlling chromophore shape, *Science*, 288, 119–122, 2000.
117. S. Ermer, D.G. Girton, L.S. Dries, R.E. Taylor, W. Eades, T.E. van Eck, A.S. Moss, and W.W. Anderson, Low-voltage electro-optic modulation using amorphous polycarbonate host material, *Proceedings of SPIE*, vol. 3949, pp. 148–155, 2000.
118. M.-C. Oh, H. Zhang, A. Szep, V. Chuyanov, W.H. Steier, C. Zhang, L.R. Dalton, H. Erlig, B. Tsap, and H.R. Fetterman, Electro-optic polymer modulators for 1.55 m wavelength using phenyltetraene bridged chromophore in polycarbonate, *Appl. Phys. Lett.*, 76(24), 3525–3527, 2000.
119. I. Liakatas, C. Cai, M. Bösch, M. Jäger, Ch. Bosshard, P. Günter, C. Zhang, and L.R. Dalton, Importance of intermolecular interactions in the nonlinear optical properties of poled polymers, *Appl. Phys. Lett.*, 76(11), 1368–1370, 2000.
120. C.-C. Teng, Precision measurements of the optical attenuation profile along the propagation path in thin-film waveguides, *Appl. Opt.*, 32(7), 1051–1054, 1993.
121. R.A. Soref, J. Schmidtchen, and K. Petermann, Large single-mode rib waveguides in GeSi–Si and Si-on-SiO, *IEEE J. Quantum Electron.*, 27(8), 1971–1974, 1991.
122. S.P. Pogossian, L. Vescan, and A. Vonsovici, The single-mode condition for semiconductor rib waveguides with large cross section, *J. Lightwave Technol.*, 16(10), 1851–1853, 1998.
123. U. Fisher, T. Zinke, J.-R. Kropp, F. Arndt, and K. Petterman, 0.1 dB/cm waveguide losses in single-mode SOI rib waveguides, *IEEE Photon. Technol. Lett.*, 8(5), 647–648, 1996.
124. K.H. Hahn, D.W. Dolfi, R.S. Moshrefzadeh, P.A. Pedersen, and C.V. Francis, Novel two-arm microwave transmission line for high-speed electro-optic polymer modulators, *Electron. Lett.*, 30, 1220–1222, 1994.
125. S.K. Mohapatra, C.V. Francis, K. Hahn, and D.W. Dolfi, Microwave loss in nonlinear optical polymers, *J. Appl. Phys.*, 73, 2569–2571, 1993.
126. C. Zhang, G. Todorova, C. Wang, T. Londergan, and L.R. Dalton, Synthesis of new second-order nonlinear optical chromophores: Implementing lessons learned from theory and experiment, *Proceedings of SPIE*, vol. 4114, pp. 77–87, 2000.
127. A. Otomo, G.I. Stegeman, W.H.G. Horsthuis, and G.R. Möhlmann, Strong field, in-plane poling for nonlinear optical devices in highly nonlinear side chain polymers, *Appl. Phys. Lett.*, 65, 2369–2371, 1994.
128. P.T. Dao and D.J. Williams, Constant current corona charging as a technique for poling organic nonlinear optical thin films and the effect of ambient gas, *J. Appl. Phys.*, 73, 2043–2050, 1993.

2 Incoherence and Coherence in Photonic Signal Processing

This chapter explains the fundamental theory of incoherent and coherent optical signal processing, which provides the basis for later chapters. The advantages and disadvantages of incoherent and coherent optical systems and means of overcoming their limitations are outlined in Section 2.1. The characteristics of the fundamental components of incoherent fiber-optic signal processors (Section 2.2) and coherent integrated-optic signal processors (Section 2.3) are then described.

2.1 INTRODUCTION

The phase of the optical signal is sensitive to environmental fluctuations, such as temperature and pressure changes and acoustic vibrations, as well as frequency fluctuations of the optical source. The inherently high sensitivity of the optical phase to environmental effects has made it attractive for sensor applications but unattractive for signal processing operations in which stability is essential. Obviously, these effects can be obviated by discarding the optical phase through the use of an incoherent light.

Incoherent optical signal processors require the coherence time of the optical source to be much shorter than the basic time delay (or sampling period) to avoid undesirable effects of optical interference. Hence, incoherent systems use intensity variations on optical carriers for performing signal processing operations. In incoherent systems, single-mode optical fibers can be used as a promising delay-line medium for processing broadband signals because of the large bandwidth of optical fibers. Typically, the basic delay-line length* of an incoherent fiber-optic signal processor is in the meter-order and is at least several orders of magnitude greater than the coherence length of the optical source, depending on the frequency of operation. For this reason, changes in the basic delay-line length, due to environmental effects and errors in cutting the fiber length, can be tolerated without causing significant degradation of the system performance. Although incoherent fiber-optic signal processors are stable and robust, they can only perform positive-valued but not

* The basic delay-line length has a delay corresponding to the basic time delay or sampling period of the system.

bipolar or complex-valued signal processing operations and hence have limited applications. This serious limitation can clearly be overcome with a coherent light, if the instability can be found.

In contrast to the incoherent case, coherent optical signal processors require the coherence time of the optical source to be much longer than the basic time delay to achieve coherent interference of the delayed signals. Coherent systems are thus capable of performing complex-valued signal processing operations because both the phase and amplitude of the optical signal are retained in the processed information. As pointed out above, coherent systems cannot operate stably unless the frequency fluctuations of the optical source and environmental effects can be prevented. The frequency can be stabilized by using highly coherent semiconductor lasers, which are commercially available. The environmental effects can be suppressed by using integrated optical waveguides (instead of optical fibers) as a comparatively small delay-line medium for broadband signal processing because of their large bandwidth. Coherent integrated-optic signal processors can operate stably because the waveguide length, which is in the centimeter- or millimeter-order, can be accurately fabricated to the precision of the wavelength order and the phase of the optical signal can be conveniently controlled to the precision of the wavelength order.

In this book, optical fibers and integrated optical waveguides have been considered as the delay-line medium of choice for incoherent and coherent optical signal processing, respectively. The fundamental theories of both incoherent fiber-optic signal processing and coherent integrated-optic signal processing are presented in the next sections.

2.2 INCOHERENT FIBER-OPTIC SIGNAL PROCESSING

The potentially large bandwidth of optical fiber has made it an attractive delay-line medium for incoherent processing of broadband signals. This section describes the theory of incoherent fiber-optic signal processing given in Refs. [1,2].

In incoherent optical signal processors, the information signal (e.g., RF or microwave) to be processed is modulated as intensity variations onto an optical carrier whose coherence time is much shorter than the basic time delay in the system. The optical source can be a broad-linewidth semiconductor laser diode, which can be directly modulated at speeds up to several gigahertz. In the time domain, the modulated wideband signals do not interfere with each other but are appropriately delayed and incoherently combined at the system output. In the frequency domain, the frequency response of the incoherent system depends on the interference of the modulation frequency (RF or microwave) but not the optical carrier frequency. In other words, an incoherent system is incoherent at the optical carrier frequency but coherent at the modulation frequency. Thus, the phase of the optical carrier can be discarded and the signals add on an intensity basis. As a result, although incoherent systems are stable and robust, they can only perform positive-valued signal processing operations because intensity cannot be negative, and hence have limited applications. Using the theory of positive systems, it has been shown that the impulse response of an incoherent system is real and positive valued [1]. In addition, the magnitude of the frequency response of an incoherent system is greatest at the origin of the frequency axis. Consequently, incoherent systems can only be designed to have a limited number of lowpass characteristics but not highpass or bandpass characteristics.

Incoherent fiber-optic signal processing was initiated by the research group at Stanford University in the 1980s [1,2]. Optical fibers and tunable fiber-optic directional couplers have been used in the analysis, design, and construction of a number of incoherent FIR* (finite impulse response) and IIR† (infinite impulse response) fiber-optic signal processors that can perform a variety of linear signal processing functions. These include convolution, correlation, analog matrix operations, frequency filtering, pulse-train generation, data-rate transformation, and code generation. Considerable research effort has produced a number of new concepts, techniques, and applications as a result of the advanced development of fiber-optic signal processors [3–22]. Optical amplifiers, in particular erbium-doped fiber amplifiers (EDFAs), have been used to overcome losses as well as to provide greater flexibility in the analysis, synthesis, and construction of incoherent fiber-optic signal processors for various filtering applications [7–22]. The resulting amplified fiber-optic signal processors have better performances and hence more applications than the unamplified processors [1–6]. Adaptive techniques have also been proposed to provide dynamic weighting of the filter coefficients as well as reconfiguration of the filter delays [4–6,9–12].

The limitation of the incoherent (or positive) fiber-optic signal processors may be reduced by using an electronic differential detection scheme, which can have negative filter coefficients but at the expense of increased system complexity [1,23,24]. It has been claimed that this synthesis technique can implement not only lowpass filters but also highpass and bandpass filters [23]. The performance of the synthesized filter can only approximate that of the desired filter because the synthesis method is based on the least squares approach. Nevertheless, impressive performances of the synthesized lowpass and highpass filters have been experimentally demonstrated [24]. However, the synthesis technique can only handle bipolar numbers but not complex numbers, which must be operated by coherent systems.

It is well known that the basic elements required for implementation of the FIR and IIR digital signal processors are delays, adders, and multipliers [25]. As a result, the basic components required for the realization of the FIR and IIR incoherent fiber-optic signal processors are fiber-optic delay lines, fiber-optic directional couplers, and fiber-optic (or semiconductor) amplifiers. These are described in the following sections.

2.2.1 Fiber-Optic Delay Lines

The low loss and broad bandwidth of optical fibers have made them an attractive delay-line medium for incoherent processing of high-speed broadband signals directly in the optical domain. The loss of optical fibers is about 0.5 dB/km at 1300 nm and about 0.2 dB/km at 1550 nm. The bandwidth–distance product of optical

---

* Note that FIR filters have no feedback loops and are also known as transversal, nonrecursive, or tapped delay-line filters.

† Note that IIR filters have at least one feedback loop and are also known as recursive or recirculating delay-line filters. All-pole and all-pass filters are special types of IIR filters.

fibers is about 32 THz km at 1300 nm and about 100 GHz km at 1550 nm [26]. As a result, the time–bandwidth product of optical fibers exceeds 10^7 at 1300 nm and 10^5 at 1550 nm, assuming a delay per unit length of 5 μs/km.

2.2.2 Fiber-Optic Directional Couplers

One of the fundamental elements in incoherent fiber-optic signal processors is a fiber-optic directional coupler, which performs signal collection (or addition) or signal distribution (or tapping).

The 2×2 fiber-optic directional coupler (see Figure 2.1) is a symmetrical and reciprocal four-port device, which can be designed to have fixed or tunable coupling coefficient. The underlying principle is based on the interaction of the evanescent fields between two parallel fiber cores placed sufficiently close to each other [27,28]. The coupler exhibits very little dependence on the state of polarization of the input fields even though the polarization effect can be easily overcome in practice by means of a fiber polarization controller. An alternative and more promising approach is to use polarization maintaining fibers in the construction of the couplers.

In the incoherent operating regime, the intensity transfer matrix of the 2×2 tunable fiber-optic directional coupler can be described by [1]

$$\begin{bmatrix} I_3 \\ I_4 \end{bmatrix} = (\gamma) \begin{bmatrix} 1-\kappa & \kappa \\ \kappa & 1-\kappa \end{bmatrix} \begin{bmatrix} I_1 \\ I_2 \end{bmatrix} \tag{2.1}$$

where

$\{I_1, I_2\}$ and $\{I_3, I_4\}$ are the intensities at the input and output ports, respectively
κ $(0 \leq \kappa \leq 1)$ is the cross-coupled intensity coefficient
γ (typically $0.95 < \gamma < 1$) is the intensity transmission coefficient

Equation 2.1 means that, when the input port {2} is not excited, the signal intensity at the input port {1} is directly coupled to the output port {3} with an intensity coupling coefficient of $\gamma(1 - \kappa)$ and cross coupled to the output port {4} with an intensity coupling coefficient of $\gamma\kappa$. Similarly, when the input port {1} is not excited, the signal intensity at the input port {2} is directly coupled to the output port {4} with an intensity coupling coefficient of $\gamma(1 - \kappa)$ and cross coupled to the output port {3} with an intensity coupling coefficient of $\gamma\kappa$.

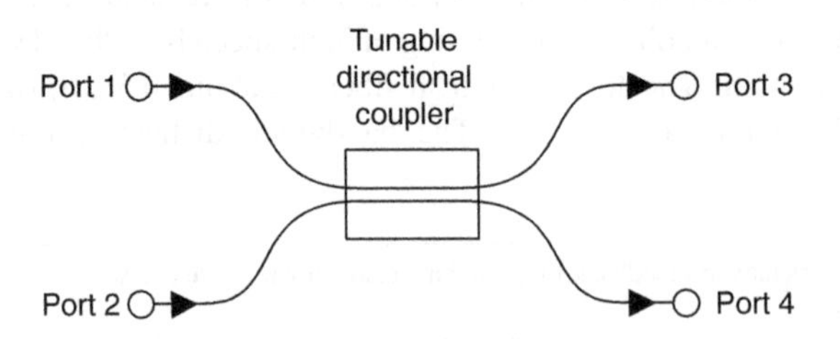

FIGURE 2.1 Schematic diagram of a tunable fiber-optic directional coupler.

2.2.3 Fiber-Optic and Semiconductor Amplifiers

An optical amplifier is an important component in incoherent fiber-optic signal processors because it can compensate for optical losses as well as providing design flexibility resulting in potential applications.

Optical amplifiers can provide signal amplification directly in the optical domain. The operational principles, characteristics, and performances of the optical amplifiers described here have been obtained from Refs. [26,29]. The gain of an optical amplifier is generated by the processes of stimulated scattering induced by nonlinear scattering in an optical fiber, or stimulated emission caused by a population inversion in a lasing medium. The former process is utilized by stimulated Raman scattering fiber amplifiers and stimulated Brillouin scattering fiber amplifiers, which are of little interest in this investigation because they are based on nonlinear effects in fibers. The latter process is employed by semiconductor laser amplifiers (SLAs) or rare-earth-doped fiber amplifiers.

Semiconductor lasers can be designed to act as amplifiers and hence the acronym SLAs. SLAs can be categorized, according to biasing condition and structure, into three types: injection-locked (IL), Fabry–Perot (FP), and traveling-wave (TW) SLAs. IL-SLA and FP-SLA, which are based on resonance effects, require the biasing of the semiconductor laser above and below the lasing threshold, respectively. By contrast, TW-SLA, which exploits single-pass amplification, requires both facets of the semiconductor laser to have antireflection coating. Considerable research attention was initially paid to IL-SLA and FP-SLA with a view of improving the inferior antireflection coating methods. However, TW-SLA has recently attracted the most attention because of its superior performance (saturation output, noise, and bandwidth, to mention a few) and the considerable progress with coating techniques. However, note that it is difficult to differentiate between FP-SLA and TW-SLA because complete zero reflectivity cannot be easily achieved by actual antireflection coating techniques. Thus, it is generally accepted that TW-SLA and FP-SLA have reflectivities less than 0.1%–1% and more than 30%, respectively. For these reasons, the TW-SLAs have been chosen in this study for application in incoherent fiber-optic signal processors.

The rare-earth-doped fiber amplifiers make use of ions, such as erbium, neodymium, and praseodymium, to name a few, as the gain medium to provide optical amplification. In recent years, EDFAs have been the main subject of research simply because they operate near 1550 nm, the wavelength region in which the fiber loss is minimum and hence, the wavelength window of interest for next-generation lightwave systems. For these reasons, EDFAs have also been considered in this investigation.

Because the TW-SLAs and EDFAs have been considered in this research, it is necessary to understand their characteristics and performances, which are summarized in Table 2.1 [29]. Compared with TW-SLAs, the advantages of EDFAs are higher unsaturated (or small-signal) gain, higher saturation output power, lower noise figure, lower fiber coupling loss, and polarization independence. However, TW-SLAs generally have larger bandwidth and are more compact than EDFAs. The high gain, high saturation output, large bandwidth, low noise, low fiber coupling loss, and polarization independence of EDFAs make them an ideal choice for application in the incoherent fiber-optic signal processors. In addition to the

TABLE 2.1
Comparison of the Characteristics and Performances of TW-SLAs and EDFAs

Characteristics/Performances	TW-SLAs	EDFAs
Signal wavelength	Various wavelengths	Currently 1550 nm band only
Unsaturated gain	15 ~ 20 dB	40 ~ 50 dB
Saturation output power	0 ~ 3 dBm	10 ~ 20 dBm
Bandwidth	>3 THz	1 ~ 4 THz
Polarization dependence	Yes	No
Noise figure	6 ~ 9 dB	3 ~ 5 dB
Fiber coupling loss	Large loss (9 ~ 10 dB)	Low loss (<0.5 dB)
Switching speed	Fast switching (<1 ns) Short carrier lifetime	Slow switching (0.2 ~ 10 ms) Long carrier lifetime
Size/length	Small, amplifier length of less than 1.0 mm	Several meters to several 100 m of fiber length

enormous bandwidth and compactness, the relatively fast switching speed of TW-SLAs makes them more attractive than EDFAs for application in adaptive (or programmable) incoherent fiber-optic signal processors, where the gains of the TW-SLAs can be altered by varying the injection current sources driving the semiconductor lasers. Programmable incoherent fiber-optic signal processors incorporating TW-SLAs must use polarization-preserving fibers and couplers because of the polarization dependence of TW-SLAs.

In this chapter, EDFAs and TW-SLAs have, thus, been considered as the amplifiers of choice for nonprogrammable and programmable incoherent fiber-optic signal processors, respectively.

2.3 COHERENT INTEGRATED-OPTIC SIGNAL PROCESSING

As described in Section 2.1, integrated optical waveguides can be used as an attractive delay-line medium for coherent processing of high-speed broadband signals because of the high precision (and hence stability) and large bandwidth of the waveguides.

In coherent optical signal processors, the information signal (RF, microwave or millimeter-wave) to be processed is impressed onto an optical carrier whose coherence time is much longer than the basic time delay in the system. The optical source must be a frequency-stabilized highly coherent semiconductor laser with a very narrow linewidth to suppress the frequency instability. Thus, the optical source must be externally (rather than directly) modulated so that its high degree of coherence can be maintained. This is because, the direct current modulation of the injection lasers causes a dynamic shift of the peak emission wavelength resulting in broadening the spectral width [30]. In addition, external modulation (e.g., using titanium-diffused lithium niobate [Ti:$LiNbO_3$] waveguide modulators) of the optical source permits high-speed modulation, which is ideal for high-speed signal processing. Thus, the use of external modulators allows the optical source to be optimized for spectral quality as well as for obtaining high modulation bandwidth. In the time domain, the modulated signals constructively or destructively interfere with each other, depending on their relative phases. In the frequency domain, the frequency response depends on the interference of the optical carrier frequency. Thus, coherent optical signal processors using integrated-optic waveguides can stably perform high-speed complex-valued signal processing operations because both the phase and amplitude of the optical carrier are retained in the processed information.

The term "integrated optics" was suggested by Miller in 1969 of Bell Laboratory as the lightwave equivalent of "integrated electronics" [31]. Since then, research in integrated optics has begun and gained momentum at about the same time as the development of low-loss optical fibers and semiconductor lasers. The concept of integrated optics involves the use of thin film and microfabrication technologies in the development of a large number of individually fabricated miniature optical components, which may be integrated (or individually interconnected) onto a single chip in a similar fashion to the one that had taken place with integrated electronic circuits [31–35]. Enormous progress has been made in this field with advances in material developments, design techniques, fabrication processes, and component developments. Developments have now reached the stage where integrated optical circuits can be realized to perform various all-optical signal processing and switching functions. In fact, recent advances in lightwave technology and networks have further accelerated the pace for the development of compact, rugged, stable, and economical integrated optical circuits for flexible processing and switching of high-speed broadband signals directly in the optical domain.

The integrated optical circuit has several advantages over its counterpart, the integrated electronic circuit, or over conventional bulk-optic systems consisting of relatively large discrete components [35]. When compared with bulk-optic systems, integrated optical circuits share the same advantages as those of the integrated electronic circuits such as smaller size and weight as well as improved reliability and batch fabrication economy. However, the integrated optical circuit, which uses a high carrier frequency for information processing, inherently has a higher processing speed than the integrated electronic circuit. As with any other new technology, a high development cost of integrated-optic technology (e.g., developing new fabrication technology) is initially required. Nevertheless, the great potential of integrated optics will clearly justify the high development cost in the long run.

Integrated optical circuits can be fabricated on several different materials, each with its own particular features. The choice of a substrate material depends very much on the function to be performed by the circuit. Commonly used substrate materials are glass, $LiNbO_3$, silicon (Si), III–V* semiconductors, such as gallium arsenide (GaAs) and indium phosphide (InP) [32–35]. For example, the InGaAsP/InP material system has been used for the development of InGaAsP/InP 1.55 μm distributed feedback lasers because the InP substrate is capable of emitting light in the 1.3–1.6 μm spectral region, which is important for lightwave systems [36]. The $LiNbO_3$

* Note that compounds composed of elements (e.g., gallium and arsenic) found in the third and fifth columns of the periodic table are called III–V semiconductors.

dielectric material has been widely used for the development of Ti:$LiNbO_3$ waveguide modulators because of its linear electro-optic or Pockel's effect, and that the optical waveguide can be easily formed by diffusing a thin film of titanium into the $LiNbO_3$ substrate [32–35]. Ge, InGaAs/InP, and InGaAsP/InP materials have been used for the fabrication of avalanche photodiodes because of their high absorption coefficients (or responsivity) in the 1.1–1.6 μm low-loss wavelength region [36].

The fundamental component of any integrated optical circuit is the waveguide. Compared with waveguides made of other materials, single-mode silica-based waveguides, which have almost the same composition as that of single-mode optical fibers, are more compatible with optical fibers and hence have lower fiber coupling loss [37–39]. Two major processes have been used for the fabrication of single-mode silica-based waveguides on planar silicon substrates: chemical vapor deposition and flame hydrolysis* deposition. The single-mode waveguide patterns are then defined by photolithographic pattern definition processes followed by reactive ion etching. The glass systems commonly used for the silica-based waveguides are phosphorous-doped silica (SiO_2–P_2O_5), which is formed by chemical vapor deposition [40], and titanium-doped silica (SiO_2–TiO_2) and germanium-doped silica (SiO_2–GeO_2), which are formed by flame hydrolysis deposition [37,38]. It has been claimed by research groups at NTT Japan laboratories that the combination of flame hydrolysis deposition and reactive ion etching can produce low-loss silica-based waveguides, which are best matched to optical fibers [37,38]. A variety of passive integrated optical circuits, which are also known as planar lightwave circuits (PLCs), using single-mode silica-based waveguides on silicon substrates fabricated by a combination of flame hydrolysis deposition and reactive ion etching have been demonstrated as splitters [37], low-speed optical switches [37,41–43], optical wavelength-division multi/demultiplexers [37,44–46], optical frequency-division multi/demultiplexers [37,47–50], tunable optical filters [50–52], and optical dispersion compensators [53,54].

Integrated optical circuits can be realized by two different approaches: hybrid integration, where several devices are fabricated on different materials, with each optimized in a given material, and combined on a common substrate, and monolithic integration, where all devices are fabricated on a common substrate [32–35]. Although the monolithic integration is economically attractive because mass production of the circuit can be achieved by automatic batch processing, there is no single substrate material that is ideal in all respects, as described above. In recent years, hybrid integration has been considered as a practical and promising approach for combining many desired functions on a common substrate, and silicon is an ideal substrate material [37,38,55–59]. The compatibility of silica-based waveguides on silicon with the optical fibers, the high thermal conductivity of silicon (and hence a good heat sink), and the good mechanical stability of silicon make silicon an attractive substrate not only for passive PLCs but also as a platform (or motherboard) for hybrid integration of optoelectronic devices [55]. Hybrid integration platforms

* Flame hydrolysis is a method originally developed for fiber preform fabrication.

have been successfully developed to enable the integration of silica-based waveguides and laser diode chips all on the same silicon substrate [55–59]. The present challenging task is to incorporate integrated-optic waveguide amplifiers (e.g., using erbium-doped silica-based waveguides [60–62]) into the hybrid integration platform to provide greater functionality for next-generation optical networks. In addition, the well-developed silicon technology in the microelectronics industry can be applied to the mass production of hybrid integration of optoelectronic devices, such as optical sources, optical amplifiers, and optical functional circuits, all on a general-purpose silicon platform at a potentially low cost.

In this investigation, low-loss single-mode silica-based waveguides embedded on silicon substrates, whose advantages have been described above, have been chosen as the integrated-optic technology for the design of coherent integrated-optic signal processors. However, the methodology and results are applicable to optical signal processors using other waveguide materials. Similar to the incoherent fiber-optic signal processors described in Section 2.2, the basic components required for the realization of FIR and IIR coherent integrated-optic signal processors are integrated-optic delay lines, integrated-optic phase shifters, integrated-optic directional couplers, and integrated-optic amplifiers. These are described in the following sections.

2.3.1 Integrated-Optic Delay Lines

The high precision and stability of the low-loss large-bandwidth single-mode silica-based waveguide on a silicon substrate have made the waveguide an attractive delay-line medium for processing high-speed broadband signals directly in the optical domain.

The minimum curvature (or bending) radius,* the propagation loss of the waveguides, and the waveguide–fiber coupling loss are very important characteristics in designing and fabricating PLCs. For high relative refractive index difference between the core and cladding (Δ), the waveguides have an advantage of having a small curvature radius but at the expense of having a large propagation loss and a large fiber coupling loss [37,49]. In recent years, the SiO_2–GeO_2 waveguide has been preferred to the SiO_2–TiO_2 waveguide because the former glass system has a lower propagation loss. The discussion here is thus focused on the SiO_2–GeO_2 waveguides.

Typically, SiO_2–GeO_2 waveguides with a core of $8 \times 8 \sim 6 \times 6$ μm^2 and a low Δ of 0.25% ~ 0.75% have a minimum curvature radius of 25 ~ 5 mm and a propagation loss less than 0.1 dB/cm [49]. However, waveguides of high relative refractive index difference ($\Delta = 1.5\%$) with a core of 4.5×4.5 μm^2 have a minimum curvature radius of 2 mm, a low propagation loss of 0.073 dB/cm, and a fiber coupling loss of 0.9 dB [49]. The small curvature radius of high-Δ waveguides makes the fabrication of circuits with curvatures possible, and hence permits high-density integration of circuits. For example, ring resonators require a small ring radius to have a large free spectral range [49,50]. However, high-Δ waveguides are disadvantageous in terms of poor coupling with conventional single-mode fibers because of

* The minimum curvature radius of the waveguide is the radius above which the bending loss is negligible.

the mismatch between their optical mode fields. This problem can be solved by using mode-field converters by means of a thermally expanded core technique. For example, the coupling loss was reduced from 2.0 to 0.9 dB, when thermally expanded core waveguides were used [49].

The stress-induced birefringence of the waveguide, which is caused by the difference between the thermal expansion coefficients of the silica glass layers and the silicon substrate, is unavoidable in the fabrication of PLCs. The waveguide birefringence can be eliminated by using either polarization mode converter with a polyimide half waveplate [63] or the laser trimming method [37,49].

2.3.2 Integrated-Optic Phase Shifters

A thermo-optic phase shifter (PS), which utilizes the thermo-optic effect to change the phase of the optical carrier, is an important element in PLCs because it provides an extra degree of freedom in circuit design [37].

The thermo-optic PS, which consists of a thin-film heater placed on the silica waveguide, is based on the temperature dependence of the refractive index of the waveguide. When an electric voltage is applied to the thin-film heater, the refractive index of the heated waveguide increases, thus changing the optical path length by $(\mathrm{d}n/\mathrm{d}T)L\Delta T$ where $\mathrm{d}n/\mathrm{d}T = 1 \times 10^{-5}$ is the thermo-optic constant of silica waveguide, L is the heated waveguide length, and ΔT is the increase in temperature. For example, when a 5 mm long waveguide is heated at 30°C, the optical path length changes by 1.5 μm, which corresponds to a phase shift of 2π for a 1.5 μm lightwave [64].

2.3.3 Integrated-Optic Directional Couplers

One of the fundamental elements in PLCs is an integrated-optic waveguide directional coupler, which performs signal collection (addition) or signal distribution (tapping). It is, thus, useful to mathematically characterize both the nontunable and tunable directional couplers. For analytical simplicity, the insertion loss, propagation delay, and waveguide birefringence of the directional coupler are not considered.

The electric field transfer matrix of the lossless nontunable waveguide directional coupler (the left directional coupler [DC] of Figure 2.2) can be described by Equation 2.2 [65].

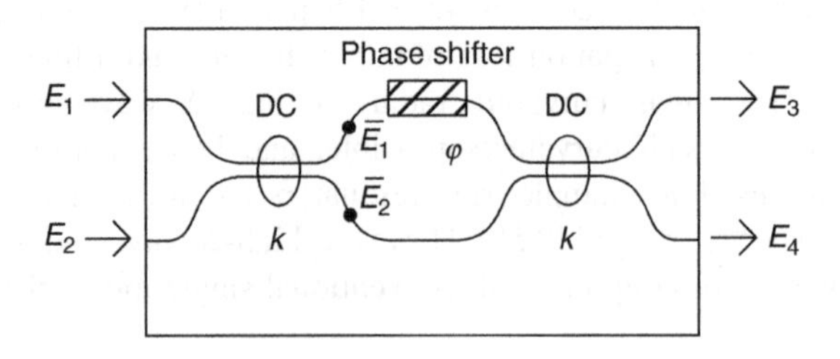

FIGURE 2.2 Schematic diagram of the symmetrical Mach–Zehnder interferometer, which is used as a tunable coupler (TC). DC represents the nontunable directional coupler.

$$\begin{bmatrix} \overline{E}_1 \\ \overline{E}_2 \end{bmatrix} = \begin{bmatrix} \sqrt{1-k} & -j\sqrt{k} \\ -j\sqrt{k} & \sqrt{1-k} \end{bmatrix} \begin{bmatrix} E_1 \\ E_2 \end{bmatrix} \tag{2.2}$$

where

$\{E_1, E_2\}$ and $\{\overline{E}_1, \overline{E}_2\}$ are the electric field amplitudes at the input and output ports of the left DC, respectively
k $(0 \leq k \leq 1)$ is the cross-coupled intensity coefficient
$j = \sqrt{-1}$

Equation 2.2 implies that when the input port $\{2\}$ is not excited (i.e., $E_2 = 0$), the input lightwave E_1 is directly coupled to the output port $\{\bar{1}\}$ with an amplitude coupling coefficient of $\sqrt{1-k}$ and cross coupled to the output port $\{\bar{2}\}$ with an amplitude coupling coefficient of $-j\sqrt{k}$. Similarly, when the input port $\{1\}$ is not excited (i.e., $E_1 = 0$), the input lightwave E_2 is directly coupled to the output port $\{\bar{2}\}$ with an amplitude coupling coefficient of $\sqrt{1-k}$ and cross coupled to the output port $\{\bar{1}\}$ with an amplitude coupling coefficient of $-j\sqrt{k}$. Note that the cross-coupled lightwave experiences a phase shift of $-\pi/2$.

The fixed coupling coefficient of the nontunable DC restricts its application in some PLCs. Another disadvantage is that it is difficult to fabricate the nontunable DC with a precise coupling coefficient. This problem can be overcome by using the symmetrical Mach–Zehnder interferometer (see Figure 2.2), which can be designed to operate as a tunable coupler (TC) or an optical switch [64]. The TC consists of two identical nontunable DCs interconnected by two waveguide arms of equal length. A thermo-optic PS (see Section 2.3.2) placed on the upper arm induces a phase shift of φ.

When the input port $\{2\}$ is not excited (i.e., $E_2 = 0$) and using Equation 2.2, the transfer functions are given by Equations 2.3 and 2.4.

$$\begin{aligned} \left.\frac{E_3}{E_1}\right|_{E_2=0} &= \sqrt{\gamma_w}\left[\sqrt{1-k}\,\exp(j\varphi)\left(\sqrt{1-k} - j\sqrt{k}\right)\left(-j\sqrt{k}\right)\right] \\ &= \sqrt{\gamma_w}[(1-k)\exp(j\varphi) - k] \end{aligned} \tag{2.3}$$

and

$$\begin{aligned} \left.\frac{E_4}{E_1}\right|_{E_2=0} &= \sqrt{\gamma_w}\left[\sqrt{1-k}\,\exp(j\varphi)\left(-j\sqrt{k} - j\sqrt{k}\right)\left(\sqrt{1-k}\right)\right] \\ &= -j\sqrt{\gamma_w}\sqrt{k(1-k)}\,[1 + \exp(j\varphi)] \end{aligned} \tag{2.4}$$

where $\exp(j\varphi)$ is the phase shift factor of the PS and γ_w (typically $\gamma_w = 0.89$ for an insertion loss* of 0.5 dB) is the intensity transmission coefficient of the waveguide TC. Similarly, when the input port $\{1\}$ is not excited (i.e., $E_1 = 0$), the transfer functions are given by Equations 2.5 and 2.6.

* Note that the insertion loss including fiber coupling loss is 0.8 dB [64].

$$\left.\frac{E_3}{E_2}\right|_{E_1=0} = \left.\frac{E_4}{E_1}\right|_{E_2=0} \tag{2.5}$$

$$\begin{aligned}\left.\frac{E_4}{E_2}\right|_{E_1=0} &= \sqrt{\gamma_w}\left[\sqrt{1-k}\left(\sqrt{1-k} - j\sqrt{k}\right)\exp(j\varphi)\left(-j\sqrt{k}\right)\right] \\ &= \sqrt{\gamma_w}[(1-k) - k\exp(j\varphi)]\end{aligned} \tag{2.6}$$

By simple algebraic manipulation of Equations 2.3 through 2.6, the electric field transfer matrix of the TC, which was proposed by Ngo et al. [66], is given by

$$\begin{bmatrix} E_3 \\ E_4 \end{bmatrix} = \sqrt{\gamma_w}\begin{bmatrix} \sqrt{1-K}\exp(j\theta_{31}) & \sqrt{K}\exp(j\theta_{32}) \\ \sqrt{K}\exp(j\theta_{32}) & \sqrt{1-K}\exp(j\theta_{42}) \end{bmatrix}\begin{bmatrix} E_1 \\ E_2 \end{bmatrix} \tag{2.7}$$

where

$$K = 2k(1-k)(1+\cos\varphi) \tag{2.8}$$

$$0 \le K \le 4k(1-k) \quad \text{or} \quad 0.5 - 0.5\sqrt{1-K} \le k \le 0.5 + 0.5\sqrt{1-K} \tag{2.9}$$

$$\theta_{31} = \tan^{-1}\left[\frac{\sin\varphi}{\cos\varphi - k/(1-k)}\right] \tag{2.10}$$

$$\theta_{32} = -\tan^{-1}\left[\frac{(1+\cos\varphi)}{\sin\varphi}\right] \tag{2.11}$$

$$\theta_{42} = \tan^{-1}\left[\frac{\sin\varphi}{\cos\varphi - (1-k)/k}\right] \tag{2.12}$$

In Equations 2.7 through 2.12, (E_3, E_4) are the output electric field amplitudes of the TC, K is the cross-coupled intensity coefficient of the TC, and θ_{nm} is the effective phase shift from the input port m to the output port n of the TC. The maximum value of K is 1, which only occurs at $k = 0.5$ according to Equation 2.9. It is thus preferable to design both the nontunable DCs with $k \cong 0.5$ to maximize the dynamic tuning range of the TC, which is $0 \le K \le 1$. Note that $k = 0.5$ results in $\theta_{31} = \theta_{32} = \theta_{42}$, which implies that the TC is a symmetrical and reciprocal device. Figure 2.3 shows the effective intensity coupling coefficient and the effective phase shifts of the TC for $k = 0.5$ and $0 \le \varphi \le 2\pi$. It can be seen that $0 \le K \le 1$ and $-\pi/2 \le \theta_{31},\theta_{32},\theta_{42} \le +\pi/2$ for $0 \le \varphi \le 2\pi$, and that the same value of K occurs at two different values of φ because of the periodicity of K.

Note that it is difficult to precisely fabricate 3 dB ($k = 0.5$) nontunable DCs, which are important for optical communication or sensor systems. However, this problem can be overcome by using the TC, which can be made exactly 3 dB ($K = 0.5$) provided that $0.1464 \le k \le 0.8536$ (see Equation 2.9).

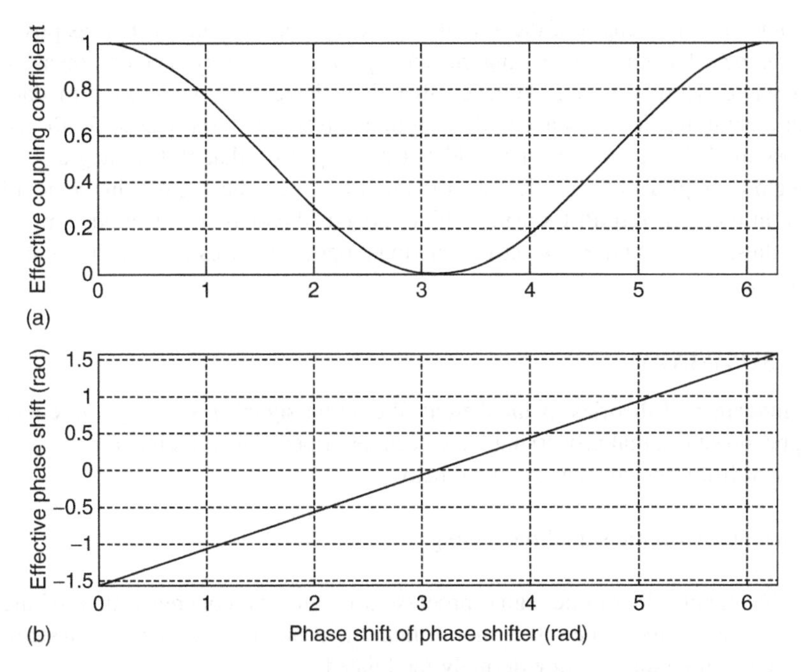

FIGURE 2.3 Output responses of the tunable coupler (TC) for $k = 0.5$ and $0 \leq \varphi \leq 2\pi$. (a) Intensity coupling coefficient K and (b) effective phase shifts $\theta_{31} = \theta_{32} = \theta_{42}$.

The TC can behave as an optical switch when $k = 0.5$ and $\varphi \in (0,\pi)$ [37,64]. When no electric power is applied to the PS ($\varphi = 0$, off state), the input signals are cross-switched according to the paths ($1 \rightarrow 4$, $2 \rightarrow 3$). With electric power corresponding to a phase shift of $\varphi = \pi$ is applied (on state), the input signals are direct-switched according to the paths ($1 \rightarrow 3$, $2 \rightarrow 4$). Typically, the power required for switching is about 0.5 W and the response time is about 1 ms [64].

The TC can operate stably against temperature variation because its operating condition is hardly influenced by environmental temperature change [37]. This is because it is the temperature difference between the two waveguide arms and not the absolute temperature of each arm that is important for tunable or switching operation.

2.3.4 Integrated-Optic Amplifiers

Integrated-optic waveguide amplifiers are important active elements for loss compensation as well as for providing design flexibility of PLCs. The resulting amplified PLCs can perform functions, which are otherwise not available with the unamplified PLCs.

A 50 cm long erbium-doped silica-based waveguide amplifier integrated with a WDM (wavelength-division multiplexing) coupler has been successfully demonstrated with a gain of 27 dB, a low noise figure of 5 dB, and a saturated output power of 4.4 dBm [61]. To fully integrate amplifier devices with other components on the same chip, the length of the amplifying waveguide must be as short as possible. This can be achieved by increasing the doping level of erbium concentration as much as possible. However, it is believed that there have been no reports to date of experimental results of waveguide amplifiers integrated on PLCs. With recent success in PLC technology, it would not be surprising that PLCs integrated with waveguide amplifiers (or semiconductor amplifiers) become a reality in the next few years. In this research, PLCs using erbium-doped silica-based waveguide amplifiers have, thus, been proposed as active functional optical devices for optical communication systems.

2.4 SUMMARY

The fundamental theories of incoherent fiber-optic signal processing and coherent integrated-optic signal processing have been described. The principal points that can be drawn from this chapter are given below.

Incoherent fiber-optic signal processing

1. Incoherent fiber-optic signal processors require the coherence time of the optical source to be much shorter than the basic time delay in the system, and can be directly or externally modulated.
2. In incoherent fiber-optic signal processors, the low loss and large bandwidth of the single-mode optical fiber have made it an attractive delay-line medium for processing broadband signals directly in the optical domain.
3. The characteristics of the fundamental fiber-optic elements (such as fiber-optic delay lines, fiber-optic directional couplers, and fiber-optic and semiconductor amplifiers) of the incoherent fiber-optic signal processors have been described.
4. Although incoherent fiber-optic signal processors are stable and robust, they can only perform positive-valued signal processing operations and thus have limited applications. They can process RF and microwave signals with speed ranging from hundreds of megahertz to a few gigahertz because the basic filter length typically ranges from a few meters to tens of meters depending on the frequency of operation.
5. The fundamental theory of incoherent fiber-optic signal processing described in this chapter is used in the analysis and design of incoherent fiber-optic signal processors, which are presented in Chapters 3 through 5.

Coherent integrated-optic signal processing

1. Coherent integrated-optic signal processors require the coherence time of the optical source to be much longer than the basic time delay in the system. The optical source must be a frequency-stabilized highly coherent semiconductor laser which must be externally modulated.
2. In coherent integrated-optic signal processors, the high precision and stability of the low-loss large-bandwidth single-mode silica-based waveguide on a

silicon substrate have made the waveguide an attractive delay-line medium for high-speed processing of broadband signals directly in the optical domain.
3. The characteristics of the fundamental integrated-optic elements (such as integrated-optic delay lines, integrated-optic phase shifters, integrated-optic directional couplers, and integrated-optic amplifiers) of the coherent integrated-optic signal processors have been described.
4. Coherent integrated-optic signal processors can stably perform complex-valued signal processing operations and thus have potential applications in optical communication systems. They can process millimeter-wave signals with speed in the range of tens of gigahertz because the basic filter length typically ranges from a few millimeters to a few centimeters depending on the frequency of operation.
5. The fundamental theory of coherent integrated-optic signal processing described in this chapter is used in the analysis and design of coherent integrated-optic signal processors presented in later chapters.

Note that the choice between an incoherent and a coherent optical signal processor depends on particular applications.

REFERENCES

1. B. Moslehi, J.W. Goodman, M. Tur, and H.J. Shaw, Fiber-optic lattice signal processing, *Proc. IEEE*, 72, 909–930, 1984.
2. K.P. Jackson, S.A. Newton, B. Moslehi, M. Tur, C.C. Cutler, J.W. Goodman, and H.J. Shaw, Optical fiber delay-line signal processing, *IEEE Trans. Microwave Theory Tech.*, MTT-33, 193–210, 1985.
3. C.C. Wang, High-frequency narrow-band single-mode fiber-optic transversal filters, *J. Lightwave Technol.*, LT-5, 77–81, 1987.
4. R.I. MacDonald, Switched optical delay-line signal processors, *J. Lightwave Technol.*, LT-5, 856–861, 1987.
5. D.M. Gookin and M.H. Berry, Finite impulse response filter with large dynamic range and high sampling rate, *Appl. Opt.*, 29, 1061–1062, 1990.
6. A. Ghosh and S. Frank, Design and performance analysis of fiber-optic infinite-impulse response filters, *Appl. Opt.*, 31, 4700–4711, 1992.
7. B. Moslehi, Fiber-optic filters employing optical amplifiers to provide design flexibility, *Electron. Lett.*, 28, 226–228, 1992.
8. B. Moslehi and J.W. Goodman, Novel amplified fiber-optic recirculating delay line processor, *J. Lightwave Technol.*, 10, 1142–1147, 1992.
9. B. Moslehi, K.K. Chau, and J.W. Goodman, Optical amplifiers and liquid-crystal shutters applied to electrically reconfigurable fiber optic signal processors, *Opt. Eng.*, 32, 974–981, 1993.
10. J. Capmany and J. Cascon, Optical programmable transversal filters using fiber amplifiers, *Electron. Lett.*, 28, 1245–1246, 1992.
11. J. Capmany and J. Cascon, Discrete time fiber-optic signal processors using optical amplifiers, *J. Lightwave Technol.*, 12, 106–117, 1994.
12. S. Sales, J. Capmany, J. Marti, and D. Pastor, Solutions to the synthesis problem of optical delay line filters, *Opt. Lett.*, 20, 2438–2440, 1995.

13. S. Sales, J. Capmany, J. Marti, and D. Pastor, Novel and significant results on the non-recirculating delay line with a fiber loop, *IEEE Photon. Technol. Lett.*, 7, 1439–1440, 1995.
14. S. Sales, D. Pastor, J. Capmany, and J. Marti, Fiber-optic delay-line filters employing fiber loops: Signal and noise analysis and experimental characterization, *J. Opt. Soc. Am. A*, 12, 2129–2135, 1995.
15. D. Pastor, S. Sales, J. Capmany, J. Marti, and J. Cascon, Amplified double coupler fiber-optic delay line filter, *IEEE Photon. Technol. Lett.*, 7, 75–77, 1995.
16. M.C. Vazquez, B. Vizoso, M. Lopez-Amo, and M.A. Muriel, Single and double amplified recirculating delay lines as fiber-optic filters, *Electron. Lett.*, 28, 1017–1019, 1992.
17. M.C. Vazquez, R. Civera, M. Lopez-Amo, and M.A. Muriel, Analysis of double-parallel amplified recirculating optical-delay lines, *Appl. Opt.*, 33, 1015–1021, 1994.
18. C. Vazquez, M. Lopez-Amo, M.A. Muriel, and J. Capmany, Performance parameters and applications of a modified amplified recirculating delay line, *Fiber Integrated Opt.*, 14, 347–358, 1995.
19. B. Vizoso, C. Vazquez, R. Civera, M. Lopez-Amo, and M.A. Muriel, Amplified fiber-optic recirculating delay lines, *J. Lightwave Technol.*, 12, 294–305, 1994.
20. B. Vizoso, I.R. Matias, M. Lopez-Amo, M.A. Muriel, and J.M. Lopez-Higuera, Design and application of double amplified recirculating ring structure for hybrid fiber buses, *Opt. Quantum Electron.*, 27, 847–857, 1995.
21. E.C. Heyde and R.A. Minasian, A solution to the synthesis problem of recirculating optical delay line filters, *IEEE Photon. Technol. Lett.*, 6, 833–835, 1994.
22. E.C. Heyde, Theoretical methodology for describing active and passive recirculating delay line systems, *Electron. Lett.*, 31, 2038–2039, 1995.
23. J. Capmany, J. Cascon, J.L. Martin, S. Sales, D. Pastor, and J. Marti, Synthesis of fiber-optic delay line filters, *J. Lightwave Technol.*, 13, 2003–2012, 1994.
24. S. Sales, J. Capmany, J. Marti, and D. Pastor, Experimental demonstration of fiber-optic delay line filters with negative coefficients, *Electron. Lett.*, 31, 1095–1096, 1995.
25. A.V. Oppenheim and R.W. Schafer, *Discrete-Time Signal Processing*, Englewood Cliffs, NJ: Prentice-Hall, 1989.
26. G.P. Agrawal, *Fiber-Optic Communication Systems*, New York: John Wiley & Sons, 1992.
27. S.K. Sheem and T.G. Giallorenzi, Single-mode fiber-optical power divider: Encapsulated etching technique, *Opt. Lett.*, 4, 29–31, 1979.
28. M.J.F. Digonnet and H.J. Shaw, Analysis of a tunable single mode optical fiber coupler, *IEEE J. Quantum Electron.*, QE-18, 746–754, 1982.
29. S. Shimada and H. Ishio (Eds.), *Optical Amplifiers and Their Applications*, chapters 1–5, New York: John Wiley & Sons, 1994.
30. J.C. Cartledge and G.S. Burley, The effect of laser chirping on lightwave system performance, *J. Lightwave Technol.*, 7, 568–573, 1989.
31. S.E. Miller, Integrated optics: An introduction, *Bell Syst. Tech. J.*, 48, 2059–2069, 1969.
32. T. Tamir (Ed.), *Integrated Optics*, New York: Springer-Verlag, 1975.
33. L.D. Hutcheson, *Integrated Optical Circuits and Components*, New York: Marcel Dekker, 1987.
34. S.E. Miller and I.P. Kaminow (Eds.), *Optical Fiber Telecommunications II*, New York: Academic Press, 1988.
35. R.G. Hunsperger, *Integrated Optics: Theory and Technology*, 3rd edn., New York: Springer-Verlag, 1991.

36. J.M. Senior, *Optical Fiber Communications: Principles and Practice*, 2nd edn., London: Prentice Hall, 1992.
37. M. Kawachi, Silica waveguides on silicon and their application to integrated-optic components, *Opt. Quantum Electron.*, 22, 391–416, 1990.
38. M. Kawachi, Recent progress in planar lightwave circuits, *Proceedings of the IOOC*, 10th International Conference on Integrated Optics and Optical Fiber Communication, Hong Kong, vol. 3, pp. 32–33, 1995.
39. C.H. Henry, Silica planar waveguides, *Proceedings of the IREE*, 19th Australian Conference on Optical Fiber Technology, Melbourne, pp. 326–328, 1994.
40. B.H. Verbeek, C.H. Henry, N.A. Olsson, K.J. Orlowsky, R.F. Kazarinov, and B.H. Johnson, Integrated four-channel Mach–Zehnder multi/demultiplexer fabricated with phosphorous doped SiO2 waveguides on Si, *J. Lightwave Technol.*, 6, 1011–1017, 1988.
41. T. Kitoh, N. Takato, K. Jinguji, M. Yasu, and M. Kawachi, Novel broad-band optical switch using silica-based planar lightwave circuit, *IEEE Photon. Technol. Lett.*, 4, 735–737, 1992.
42. M. Okuno, K. Kato, Y. Ohmori, M. Kawachi, and T. Matsunaga, Improved 8×8 integrated optical matrix switch using silica-based planar lightwave circuits, *J. Lightwave Technol.*, 12, 1597–1606, 1994.
43. R. Nagase, A. Himeno, M. Okuno, K. Kato, K. Yukimatsu, and M. Kawachi, Silica-based 8×8 optical matrix switch module with hybrid integrated driving circuits and its system application, *J. Lightwave Technol.*, 12, 1631–1639, 1994.
44. Y. Hida, N. Takato, and K. Jinguji, Wavelength division multiplexer with passband and stopband for 1.3/1.55 μm using silica-based planar lightwave circuit, *Electron. Lett.*, 31, 1377–1379, 1995.
45. K. Okamoto, K. Takiguchi, and Y. Ohmori, 16-Channel optical add/drop multiplexer using silica-based array-waveguide gratings, *Electron. Lett.*, 31, 723–724, 1995.
46. K. Okamoto, M. Ishii, Y. Hibino, Y. Ohmori, and H. Toba, Fabrication of unequal channel spacing array-waveguide grating multiplexer modules, *Electron. Lett.*, 31, 1464–1465, 1995.
47. K. Sasayama, M. Okuno, and K. Habara, Photonic FDM multichannel selector using coherent optical transversal filter, *J. Lightwave Technol.*, 12, 664–669, 1994.
48. K. Oda, S. Suzuki, H. Takahashi, and H. Toba, An optical FDM distribution experiment using a high finesse waveguide-type double ring resonator, *IEEE Photon. Technol. Lett.*, 6, 1031–1034, 1994.
49. S. Suzuki, M. Yanagisawa, Y. Hibino, and K. Oda, High-density integrated planar lightwave circuits using SiO_2–GeO_2 waveguides with a high-refractive index difference, *J. Lightwave Technol.*, 12, 790–796, 1994.
50. S. Suzuki, K. Oda, and Y. Hibino, Integrated-optic double-ring resonators with a wide free spectral range of 100 GHz, *J. Lightwave Technol.*, 13, 1766–1771, 1995.
51. E. Pawlowski, K. Takiguchi, M. Okuno, K. Sasayama, A. Himeno, K. Okamato, and Y. Ohmori, Variable bandwidth and tunable centre frequency filter using transversal-form programmable optical filter, *Electron. Lett.*, 32, 113–114, 1996.
52. K. Okamoto, M. Ishii, Y. Hibino, and Y. Ohmori, Fabrication of variable bandwidth filters using arrayed-waveguide gratings, *Electron. Lett.*, 31, 1592–1594, 1995.
53. K. Takiguchi, K. Jinguji, K. Okamato, and Y. Ohmori, Dispersion compensation using a variable group-delay dispersion equaliser, *Electron. Lett.*, 31, 2129–2194, 1995.
54. K. Takiguchi, S. Kawanishi, H. Takara, K. Okamato, K. Jinguji, and Y. Ohmori, Higher order dispersion equaliser of dispersion shifted fiber using a lattice-form programmable optical filter, *Electron. Lett.*, 32, 755–757, 1996.

55. C.H. Henry, G.E. Blonder, and R.F. Kazarinov, Glass waveguides on silicon for hybrid optical packaging, *J. Lightwave Technol.*, 7, 1530–1539, 1989.
56. Y. Yamada, A. Takagi, I. Ogawa, M. Kawachi, and M. Kobayashi, Silica-based optical waveguide on terraced silicon substrate as hybrid integration platform, *Electron. Lett.*, 29, 444–446, 1993.
57. S. Mino, K. Yoshino, Y. Yamada, M. Yasu, and K. Moriwaki, Optoelectronic hybrid integrated laser diode module using planar lightwave circuit platform, *Electron. Lett.*, 30, 1888–1890, 1994.
58. E.E.L. Friedrich, M.G. Oberg, B. Broberg, S. Nilsson, and S. Valette, Hybrid integration of semiconductor lasers with Si-based single-mode ridge waveguides, *J. Lightwave Technol.*, 10, 336–339, 1992.
59. Y. Yamada, H. Terui, Y. Ohmori, M. Yamada, A. Himeno, and M. Kobayashi, Hybrid-integrated 4×4 optical gate matrix switch using silica-based optical waveguides and LD array chips, *J. Lightwave Technol.*, 10, 383–390, 1992.
60. G. Nykolak, M. Haner, P.C. Becker, J. Shmulovich, and Y.H. Wong, Systems evaluation of an Er^{3+}-doped planar waveguide amplifier, *IEEE Photon. Technol. Lett.*, 5, 1185–1187, 1993.
61. K. Hattori, T. Kitagawa, M. Oguma, Y. Ohmori, and M. Horiguchi, Erbium-doped silica-based waveguide amplifier integrated with a 980/1480 nm WDM coupler, *Electron. Lett.*, 30, 856–857, 1994.
62. R.N. Ghosh, J. Shmulovich, C.F. Kane, M.R.X. de Barros, G. Nykolak, A.J. Bruce, and P.C. Becker, 8-mW threshold Er^{3+}-doped planar waveguide amplifier, *IEEE Photon. Technol. Lett.*, 8, 518–520, 1996.
63. K. Takiguchi, K. Okamoto, Y. Inoue, M. Ishii, K. Moriwaki, and S. Ando, Planar lightwave circuit dispersion equalizer module with polarization insensitive properties, *Electron. Lett.*, 31, 57–58, 1995.
64. N. Takato, K. Jinguji, M. Yasu, H. Toba, and M. Kawachi, Silica-based single-mode waveguides on silicon and their application to guided-wave optical interferometers, *J. Lightwave Technol.*, 6, 1003–1010, 1988.
65. K. Oda, N. Takato, H. Toba, and K. Nosu, A wide-band guided-wave periodic multi/demultiplexer with a ring resonator for optical FDM transmission systems, *J. Lightwave Technol.*, 6, 1016–1023, 1988.
66. Q.N. Ngo and L.N. Binh, Novel realisation of monotonic butterworth-type lowpass, highpass and bandpass optical filters using phase-modulated fibre-optic interferometers and ring resonators, *IEEE J. Lightwave Technol.*, 12, 827–841, 1994.

3 Photonic Computing Processors

In this chapter, incoherent fiber-optic systolic array processors (FOSAPs), which employ a digital-multiplication-by-analog-convolution (DMAC) algorithm and the extension of the DMAC algorithm, are proposed for real-valued digital matrix computations. The important role of optics in optical computing and a variety of existing optical architectures using the DMAC algorithm are described in Section 3.1.2. Section 3.1.2.4 presents mathematical formulations of the DMAC algorithm and the twos-complement binary (TCB) arithmetic, while Section 3.1.3 outlines the operational principles of the elemental processors of the FOSAP architectures. The performances of the FOSAP multipliers (Section 3.1.5) are compared with the performances of digital electronic multipliers and other optical DMAC multipliers. Means of overcoming the limitation of the FOSAP architectures are discussed in Section 3.1.6. The theoretical aspects of incoherent fiber-optic signal processing described in Chapter 2 are applied in this chapter where intensity-based signals are considered.

Furthermore, the design of a programmable incoherent Newton–Cotes optical integrator (INCOI) is described. The definition, existing design techniques, and application of digital integrators are described in Section 3.2. A generalized theory of the Newton–Cotes digital integrators, whose derivation is given in Appendix A, and their magnitude and impulse responses are given in Section 3.2. On the basis of this theory, algorithms for the synthesis of the INCOI processor are proposed. These algorithms are then used in the design of the programmable INCOI processor, which essentially consists of a microprocessor, fiber-optic architectures, optical switches, and optical and semiconductor amplifiers. Several types of input pulse sequences are chosen as examples for illustration of the incoherent processing accuracy of the programmable INCOI processor. In addition, the incoherent RFOSP is used in the design of the programmable INCOI processor. The theory of incoherent fiber-optic signal processing described in Chapter 2 is employed in this chapter where intensity-based signals are considered. Section 3.3 then outlines the theoretical development for optical differentiating processors and their implementation.

3.1 INCOHERENT FIBER-OPTIC SYSTOLIC ARRAY PROCESSORS

3.1.1 INTRODUCTION

In a digital electronic computer, a coprocessor is an important special-purpose processor, which is mainly responsible for performing specific high-speed arithmetic operations. A high-performance coprocessor is necessary for performing massive computational tasks, such as image processing, pattern recognition, and signal processing problems, which would otherwise be performed by very computationally expensive computer software. There has been considerable research interest in developing optical coprocessors for general-purpose digital electronic computers because of the massive parallelism, high processing speed, and high spatial- and temporal-bandwidth of optics compared with electronics.

Optical systolic array processors* [1] have been proposed as optical coprocessors [2]. Initially, they were proposed for performing analog matrix–vector and matrix–matrix operations using acousto-optic [1–4] and fiber-optic architectures [5,6]. However, the main drawback of these analog optical processors is their restriction to operating in the low accuracy range (13–16 bits). The accuracy is limited primarily by the linear dynamic range of the devices used in the system, e.g., an erbium-doped fiber amplifier (EDFA) can provide an intensity gain of 40–50 dB, which translates to 13–16 bits of accuracy.

Various algorithms have been described that enable optical systolic array processors to compute with high (or digital) accuracy, with techniques such as residue arithmetic [7], modified signed-digit number representation [8], redundant number representation [9], and symbolic substitution [10]. The most commonly used technique is the DMAC algorithm [11], which is employed in this chapter, because convolution can be easily performed in optics.

Whitehouse and Speiser [11] proposed that the digital multiplication of two binary numbers (originally known as the Swartzlander multiplier [12]) is equivalent to the analog convolution of these numbers, provided that the analog result of the digital product is represented in the mixed-binary format. This DMAC technique is known as the DMAC algorithm [13]. The mixed-binary representation of the digital product can be converted to the standard binary representation of the digital product by an analog-to-digital converter (ADC) and a shift-and-add (S/A) circuit in the postprocessing unit. This basic idea has been applied to digital matrix–vector and matrix–matrix operations using time-integrating (TI) and space-integrating (SI) acousto-optic architectures [14–22], magneto-optic spatial light modulators [23], nonlinear optical devices [24,25], and logic counters [26].

In this chapter, incoherent FOSAPs employing the DMAC algorithm are proposed for real-valued digital matrix multiplications. Most of the work presented here has been described by Ngo and Binh [27].

* Although there are various architectures of optical systolic array processors, they all have one general feature in that the input data flow in a pulsating fashion, and hence the name "systolic," through one- or two-dimensional identical array processing elements where computation is performed on the data currently present.

3.1.2 Digital-Multiplication-by-Analog-Convolution Algorithm and Its Extended Version

The DMAC algorithm, its extended version, and the TCB arithmetic are described. For the sake of hardware simplicity, the binary words are assumed to be of the same length, although the DMAC algorithm is generally valid for binary data of any length. Unless otherwise stated, variables with brackets denote binary sequences. For example, f represents the analog value whose binary sequence is denoted by $\{f\}$.

3.1.2.1 Multiplication of Two Digital Numbers

Multiplication is a standard operation in matrix computations. It is shown here that the digital multiplication of two binary numbers can be determined by performing the analog convolution of these binary numbers followed by the function of the postprocessor.

The standard binary representation of two positive integers f and g is given by the sequences $\{f\} = \{f_{n-1} \ldots f_1 f_0\}$ and $\{g\} = \{g_{n-1} \ldots g_1 g_0\}$, where n is the number of bits in a binary number, $f_i, g_i \in (0,1)$ for $0 \leq i \leq n-1$, f_0 and g_0 are the least significant bits (LSBs), and f_{n-1} and g_{n-1} are the most significant bits (MSBs). The mathematical descriptions of these positive n-bit words are given by

$$f = \sum_{i=0}^{n-1} f_i 2^i \tag{3.1a}$$

$$g = \sum_{j=0}^{n-1} g_j 2^j \tag{3.1b}$$

The product of these binary numbers is given by [13]

$$f \cdot g = \sum_{k=0}^{2n-2} 2^k y_k \tag{3.2}$$

where

$$y_k = \sum_{i=0}^{k} f_i g_{k-i}, \quad (0 \leq k \leq 2n-2) \tag{3.3}$$

with $f_i = g_i = 0$ for $i < 0$ or $i > n-1$.

Equation 3.3 is easily recognized as the discrete convolution of two sequences, i.e., $\{f\} * \{g\}$ where $*$ designates the convolution operation. The analog result of the discrete convolution is in the mixed-binary format. Note that the digital multiplication of two binary numbers without carries, i.e., $\{f\} \cdot \{g\}$, is also in the mixed-binary format. Furthermore, the kth analog value y_k of the convolution result represents the sum of partial products (without carries) in the kth column of the product $\{f\} \cdot \{g\}$. This process of digital multiplication by analog convolution is known as the DMAC algorithm.

Evaluation of Equation 3.2 requires an optical convolver, an optical detector, an ADC, and an S/A circuit. At a particular discrete time k, an ADC* is used to convert the kth analog value y_k into its binary representation, which is then upshifted to the left by k bits by a shift register. This process is equivalent to evaluating the partial product $2^k y_k$ of Equation 3.2. The last step for evaluation of Equation 3.2 requires the addition of the partial products (with carries) by a binary adder.

The DMAC algorithm is best illustrated by means of a numerical example of multiplying two positive n-bit words whose standard binary representations are $\{f\} = \{110\}$ and $\{g\} = \{011\}$, which correspond to the analog values $f = 6$ and $g = 3$. Figures 3.1a and b show the discrete-time representation of these sequences, where $f[i]$ is the impulse response of the optical convolver, $g[i]$ represents the modulated laser pulses (rectangular profile assumed) of unity height to be launched into the convolver, T_p is the pulse width, and T is the bit period of the input pulse sequence or the sampling period of the optical convolver. The convolver pulse response, which is simply the discrete convolution of its impulse response $f[i]$ with the input pulse sequence $g[i]$ is given, from Equation 3.3, as

$$y[k] = \sum_{i=0}^{k} f[i]g[k-i], \quad (0 \leq k \leq 2n-2) \tag{3.4}$$

Equation 3.4 can be evaluated graphically as shown in Figures 3.1c through g for $n = 3$. In Figure 3.1c, $g[i]$ is folded about $i = 0$ to become $g[-i]$ and is slid past (with the LSB first) the digits of $f[i]$. The sequence $g[k-i]$ is simply the folded sequence of $g[i]$ shifted to the right by k units of delay T, as shown in Figures 3.1c through g for $k = 0, 1, 2, 3, 4$, respectively. The pulse shown by the broken curve corresponds to the LSB of the next word. t_0 and t_1 denote the starting times of the LSB of the first and second words, respectively. The word strobe period $T_w = t_{j+1} - t_j$ is the time separation between the LSB of the previous word and the LSB of the next word and is given by $T_w = (2n-1)T$. To avoid overflow, $n-1$ zeros are required for padding between the words. The analog result $y[k]$ is shown in Figure 3.1h, where T_{conv} is the propagation delay of the optical convolver, which has been ignored during the convolution process in Figures 3.1c through g for the sake of clarity. These mixed-binary pulses are converted into their binary representations by a 2-bit ADC and shifted and added by a 2-bit S/A circuit in the postprocessing unit, as shown in Figure 3.1i. The standard binary representation $\{010010\}$ corresponds to the integer 18, which is expected from the product (6)(3).

3.1.2.2 High-Order Digital Multiplication

High-order multiplication is required for high-order matrix computations [28–30]. It is shown here that the digital multiplication of $\hat{N}$ binary numbers can be determined

* The ADC requires $\log_2 n$ bits accuracy because n is the maximum analog value as a result of the convolution of two n-bit words.

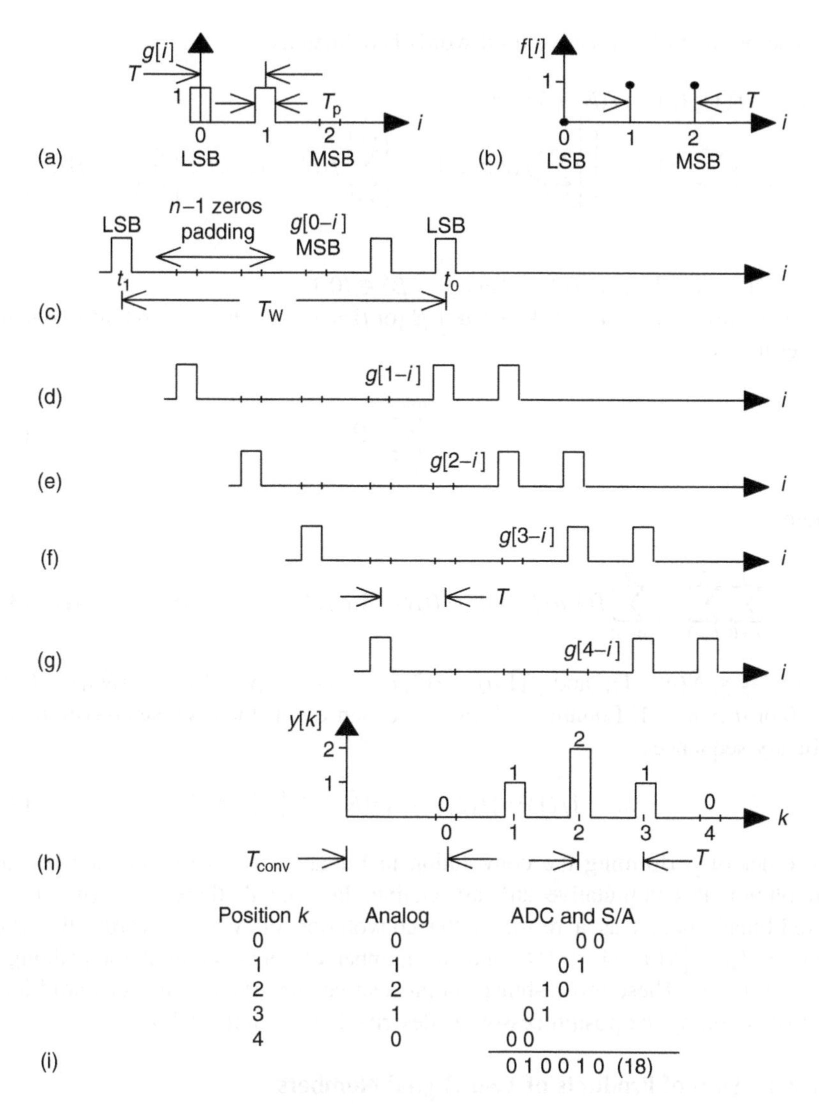

Position k	Analog	ADC and S/A
0	0	0 0
1	1	0 1
2	2	1 0
3	1	0 1
4	0	0 0
		0 1 0 0 1 0 (18)

FIGURE 3.1 Graphical illustration of the DMAC technique. (a) The input pulse sequence $g[i]$, (b) the convolver impulse response $f[i]$, (c–g) the convolution operation, (h) the convolution output $y[k]$ is in the mixed-binary format, and (i) the analog output is operated by the postprocessing unit to obtain the expected standard binary representation of the decimal number 18.

by performing the analog convolution of these $\hat{N}$ binary numbers followed by the function of the postprocessor. This process of "high-order digital-multiplication-by-analog-convolution" is referred to as the HO-DMAC algorithm.

The product of $\hat{N}$ positive n-bit words is defined as

$$P_{\hat{N}} = f(1)f(2)\ldots f(\hat{N}-1)f(\hat{N})$$
$$= \left[\sum_{a=0}^{n-1} f(1,a)2^a\right]\left[\sum_{b=0}^{n-1} f(2,b)2^b\right]\ldots\left[\sum_{\alpha=0}^{n-1} f(\hat{N}-1,\alpha)2^\alpha\right]\left[\sum_{\beta=0}^{n-1} f(\hat{N},\beta)2^\beta\right] \tag{3.5}$$

where $f(1,a), f(2,b), \ldots, f(\hat{N}-1,\alpha), f(\hat{N},\beta) \in (0,1)$.

Substitution of $\gamma = a + b + \cdots + \alpha + \beta$ (or $\beta = \gamma - a - b - \cdots - \alpha$) into Equation 3.5 results in

$$P_{\hat{N}} = \sum_{\gamma=0}^{\hat{N}(n-1)} 2^\gamma y_\gamma \tag{3.6}$$

where

$$y_\gamma = \sum_{a=0}^{\gamma}\sum_{b=0}^{\gamma}\cdots\sum_{\alpha=0}^{\gamma} f(1,a)f(2,b)\cdots f(\hat{N}-1,\alpha)f(\hat{N},\gamma-a-b-\cdots-\alpha) \tag{3.7}$$

for $0 \le \gamma \le \hat{N}(n-1)$, and $f(1,a) = f(2,a) = \cdots = f(\hat{N}-1,a) = f(\hat{N},a) = 0$ for $a < 0$ or $a > n-1$. Equation 3.7 can be recognized as the discrete convolution of $\hat{N}$ binary sequences

$$y_\gamma = \{f(1)\}*\{f(2)\}\cdots\{f(\hat{N}-1)\}*\{f(\hat{N})\} \tag{3.8}$$

The order of performing the convolution in Equation 3.8 is unimportant because convolution is commutative and associative. In general, there are $\hat{N}(n-1)+1$ mixed-binary points as a result of the convolution of $\hat{N}$ n-bit words, the word cycle is $T_w = [\hat{N}(n-1)+1]T$, and the number of zeros required for padding is $(\hat{N}-1)(n-1)$. These mixed-binary points can be converted to the standard binary representation by the postprocessor, as described in Section 3.1.2.1.

3.1.2.3 Sum of Products of Two Digital Numbers

The sum of products of two digital numbers, which is a vector inner-product operation, is also a standard operation in matrix computations. It is shown here that the digital summation of products of two binary numbers can be determined by performing the analog summation of convolutions of these binary numbers followed by the function of the postprocessor. This process of "sum of digital-multiplication-by-analog-convolution" is referred to as the S-DMAC algorithm.

The standard binary representations of the pth positive integers $f(p)$ and $g(p)$ are given by

$$f(p) = \sum_{i=0}^{n-1} f_i(p)2^i, \quad g(p) = \sum_{j=0}^{n-1} g_j(p)2^j \tag{3.9}$$

where $f_i(p),\ g_j(p) \in (0,1)$. The sum of products of these positive n-bit words is given by

$$S_{\hat{M}} = \sum_{p=1}^{\hat{M}} f(p)g(p) \tag{3.10}$$

where $\hat{M}$ is the number of products of such two n-bit words. Equation 3.10 can be shown to be [20]

$$S_{\hat{M}} = \sum_{k=0}^{2n-2} 2^k \left\{ \sum_{p=1}^{\hat{M}} y_k(p) \right\} \tag{3.11}$$

where

$$y_k(p) = \sum_{i=0}^{k} f_i(p) g_{k-i}(p), \quad (0 \le k \le 2n-2) \tag{3.12}$$

Equation 3.12 can be recognized as the discrete convolution of two binary numbers. Evaluation of Equation 3.11 requires $\hat{M}$ optical convolvers, an optical combiner, an optical detector, an ADC, and an S/A circuit. At a particular discrete time k, the term in brackets in Equation 3.11, which represents the sum of $\hat{M}$ analog values of $y_k(p)$ for $1 \le p \le \hat{M}$, can be performed by an optical combiner. The analog result of this summation is then passed to the ADC for digitization, followed by the S/A circuit to obtain the standard binary representation. The advantage of the S-DMAC algorithm lies in the fact that a combination of product and summation operations can be performed.

3.1.2.4 Twos-Complement Binary Arithmetic

The TCB representation is a powerful encoding scheme that permits both positive and negative numbers to be represented in binary form. The encoding scheme requires a sign bit (SB) to be attached to the leftmost bit of the binary number, i.e., SB = 0 for positive numbers and SB = 1 for negative numbers. For example, the TCB sequence of the positive number* +5.5 is {0 1 0 1.1}, and the negative number −13.375 is {1 0 0 1 0.1 0 1}.

On the basis of the DMAC algorithm as described in Section 3.1.2.1, the multiplication of two real numbers using the TCB arithmetic requires that the input numbers be represented by the same number of bits as the output number [29]. For example, the TCB representation of the product (+5.5)(−13.375) = −73.5625 is {1 0 1 1 0 1 1 0.0 1 1 1}, which is a 12-bit word. The 12-bit TCB sequence of the input numbers can thus be obtained by inserting seven zeros to the left of the SB of

---

* Implementation of the radix (or decimal) point shifting is not discussed here because it is not a hardware issue.

+5.5 to become {0 0 0 0 0 0 0 0 1 0 1.1} and four ones to the left of the SB of −13.375 to give {1 1 1 1 1 0 0 1 0.1 0 1}. The discrete convolution of these two sequences results in the mixed-binary sequence {1 1 1 2.1 2 0 2 2 2 3 3 2 1 1 0 0 0 0 0 0 0 0}, in which the last 13 analog bits are discarded. The standard TCB format of the first 12 chosen analog bits (the boldface bits) is {1 0 1 1 0 1 1 0.0 1 1 1} after the ADC and S/A operations, which is expected for the negative number −73.5625. The TCB arithmetic is not only applicable to the DMAC algorithm but also can be used in conjunction with the HO-DMAC (see Section 3.1.2.2) and S-DMAC (see Section 3.1.2.3) algorithms.

The main disadvantage of any optical DMAC processor employing the TCB arithmetic is the reduction in the preprocessing speed, and this can be shown as follows. In the following discussion, primes are used to denote the TCB variables. The convolution of $\hat{N}$ n'-bit TCB numbers generates $[\hat{N}(n'-1)+1]$ mixed-binary points, in which only the first n' analog bits are useful for decoding into the binary format, and the word cycle is $T'_{\mathrm{w}} = [\hat{N}(n'-1)+1]T$. If the first useful n' analog bits of the TCB convolution are equal to the $[\hat{N}(n-1)+1]$ analog bits of the unsigned convolution (see Section 3.1.2.2), then the following relationship is obtained:

$$T'_{\mathrm{w}} = \hat{N}T_{\mathrm{w}} + (1-\hat{N})T \tag{3.13}$$

in which the first term is dominant for large word length. Thus, the processing power of the optical DMAC processor incorporating the TCB arithmetic is approximately reduced by a factor of $\hat{N}$ (i.e., $T'_{\mathrm{w}} \approx \hat{N}T_{\mathrm{w}}$ where $\hat{N}$ is the number of integers [or matrices] to be multiplied) as compared with its unsigned counterpart. Optical DMAC processors based on the TCB arithmetic also suffer from an increase in the resolution bits of the ADC and the S/A circuit because all the bits representing the output number are not fully utilized. This has the effect of reducing the processing speed because a higher-resolution ADC operates at a much lower speed. Nevertheless, optical DMAC processors incorporating the TCB representations can operate on real numbers.

3.1.3 Elemental Optical Signal Processors

This section describes three elemental optical signal processors: an optical splitter, an optical combiner, and a binary programmable incoherent fiber-optic transversal filter. These processors are the building blocks of the FOSAP matrix multipliers, which are to be outlined in Section 3.1.4.

It is assumed that the optically encoded signals to be processed are modulated onto an optical carrier whose coherence time T_{p} is very short compared to the basic time delay T of the incoherent fiber-optic transversal filter, i.e., $T_{\mathrm{p}} \ll T$. In other words, incoherent fiber-optic signal processing (see Chapter 2) is considered here where signals add on an intensity basis. It is further assumed that the modulated laser pulse has a pulse width T_{p} and a height of one unit.

3.1.3.1 Optical Splitter and Combiner

Optical splitter and combiner are useful for signal distribution and collection, respectively. The $1 \times n$ optical splitter (see Figure 3.2) and the $n \times 1$ optical combiner

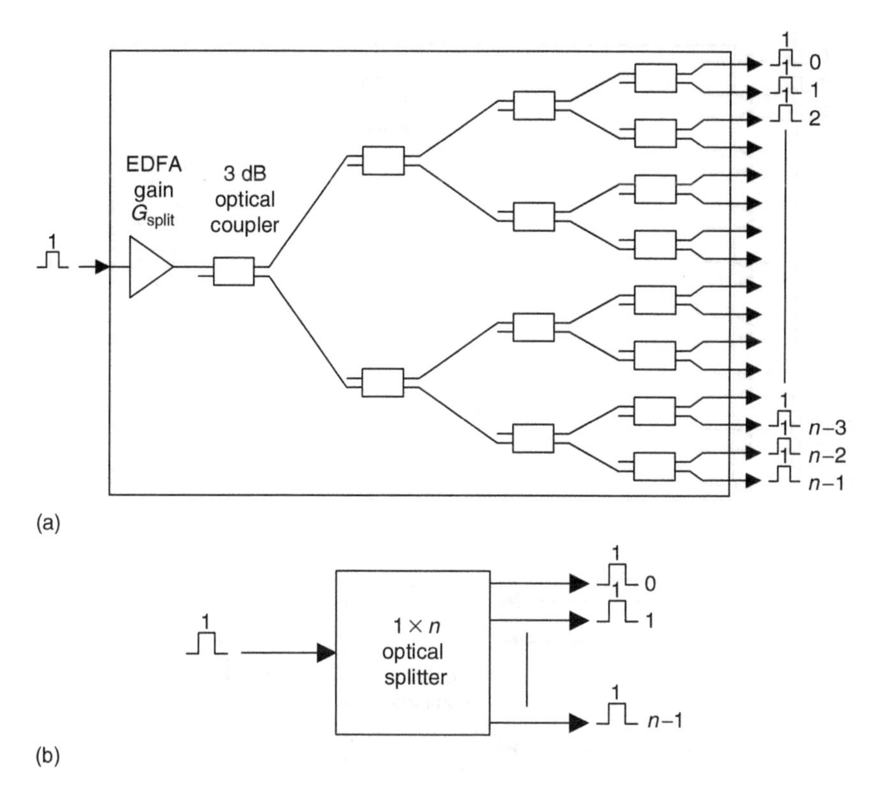

FIGURE 3.2 $1 \times n$ Optical splitter. (a) Schematic diagram and (b) block diagram.

(see Figure 3.3) can be constructed from $n - 1$ 3 dB fiber-optic directional couplers arranged in a binary tree structure. The $1 \times n$ splitter distributes the incoming signal intensity evenly to n output ports, while the $n \times 1$ combiner collects signal intensities from n input ports into a single output port.

The signal intensity coming into the splitter will experience a 3 dB coupling loss at every stage of the tree, and this results in a common intensity loss of $10 \log n$ (dB) at each of the n output ports. An EDFA, which has been described in Chapter 2, of intensity gain G_{split} is incorporated at the start of the tree to compensate for such a loss. Thus, the signal intensity at each of the output ports is equal to the incoming signal intensity provided that $G_{split} = 10 \log n$ (dB). Likewise, the incoming signal intensity at each of the n input ports of the combiner will also experience a common intensity loss of $10 \log n$ (dB) at the end of the tree. The signal intensity of each input port can be recovered at the output port, provided that an EDFA of intensity gain $G_{comb} = 10 \log n$ (dB) is incorporated at the output of the device. For a typical 30 dB gain of the EDFA, the value n can be as large as 1000. Note that the propagation delays of the splitter (T_{split}) and combiner (T_{comb}) are mainly due to the amplifier lengths of the EDFAs.

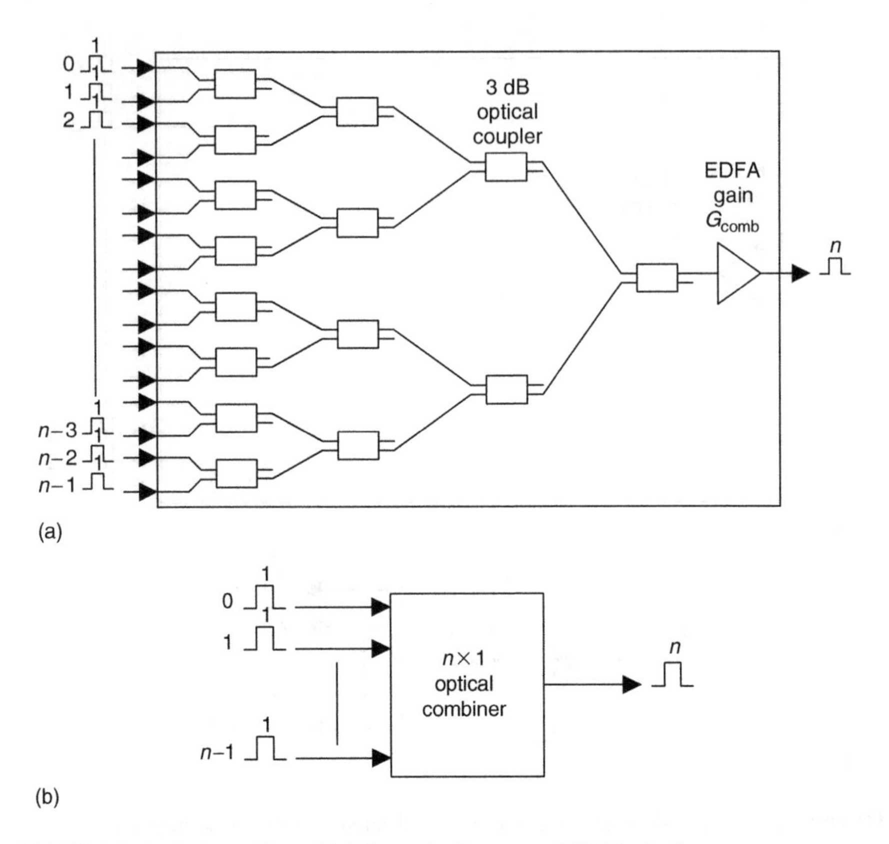

FIGURE 3.3 Optical combiner. (a) Schematic diagram and (b) block diagram.

3.1.3.2 Binary Programmable Incoherent Fiber-Optic Transversal Filter

An optical transversal filter (or optical convolver) is useful for performing the convolution operation. A binary programmable incoherent fiber-optic transversal filter has the advantage of being adaptive, enabling an arbitrary binary intensity impulse response to be obtained by means of external optical or electronic control.

Figures 3.4a and b show the schematic and block diagrams of the binary programmable incoherent fiber-optic transversal filter, which essentially consists of an optical splitter, an optical combiner, optical switches, and fiber delay lines [31]. The binary-code selector is word-parallel loaded into the optical convolver, and is used to set the binary impulse response of the optical convolver by simultaneously controlling the binary states of the 1×2 (or 2×2) optical switches. The binary states of the optical switches correspond to the code word $b_i = \{s_{n-1}\ s_{n-2} \cdots s_0\}$, where $s_i \in (0,1)$, s_0 and s_{n-1} designate the LSB and MSB, respectively, and n is the number of bits in a binary word b_i. Each optical switch permits the signal intensity from the optical splitter to either connect (binary state $s_i = 1$) or bypass (binary state $s_i = 0$) a particular fiber delay path. The optical combiner collects the signal intensities according to the binary states of the optical switches. The optical splitter and

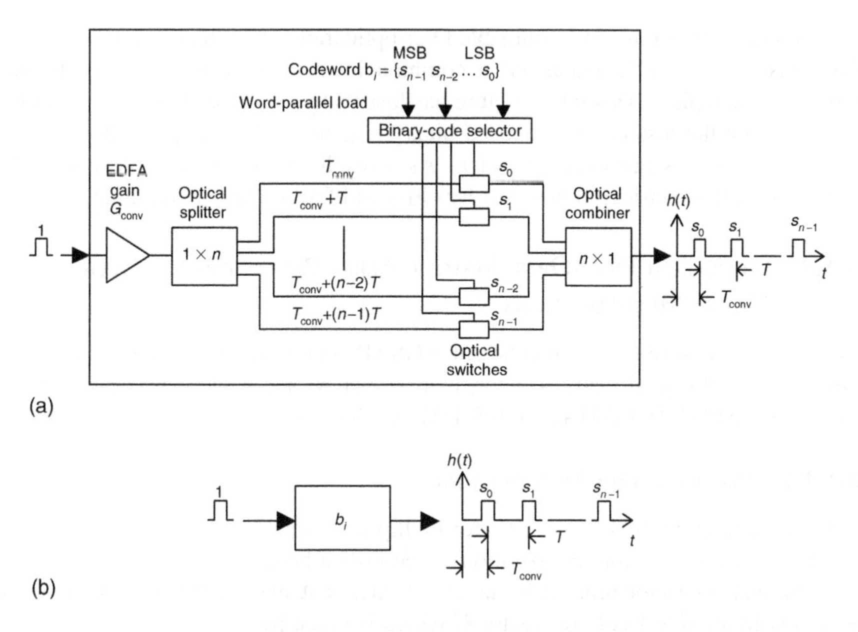

FIGURE 3.4 Binary programmable incoherent fiber-optic transversal filter (or optical convolver). (a) Schematic diagram and (b) block diagram.

combiner are those of Figures 3.2 and 3.3, except that no EDFAs are used. Instead, an EDFA of intensity gain G_{conv} is placed at the start of the splitter to compensate for the losses* of the splitter and combiner. That is, the signal intensity at the output of the optical convolver is the same as the signal intensity at its input provided that $G_{\text{conv}} = 20\log n$ (dB). The convolver impulse response is given by, provided that $G_{\text{conv}} = 20\log n$ (dB),

$$h(t) = \sum_{i=0}^{n-1} s_i \delta(t - iT - T_{\text{conv}}) \tag{3.14}$$

where

$\delta(t)$ is the delta function

T_{conv} the filter propagation delay, which is mainly due to the amplifier length of the EDFA

T the time delay difference between the fiber delay lines and the sampling period of the filter

$1/T$ is the sampling frequency of the filter

---

* Other minor losses associated with the fiber delay lines, such as the insertion loss of the optical switch and connector (or splice) loss, are not considered here for the sake of analytical simplicity.

Another form of the programmable fiber-optic transversal filter is in the forward-flow bus structure [7], which has a common intensity loss of $3(n + 1)$ dB and requires $2n$ couplers. This bus structure has loss that varies linearly with the number of bits n, but the loss of the proposed binary structure, which requires $2 \times (n - 1)$ couplers, varies as the logarithm of n. As a result, the programmable fiber-optic transversal filter considered here has less loss and fewer couplers required.

3.1.4 Incoherent Fiber-Optic Systolic Array Processors for Digital Matrix Multiplications

This section describes the incoherent FOSAP architectures for computation of positive-valued digital matrix–vector, matrix–matrix, and triple-matrix products by using the DMAC, HO-DMAC, and S-DMAC algorithms.

3.1.4.1 Matrix–Vector Multiplication

A high-accuracy FOSAP matrix–vector multiplier based on the S-DMAC algorithm and the vector inner-product operations is described here.

The matrix–vector product of an $M \times N$ matrix **A** and an $N \times 1$ column vector **B** results in an $M \times 1$ column vector **C** which is given by

$$\begin{bmatrix} c_1 \\ c_2 \\ \vdots \\ c_M \end{bmatrix} = \begin{bmatrix} a_{11} & a_{12} & \cdots & a_{1N} \\ a_{21} & a_{22} & \cdots & a_{2N} \\ \vdots & \vdots & \cdots & \vdots \\ a_{M1} & a_{M2} & \cdots & a_{MN} \end{bmatrix} \begin{bmatrix} b_1 \\ b_2 \\ \vdots \\ b_N \end{bmatrix} = \begin{bmatrix} a_{11}b_1 + a_{12}b_2 + \cdots + a_{1N}b_N \\ a_{21}b_1 + a_{22}b_2 + \cdots + a_{2N}b_N \\ \vdots \\ a_{M1}b_1 + a_{M2}b_2 + \cdots + a_{MN}b_N \end{bmatrix} \tag{3.15}$$

where the words of vector **C** are given by

$$c_i = \sum_{j=1}^{N} a_{ij}b_j \quad (i = 1, 2, \ldots, M), \tag{3.16}$$

with a_{ij} and b_j being the n-bit words of matrix **A** and vector **B**, respectively. Note that the matrix–vector product can be decomposed into parallel vector inner-product operations, as shown in Equation 3.15.

Figure 3.5 shows the block diagram of the high-accuracy FOSAP matrix–vector multiplier based on the S-DMAC algorithm and the vector inner-product operations. At time t_0, the N n-bit binary words $(b_1\ b_2 \cdots b_N)$ of vector **B** are word-parallel loaded into the corresponding N optical convolvers by selection of the appropriate optical switches. This process is equivalent to setting the desired binary impulse responses of the optical convolvers, as described above. At the same time, the LSBs of the N n-bit words $(a_{11}\ a_{12} \cdots a_{1N})$ on the first row of matrix **A** are bit-serial fed into

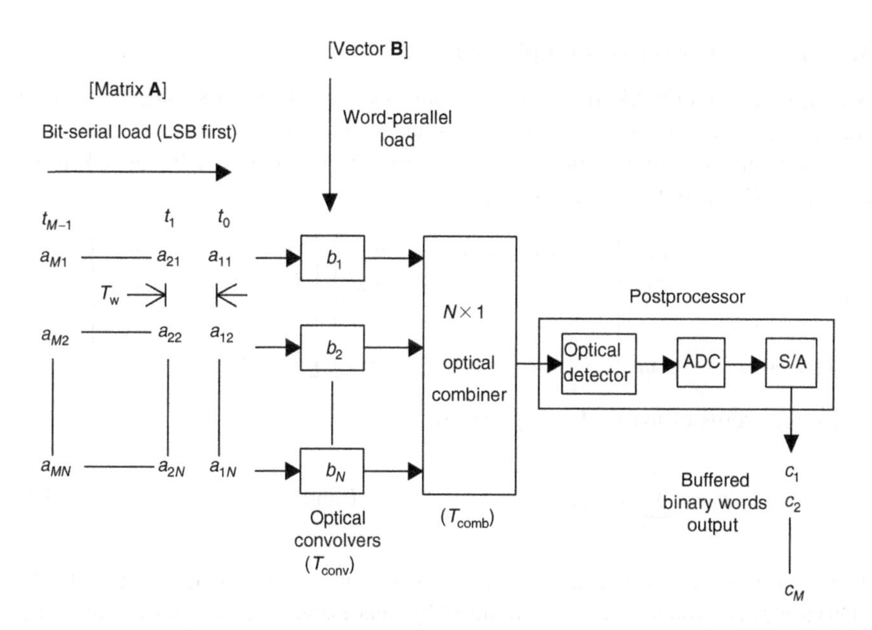

FIGURE 3.5 The high-accuracy FOSAP matrix–vector multiplier based on the S-DMAC algorithm and the vector inner-product operations.

the appropriate inputs of the N optical convolvers. The N analog output signals of the optical convolvers are then passed to the $N \times 1$ optical combiner, where analog summations begin. The resulting analog optical signal is then passed to an optical detector and converted into its binary representation by a $\log_2 Nn$-bit ADC* and a $\log_2 Nn$-bit S/A circuit. At a later bit time $t_0 + T$, the second bits of the same words $(a_{11}\, a_{12} \cdots a_{1N})$ are bit-serial loaded into the optical convolvers, and the analog signals are then combined by the optical combiner. The detected analog signal is converted into its binary format by an ADC, which is then upshifted by one position to the left and binary added to the previous $\log_2 Nn$ binary bits by the S/A circuit to form a new binary sequence. This process continues until the computation of the first row of matrix **A** is completed, which takes one word cycle T_w where $T_w = (2n - 1)T$. At time $t_1 + T$, the first element c_1 of the column vector **C** is computed and stored in an output buffer. At the same time, the LSBs of the N words $(a_{21}\, a_{22} \cdots a_{2N})$ on the second row of the matrix **A** are convolved with the LSBs of the N optical convolvers. This computation process is repeated as before, and the second element c_2 of the vector **C** is available after two word cycles $2T_w$ and stored in the output buffer. There are $n - 1$ zeros for padding between the words to avoid overflow. The computation time of the high-accuracy FOSAP matrix–vector multiplier, which takes M word cycles, is thus, provided that $T_{ADC} < T$, given by

$$T_{MV} = (2n - 1)MT + T_{conv} + T_{comb} \tag{3.17}$$

* It is assumed that the conversion time of the ADC (T_{ADC}) is smaller than the bit time T.

3.1.4.2 Matrix–Matrix Multiplication

A high-accuracy FOSAP matrix–matrix multiplier based on the S-DMAC algorithm and the vector outer-product operations is described here.

The matrix–matrix product of an $M \times N$ matrix **A** and an $N \times P$ matrix **B** results in an $M \times P$ matrix **C** which is given by

$$\begin{bmatrix} c_{11} & c_{12} & \cdots & c_{1P} \\ c_{21} & c_{22} & \cdots & c_{2P} \\ \vdots & \vdots & \cdots & \vdots \\ c_{M1} & c_{M2} & \cdots & c_{MP} \end{bmatrix} = \begin{bmatrix} a_{11} & a_{12} & \cdots & a_{1N} \\ a_{21} & a_{22} & \cdots & a_{2N} \\ \vdots & \vdots & \cdots & \vdots \\ a_{M1} & a_{M2} & \cdots & a_{MN} \end{bmatrix} \begin{bmatrix} b_{11} & b_{12} & \cdots & b_{1P} \\ b_{21} & b_{22} & \cdots & b_{2P} \\ \vdots & \vdots & \cdots & \vdots \\ b_{N1} & b_{N2} & \cdots & b_{NP} \end{bmatrix} \quad (3.18)$$

where the words of matrix **C** are given by

$$c_{ij} = \sum_{k=1}^{N} a_{ik} b_{kj}, \quad \text{for } i = 1, 2, \ldots, M \text{ and } j = 1, 2, \ldots, P \quad (3.19)$$

The n-bit words a_{ik} and b_{kj} are the elements of matrices **A** and **B**, respectively. The matrix–matrix product can be computed by successive vector outer-product operations [14], as in the following illustration:

$$\begin{bmatrix} c_{11} & c_{12} & c_{13} \\ c_{21} & c_{22} & c_{23} \\ c_{31} & c_{32} & c_{33} \end{bmatrix} = \begin{bmatrix} a_{11} \\ a_{21} \\ a_{31} \end{bmatrix} \cdot \begin{bmatrix} b_{11} & b_{12} & b_{13} \end{bmatrix} + \begin{bmatrix} a_{12} \\ a_{22} \\ a_{32} \end{bmatrix} \cdot \begin{bmatrix} b_{21} & b_{22} & b_{23} \end{bmatrix} + \begin{bmatrix} a_{13} \\ a_{23} \\ a_{33} \end{bmatrix} \cdot \begin{bmatrix} b_{31} & b_{32} & b_{33} \end{bmatrix} \quad (3.20)$$

Note that the vector outer-product operations, e.g., the first term on the right of Equation 3.20, are defined as

$$\begin{bmatrix} a_{11} \\ a_{21} \\ a_{31} \end{bmatrix} \cdot \begin{bmatrix} b_{11} & b_{12} & b_{13} \end{bmatrix} = \begin{bmatrix} a_{11}b_{11} & a_{11}b_{12} & a_{11}b_{13} \\ a_{21}b_{11} & a_{21}b_{12} & a_{21}b_{13} \\ a_{31}b_{11} & a_{31}b_{12} & a_{31}b_{13} \end{bmatrix} \quad (3.21)$$

Figure 3.6 shows the block diagram of the high-accuracy FOSAP matrix–matrix multiplier based on the S-DMAC algorithm and the vector outer-product operations. At time t_0, all the n-bit binary words $(b_{11} \cdots b_{NP})$ of matrix **B** are word-parallel loaded into the optical convolvers by selection of the appropriate optical switches. At the same time, the N n-bit words $(a_{11}a_{12} \cdots a_{1N})$ on the first row of matrix **A** are bit-serial fed (with the LSBs first) into the N $1 \times P$ optical splitters. The LSBs of these words are then bit-serial fed into the inputs of the appropriate optical convolvers, where convolution computations begin. The analog output signals of the optical convolvers are then passed to the P $N \times 1$ optical combiners, where analog additions begin. The P postprocessors, as shown in Figure 3.5, perform the ADC and S/A

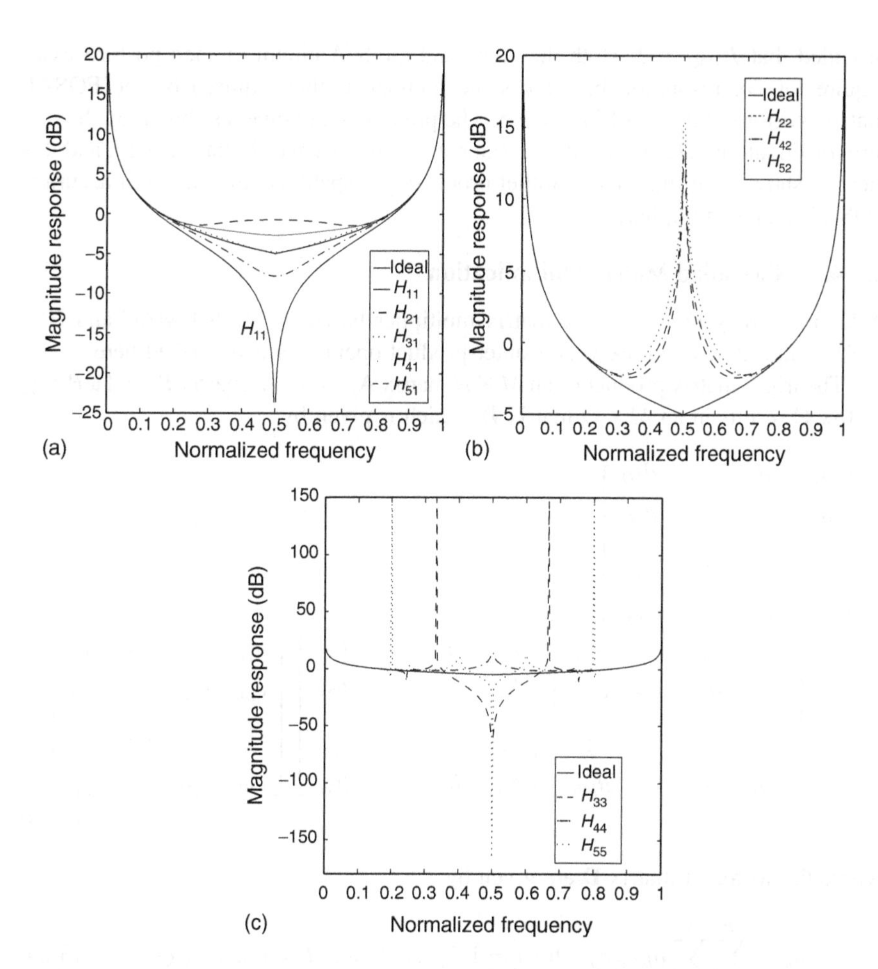

FIGURE 3.6 Magnitude responses of several families of the Newton–Cotes digital integrators. Here, the normalized frequency corresponds to $\omega T/2\pi$, H_{mp} means $H_{mp}(z)$, and the sampling period is assumed to be $T=1$ without loss of generality. (a) Trapezoidal family $H_{m1}(z;\ p=1;\ m=1,2,3,4,5)$; (b) Simpson's 1/3 family $H_{m2}(z;\ p=2;\ m=2,4,5)$; and (c) Simpson's 3/8 integrator $H_{33}(z)$, Boole's integrator $H_{44}(z)$, and the fifth-order integrator $H_{55}(z)$.

operations to yield the standard binary sequences for the first row of matrix **C**, i.e., $(c_{11}c_{12}\cdots c_{1P})$, which takes one word cycle T_w, where $T_w=(2n-1)T$. At time t_1+T, the LSBs of the N n-bit elements $(a_{21}a_{22}\cdots a_{2N})$ on the second row of matrix **A** are sequentially fed into the N $1\times P$ optical splitters, and the process repeats as before until time t_{M-1}, which takes M word cycles.

The computation time of the high-accuracy FOSAP matrix–matrix multiplier, which takes M word cycles, is thus given by

$$T_{MM}=(2n-1)MT+T_{split}+T_{conv}+T_{comb} \tag{3.22}$$

provided that $T_{ADC} < T$. Both the ADC and the S/A circuit in each postprocessor require $\log_2 Nn$ resolution bits, the same number as that required by the FOSAP matrix–vector multiplier of Figure 3.5. The preprocessing time, as shown by the first term of Equation 3.22, is exactly the same as the first term of Equation 3.17, and this clearly shows the massive parallel-processing capability of the high-accuracy FOSAP matrix multiplier.

3.1.4.3 Cascaded Matrix Multiplication

A high-accuracy FOSAP triple-matrix multiplier based on the S-DMAC and HO-DMAC algorithms and the vector outer-product operations is described here.

The triple-matrix product of an $M \times N$ matrix **A**, an $N \times P$ matrix **B**, and a $P \times Q$ matrix **C** results in an $M \times Q$ matrix **D**, which is given by

$$\begin{bmatrix} d_{11} & d_{12} & \cdots & d_{1Q} \\ d_{21} & d_{22} & \cdots & d_{2Q} \\ \vdots & \vdots & \cdots & \vdots \\ d_{M1} & d_{M2} & \cdots & d_{MQ} \end{bmatrix} = \begin{bmatrix} a_{11} & a_{12} & \cdots & a_{1N} \\ a_{21} & a_{22} & \cdots & a_{2N} \\ \vdots & \vdots & \cdots & \vdots \\ a_{M1} & a_{M2} & \cdots & a_{MN} \end{bmatrix} \begin{bmatrix} b_{11} & b_{12} & \cdots & b_{1P} \\ b_{21} & b_{22} & \cdots & b_{2P} \\ \vdots & \vdots & \cdots & \vdots \\ b_{N1} & b_{N2} & \cdots & b_{NP} \end{bmatrix} \begin{bmatrix} c_{11} & c_{12} & \cdots & c_{1Q} \\ c_{21} & c_{22} & \cdots & c_{2Q} \\ \vdots & \vdots & \cdots & \vdots \\ c_{P1} & c_{P2} & \cdots & c_{PQ} \end{bmatrix} \quad (3.23)$$

where the words of matrix **D** are given by

$$d_{il} = \sum_{j=1}^{P} \sum_{k=1}^{N} a_{ik} b_{kj} c_{jl} \quad \text{for } i = 1, 2, \ldots, M \text{ and } l = 1, 2, \ldots, Q \quad (3.24)$$

The n-bit binary words a_{ik}, b_{kj}, and c_{jl} are the elements of matrices **A**, **B**, and **C**, respectively.

Figure 3.7 shows the block diagram of the high-accuracy FOSAP triple-matrix multiplier based on the S-DMAC and HO-DMAC algorithms, and the vector outer-product operations. The operating principle of the FOSAP triple-matrix multiplier is very similar to that of the FOSAP matrix–matrix multiplier. The analog outputs of the P $N \times 1$ optical combiners are due to the convolution of the elements of matrix **A** with the elements of matrix **B**, i.e., **A**∗**B**. At this point, the preprocessing architecture is exactly the same as that of the FOSAP matrix–matrix multiplier of Figure 3.8. The P analog outputs of **A**∗**B** are sequentially fed to the $P 1 \times Q$ optical splitters and then to the appropriate optical convolvers, whose binary impulse responses correspond to the words of matrix **C**. The Q $P \times 1$ optical combiners perform the analog summations to generate the Q analog outputs of **(A**∗**B)**∗**C**, which are then fed to the postprocessors for digitization to yield the standard binary sequences of the elements of matrix **D**.

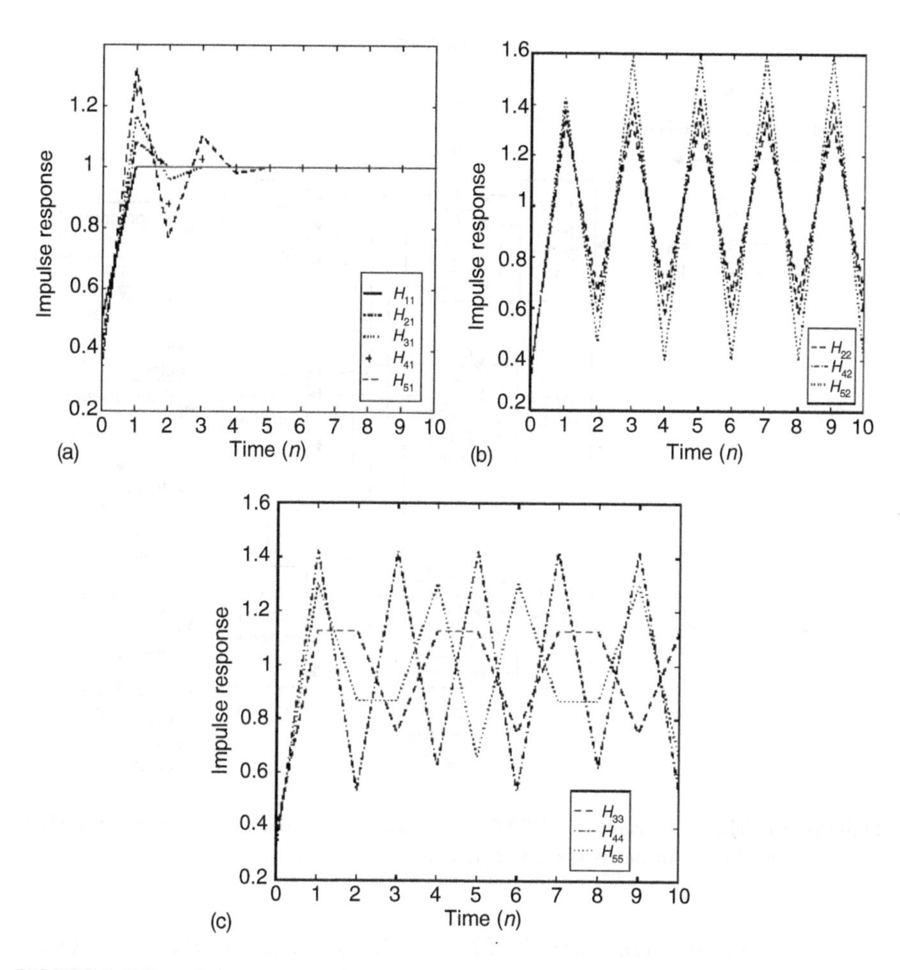

FIGURE 3.7 Impulse responses of several families of the Newton–Cotes digital integrators. Here, H_{mp} means $H_{mp}(z)$, time n means nT, and the sampling period T is assumed to be $T = 1$ without loss of generality. (a) Trapezoidal family $H_{m1}(z; p = 1; m = 1, 2, 3, 4, 5)$; (b) Simpson's 1/3 family $H_{m2}(z;\ p = 2;\ m = 2, 4, 5)$; and (c) Simpson's 3/8 integrator $H_{33}(z)$, Boole's integrator $H_{44}(z)$, and the fifth-order integrator $H_{55}(z)$.

The word cycle T_w is given by $T_w = (3n - 2)T$, and the number of zeros required for padding between the words is $2(n \times 1)$. The largest analog value of the convolution of 3 n-bit words (i.e., $a_{ik} * b_{kj} * c_{jl}$) in which all binary bits have logic one is given by $0.75n^2$ when n is even and $0.25(3n^2 + 1)$ when n is odd. Thus, the largest analog bit of d_{il} in Equation 3.24 is given by $0.25NP(3n^2 + 1)$, where N and P are the number of summations performed by the $N \times 1$ and $P \times 1$ optical combiners to generate $(\mathbf{A}*\mathbf{B})$ and $(\mathbf{A}*\mathbf{B})*\mathbf{C}$, respectively. Thus, both the ADC and S/A circuit in the postprocessor require $\log_2[0.25NP(3n^2 + 1)]$ resolution bits. The computation time of the high-accuracy FOSAP triple-matrix multiplier, which takes M word cycles, is thus given by

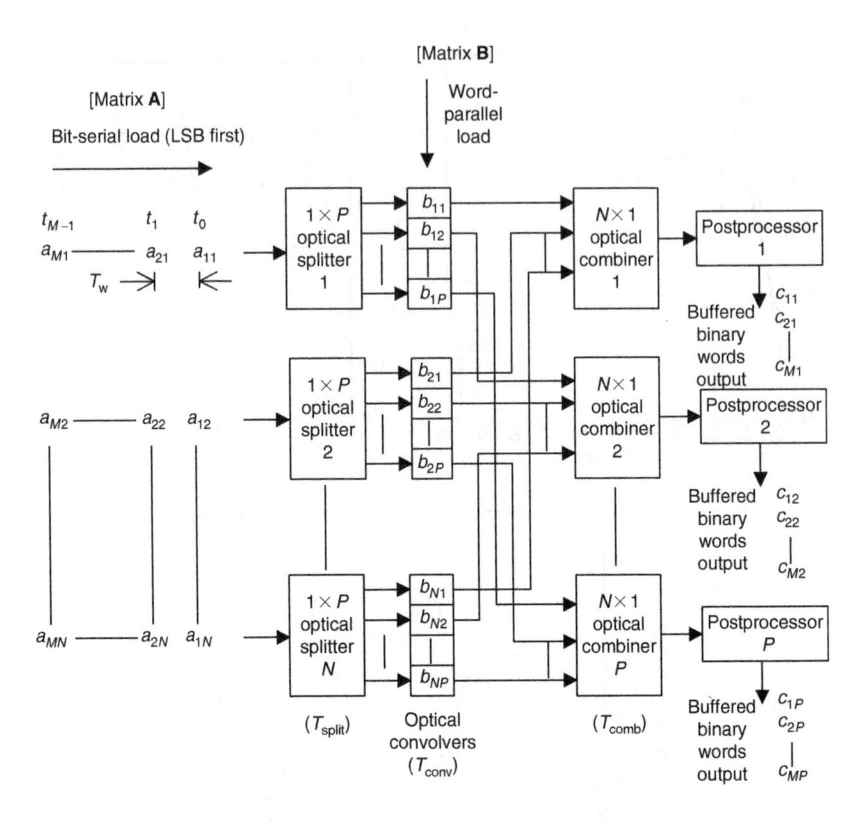

FIGURE 3.8 The high-accuracy FOSAP matrix–matrix multiplier based on the S-DMAC algorithm and the vector outer-product operations.

$$T_{MMM} = (3n-2)MT + 2T_{split1} + 2T_{conv} + T_{comb1} + T_{split2} \tag{3.25}$$

provided that $T_{ADC} < T$ and $T_{split1} = T_{comb2}$.

The major advantage of the S-DMAC and HO-DMAC algorithms lies in the fact that a combination of convolution and addition operations can be optically performed in the preprocessing unit. Higher-order matrix operations can be performed by cascading the basic building blocks, as highlighted in Figure 3.8, where each block corresponds to the elements of a matrix.

3.1.5 Performance Comparison

In this section, the performance of the FOSAP architecture using nonbinary data is described. The performances of the FOSAP matrix–vector and matrix–matrix multipliers are compared with those of the digital electronic multipliers and other optical multipliers. Also described is the performance of the high-order FOSAP architecture. For analytical simplicity, the propagation delays T_{split}, T_{comb}, and T_{conv} of the FOSAP architectures are assumed to be negligible compared with the word cycles because the matrix dimensions considered here are large.

3.1.5.1 Fiber-Optic Systolic Array Processors Using Nonbinary Data

Using the Psaltis–Athale performance ratio as a performance measure [13,14], the performance of the FOSAP matrix multiplier, in which the FOSAP matrix–vector multiplier of Figure 3.5 is chosen as an example, is described here for nonbinary-encoded data.

The number of bits n required for representing a base-b m-bit number is given by

$$n \geq \log_b 2^m \tag{3.26}$$

The Psaltis–Athale performance ratio R is defined as the number of multiplications per ADC operation, and can be obtained by dividing the number of operations* per second by the number of ADC operations per second (the number of ADCs times the clock rate) [20]:

$$\begin{aligned} \mathrm{R} &= \text{number of multiplications per ADC operation} \\ &= \frac{\text{number of operations per second}}{\text{number of ADCs} \times \text{clock rate}} \end{aligned} \tag{3.27}$$

The Psaltis–Athale performance ratio provides a direct comparative estimate of an optical implementation versus an electronic implementation. The performance ratio R is independent of the clock rate and must exceed unity to keep the complexity of the electronics to a minimum. As a result, optical DMAC processors are only superior to the digital electronic processors if $R > 1$. If, e.g., $R = 1$ then only one binary multiplication is being performed by the optical system per ADC operation. Because it is about equally difficult to perform multiplications and ADC operations electronically, this example shows that optics offers no advantage over electronics.

The maximum value of the matrix–vector convolution, in which the elements are base-b n-bit numbers, is given by $Nn(b-1)^2$, which must not exceed the dynamic range of the ADC of $2^{N_{\mathrm{ADC}}}$, where N_{ADC} is the ADC resolution bits. Thus, the matrix dimension N of the FOSAP matrix–vector multiplier of Figure 3.5 is limited by the dynamic range of the ADC according to

$$N \leq \frac{2^{N_{\mathrm{ADC}}}}{n(b-1)^2} \tag{3.28}$$

The FOSAP matrix–vector multiplier, which performs MN operations in time MT_{w}, has the processing speed given by

$$\mathrm{MOPS} = \frac{MN}{MT_{\mathrm{w}}} = \frac{N}{(2n-1)T} \tag{3.29}$$

* Here, one operation is considered to be equivalent to one multiplication and one addition.

where MOPS stands for mega operations per second and $1/T$ is the clock rate or ADC speed in megahertz. The performance ratio of the FOSAP matrix–vector multiplier is thus given, from Equation 3.27, as

$$R = \frac{\text{MOPS}}{1 \times 1/T} = \frac{N}{(2n - 1)} \quad (3.30)$$

Table 3.1 shows the operational parameters for various case studies of the FOSAP matrix–vector multiplier of Figure 3.5 using Equations 3.26 through 3.30. Cases 1–3, where the resolution bits N_{ADC} and speed $1/T$ of the 12-bit 100 MHz ADC are fixed, show that increasing the base b results in decreasing the values of n and N. Consequently, the values of MOPS and R are significantly reduced, and the FOSAP matrix–vector multiplier performs best with binary-encoded data (case 1). Cases 4–6 correspond to a fixed matrix dimension $N = 128$. Increasing the base b results in increasing N_{ADC}, which greatly reduces the processing speed of MOPS because higher-resolution ADCs operate at much lower speeds. However, the FOSAP matrix–vector multiplier still outperforms its digital electronic counterpart because of the large value of R but at the expense of lower accuracy of n. Hence, the overall performance of the FOSAP matrix–vector multiplier is best achieved with binary-encoded data because of the desirable values of MOPS and R, as shown in case 4 which is in fact case 1.

Similarly, the FOSAP matrix–matrix multiplier of Figure 3.8 also has the same operational parameters as the parameters in Table 3.1 except that it has P times the values of the MOPSs. The analysis here also applies to higher-order FOSAP matrix multipliers, and it can be deduced that the FOSAP matrix multipliers perform better with binary-encoded data.

TABLE 3.1
Operational Parameters for Various Case Studies of the FOSAP Matrix–Vector Multiplier of Figure 3.5 for 32-Bit ($m = 32$) Multiplications

Case	b	n	N	N_{ADC}	$1/T$ (MHz)	MOPS	R	Remarks
1	2	32	128	12	100	203.2	2.032	Constant clock rate $1/T$,
2	4	16	28	12	100	90.3	0.903	constant ADC bits N_{ADC}
3	8	11	7	12	100	33.3	0.333	
4	2	32	128	12	100	203.2	2.032	Constant matrix dimension N
5	4	16	128	16	20	82.6	4.13	
6	8	11	128	18	0.20	1.22	6.1	

Note: b, the base used; n, the digits of accuracy; N, the matrix dimension; N_{ADC}, the ADC resolution bits; $1/T$, the ADC speed or clock rate; MOPS, mega operations per second; and R, the Psaltis–Athale performance ratio.

3.1.5.2 High-Order Fiber-Optic Systolic Array Processors

From Section 3.1.4.2, the FOSAP matrix–matrix multiplier achieves the performance ratio $R = 2.032$, which indicates its superiority over other nonfiber and digital electronic processors. In this section, the performance of the positive-valued high-order FOSAP matrix multiplier is compared with that of its digital electronic counterpart. For analytical simplicity, the matrix is assumed to be square and has dimension M, and its elements are n-bit words.

The high-order FOSAP matrix multiplier requires M word cycles to perform the product of x $M \times M$ matrices, and its processing time (or the number of clock cycles) is thus given by

$$T_x = M[x(n-1)+1]T \tag{3.31}$$

The number of operations (NO) involved in the product of x $M \times M$ matrices is

$$\mathrm{NO} = (x-1)M^3 \tag{3.32}$$

The number of operations per second (NOPS) performed by the high-order FOSAP matrix multiplier is thus given by

$$\mathrm{NOPS} = \frac{\mathrm{NO}}{T_x} = \frac{(x-1)M^2}{[x(n-1)+1]T} \tag{3.33}$$

The number of ADCs required by the high-order FOSAP matrix multiplier is always M. Using Equation 3.27, the Psaltis–Athale performance ratio of the positive-valued high-order FOSAP matrix multiplier is thus given by

$$R = \frac{\mathrm{NOPS}}{M \times 1/T} = \frac{(x-1)M}{[x(n-1)+1]} \approx \left(\frac{M}{n}\right)\left(\frac{x-1}{x}\right) \tag{3.34}$$

which is greater than unity if

$$M > n\left(\frac{x}{x-1}\right) \tag{3.35}$$

The performance ratio R in Equation 3.34 is expected to exceed unity because the matrix dimension M is, for many practical high-order matrix operations, usually a few orders of magnitude larger than the word length n. This shows the superiority of the high-order FOSAP matrix multiplier over its digital electronic counterpart. Thus, the high-order FOSAP matrix multiplier may be used to perform various linear algebra operations, such as solutions of algebraic equations, 2-D mathematical transform, matrix-inversions, and pattern recognition, which require high-order matrix operations [15].

3.1.6 Remarks

The FOSAP matrix multipliers may be used to perform several linear algebra operations, which require basic matrix–vector and matrix–matrix operations, for advanced signal processing tasks such as pattern recognition and image processing. For example, the following linear algebra operations involve matrix operations: LU factorization, QR factorization, singular value decomposition, solution of simultaneous algebraic and differential equations, least squares solution, matrix inversion, and solutions to eigenvalue problems [2].

The main limitation of any optical DMAC processor is the slow processing speed of the electronic postprocessor, in which the ADC is often the slowest component, with a bit-time limit of T. This limitation means that the overall performance of any optical DMAC system is highly compromised. It has been predicted that an electronic 8-bit ADC can operate at 1.5 GHz, and that future development of a 6-bit ADC at 6 GHz is feasible [32]. Electro-optic 2-bit and 4-bit ADCs were experimentally demonstrated to operate in the gigahertz range [33]. However, future development of high-speed and high-resolution optical ADCs, with speed in the gigahertz range and resolution bits greater than 12, will enable the proposed FOSAP architectures to process more information at a faster rate than the digital electronic architectures.

- Using the DMAC, HO-DMAC, and S-DMAC algorithms, the incoherent FOSAP matrix multipliers have been designed using optical splitters, optical combiners, programmable fiber-optic transversal filters, and postprocessors consisting of optical detectors, ADCs, and S/A circuits.
- The FOSAP multipliers, which perform best with binary-encoded data, can perform real-valued digital matrix–vector, matrix–matrix, triple-matrix, and higher-order matrix computations.
- The positive-valued FOSAP matrix–vector and matrix–matrix multipliers have higher computational power than the digital electronic multipliers and other optical DMAC multipliers, showing their massive parallel-processing capability.
- The computational power of any real-valued optical DMAC system incorporating TCB arithmetic is reduced by a factor approximately equal to the number of matrices to be multiplied compared with the positive-valued optical DMAC systems.
- The positive-valued high-order FOSAP multipliers have higher computational power than the digital electronic multipliers when the matrix dimension is greater than the word length by a few orders of magnitude, which is often the case for many signal processing applications. However, the real-valued high-order FOSAP multipliers using TCB arithmetic are unlikely to offer any advantage over the digital electronic multipliers.
- The FOSAP architectures have a number of advantages including massive pipeline capability, high computational power, modularity (construction of a larger architecture from several smaller architectures), and size scalability (the size of the architecture can be increased with nominal changes in the existing architecture with a comparable increase in performance).
- The FOSAP multipliers may be used as high-performance coprocessors in a general-purpose digital electronic computer to perform various linear algebra operations, such as those in pattern recognition and image processing.

The optical transversal filter structure described in this section will be used in the design of incoherent optical integrators in Section 3.2 and in the design of coherent optical differentiators in Section 3.3.

3.2 PROGRAMMABLE INCOHERENT NEWTON–COTES OPTICAL INTEGRATOR

In this section, the design of a programmable incoherent Newton–Cotes optical integrator (INCOI) is described. The definition, existing design techniques, and application of digital integrators are described in Section 3.2.1. A generalized theory of the Newton–Cotes digital integrators, whose derivation is given in Appendix A, and their magnitude and impulse responses are given in Section 3.2.2. On the basis of this theory, algorithms for the synthesis of the INCOI processor are proposed in Section 3.2.2.2. These algorithms are then used in the design of the programmable INCOI processor, which essentially consists of a microprocessor, fiber-optic architectures, optical switches, and optical and semiconductor amplifiers (Section 3.2.2.3). Several types of input pulse sequences are chosen as examples for illustration of the incoherent processing accuracy of the programmable INCOI processor (Section 3.2.3). In addition, the incoherent RFOSP described in Section 3.1.4 is used in the design of the programmable INCOI processor. The theory of incoherent fiber-optic signal processing described in Chapter 2 is employed henceforth where intensity-based signals are considered.

3.2.1 INTRODUCTORY REMARKS

While digital signal processing was originally used in the 1950s as a technique for simulating continuous-time systems using discrete-time computations, it has since become a field of study in its own right. One example is a discrete-time (or digital) integrator, which can be used to simulate the behavior of a continuous-time (or analog) integrator. A digital integrator forms a fundamental part of many practical signal processing systems because the time integral of signals is sometimes required for further use or analysis. For example, digital integrators have been used in the design of compensators for control systems [34] and for measuring the cardiac output, the volume of blood pumped by the heart per unit time [35].

A digital integrator is a processor whose output pulse sequence is obtained by approximating the integral of a continuous-time signal from the samples of that signal. A continuous-time signal $x(t)$, whose values are known at the discrete time $t = nT$ for $n = 0, 1, 2, \ldots$, where T is the period between successive samples, can be integrated by a digital integrator. The frequency response of an ideal digital integrator is given by [35]

$$H_I(\omega) = \begin{cases} \dfrac{1}{j\omega T}, & 0 \leq \omega T/(2\pi) \leq 1/2 \\ \dfrac{1}{j(2\pi - \omega T)}, & 1/2 < \omega T/(2\pi) \leq 1 \end{cases} \tag{3.36}$$

where

$j = \sqrt{-1}$

ω is the angular frequency

T is the sampling period of the integrator

Digital integrators may be designed by using one of the many classical numerical integration techniques such as the Newton–Cotes, Lagrange, Romberg, and Gauss–Legendre formulas [34–40]. Of these, the Newton–Cotes integration scheme has been extensively used. For example, the well-known trapezoidal, Simpson's 1/3, Simpson's 3/8, and Boole's integrators all belong to the family of the Newton–Cotes digital integrators [34–39]. The underlying principle of the Newton–Cotes integration scheme is to fit a continuous-time interpolation polynomial $x(t)$ to a given input pulse sequence $f(nT)$ where $f(nT) = x(nT)$. The continuous-time interpolation polynomial $x(t)$ is then integrated resulting in $y(t) = \int_0^t x(t)\mathrm{d}t$. Sampling the integrated continuous-time polynomial $y(t)$ by a digital integrator at the sampling period yields the output pulse sequence $y(nT)$. Thus, the output pulse sequence of a digital integrator effectively approximates the integral of a continuous-time signal according to

$$y(nT) = \int_0^{nT} x(t)\,\mathrm{d}t \tag{3.37}$$

The magnitude responses of the Newton–Cotes digital integrators generally approximate the ideal magnitude response reasonably well over the lower frequency band of $0 \le \omega T/(2\pi) \le 0.2$. The Newton–Cotes digital integrators may thus be referred to as narrowband integrators.

Unlike the well-known Newton–Cotes digital integrators, the concept of optical integration is still new in the area of optical signal processing. In this section, a programmable INCOI processor is described. Most of the work presented here has been described by Ngo and Binh [40]. The derivation of a generalized theory of the Newton–Cotes digital integrators, which was not given in Ref. [41], is shown in Appendix A.

3.2.2 Newton–Cotes Digital Integrators

3.2.2.1 Transfer Function

The transfer function of the pth-order Newton–Cotes digital integrators can be generally expressed as*

$$\begin{aligned} H_{mp}(z) &= \frac{T}{1-z^{-p}} \sum_{k=0}^{m} C_k(p)\Delta^k D(z) \\ &= \frac{T}{1-z^{-p}} \left[C_0(p) + C_1(p)\Delta D(z) + C_2(p)\Delta^2 D(z) + \cdots + C_m(p)\Delta^m D(z)\right] \end{aligned} \tag{3.38}$$

* The derivation of Equations 3.38 through 3.40 is given in Appendix A where Equations A.3.32 through A.3.34 correspond to Equations 3.38 through 3.40.

where the kth coefficient is given by*

$$C_k(p) = \int_0^p \binom{\eta}{k} d\eta \quad (3.39)$$

and the kth difference equation is given by

$$\Delta^k D(z) = (-1)^k (1 - z^{-1})^k \quad (3.40)$$

where $k = 0, 1, \ldots, m$, $1 \leq p \leq m$, and $z = \exp(j\omega T)$ is the z-transform parameter [22]. Note that these digital integrators are marginally stable because of p poles on the unit circle in the z-plane.

Equation 3.38 can be generally expressed as the product of the transfer function of the FIR (finite impulse response) filter and the transfer function of the IIR (infinite impulse response) filter according to

$$H_{mp}(z) = \frac{T \cdot \sum_{k=0}^{m} b_k z^{-k}}{1 - az^{-p}} \quad (3.41)$$

where

$a = 1$ is the pole value
b_k is the tap coefficient of the FIR filter
$b_k = b_{m-k}$ is positive for $m = p$
b_k is real for $m > p$

The transfer functions of several families of the Newton–Cotes digital integrators are tabulated in Table 3.2. Note that $H_{11}(z)$, $H_{22}(z)$, $H_{33}(z)$, and $H_{44}(z)$ are, respectively, the well-known transfer functions of the trapezoidal, Simpson's 1/3, Simpson's 3/8, and Boole's integrators [34–37].

Figures 3.6a through c show the magnitude responses of several families of the Newton–Cotes digital integrators. Figure 3.6a shows that the magnitude response of the trapezoidal integrator $H_{31}(z)$ approximates that of the ideal integrator much better than other integrators of the same or a different family. This does not necessarily mean that the trapezoidal integrator is superior to other Newton–Cotes integrators because the time-domain performance, which is described in Section 3.2.3, needs to be considered.

Figures 3.7a through c show that the impulse responses of several families of the Newton–Cotes digital integrators are real and positive and hence characterize incoherent systems (see Chapter 2). As a result, incoherent Newton–Cotes optical integrators can be synthesized from the Newton–Cotes digital integrators.

* The binomial coefficient is defined as $\binom{\eta}{k} = \frac{\eta(\eta-1)\cdots(\eta-(k-1))}{k!} = \frac{\eta!}{(\eta-k)!k!}$.

TABLE 3.2
Digital Tap Coefficients of Several Families of the Newton–Cotes Digital Integrators, as Computed from Equations 3.38 and 3.39, with Transfer Functions Expressed in the Form of Equation 3.41

Filter Family	$H_{mp}(z)$	b_0	b_1	b_2	b_3	b_4	b_5
Trapezoidal	$2T^{-1}H_{11}(z)$	1	1	0	0	0	0
	$12T^{-1}H_{21}(z)$	5	8	−1	0	0	0
	$24T^{-1}H_{31}(z)$	9	19	−5	1	0	0
	$720T^{-1}H_{41}(z)$	251	646	−264	106	−19	0
	$1440T^{-1}H_{51}(z)$	475	1427	−798	482	−173	27
Simpson's 1/3	$[3T^{-1}H_{22}(z)=3T^{-1}H_{32}(z)]$	1	4	1	0	0	0
	$90T^{-1}H_{42}(z)$	29	124	24	4	−1	0
	$90T^{-1}H_{52}(z)$	28	129	14	14	−6	1
Simpson's 3/8	$(8/3)T^{-1}H_{33}(z)$	1	3	3	1	0	0
Boole	$(45/2)T^{-1}H_{44}(z)$	7	32	12	32	7	0
Fifth-order	$288T^{-1}H_{55}(z)$	95	375	250	250	375	95

3.2.2.2 Synthesis

One common approach to the optical synthesis of Equation 3.41 is to cascade the FIR fiber-optic signal processor (FOSP) with the IIR FOSP. This approach can only be used for the synthesis of incoherent FOSPs comprising positive tap coefficients, e.g., the tap coefficients b_{ks} of $H_{mp}(z)\big|_{m=p}$ are all positive (see Tables 3.1 and 3.2). However, such a technique cannot be used for the case $H_{mp}(z)\big|_{m>p}$ where some of the negative tap coefficients cannot be optically implemented with incoherent FOSPs. An alternative optical synthesis method, which enables the negative tap coefficients of Equation 3.41 to be implemented with incoherent FOSPs, is described here. In the following analysis, the transfer function $H_{mp}(z)$ and parmeters a and b_k are associated with any generic digital filters, while $\hat{H}_{mp}(z)$, $\hat{a}$, and $\hat{b}_k$ are the variables of their optical counterparts.

When p incoherent IIR FOSPs are incorporated into p higher-order delay lines of an incoherent FIR FOSP, the overall transfer function of the pth-order INCOI can be described by

$$\hat{H}_{mp}(z) = T \cdot \sum_{k=0}^{m} \left[\frac{\hat{b}_k z^{-k}}{1 - \hat{a} z^{-p}} \right] \tag{3.42}$$

where

$$\hat{a} = \begin{cases} 0 & \text{for } k = 0, 1, \ldots, m-p \\ a = 1 & \text{for } k = m-(p-1), \ldots, m \end{cases}$$

Equation 3.42 can be rewritten as

$$\hat{H}_{mp}(z) = \frac{T \cdot \sum_{k=0}^{m} (\hat{b}_k - \hat{b}_{k-p}\hat{a}) z^{-k}}{1 - \hat{a} z^{-p}} \quad (\hat{b}_i = 0 \text{ for } i < 0) \tag{3.43}$$

TABLE 3.3
Optical Tap Coefficients of Several Families of the Synthesized INCOI Processor as Computed from Equations 3.44b and 3.44c

INCOI Family	$\hat{H}_{mp}(z)$	$\hat{b}_0$	$\hat{b}_1$	$\hat{b}_2$	$\hat{b}_3$	$\hat{b}_4$	$\hat{b}_5$
Trapezoidal	$T^{-1}\hat{H}_{11}(z)$	0.50	1.0	0	0	0	0
	$T^{-1}\hat{H}_{21}(z)$	0.4167	1.0833	1.0	0	0	0
	$T^{-1}\hat{H}_{31}(z)$	0.375	1.1667	0.9583	1.0	0	0
	$T^{-1}\hat{H}_{41}(z)$	0.3486	1.2458	0.8792	1.0264	1.0	0
	$T^{-1}\hat{H}_{51}(z)$	0.3299	1.3208	0.7667	1.1014	0.9812	1.0
Simpson's 1/3	$T^{-1}\hat{H}_{22}(z)$ $= T^{-1}\hat{H}_{32}(z)$	0.3333	1.3333	0.6667	0	0	0
	$T^{-1}\hat{H}_{42}(z)$	0.3222	1.3778	0.5889	1.4222	0.5778	0
	$T^{-1}\hat{H}_{52}(z)$	0.3111	1.4333	0.4667	1.5889	0.40	1.6
Simpson's 3/8	$T^{-1}\hat{H}_{33}(z)$	0.375	1.125	1.125	0.50	0	0
Boole	$T^{-1}\hat{H}_{44}(z)$	0.3111	1.4222	0.5333	1.4222	0.6222	0
Fifth-order	$T^{-1}\hat{H}_{55}(z)$	0.3299	1.3021	0.8681	0.8681	1.3021	0.6597

The synthesis technique requires Equation 3.43 to be equal to Equation 3.41 so that the following necessary and sufficient conditions can be obtained:

$$\hat{a} = a = 1 \tag{3.44a}$$

$$\hat{b}_k = b_k > 0 \quad (k = 0, 1, \ldots, p-1) \tag{3.44b}$$

$$\hat{b}_k = \sum_{q=0}^{k} b_{k-pq}a^q > 0 \quad (k = p, p+1, \ldots, m) \tag{3.44c}$$

where $b_i = 0$ for $i < 0$. The advantage of this synthesis method is evident from Equation 3.44, where the new optical tap coefficient $(\hat{b}_k - \hat{b}_{k-p}\hat{a})$ can be made positive or negative depending on the design requirements. Table 3.3 shows that the computed optical tap coefficients are positive, showing the effectiveness of the synthesis method. As a result, the requirement for the synthesis of the incoherent INCOI processor is met. Thus, the optical synthesis technique described here provides greater design flexibility than the conventional approach.

3.2.2.3 Design of a Programmable Optical Integrating Processor

On the basis of the optical synthesis technique described in the previous section, the design of a programmable INCOI processor is outlined in this section.

3.2.2.3.1 Adaptive Algorithm for the INCOI Processor

One possible approach for overcoming the marginal stability of the INCOI processor is to replace the IIR filter, as described by Equation 3.41, by another FIR filter knowing that

$$\frac{1}{1 - az^{-p}} = \sum_{i=0}^{\infty} (az^{-p})^i \tag{3.45}$$

This approach is not practically feasible because of the infinite number of taps involved in the FIR filter. However, if the order of the INCOI processor is low (e.g., $p = 1$) and the duration of the input pulse sequence to be processed is short (e.g., a 20 tap INCOI processor is sufficient to process a signal with a duration of 20 sampling intervals), then such a technique can be useful for this specific requirement. The drawback of this method is that only the positive tap coefficients of the two cascaded FIR filters can be optically implemented with incoherent FOSPs. As a result, the technique is restricted to the INCOI processor with $m = p$.

An adaptive algorithm for overcoming the marginal stability of the INCOI processor is described here. The INCOI impulse response $\hat{h}_{mp}[n]$ is given by the inverse z-transform of Equation 3.42 as

$$T^{-1}\hat{h}_{mp}[n] = \hat{b}_0\delta[n] + \cdots + \hat{b}_{m-p}\delta[n-(m-p)] + \sum_{k=m-(p-1)}^{m} \hat{b}_k\hat{a}^{(n-k)/p}\sum_{q=0}^{\infty}\delta[n-(k+pq)] \tag{3.46}$$

where the unit-sample sequence is defined as [22]

$$\delta[n] = \begin{cases} 1 & \text{for } n = 0 \\ 0 & \text{for } n \neq 0 \end{cases} \tag{3.47}$$

and the delayed unit-sample sequence is defined as

$$\delta[n-i] = \begin{cases} 1 & \text{for } n = i \\ 0 & \text{for } n \neq i \end{cases} \tag{3.48}$$

where n is the discrete-time index, i.e., $n = 1$ means one unit-time delay T. Optical implementation of the positive tap coefficients in Equation 3.46 requires adaptive control of the optical gain values such that

$$T^{-1}\hat{h}_{mp}[n] = G_0[0] + \cdots + G_{m-p}[m-p] + \sum_{k=m-(p-1)}^{m} G_k[n]\hat{a}^{(n-k)/p} \tag{3.49}$$

where the time-variant optical gains are given by

$$G_k[n]|_{n=k} = \hat{b}_k \quad (k = 0, 1, \ldots, m-p) \tag{3.50a}$$

$$G_k[n] = \hat{b}_k\hat{a}^{-(n-k)/p}\sum_{q=0}^{\infty}\delta[n-(k+pq)] \quad (k = m-(p-1), \ldots, m) \tag{3.50b}$$

where $0 < \hat{a} < 1$. The summation term in Equations 3.46 and 3.50b shows that the pole value $\hat{a}$ is not strictly required to be equal to 1, but can take any value in the range of $0 < \hat{a} < 1$ while still maintaining the characteristics and overcoming the marginal stability of the INCOI processor. The dynamic range of the time-variant optical gain in Equation 3.50b can be increased by setting the pole value $\hat{a}$ as close to unity as possible (e.g., $\hat{a} = 0.95$) but not so close to unity that the original problem of marginal stability reappears.

3.2.2.3.2 Analysis of the Programmable INCOI Processor

The traveling-wave semiconductor laser amplifiers (TW-SLAs) [42,43] are considered to be the gain elements in the programmable INCOI processor because of their fast switching speed (see Chapter 2). The TW-SLA can operate at either 1300 or 1550 nm. The TW-SLA gain depends on both the injection current and the injected light power; it increases with increasing injection current but decreases with increasing injected power. Thus, the injected power level must be chosen to within the range over which the TW-SLA gate can be driven into saturation where the gain does not differ significantly [42]. With this fixed level of input power at a particular operating wavelength, the required TW-SLA gain can be obtained from the injection current source driving the gate. The effects of polarization sensitivity, arising from different TE and TM mode confinement factors, and the amplifier spontaneous emission (ASE) noise of the TW-SLAs are ignored here because they can be easily overcome in practice. A polarization insensitive TW-SLA has been demonstrated to be capable of achieving an effective gain of up to 20 dB, with less than 1 dB polarization sensitivity, less than 1 dB spectral gain ripple, and 3 dB bandwidth of 55 nm. The ASE noise can be minimized by means of a tunable optical filter with sufficient narrow bandwidth [44].

Figure 3.9 shows a block diagram of the proposed microprocessor (μP) controlled INCOI processor. The following discussion focuses on the μP-controlled TW-SLAs. The software-controlled μP chip is used to control the injection current sources driving the TW-SLAs to provide the required time-variant optical gains according to Equation 3.50.

Figure 3.10 shows a graphical illustration of the performance of the programmable trapezoidal INCOI processor $\hat{H}_{11}(z)$ in processing a rectangular input pulse sequence, $x[n] = 1$ for $n \geq 0$. Figure 3.10a and b shows, respectively, the profiles of the time-variant TW-SLA gains $G_0[n]$ and $G_1[n]$. From Equation 3.50a, the time-variant TW-SLA gain $G_k[n]\big|_{n=k}$ is only active at the appropriate discrete time n. This is shown in Figure 3.10a where $G_0[n]\big|_{n=0} = 0.5$ is only applied at time $n = 0$. It is assumed that $G_0[n]$ has already reached its steady-state value when the first pulse of the input pulse sequence arrives. Figure 3.10b shows that the time-variant TW-SLA gain $G_1[n]$ is active at every clock period T_{μ} of the μP and takes T_{SLA}, the switching time of the TW-SLA, to reach its steady state and thereby provides the required gain $G_1[n]$ according to Equation 3.50b.

From Figures 3.10a through c, the following timing requirements must be met in order to obtain optimum performance of the programmable INCOI processor:

$$\tau_{\text{W}} \ll \tau \tag{3.51a}$$

$$\tau = qT_{\text{SLA}} \tag{3.51b}$$

$$T = q\tau, \quad q = 1, 2, \ldots \tag{3.51c}$$

$$T_{\mu} = T \tag{3.51d}$$

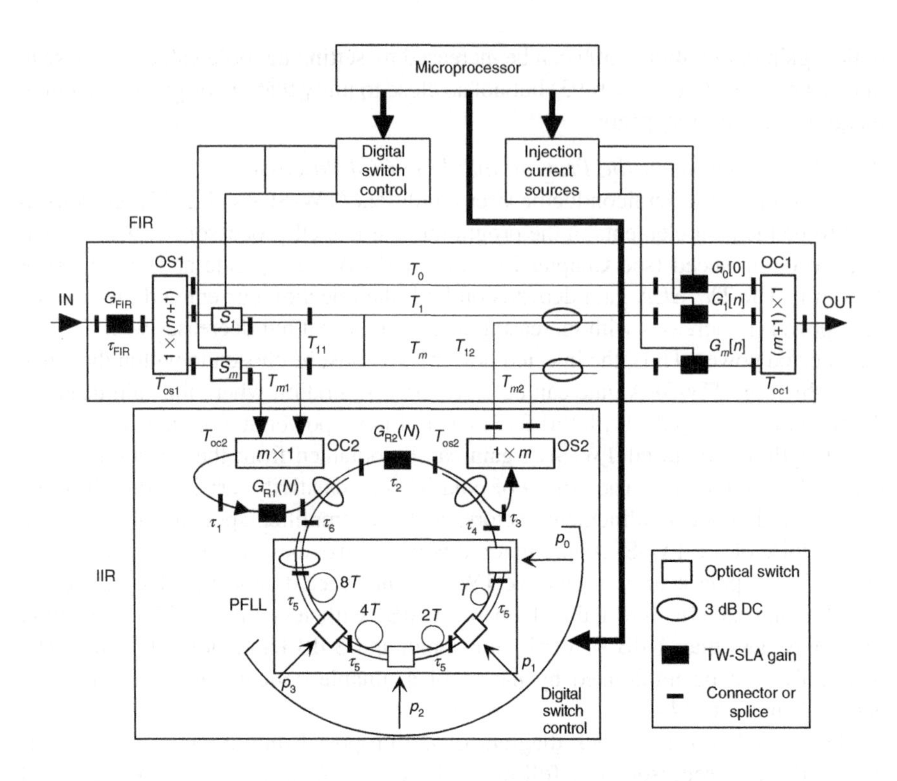

FIGURE 3.9 Block diagram of the proposed programmable INCOI processor.

Equation 3.51a indicates that the full-width half-maximum (FWHM) of the input pulse τ_w must be very much less than the bit period τ of the input pulse sequence in order to satisfy the incoherent requirement of the INCOI processor. Equation 3.51b requires τ to be a multiple integer of T_{SLA}, e.g., Figure 3.10 shows the case where $\tau = T_{SLA}$. Equation 3.51c shows that the multiple integer of τ must be equal to the unit-time delay T, e.g., Figure 3.10 shows the case where $T = 3\tau$. Equation 3.51d requires the clock period T_μ of the μP to be equal to the unit-time delay T of the INCOI processor in order to achieve system synchronization. The conditions described by Equations 3.51b through 3.51d are necessary for obtaining high processing accuracy. The deviation of the unit-time delay from below ($-\Delta T$) or above ($+\Delta T$) its nominal value T by $/\Delta T/$, as shown in Figure 3.10b, will not cause significant performance degradation on the programmable INCOI processor provided that the following condition is satisfied:

$$|\Delta T| \leq \tau_w/2 \tag{3.52}$$

The output pulse sequence is shown in Figure 3.10d. Note that the speed ($1/T$) of the programmable INCOI processor is ultimately limited by the μP speed ($1/T_\mu$), the clock rate at which the μP sequentially fetches and executes one instruction after

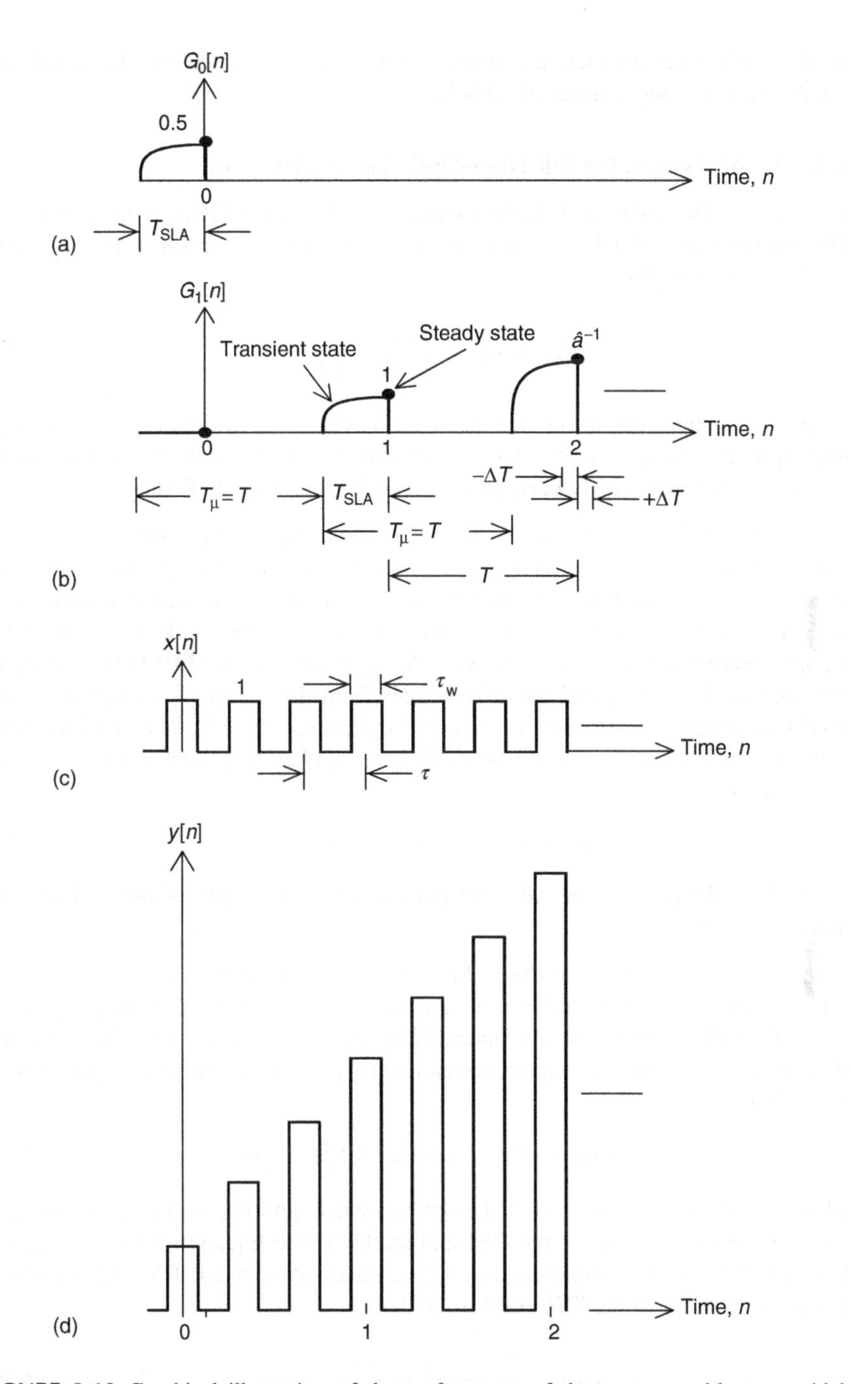

FIGURE 3.10 Graphical illustration of the performance of the programmable trapezoidal INCOI processor $\hat{H}_{11}(z)$ in processing a rectangular input pulse sequence. (a) TW-SLA gain profile of $G_0[n]$, (b) TW-SLA gain profile of $G_1[n]$, (c) rectangular input pulse (rectangular profile assumed) sequence $x[n] = 1$ for $n \geq 0$, and (d) output pulse sequence $y[n]$.

another until a halt instruction is processed. Commercially available 32-bit μPs are capable of achieving a speed of 60 MHz [45].

3.2.2.4 Analysis of the FIR Fiber-Optic Signal Processor

The delay and loss of the FIR FOSP block, as highlighted in Figure 3.9, are analyzed. The insertion loss of a $1 \times (m+1)$ optical splitter (L_{os}) or an $(m+1) \times 1$ optical combiner (L_{oc}) is given by

$$L_{os} = L_{oc} = \frac{(L_{s/c} L_{dc})^{\log_2 (m+1)}}{(m+1) L_{s/c}} \tag{3.53}$$

where $L_{s/c}$ is the splice (or connector) loss and L_{dc} is the insertion loss of the 3 dB fiber-optic directional coupler (DC). Note that the structures of the optical splitter and combiner have already been described in Section 3.1 (see Figures 3.2 and 3.3).

3.2.2.4.1 Delay Analysis of the FIR Fiber-Optic Signal Processor

The delay analysis considered here is associated only with the optical components inside the FIR FOSP block. In the following analysis, it is assumed that the cross talks of the optical switches are acceptably low, e.g., <-50 dB. It is presumed that the μP-controlled optical switches, $S_1 \cdots S_m$, $S_k \in (0,1)$, of the FIR FOSP are in the bar state so that the optical intensity signals from the $1 \times (m+1)$ optical splitter (OS1) are routed directly to the $(m+1) \times 1$ optical combiner (OC1). The FIR FOSP requires the differential delay between neighboring fiber delay lines to be exactly one unit-time delay T, i.e.,

$$T_k = T_0 + kT \quad (k = 0, 1, \ldots, m) \tag{3.54}$$

where T_k is the delay of the kth fiber path between the output of the OS1 and the input of the OC1.

3.2.2.4.2 Loss Analysis of the FIR Fiber-Optic Signal Processor

The loss analysis considered here is associated only with the optical components inside the FIR FOSP block. An optical intensity signal coming into the input port (IN) of the FIR FOSP will experience an intensity loss L_{FIR} at its output port (OUT) according to

$$L_{FIR} = L_{s/c}^{7} \cdot L_{os1} \cdot L_{sw} \cdot (0.5 L_{dc}) \cdot L_{oc1} \tag{3.55}$$

where L_{sw} is the insertion loss of the optical switch, and L_{os1} and L_{oc1} are, respectively, the insertion losses of the OS1 and the OC1 (see Equation 3.53). The loss in Equation 3.55 can be compensated by a time-invariant (or fixed) TW-SLA gain G_{FIR} placed at the front of the FIR FOSP such that

$$G_{FIR} = \frac{2(m+1)^2 T}{L_{s/c}^{5} L_{dc} L_{sw} (L_{s/c} L_{dc})^{\log_2 (m+1)^2}} \tag{3.56}$$

Equation 3.56 has been obtained by assuming that all the fiber paths of the FIR FOSP have the same intensity loss. However, the loss of the upper fiber path (where $k=0$)

is less than those on the lower fiber paths (where $k = 0, 1, 2, \ldots$) because of the absence of one splice (or connector), one optical switch, and one 3 dB DC. The absence of this loss must be incorporated into the time-variant TW-SLA gain $G_0[0]$ so that every fiber path of the FIR FOSP has the same intensity loss. As a result, the new $G_0[0]$ in Equation 3.51a now becomes

$$G_0[0] = (0.5L_{s/c}L_{sw}L_{dc})\hat{b}_0 \tag{3.57}$$

3.2.2.5 Analysis of the IIR Fiber-Optic Signal Processor

The programmable fiber loop length (PFLL), the delay and loss of the IIR FOSP block as highlighted in Figure 3.9, are described here. From Equation 3.41, p incoherent IIR FOSPs are required to be incorporated into the p higher-order delay lines of the FIR FOSP. This requirement can be alternatively achieved, as shown inside the IIR FOSP block of Figure 3.9, with an $m \times 1$ optical combiner (OC2), a $1 \times m$ optical splitter (OS2), and only one (instead of p) IIR FOSP. It is clear that this IIR FOSP block is structurally simpler and yet performs the same function as if p IIR FOSPs are introduced into the p higher-order fiber paths of the FIR FOSP. Note that the IIR FOSP, which is an all-pole filter, has already been described in Chapter 2 where it was referred to as the incoherent RFOSP.

3.2.2.5.1 Programmable Fiber Loop Length

The PFLL is schematically shown and highlighted inside the IIR FOSP block (see Figure 3.9). The intensity signal coming into the PFLL is optionally routed through N fiber segments whose lengths are arranged in a binary sequence of lengths giving delays of 2^iT $(i = 0, 1, \ldots, N-1)$ so that $2^N - 1$ different delays with a resolution of T can be selected [46]. Selection of the appropriate fiber segments is by the μP-controlled optical switches, which permit the signal either to connect or bypass a particular fiber segment according to the state $p_i \in (0,1)$. The PFLL requires N optical switches and one 3 dB DC, e.g., Figure 3.9 shows the case where $N = 4$. The required fiber loop delay $T_{p,\mathrm{IIR}}$ of the pth-order IIR FOSP is given by

$$T_{p,\mathrm{IIR}} = pT \quad (1 \le p \le m) \tag{3.58}$$

as required by the loop delay z^{-p} in the denominator of Equation 3.41. The integer value p in Equation 3.58 takes the binary representation of the N fiber segments as

$$p = \sum_{i=0}^{N-1} p_i 2^i, \quad p_i \in (0,1) \tag{3.59}$$

which lies in the range of $1 \le p \le 2^N - 1$. Thus, the number of optical switches required to achieve p unit-time delays is given by

$$N \ge \log_2(p+1) \tag{3.60}$$

which is relatively small because of the logarithmic relation.

3.2.2.5.2 Delay Analysis of the IIR Fiber-Optic Signal Processor

The delay analysis considered here is associated only with the optical components inside the IIR FOSP block. It is presumed that the μP-controlled optical switches $S_1 \ldots S_m$ of the FIR FOSP are now in the cross state so that the intensity signals coming into these switches are routed through the OC2. Selection of the states of the switches is governed by Equation 3.59. There are $m + 1$ possible paths through which the signal at the input port can propagate to the output port. Each of these paths introduces a delay D_k between the input and output ports according to

$$D_0 = \tau_{\mathrm{FIR}} + 2T_{\mathrm{os1}} + T_0 \tag{3.61a}$$

$$D_k = \tau_{\mathrm{FIR}} + 2T_{\mathrm{os1}} + 2T_{\mathrm{os2}} + \tau_1 + \tau_2 + \tau_3 + T_{k1} + T_{k2} \quad (k = 1, 2, \ldots, m) \tag{3.61b}$$

where $T_{\mathrm{os1}} = T_{\mathrm{oc1}}$ and $T_{\mathrm{os2}} = T_{\mathrm{oc2}}$ have been used. The parameters of Equation 3.61 are defined as follows: T_{os1}, the propagation delay of the OS1; T_{oc1}, the propagation delay of the OC1; T_{os2}, the propagation delay of the OS2; T_{oc2}, the propagation delay of the OC2; τ_{FIT}, the delay of the fiber path between the input port IN and the input of the OS1; τ_1, the delay of the fiber path between the output of the OC2 and the input of the IIR FOSP via the TW-SLA with gain $G_{\mathrm{R1}}(N)$; τ_2, the delay of the fiber path between the two 3 dB DCs via the TW-SLA with gain $G_{\mathrm{R2}}(N)$; τ_3, the delay of the fiber path between the output of the IIR FOSP and the input of the OS2; T_{k1}, the delay of the kth fiber path between the output of the OS1 and the input of the OC2 via the kth optical switch; T_{k2}, the delay of the kth fiber path between the output of the OS2 and the input of the OC1 via the 3 dB DC and the kth time-variant TW-SLA with gain $G_k[n]$. The delay D_0 in Equation 3.61a is the pure propagation delay of the INCOI processor.

Again, the FIR FOSP requires the differential delay between neighboring fiber delay lines to be exactly one unit-time delay T, i.e.,

$$D_k - D_0 = kT \quad (k = 0, 1, \ldots, m) \tag{3.62a}$$

or

$$T_k = 2T_{\mathrm{os2}} + \tau_1 + \tau_2 + \tau_3 + T_{k1} + T_{k2} \quad (k = 1, 2, \ldots, m) \tag{3.62b}$$

when Equations 3.54 and 3.61 have been substituted into Equation 3.62a. The μP-controlled optical switches $S_1 \ldots S_m$ must be active at the instant when the first pulse of the input pulse sequence arrives. This timing requirement can be met by having

$$\tau_{\mathrm{FIR}} + T_{\mathrm{os1}} = T_{\mathrm{sw}} + T_{\mu} \tag{3.63}$$

where T_{sw} is the switching time of the optical switches.

3.2.2.5.3 Loss Analysis of the IIR Fiber-Optic Signal Processor

The loss analysis considered here is associated only with the optical components inside the IIR FOSP block. The loop gain $G_{\mathrm{loop}}(\omega)$ of the IIR FOSP is given by

$$G_{\mathrm{loop}}(\omega) = (0.5L_{\mathrm{dc}})^3 \cdot L_{\mathrm{s/c}}^{N+4} \cdot L_{\mathrm{sw}}^{N} \cdot G_{\mathrm{R2}}(N) \cdot \exp[-j\omega(pT + \tau_2 + \tau_4 + \tau_6)] \tag{3.64}$$

where τ_4 is the delay of the fiber path between the 3 dB DC and the optical switch p_0 and τ_6 is the delay of the fiber path between the 3 dB DC of the PFLL and another 3 dB DC. In the PFLL, the effect of the delay of the lower fiber line between neighboring optical switches, i.e., τ_5 has been ignored in Equation 3.64 because the states of the optical switches are not known in advance to the designer. Comparison of Equations 3.54 and 3.58 requires $\tau_2 = \tau_4 = \tau_6 = 0$, which is not practically attainable. However, the effects of the delays τ_2, τ_4, τ_6, and τ_5 can be ignored without much deterioration of the performance of the programmable INCOI processor if they can be made sufficiently small compared to T, i.e., $(\tau_2, \tau_4, \tau_6, \tau_5) \ll T$. The absolute value of the loop gain in Equation 3.64 is required to be equal to $\hat{a}$ according to Equation 3.41 such that

$$G_{R2}(N) = \frac{8\hat{a}}{L_{dc}^3 L_{sw}^N L_{s/c}^{N+4}} \quad (0 < \hat{a} < 1) \tag{3.65}$$

In Equations 3.41 and 3.45, the numerator of the IIR FOSP is equal to unity, and this requires the forward-path gain of the kth fiber path between the input of the OC2 and the output of the OS2 to be equal to unity, i.e.,

$$L_{os2}^2 \cdot L_{s/c}^6 \cdot (0.5L_{dc})^2 \cdot G_{R1}(N) \cdot G_{R2}(N) = 1 \tag{3.66a}$$

or

$$G_{R1}(N) = \frac{L_{dc} L_{sw}^N L_{s/c}^{N-2}}{2\hat{a} L_{os2}^2} \tag{3.66b}$$

when Equation 3.65 has been placed into Equation 3.66a and $L_{os2} = L_{oc2}$ has been used.

3.2.2.5.4 *Timing Requirements for Multiple Input Pulse Sequences*

The timing requirements for sequential processing of multiple input pulse sequences are described here. The INCOI processor time, T_{INCOI}, the time required to process the input pulse sequence of duration, T_{pulse}, is given by

$$T_{INCOI} = 2T_{pulse} - T \tag{3.67}$$

where $T_{pulse} \gg T$. The next timing requirement assumes that each input pulse sequence is processed by a different family of the INCOI processor. The pulse strobe period, T_{psp}, defined as the time separation between the first pulse of the present input pulse sequence and the first pulse of the next input pulse sequence, and the programmability period, T_{pr}, defined as the time separation between the setup of two consecutive programs, must be equal to the INCOI processor time, i.e.,

$$T_{psp} = T_{pr} = T_{INCOI} \tag{3.68}$$

3.2.3 Remarks

This section describes the processing accuracy of the ideal and nonideal INCOI processor in the time domain. For analytical simplicity, the TIME n means nT and the sampling period is assumed to be $T = 1$.

The processing accuracy of the INCOI processor is evaluated as

$$\text{Error response} = \frac{\text{true integral} - \text{output pulse sequence}}{\text{true integral}} \times 100\% \quad (3.69)$$

where the true integral corresponds to the true integral of the input pulse sequence $x[n]$, and the output pulse sequence corresponds to the convolution of the impulse response $\hat{h}_{mp}[n]$ of the INCOI processor with the input pulse sequence $x[n]$.

There are two major error sources in the implementation of the programmable INCOI processor. The first arises from the deviation of the optical tap coefficient, that is $\Delta\hat{b}_k$, caused by the inaccuracy in the time-variant TW-SLA gain $G_k[n]$, from its nominal value $\hat{b}_k$. However, the effect of the deviation of the fiber loop gain value (i.e., $\Delta\hat{a}$), caused by the inaccuracy in the TW-SLA gain $G_{R2}[n]$, from its nominal value $\hat{a}$ can easily be included in $G_k[n]$. The second error source is due to the deviation of the unit-time delay from its nominal value T, i.e., ΔT, caused by the inaccurate cutting of fiber lengths. The effects of the detector noise and cross talks of the optical switches are assumed to be negligible here.

The performance of the trapezoidal INCOI processor $\hat{H}_{11}(z)$ in processing the linear input pulse sequence is now considered in detail. The basic time delay of the INCOI processor is defined as $T_1 - T_0 = T$, which is used as a basis for the higher-order delays (i.e., $T_k - T_0 = kT$, $k = 2, 3, \ldots, m$) of the FIR FOSP for $m \geq 2$ and for the fiber loop delay $T_{p\text{IIR}} = pT$ of the IIR FOSP. Thus, the trapezoidal integrator requires the nominal value of the fiber loop delay $T_{1,\text{IIR}}$ of the IIR FOSP to be exactly T (i.e., $T_{1,\text{IIR}} = T$) and any deviation from this nominal value is denoted as $\Delta T_{1,\text{IIR}}$.

Table 3.4 shows three cases of the $\pm 5\%$ deviation of the fiber loop delay $\Delta T_{1,\text{IIR}}/T_{1,\text{IIR}}$ and their corresponding $\pm 5\%$ deviations of all the possible sets of the optical tap coefficients $\{\Delta\hat{b}_0/\hat{b}_0, \Delta\hat{b}_1/\hat{b}_1\}$. The percent relative errors (REs) at $n = 20$ are also shown for each case.

Figures 3.11 through 3.13 show the percentage error responses of the trapezoidal integrator for the three cases. Curves Ai, Bi, and Ci ($i = 1, \ldots, 9$) show the error responses of cases A, B, and C with the corresponding ith set of optical tap coefficients. The solid curve A1 in Figure 3.12a shows the performance of the ideal trapezoidal integrator, where there are no deviations of the fiber loop length (i.e., $\Delta T_{1,\text{IIR}} = 0$) and the optical tap coefficients (i.e., $\Delta\hat{b}_0 = \Delta\hat{b}_1 = 0$) from their nominal values. Curve A1 shows that the trapezoidal integrator has zero processing error at the sampling instants (i.e., $n = 0, 1, 2, \ldots$) in processing the linear input pulse sequence. This is as expected from the numerical integration scheme where the trapezoidal rule is equivalent to approximating the area of the trapezoid under the straight line [39].

However, Figures 3.11 through 3.13 show that the processing errors are large between the sampling instants and eventually reach their steady states after $n = 20$

TABLE 3.4
Relative Errors (REs) (%) at $n=20$ of the Trapezoidal INCOI Processor $\hat{H}_{11}(z)$, with ±5% Deviations of the Optical Tap Coefficients and ±5% Deviation of the Fiber Loop Delay, in Processing the Linear Input Pulse Sequence $x[n]=n$ for $n \geq 0$

	Case A			Case B			Case C		
	$\frac{\Delta T_{1,\text{IIR}}}{T_{1,\text{IIR}}}=0$			$\frac{\Delta T_{1,\text{IIR}}}{T_{1,\text{IIR}}}=+0.05$			$\frac{\Delta T_{1,\text{IIR}}}{T_{1,\text{IIR}}}=-0.05$		
Set	$\frac{\Delta\hat{b}_0}{\hat{b}_0}$	$\frac{\Delta\hat{b}_1}{\hat{b}_1}$	RE (%)	$\frac{\Delta\hat{b}_0}{\hat{b}_0}$	$\frac{\Delta\hat{b}_1}{\hat{b}_1}$	RE (%)	$\frac{\Delta\hat{b}_0}{\hat{b}_0}$	$\frac{\Delta\hat{b}_1}{\hat{b}_1}$	RE (%)
1	0	0	0	0	0	+0.463	0	0	−0.488
2	0	+0.05	−4.75	0	+0.05	−4.625	0	+0.05	5.261
3	0	−0.05	+4.75	0	−0.05	+5.19	0	−0.05	+4.286
4	−0.05	0	−0.25	+0.05	0	+0.123	+0.05	0	−0.738
5	+0.05	+0.05	−0.5	+0.05	+0.05	−4.514	+0.05	+0.05	−5.512
6	+0.05	−0.05	+4.5	−0.05	−0.05	+4.491	+0.05	−0.05	+4.036
7	−0.05	0	+0.25	−0.05	0	+0.712	−0.05	0	−0.237
8	−0.05	+0.05	−4.5	−0.05	+0.05	−4.016	−0.05	+0.05	−5.011
9	−0.05	−0.05	+5.0	−0.05	−0.05	5.439	−0.05	−0.05	+4.537

sampling intervals, beyond which they do not differ significantly. The steady-state errors of Figures 3.11 and 3.13 are tabulated in Table 3.4 which shows that cases {A1,A4,A7}, {B1,B4,B7}, and {C1,C4,C7}, where $\Delta\hat{b}_1 = 0$, have smaller processing errors than other sets belonging to the same cases. Thus, the performance of the trapezoidal integrator is greatly affected by large deviations of $\hat{b}_1$. In addition, the

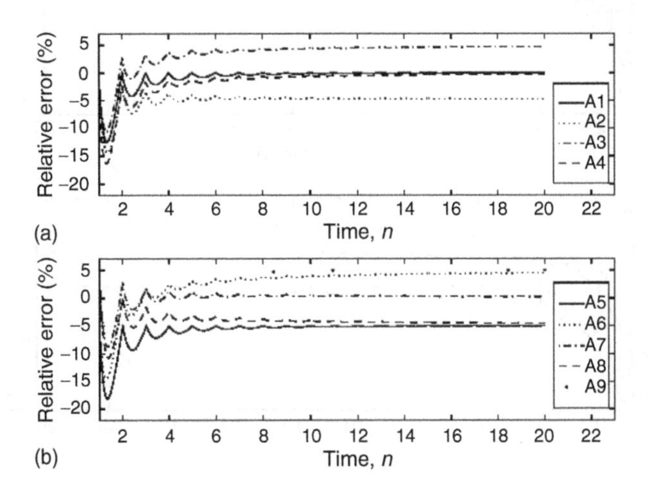

FIGURE 3.11 Relative error response (%) of the trapezoidal INCOI processor $\hat{H}_{11}(z)$ in processing the linear input pulse sequence $x[n]=n$ for $n \geq 0$. (a) Cases A1–A4 and (b) cases A5–A9 correspond to the conditions given in Table 3.4.

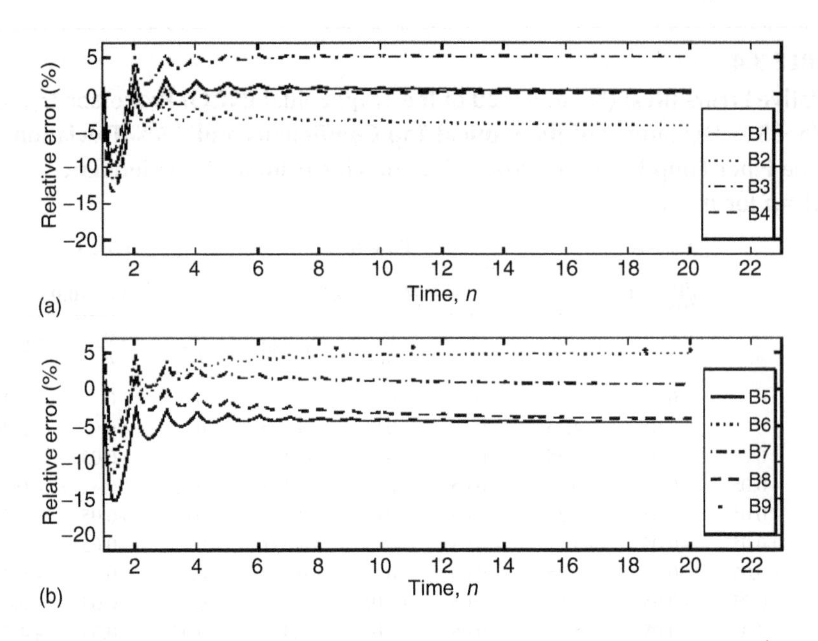

FIGURE 3.12 Relative error response (%) of the trapezoidal INCOI processor $\hat{H}_{11}(z)$ in processing the linear input pulse sequence $x[n]=n$ for $n \geq 0$. (a) Cases B1–B4 and (b) cases B5–B9 correspond to the conditions given in Table 3.4.

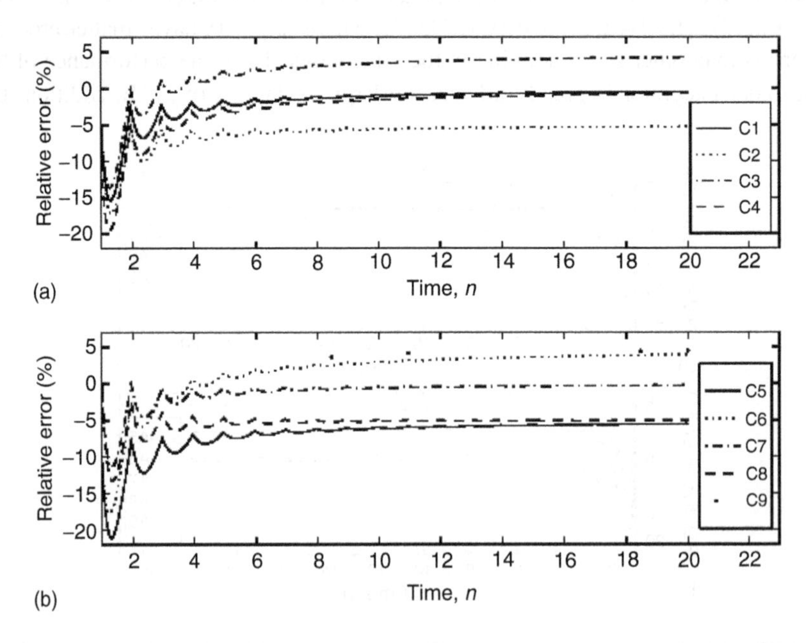

FIGURE 3.13 Relative error response (%) of the trapezoidal INCOI processor $\hat{H}_{11}(z)$ in processing the linear input pulse sequence $x[n]=n$ for $n \geq 0$. (a) Cases C1–C4 and (b) cases C5–C9 correspond to the conditions given in Table 3.4.

absolute values of the steady-state errors for all cases in Tables 3.1 and 3.2 are less than 6%. Thus, the ±5% parameter deviation may be considered as an acceptable upper bound value in the implementation of the programmable INCOI processor.

Next, the performances of several families of the ideal INCOI processor are analyzed for several types of input pulse sequences. It was found that the ideal Simpson's 1/3 integrator $\hat{H}_{22}(z)$ has zero processing error of the parabolic input pulse sequence (i.e., $x[n]=n^2$, $n \geq 0$) at the sampling instants. This is also expected because the numerical integration scheme was employed [39].

From Table 3.5, the Simpson's 1/3 integrator $\hat{H}_{22}(z)$ has higher processing accuracy of the cubic input pulse sequence (i.e., $x[n]=n^3$, $n \geq 0$) than other families of the INCOI processor, and the error rapidly decreases with time and eventually converges to zero in the steady state. For the fourth-order polynomial input pulse sequence (i.e., $x[n]=n^4$, $n \geq 0$), Table 3.6 shows that the integrator $\hat{H}_{52}(z)$ outperforms other families of the INCOI processor and has negligibly small processing error for $n \geq 3$. It was also found that the overall performance of the ideal Boole's integrator $\hat{H}_{44}(z)$ is superior to other families of the INCOI processor in processing the rth-order polynomial input pulse sequence (i.e., $x[n]=n^r$, $r \geq 5$ for $n \geq 0$). From Tables 3.4 and 3.5, as the order of the polynomial input pulse sequence is increased (i.e., $r \geq 2$), the higher-order trapezoidal integrator $\hat{H}_{m1}(z)\big|_{m\geq 2}$ outperforms the lowest-order trapezoidal integrator $\hat{H}_{11}(z)$. This shows the advantage of using a higher-order trapezoidal integrator.

For the exponential input pulse sequence, i.e., $x[n]=\exp(-n/w)$ for $n \geq 0$, with the FWHM pulse width $w=50$, Table 3.7 shows that the ideal trapezoidal integrator

TABLE 3.5
Relative Errors (%) of Several Families of the Ideal INCOI Processor, at the Sampling Instants, in Processing the Cubic Input Pulse Sequence $x[n]=n^3$ for $n \geq 0$

	Time, n					
$\hat{H}_{mp}(z)$	1	2	3	4	5	200
$\hat{H}_{11}(z)$	−100	−25	−11.1	−6.25	−4.0	−2.5E–03
$\hat{H}_{21}(z)$	−66.7	−10.4	−3.29	−1.43	−0.75	−1.25E–05
$\hat{H}_{31}(z)$	−50	−4.17	−0.82	−0.26	−0.107	−4.17E–08
$\hat{H}_{41}(z)$	−39.4	−0.87	−0.041	−0.013	−0.0053	−2.08E–09
$\hat{H}_{51}(z)$	−31.9	1.007	0.051	−0.013	−0.0053	−2.08E–09
$[\hat{H}_{22}(z)=\hat{H}_{33}(z)]$	−33.3	0	−0.41	1.11E–14	−0.053	0
$\hat{H}_{42}(z)$	−28.9	1.11	−0.302	0.0694	−0.0391	1.11E–08
$\hat{H}_{52}(z)$	−24.4	1.94	−0.41	0.104	−0.053	1.67E–08
$\hat{H}_{33}(z)$	−50	−3.13	0	−0.195	−0.08	−3.13E–08
$\hat{H}_{44}(z)$	−24.4	2.22	−0.302	0	−0.0391	0
$\hat{H}_{55}(z)$	−31.9	1.48	0.29	−0.125	−1.82E–14	0

Note: Error at time $n=0$ is undefined.

TABLE 3.6
Relative Errors (%) of Several Families of the Ideal INCOI Processor, at the Sampling Instants, in Processing the Fourth-Order Polynomial Input Pulse Sequence $x[n]=n^4$ for $n \geq 0$

	Time, *n*					
$\hat{H}_{mp}(z)$	1	2	3	4	5	200
$\hat{H}_{11}(z)$	−150	−40.6	−18.3	−10.4	−6.64	−4.2E–03
$\hat{H}_{21}(z)$	−108	−21.1	−7.17	−3.23	−1.72	−3.1E–05
$\hat{H}_{31}(z)$	−87.5	−11.9	−2.88	−0.99	−0.427	−1.97E–07
$\hat{H}_{41}(z)$	−74.3	−6.62	−0.926	−0.219	−0.072	−7.02E–10
$\hat{H}_{51}(z)$	−64.9	−3.1	−0.0386	1.39E–14	1.82E–14	−7.51E–13
$[\hat{H}_{22}(z)=\hat{H}_{32}(z)]$	−66.7	−4.17	−0.823	−0.26	−0.107	−4.17E–08
$\hat{H}_{42}(z)$	−61.1	−2.08	−0.274	−0.065	−0.0213	−2.08E–10
$\hat{H}_{52}(z)$	−55.6	0.174	0	0	0	−2.03E–13
$\hat{H}_{33}(z)$	−87.5	−11.3	−1.85	−0.525	−0.26	−9.39E–08
$\hat{H}_{44}(z)$	−55.6	0	0.229	−1.39E–14	−0.0178	−1.91E–13
$\hat{H}_{55}(z)$	−64.9	−2.81	0.37	0.0634	0	0

Note: Error at time $n=0$ is undefined.

$\hat{H}_{11}(z)$ has the smallest processing error among other families of the INCOI processor, and its steady-state error converges to -1.006% at $n=300$. Its processing error was also found to decrease with increasing pulse width, e.g., when $w=100$ the

TABLE 3.7
Relative Errors (%) of Several Families of the Ideal INCOI Processor, at the Sampling Instants, in Processing the Exponential Input Pulse Sequence $x[n]=\exp(-n/w)$ for $n \geq 0$ Where $w=50$

	Time, *n*					
$\hat{H}_{mp}(z)$	1	2	3	4	5	300
$\hat{H}_{11}(z)$	−50.51	−25.51	−17.18	−13.01	−10.51	−1.006
$\hat{H}_{21}(z)$	−50.67	−25.59	−17.22	−13.05	−10.54	−1.006
$\hat{H}_{31}(z)$	−54.96	−25.59	−17.23	−13.05	−10.54	−1.006
$\hat{H}_{41}(z)$	−60.35	−24.22	−17.23	−13.05	−10.54	−1.006
$\hat{H}_{51}(z)$	−66.06	−21.31	−17.89	−13.05	−10.54	−1.006
$[\hat{H}_{22}(z)=\hat{H}_{33}(z)]$	−67.67	−17.00	−23.01	−8.67	−14.08	−0.668
$\hat{H}_{42}(z)$	−71.06	−14.71	−24.55	−7.503	−15.02	−0.578
$\hat{H}_{52}(z)$	−75.57	−10.71	−27.63	−5.168	−16.91	−0.398
$\hat{H}_{33}(z)$	−50.76	−32.01	−12.89	−13.07	−13.19	−0.752
$\hat{H}_{44}(z)$	−74.45	−13.56	−23.79	−8.093	−15.49	−0.624
$\hat{H}_{55}(z)$	−64.17	−25.54	−12.67	−17.52	−6.933	−0.661

Note: Error at time $n=0$ is undefined.

steady-state error was reduced to -0.527% at $n = 300$, which is about half of that for the case $w = 50$.

The processing errors of the families of the INCOI processor are large over the initial time interval but eventually converge to a negligibly small, if not zero, value in the steady state. Note that if the deviation of the basic time delay satisfies the condition given in Equation 3.52, then the performance of the INCOI processor is not greatly affected by this factor. It is worth mentioning that the rectangular integrator, whose transfer function is given by $\hat{H}_R(z) = z^{-1}/(1 - z^{-1})$, has zero processing error of the constant (or rectangular) input pulse sequence, such as that shown in Figure 3.10c, at the sampling instants.

3.2.3.1 Conclusions

1. A generalized theory of the Newton–Cotes digital integrators has been developed, based on which a programmable INCOI processor has been designed using a microprocessor, fiber-optic architectures, optical switches, and optical and semiconductor amplifiers.
2. The pth-order programmable INCOI processor has high processing accuracy of the pth-order polynomial input pulse sequence (i.e., $x[n] = n^p, p \geq 1$), while the trapezoidal processor, $\hat{H}_{11}(z)$, outperforms other higher-order processors in processing the exponential pulse (i.e., $x[n] = \exp(-n/w)$, $n \geq 0$). In general, the incoherent processing accuracy of a particular processor order depends very much on the type of input pulse.
3. Optical integration is a new concept with many potential applications. One example is to apply the trapezoidal optical integrator described here to the design of an optical dark-soliton generator, which is outlined in Chapter 4.

3.3 HIGHER-DERIVATIVE FIR OPTICAL DIFFERENTIATORS

In this section, a theory of higher-derivative FIR optical differentiators is proposed. Section 3.3.1 describes the underlying principle of digital differentiators and their design techniques and applications. Section 3.3.2 presents a theory of the qth-order pth- derivative FIR digital differentiator whose derivation is given in Appendix B. Section 3.3.3 describes the synthesis of a qth-order pth-derivative FIR optical differentiator using an optical transversal filter structure, as described in Section 3.1.3.2, which consists of integrated-optic components. The theory of coherent integrated-optic signal processing described in Chapter 2 is employed in this chapter where electric-field amplitude signals are considered.

3.3.1 Introduction

Section 3.2 has established that a digital integrator can be used to simulate the behavior of an analog integrator. In this section, a digital differentiator is also used to simulate the behavior of an analog differentiator. Like the digital integrator, a digital differentiator also forms an integral part of many practical signal processing systems because the time derivative of signals is sometimes required for further use or analysis. Digital differentiators are useful in modifying the shape of the signal; they can be used to find positive-going or negative-going slopes, maxima or minima, or point of greatest slope, where the differentiated output would be positive or negative, zero, or maximum, respectively. First-derivative and second-derivative digital differentiators have been used in the design of compensators for control systems [34], for monitoring electrocardiogram (ECG) signals [3], in the study of velocity and acceleration in human locomotion [47], in the analysis of radar signals in radar systems [48], and for the calculation of geometric moments in optical systems [49].

A digital differentiator is a processor whose output pulse sequence is obtained by approximating the derivative of a continuous-time signal from the samples of that signal. A continuous-time signal $x(t)$, whose values are known at the discrete time $t = nT$ for $n = 0, 1, 2 \ldots$ where $T > 0$ is the period between successive samples, can be differentiated by a digital differentiator. The frequency response of an ideal pth-derivative digital differentiator is given by [50]

$$H_{\mathrm{d}}(\omega) = \begin{cases} (j\omega T)^p, & 0 \leq \omega T/(2\pi) \leq 1/2 \\ [j(2\pi - \omega T)]^p, & 1/2 < \omega T/(2\pi) \leq 1 \end{cases} \tag{3.70}$$

where

$j = \sqrt{-1}$

$p = 1, 2, \ldots$

ω is the angular frequency

T is the sampling period of the differentiator

The output pulse sequence of the ideal pth-derivative digital differentiator $y^{(p)}(nT)$ approximates the true pth-derivative of the continuous-time signal $x(t)$ according to

$$y^{(p)}(nT) = \left. \frac{\mathrm{d}^p x(t)}{\mathrm{d}t^p} \right|_{t=nT} \tag{3.71}$$

Digital differentiators can be designed in two ways, i.e., the time- and the frequency-domain approaches as follows.

In the time-domain approach, first-derivative FIR digital differentiators can be designed by using one of the many classical numerical differentiation algorithms such as the Newton, Bessel, Everett, Stirling, and Lagrange formulas [35–39]. The underlying principle of these numerical differentiation algorithms is to fit a continuous-time interpolation polynomial $x(t)$ to a given input pulse sequence $f(nT)$, where $f(nT) = x(nT)$, which is then differentiated to give $y(t) = \mathrm{d}x(t)/\mathrm{d}t$. Sampling the differentiated continuous-time polynomial $y(t)$ (by a digital differentiator) at the discrete time $t = nT$ yields the output pulse sequence $y(nT)$, which effectively approximates the true derivative of a continuous-time signal, i.e.,

$$y(nT) = \left. \frac{\mathrm{d}x(t)}{\mathrm{d}t} \right|_{t=nT} \tag{3.72}$$

The magnitude responses of these digital differentiators generally approximate the ideal magnitude response reasonably well over the lower frequency band

$0 \leq \omega T/(2\pi) \leq 1/4$. Thus, these differentiators are normally referred to as narrow-band digital differentiators.

In the frequency-domain approach, first-derivative and higher-derivative FIR digital differentiators satisfying prescribed specifications of the ideal frequency response can be designed by using the minimum method [50,51], the Fourier series method in conjunction with the Kaiser window function [52–54], the Fourier series method in conjunction with accuracy constraints [55–59], the eigenfilter method [60,61], and the least squares methods [62,63]. Because of the constraints imposed on the frequency responses of these digital differentiators, they are normally referred to as frequency-selective differentiating filters that, in addition to performing the function of signal differentiation, are capable of passing as well as rejecting certain frequency components of the signal. That is, they can be designed to have a narrowband* [55,56], mid-band [57,58], or wideband [50–54,59–62,64] magnitude frequency response, depending on the application. Although impressive frequency-domain performances can generally be achieved by these frequency-selective differentiating filters, the required filter order is generally very high (e.g., 30–40 taps are common).

In both the time- and frequency-domain approaches, the performances of the digital differentiators have usually been evaluated in the frequency domain by using the ideal frequency response (usually magnitude but not phase response) as a basis. This is because the differentiating accuracy of these digital differentiators in the time domain is difficult to assess as this would depend on a specific application, and hence the actual shape of the signal. It is obvious that a good digital differentiator must be capable of achieving high differentiation accuracy in the time domain while still able to reject unwanted frequency components. For example, the high frequency components of a digital signal are often corrupted with wideband noise. A narrowband digital differentiator would be useful in this case [35,47–49].

Unlike the digital differentiators that have been studied for some time, optical differentiation is still a new concept in the area of optical signal processing. Although a three tap coherent optical transversal (or FIR) filter using silica-based wave-guides integrated on a silicon substrate has been experimentally demonstrated as a second-derivative optical differentiator, no theoretical background was given [63]. In addition, a fiber-optic ring resonator has been claimed as a first-derivative optical differentiator under the resonance condition but it is not a true differentiator and hence would suffer from low processing accuracy [65].

In this section, a theory of higher-derivative FIR optical differentiators using integrated-optic structures is described. Most of the work presented here has been described by Ngo and Binh [66]. The derivation of a theory of higher-derivative FIR digital differentiators, which was not presented in Ref. [66], is given in Appendix A.

* A narrowband, mid-band, or wideband digital differentiator has a magnitude response that closely matches the ideal magnitude response over the frequency band of $0 \leq \omega T/(2\pi) \leq 1/4$, $1/8 \leq \omega T/(2\pi) \leq 3/8$, or $0 \leq \omega T/(2\pi) \leq 1/2$, respectively.

3.3.2 Higher-Derivative FIR Digital Differentiators

The transfer function of the qth-order pth-derivative FIR digital differentiator (i.e., $H_q^{(p)}(z)$) can be generally expressed as*

$$\begin{bmatrix} -a_{11} & a_{12} & -a_{13} & \cdots & (-1)^M a_{1M} \\ -a_{21} & a_{22} & -a_{23} & \cdots & (-1)^M a_{2M} \\ \vdots & \vdots & \vdots & & \vdots \\ -a_{M1} & a_{M2} & -a_{M3} & \cdots & (-1)^M a_{MM} \end{bmatrix} \begin{bmatrix} TH_M^{(1)}(z) \\ T^2 H_M^{(2)}(z) \\ \vdots \\ T^M H_M^{(M)}(z) \end{bmatrix} \cong \begin{bmatrix} z^{-1}-1 \\ z^{-2}-1 \\ \vdots \\ z^{-M}-1 \end{bmatrix} \tag{3.73}$$

where

$$a_{pq} = \frac{p^q}{q!}, \quad p,q = 1,2,\ldots,M \tag{3.74}$$

and $z = \exp(j\omega T)$ is the z-transform parameter [52]. The pulse response of the qth-order pth-derivative FIR digital differentiator is defined as

$$y_q^{(p)}(nT) = x(nT) * h_q^{(p)}(nT) \tag{3.75}$$

which approximates the true pth-derivative of the input pulse sequence $x(nT)$ as

$$y_q^{(p)}(nT) \cong \left.\frac{\mathrm{d}^p x(t)}{\mathrm{d}t^p}\right|_{t=nT} \tag{3.76}$$

where $h_q^{(p)}(nT)$ is the impulse response of $H_q^{(p)}(z)$.

The transfer function of the qth-order pth-derivative FIR digital differentiator takes the general form of

$$H_q^{(p)}(z) = \left[\frac{b_{\max}}{T^p}\right] \sum_{k=0}^{q} b(k) z^{-k}, \quad p,q = 1,2,\ldots,M \tag{3.77}$$

where $-1 \le b(k) \le 1$ is the normalized tap coefficient and $b_{\max} \ge 1$ is the normalization factor. The transfer functions of several families of the digital differentiators, as computed from Equation 3.72 for $M = 4$, are tabulated in Table 3.8. Note that the signs of the tap coefficients alternate such that even are positive and odd negative. For the special case where $q = p$, the transfer function of the qth-order pth-derivative FIR digital differentiator is generally given by

$$T^p H_q^{(p)}(z)\Big|_{q=p} = \left(1 - z^{-1}\right)^p \tag{3.78}$$

* The derivation of Equation 3.73 is given in Appendix A where Equation A.3.12a corresponds to Equation 3.73.

TABLE 3.8
Normalized Tap Coefficients and Normalization Factor, as Computed from Equation 6.4 for $M=4$, of the Several Families of the FIR Digital Differentiators with Transfer Functions Expressed in the Form of Equation 3.77

$T^pH_q^{(p)}(z)$	b_{max}	$b(0)$	$b(1)$	$b(2)$	$b(3)$	$b(4)$
$TH_1^{(1)}(z)$	1	1	−1	0	0	0
$TH_2^{(1)}(z)$	2	0.75	−1	0.25	0	0
$TH_3^{(1)}(z)$	3	0.6111	−1	0.5	−0.1111	0
$TH_4^{(1)}(z)$	4	0.5208	−1	0.75	−0.3333	0.0625
$T^2H_2^{(2)}(z)$	2	0.5	−1	0.5	0	0
$T^2H_3^{(2)}(z)$	5	0.4	−1	0.8	−0.2	0
$T^2H_4^{(2)}(z)$	9.5	0.3070	−0.9123	1	−0.4912	0.0965
$T^3H_3^{(3)}(z)$	3	0.3333	−1	1	−0.3333	0
$T^3H_4^{(3)}(z)$	12	0.2083	−0.75	1	−0.5833	0.125
$T^4H_4^{(4)}(z)$	6	0.1667	−0.6667	1	−0.6667	0.1667

3.3.3 Synthesis of Higher-Derivative FIR Optical Differentiators

The characteristics of the higher-derivative FIR digital differentiators and the planar lightwave circuit (PLC) technology outlined in Section 3.3.2 are used to synthesize higher-derivative FIR optical differentiators.

Coherent integrated-optic signal processing of electric-field amplitude signals is considered here. The unmodulated signal of the optical source is assumed to be externally modulated by an optical intensity modulator, which minimizes laser chirp as well as permitting high-speed modulation and hence high-speed signal processing. It is also assumed that the optically encoded signals to be processed by the optical differentiator are modulated onto the optical carrier whose coherence time is much longer than the sampling period T of the optical differentiator. As a result, the pulse response of the optical differentiator depends on the coherent interference of the delayed signals.

Figure 3.14 shows the schematic diagram of the proposed $(q+1)$-*tap* FIR coherent optical filter,* which is used to synthesize the qth-order pth-derivative FIR optical differentiator. The FIR coherent optical filter essentially consists of a $1 \times (q+1)$ optical splitter, a $(q+1) \times 1$ optical combiner, and $(q+1)$ waveguide delay lines, into each of which a tunable coupler (TC) and a phase shifter (PS) are incorporated. A coherent optical signal coming into the optical splitter will be evenly distributed to $(q+1)$ signals, which are then appropriately delayed by the delay lines and weighted by the TCs and PSs. These signals are then coherently collected by the optical combiner to generate the differentiated optical signal.

The FIR coherent optical filter can be constructed using the PLC technology, namely, silica-based waveguides embedded on a silicon substrate as described in

* The FIR coherent optical filter described here has a similar structure to the incoherent fiber-optic transversal filter (see Figure 4.4) outlined in the previous section.

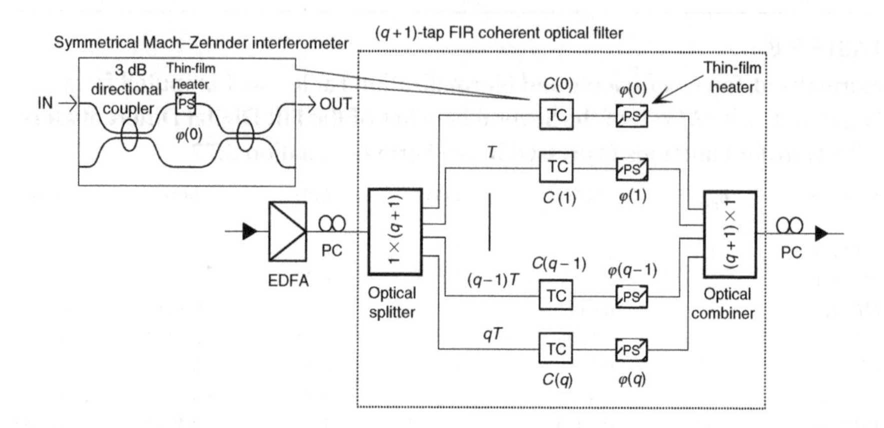

FIGURE 3.14 Schematic diagram of the proposed $(q + 1)$-tap FIR coherent optical filter used to synthesize the qth-order pth-derivative FIR optical differentiator. TC, tunable coupler; PS, phase shifter; and PC, polarization controller.

Section 3.3. The optical splitter and optical combiner can be developed using 3 dB silica-based waveguide DCs, except that no EDFAs are used here. In each delay line, the PS following the TC is a waveguide with a thin-film heater deposited on it and utilizes the thermo-optic effect to induce a carrier phase change of $\phi(k)$. The TC is a symmetrical MachZehnder interferometer (see the inset of Figure 3.14), which consists of two 3 dB DCs, two equal waveguide arms, and a thin-film heater, with a carrier phase change of $\varphi(\kappa)$, attached to one of the arms for controlling the output amplitude (see Chapter 2).

Neglecting the insertion loss of the 3 dB DCs, the propagation delay and waveguide birefringence of the TC, the kth TC transfer function, which corresponds to the transfer function E_3/E_1 in Equation 2.7, is given by

$$C(k) = |C(k)| \exp(j\angle C(k)) = 0.5\{\exp[j\varphi(k)] - 1\} \tag{3.79a}$$

where $\angle C(k)$ denotes the argument of $C(k)$,

$$|C(k)| = \sqrt{0.5 - 0.5\cos(\varphi(k))} \tag{3.79b}$$

or

$$\varphi(k) = \cos^{-1}\left[1 - 2|C(k)|^2\right] \tag{3.79c}$$

and

$$\angle C(k) = \tan^{-1}\left[\sin(\varphi(k))/(\cos(\varphi(k)) - 1)\right] \quad \text{for } k = 0, 1, \ldots, q \tag{3.79d}$$

Equation 3.79b indicates that a desired TC amplitude can be obtained by choosing an appropriate PS phase according to Equation 3.79c, and this results in the TC phase as given by Equation 3.79d. The amplitude and phase of the TC can be changed from 0 to 1 and from $-\pi/2$ to $+\pi/2$, respectively, when $\varphi(k)$ is varied from 0 to 2π (see Figure 2.3).

Neglecting the propagation delay and waveguide birefringence, the transfer function of the $(q+1)$-tap FIR coherent optical filter is given by

$$\hat{H}_q^{(p)}(z) = \sqrt{l_{\text{path}}} \cdot (q+1)^{-1} \cdot \sqrt{G} \cdot \sum_{k=0}^{q} (-1)^k \cdot |C(k)| \exp(j\angle C(k)) \cdot \exp(j\phi(k)) \cdot z^{-k} \tag{3.80}$$

where l_{path} is the intensity path loss, which takes into account all the losses associated with each delay line such as the loss of the straight and bend waveguides, the insertion loss of the 3 dB DCs in the splitter and combiner, and the insertion loss of the TC, $(q+1)^{-1}$ is the coupling loss of the splitter and combiner as a result of a 3 dB coupling loss at each stage of the structure, G is the intensity gain of the EDFA, and $(-1)^k = \exp(jk\pi)$ is the phase shift factor due to the $\pi/2$ cross-coupled phase shift of the 3 dB DCs in the splitter and combiner.

Synthesis of the qth-order pth-derivative FIR optical differentiator requires the equality of Equations 3.80 and 3.77, i.e., $\hat{H}_q^{(p)}(z) = H_q^{(p)}(z)$, such that

$$G = \frac{b_{\max}^2 (q+1)^2}{l_{\text{path}} T^{2p}} \tag{3.81a}$$

$$|C(k)| = |b(k)| \tag{3.81b}$$

$$\phi(k) = \angle b(k) - \angle(-1)^k - \angle C(k) \tag{3.81c}$$

$$\phi(k) = -\angle C(k), \quad \angle b(k) = \angle(-1)^k \tag{3.81d}$$

where $\angle b(k) = \angle(-1)^k = 0$ for even k or $\angle b(k) = \angle(-1)^k = \pi$ for odd k. Equation 3.81a shows that the required gain G is dominated by the small value of the sampling period T, and hence several EDFAs in cascade may be required at both the input and output of the optical differentiator, depending on the application. In each kth delay line, Equation 3.81b shows that the amplitude of the digital coefficient (i.e., $|b(k)|$) can be optically implemented by the TC amplitude (i.e., $|C(k)|$), and Equation 6.81d shows that the PS must provide a phase shift (i.e., φk) opposite to the TC phase (i.e., $\angle C(k)$). Because of the temperature dependence of the refractive index change of the silica waveguide, the PS can also be used to compensate for the optical path-length difference resulting from imperfect fabrication of the waveguide length. Note that if, for each delay line, a nontunable DC is used instead of the TC, then the PS is not required. However, it is difficult to practically fabricate a nontunable DC with a precise coupling coefficient as described in Chapter 2. Thus, it is preferable to use the TC which, in addition to implementing the digital coefficient, can accommodate for the deviations of the coupling coefficients of the 3 dB DCs in the splitter and combiner as a result of fabrication errors.

Because the temperature of the silicon substrate can be maintained to be within a small fraction of a degree to stabilize the refractive index of the waveguides [63], the PS can provide a very accurate phase shift. The temperature stability of the waveguides means that the optical differentiator can operate stably. Since the control of the TC amplitude and PS phase in a particular delay line is independent of those in other delay lines, the amplitude and phase of the digital coefficient can be optically implemented with high accuracy, showing the advantage of the proposed filter structure. The polarization controller (PC) placed at the input of the optical differentiator is used to counter any birefringence induced in the fiber, while the PC placed at its output is used to counter any waveguide birefringence arising from the optical differentiator. Alternatively, the waveguide birefringence may be overcome by inserting the polyamide half waveplates, acting as TE/TM mode converter, into the delay lines.

3.3.4 Remarks

The proposed qth-order pth-derivative FIR optical differentiator, $\hat{H}_q^{(p)}(z;\, p, q = 1, 2, 3, 4)$, described in Section 3.3.3, is now analyzed. For analytical simplicity, the following assumptions are used in all figures: the discrete-time index m means $m = nT$, the normalized optical frequency means $\omega T/(2\pi)$, the normalized time t/T means m, H_{pq} means $\hat{H}_q^{(p)}(z)$, the magnitude response corresponds to $|\hat{H}_q^{(p)}(z)|$, and the sampling period T is set to unity.

The processing accuracy of the qth-order pth-derivative FIR optical differentiator is evaluated by means of the Error response where

$$\text{Error response} = \frac{\text{true derivative} - \text{pulse response}}{\text{true derivative}} \times 100\% \qquad (3.82)$$

where true derivative corresponds to the true derivative of the input pulse sequence $x[m]$ and pulse response corresponds to the pulse response of the differentiator, as defined in Equation 3.77. To characterize the performance of the differentiator, the pulse response is defined as the amplitude response at the output of the optical differentiator before detection by an optical detector.

3.3.4.1 First-Derivative Differentiators

This section analyzes the performances of the qth-order first-derivative differentiators, $\hat{H}_q^{(1)}(z;\, q = 1, 2, 3, 4)$.

Figure 3.15a shows the magnitude responses of the qth-order and ideal differentiators. The magnitude response increases with increasing filter order q. The magnitudes are zero at $\omega T/(2\pi) = 0$ and $\omega T/(2\pi) = 1$ and maximum at $\omega T/(2\pi) = 0.5$ because there are at least one zero on the unit circle in the z-plane. Similar accounts can be made for the magnitude responses of the qth-order second-derivative, third-derivative, and fourth-derivative differentiators, which are shown in Figure 3.15b through d, respectively.

Figures 3.16a through d show the differentiator pulse responses when processing the Gaussian input pulse with various pulse widths* w, where $2w$ is the temporal

* The pulse width w corresponds to the normalized pulse width, i.e., w means wT.

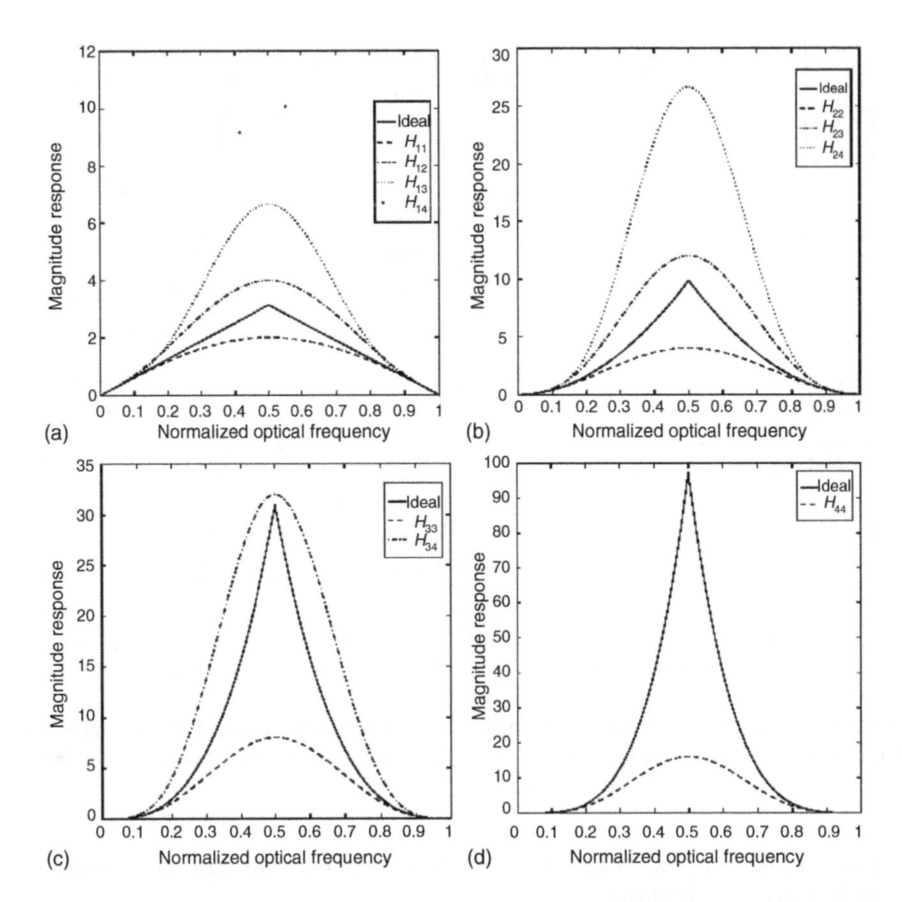

FIGURE 3.15 Magnitude responses of the optical differentiators. (a) qth-order first-derivative differentiators $\hat{H}_q^{(1)}(z;\, q = 1, 2, 3, 4)$, (b) qth-order second-derivative differentiators $\hat{H}_q^{(2)}(z;\, q = 2, 3, 4)$, (c) qth-order third-derivative differentiators $\hat{H}_q^{(3)}(z;\, q = 3, 4)$, and (d) fourth-order fourth-derivative differentiator $\hat{H}_4^{(4)}(z)$.

full-width of the intensity pulse at the 1/exp(1) points. The pulse responses closely resemble the true derivatives and the processing accuracy increases with increasing value of w.

Figures 3.17a and b show the differentiator error responses when processing the Gaussian input pulse with pulse widths $w = 5$ and $w = 20$, respectively. For $w = 5$, Figure 3.17a shows that the second-order, third-order, and fourth-order differentiators have lower processing accuracy than the first-order differentiator for time $m > 25$. However, for a larger pulse width, $w = 20$, Figure 3.17b shows that the second-order, third-order, and fourth-order differentiators have higher processing accuracy than the first-order differentiator, but at the expense of having lower processing accuracy over the initial time interval (see the enlarged curves in Figure 3.17c). A higher-order differentiator requires more hardware components than a lower-order differentiator. Thus, the first-order differentiator is considered as the

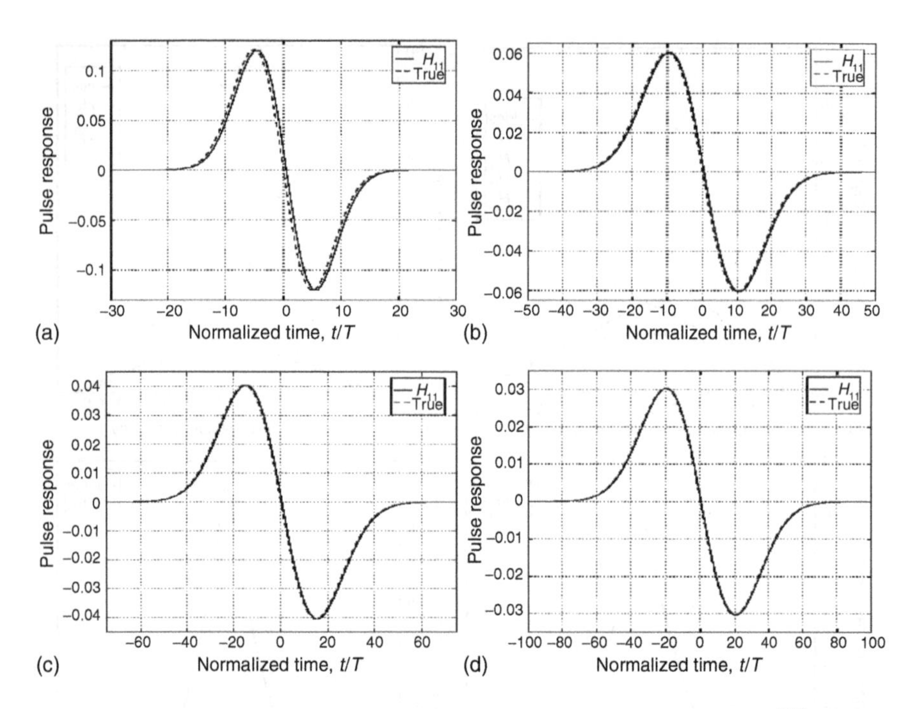

FIGURE 3.16 Pulse responses of the first-order first-derivative differentiator $\hat{H}_1^{(1)}(z)$ when processing the Gaussian input pulse (i.e., $x[m]=\exp[-m^2/(2w^2)]$) with various pulse widths w. (a) $w=5$, (b) $w=10$, (c) $w=15$, and (d) $w=20$.

optimum filter for processing a Gaussian pulse because of its structural simplicity and impressive performance.

Figures 3.18a through d show the differentiator error responses when processing the qth-order polynomial input pulse $x[m]=m^q$. The first-order, second-order, third-order, and fourth-order differentiators have higher processing accuracy of the first-order, second-order, third-order, and fourth-order input pulses, respectively, than other differentiator orders. Thus, for $x[m]=m^q$, the qth-order first-derivative differentiator has large processing error over the initial time interval $0 \leq m < q$ but has zero processing error over the longer time interval $m \geq q$. This is because the differentiators have been designed from the polynomial perspective (see Appendix A).

3.3.4.2 Second-Derivative Differentiators

This section analyzes the performances of the qth-order second-derivative differentiators, $\hat{H}_q^{(2)}(z;\ q=2,3,4)$, whose magnitude responses were shown in Figure 3.15b.

Figures 3.19a through d show the differentiator error responses when processing the qth-order polynomial input pulse. Figures 3.19a through c show that the second-order, third-order, and fourth-order differentiators have higher processing accuracy of the second-order, third-order, and fourth-order pulses, respectively, than other differentiator orders. However, Figure 3.19d shows that the fourth-order

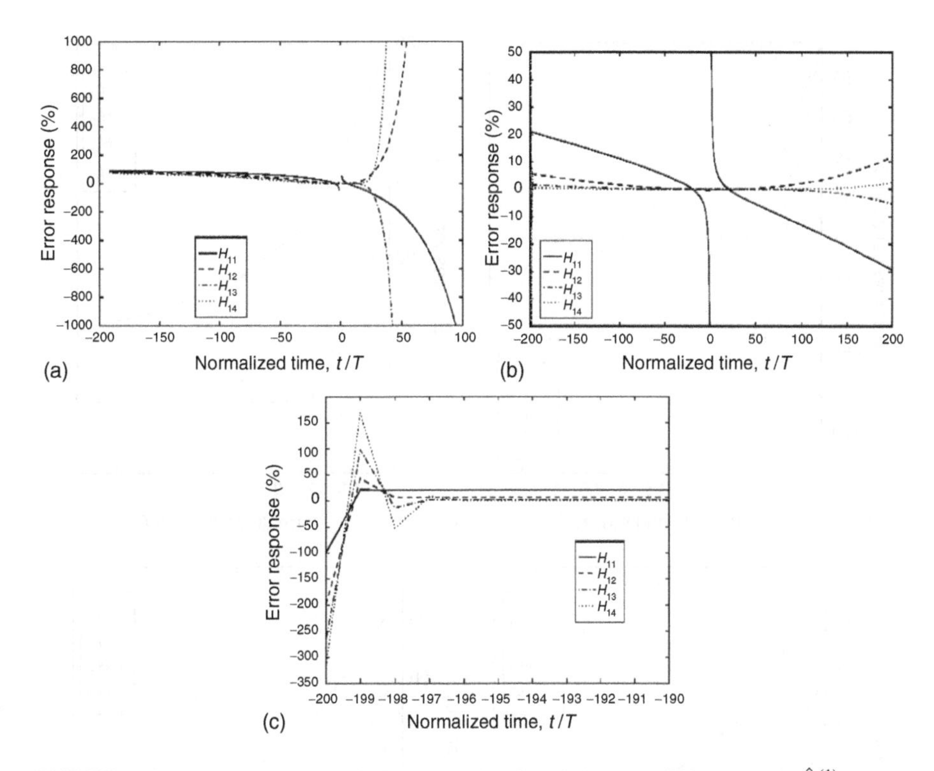

FIGURE 3.17 Error responses of the qth-order first-derivative differentiators $\hat{H}_q^{(1)}(z; q = 1, 2, 3, 4)$ when processing the Gaussian input pulse with two different pulse widths. (a) $w = 5$ and (b, c) $w = 20$ with different timescales.

differentiator has higher processing accuracy of the fifth-order pulse than the second-order and third-order differentiators. Thus, for $x[m] = m^q$, the qth-order second-derivative differentiator has large processing error over the initial time interval $0 \leq m < q$ but has zero processing error over the longer time interval $m \geq q$.

Figures 3.20a through d show the differentiator pulse responses when processing the Gaussian input pulse with various pulse widths w. The pulse responses closely resemble the true derivatives and the processing accuracy increases with increasing value of w. The second-order second-derivative differentiator analyzed here was experimentally demonstrated in Ref. [63] where a square-type input pulse was processed.

Figures 3.21a and b show the differentiator error responses when processing the Gaussian pulse with pulse widths $w = 5$ and $w = 20$, respectively. For $w = 5$, Figure 3.21a shows that the third-order and fourth-order differentiators have lower processing accuracy than the second-order differentiator for time $m > 25$. However, for a larger pulse width, $w = 20$, Figure 3.21b shows that the third-order and fourth-order differentiators have higher processing accuracy than the second-order differentiator, but at the expense of having lower processing accuracy over the initial time interval (see the enlarged curves in Figure 3.21c).

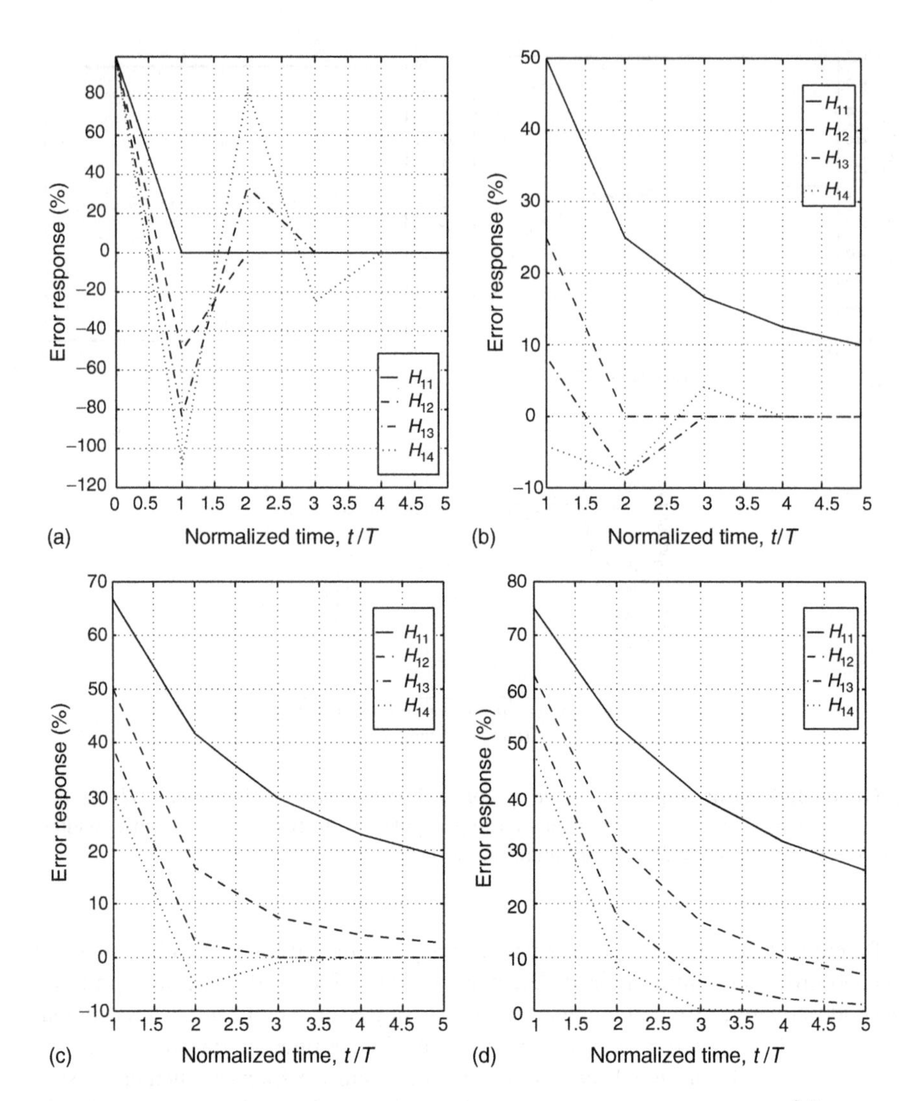

FIGURE 3.18 Error responses of the qth-order first-derivative differentiators $\hat{H}_q^{(1)}(z;\, q = 1, 2, 3, 4)$ when processing the qth-order polynomial input pulse. (a) First-order polynomial input pulse, $x[m] = m$; (b) second-order polynomial input pulse, $x[m] = m^3$; (c) third-order polynomial input pulse, $x[m] = m^3$; and (d) fourth-order polynomial input pulse, $x[m] = m^4$.

3.3.4.3 Third-Derivative Differentiators

This section analyzes the performances of the qth-order third-derivative differentiators, $\hat{H}_q^{(3)}(z;\, q = 3, 4)$, whose magnitude responses were shown in Figure 3.15c.

Figures 3.22a and b show the differentiator error responses when processing the qth-order polynomial input pulse. Figure 3.22a shows that the third-order differentiator has higher processing accuracy of the third-order pulse than the fourth-order

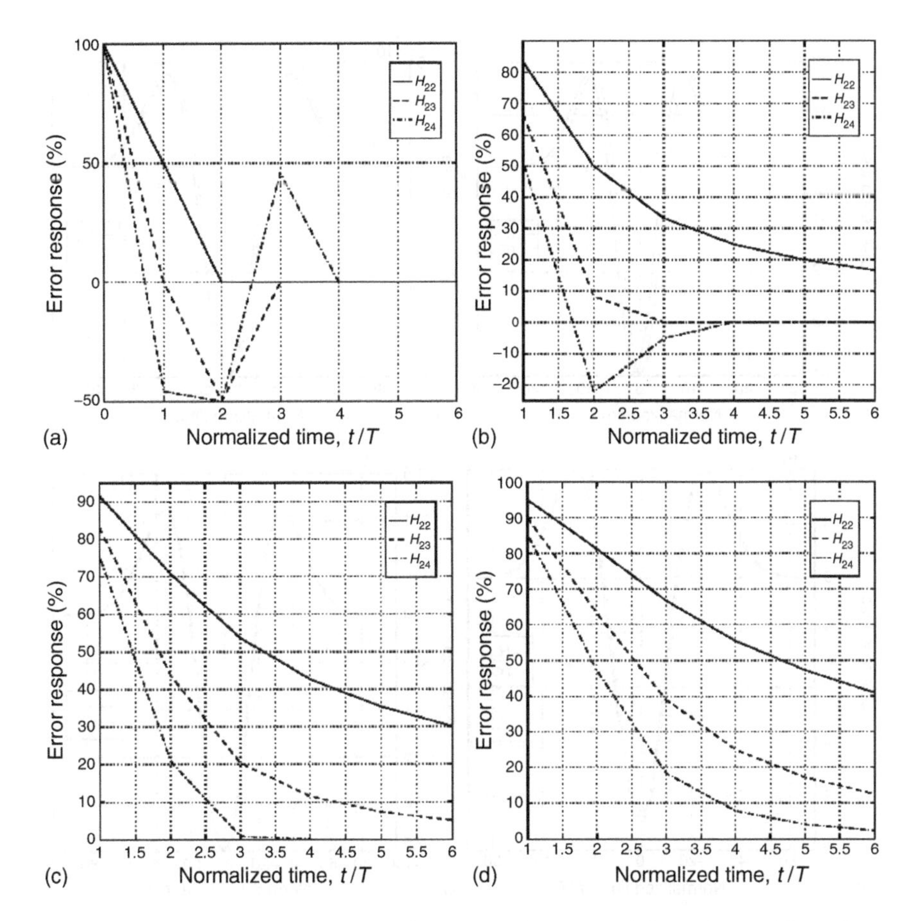

FIGURE 3.19 Error responses of the qth-order second-derivative differentiators $\hat{H}_q^{(2)}(z;\ q = 2, 3, 4)$ when processing the qth-order polynomial input pulse. (a) Second-order polynomial input pulse, $x[m] = m^2$; (b) third-order polynomial input pulse, $x[m] = m^3$; (c) fourth-order polynomial input pulse, $x[m] = m^4$; and (d) fifth-order polynomial input pulse, $x[m] = m^5$.

differentiator. However, Figure 3.22b shows that the fourth-order differentiator has higher processing accuracy of the fourth-order pulse than the third-order differentiator. Thus, for $x[m] = m^q$, the qth-order third-derivative differentiator has large processing error over the initial time interval $0 \leq m < q$ but has zero processing error over the longer time interval $m \geq q$.

Figures 3.23a through d show the differentiator pulse responses when processing the Gaussian input pulse with various pulse widths w. The pulse responses closely resemble the true derivatives, and the processing accuracy increases with increasing value of w. Figures 3.24a and b show the differentiator error responses when processing the Gaussian input pulse with pulse widths $w = 5$ and $w = 20$, respectively. For $w = 5$, Figure 3.24a shows that the fourth-order differentiator has

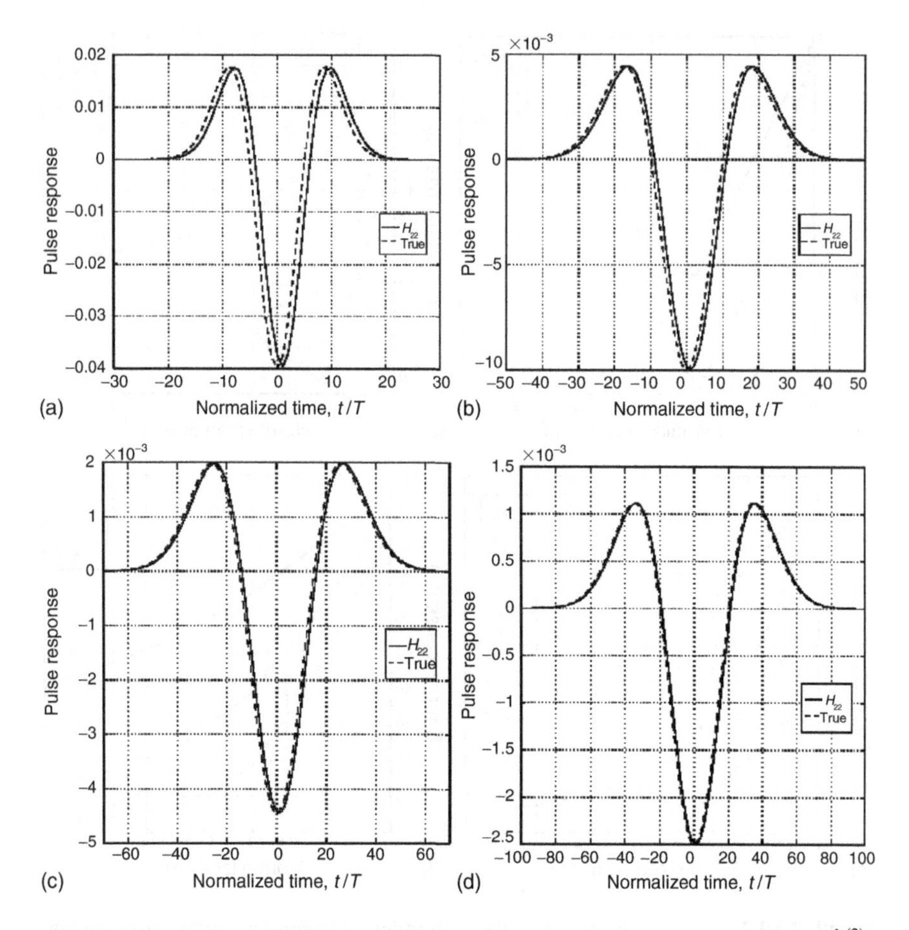

FIGURE 3.20 Pulse responses of the second-order second-derivative differentiator $\hat{H}_2^{(2)}(z)$ when processing the Gaussian input pulse with various pulse widths w. (a) $w = 5$, (b) $w = 10$, (c) $w = 15$, and (d) $w = 20$.

lower processing accuracy than the third-order differentiator for time $m > 35$. However, for a larger pulse width $w = 20$, Figure 3.24b shows that the fourth-order differentiator has higher processing accuracy than the third-order differentiator, but at the expense of having lower processing accuracy over the initial time interval (see the enlarged curves in Figure 3.24c).

3.3.4.4 Fourth-Derivative Differentiators

This section analyzes the performance of the fourth-order fourth-derivative differentiator $\hat{H}_4^{(4)}(z)$ whose magnitude response was shown in Figure 3.15d.

Figures 3.25a through d show the pulse and error responses of the differentiator when processing the fourth-order and fifth-order polynomial input pulses. For the fourth-order pulse, Figures 3.25a and b show that the fourth-order differentiator

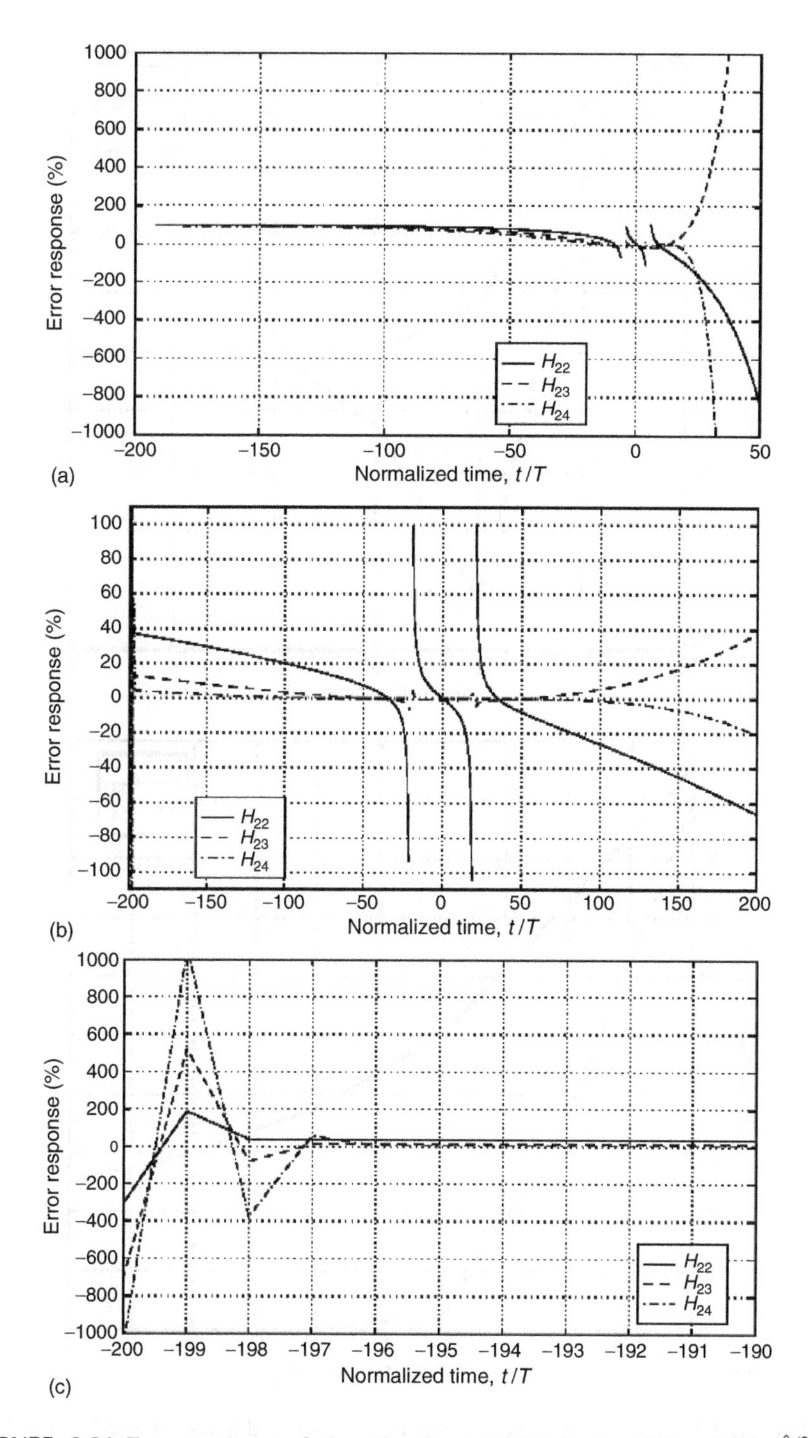

FIGURE 3.21 Error responses of the qth-order second-derivative differentiators $\hat{H}_q^{(2)}(z; q = 2, 3, 4)$ when processing the Gaussian input pulse with two different pulse widths. (a) $w = 5$ and (b, c) $w = 20$ with different timescales.

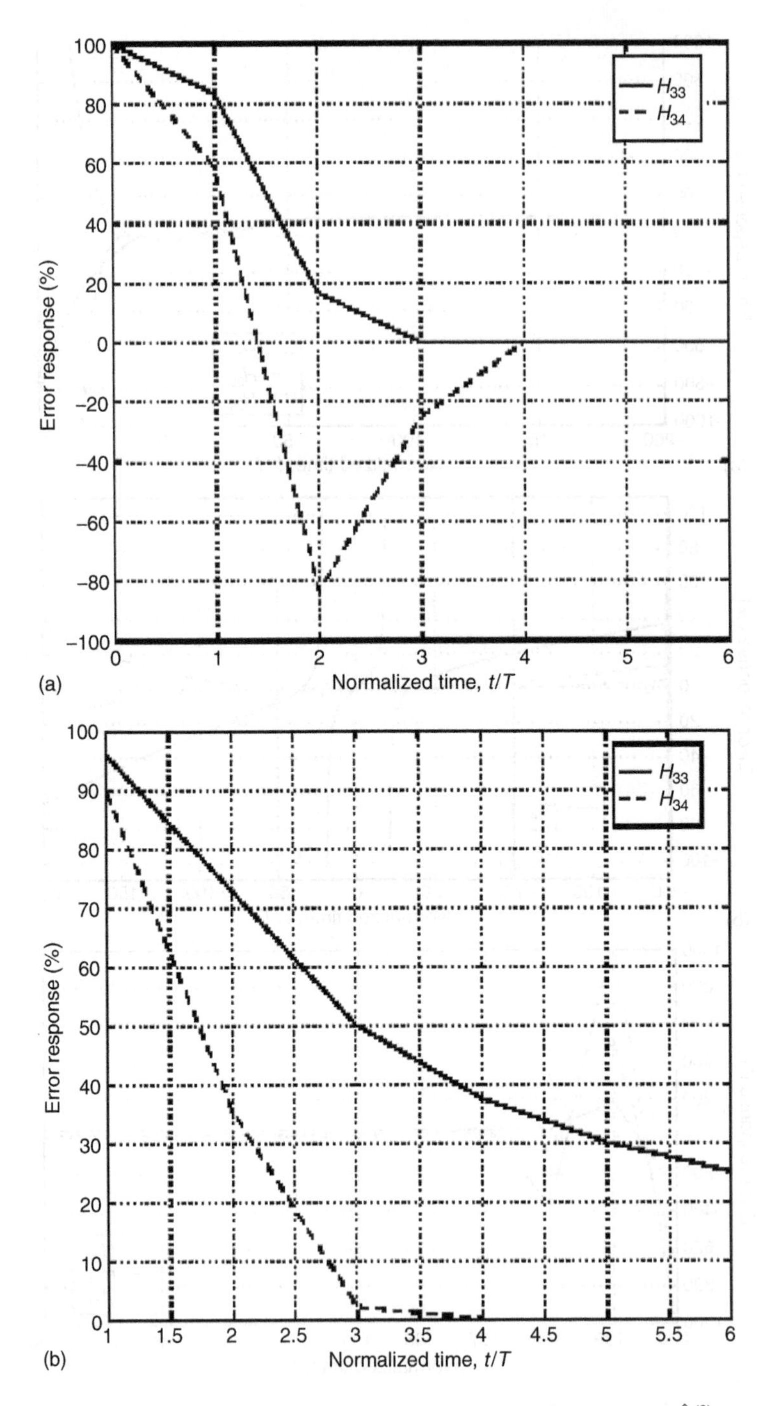

FIGURE 3.22 Error responses of the qth-order third-derivative differentiators $\hat{H}_q^{(3)}(z; q = 3, 4)$ when processing the qth-order pulse. (a) Third-order polynomial input pulse, $x[m] = m^3$, and (b) fourth-order polynomial input pulse, $x[m] = m^4$.

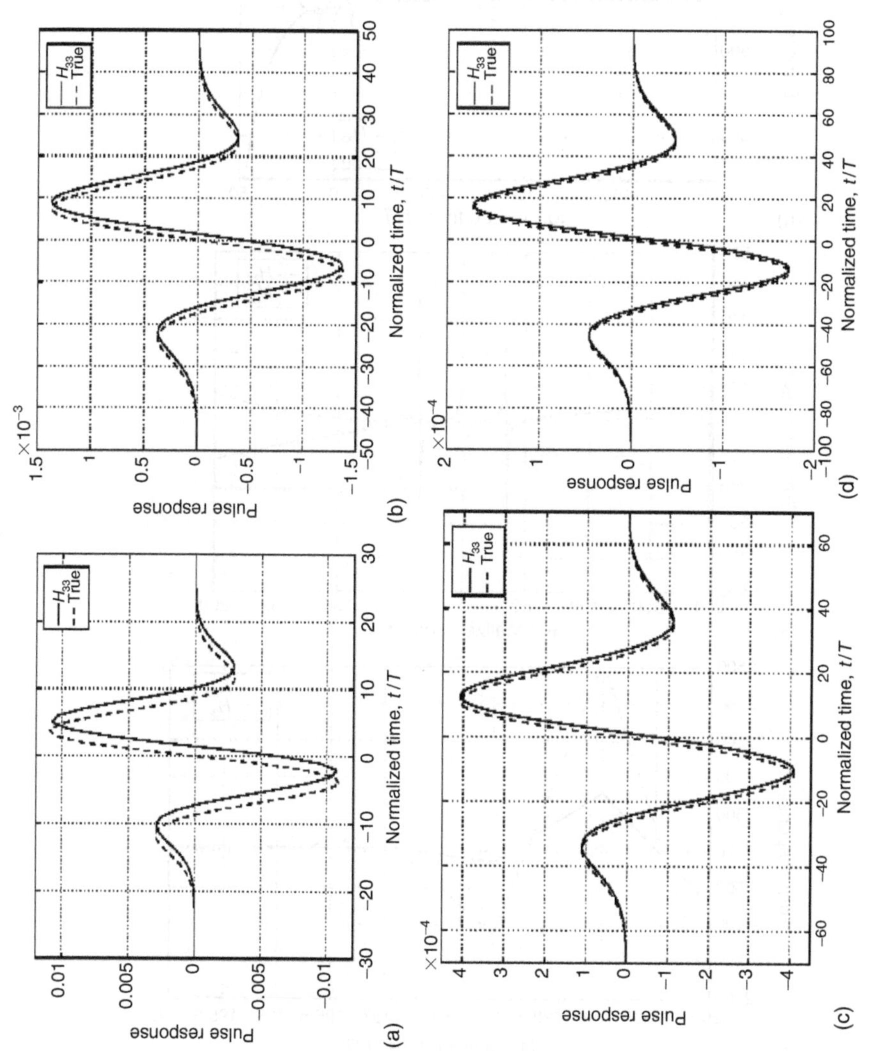

FIGURE 3.23 Pulse responses of the third-order third-derivative differentiator $\hat{H}_3^{(3)}(z)$ when processing the Gaussian input pulse with various pulse widths w. (a) $w = 5$, (b) $w = 10$, (c) $w = 15$, and (d) $w = 20$.

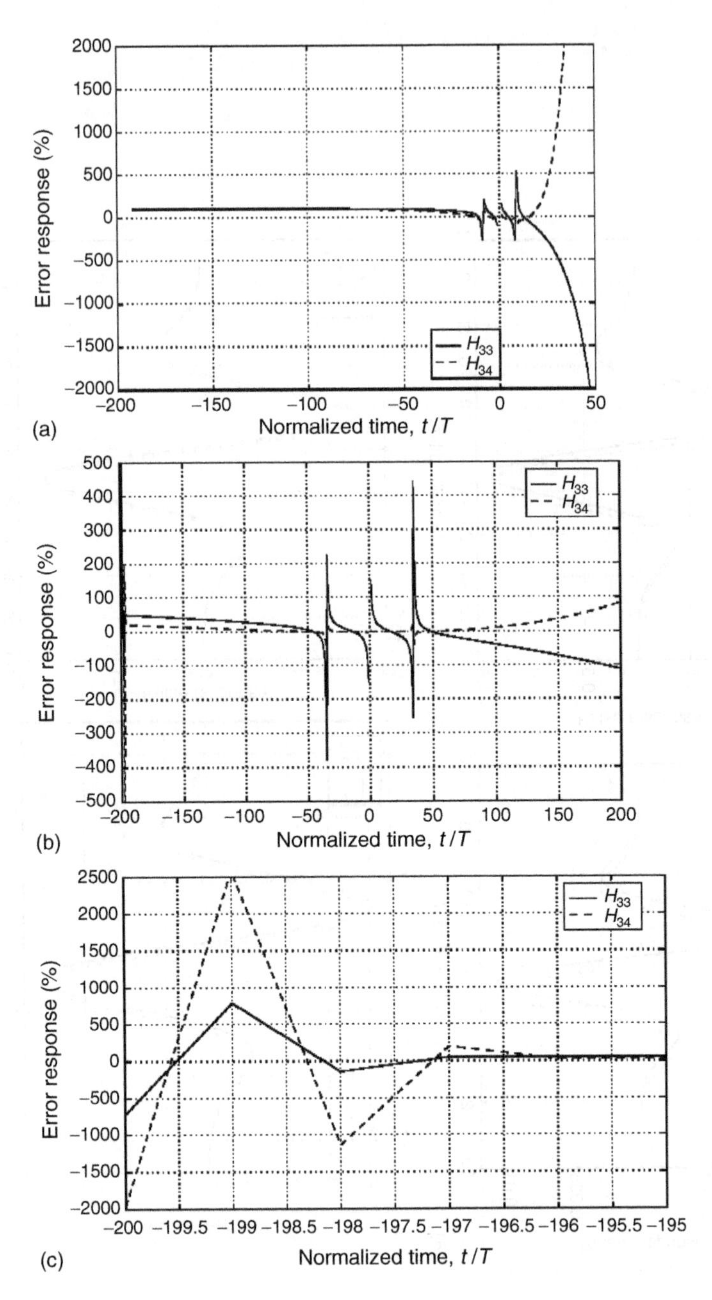

FIGURE 3.24 Error responses of the qth-order third-derivative differentiators $\hat{H}_q^{(3)}(z;\ q = 3, 4)$ when processing the Gaussian input pulse with two different pulse widths. (a) $w = 5$ and (b, c) $w = 20$ with different timescales.

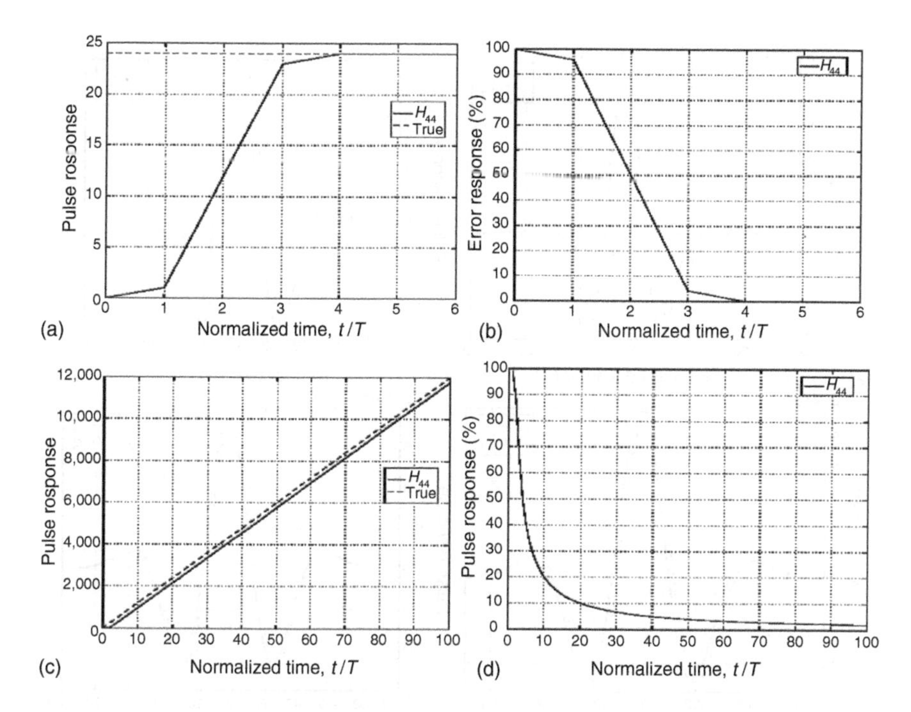

FIGURE 3.25 Pulse and error responses of the fourth-order fourth-derivative differentiator $\hat{H}_4^{(4)}(z)$ when processing the fourth-order and fifth-order polynomial input pulses. (a, b) Fourth-order polynomial input pulse, $x[m] = m^4$ and (c, d) fifth-order polynomial input pulse, $x[m] = m^5$.

has large processing error over the initial time interval $0 \leq m < 4$ but has zero processing error over the longer time interval $m \geq 4$. For the fifth-order pulse, Figures 3.25c and d show that the processing error never converges to zero. This is because the order of the differentiator is lower than the order of the input pulse. However, a fifth-order differentiator, which is not considered here, is expected to improve the processing accuracy of the fifth-order pulse.

Figures 3.26a through d show the differentiator pulse responses when processing the Gaussian input pulse with various pulse widths w. The pulse responses closely resemble the true derivatives and the processing accuracy increases with the increasing value of w.

3.3.5 Remarks

1. A theory of the qth-order pth-derivative FIR digital differentiator has been proposed, based on which the qth-order pth-derivative FIR optical differentiator has been synthesized using integrated-optic components.
2. For a qth-order polynomial input pulse (i.e., $x[m] = m^q$, $q \geq 1$), the qth-order pth-derivative FIR optical differentiator, $\hat{H}_q^{(p)}(z)$, which has higher processing accuracy than other differentiator orders, has large processing

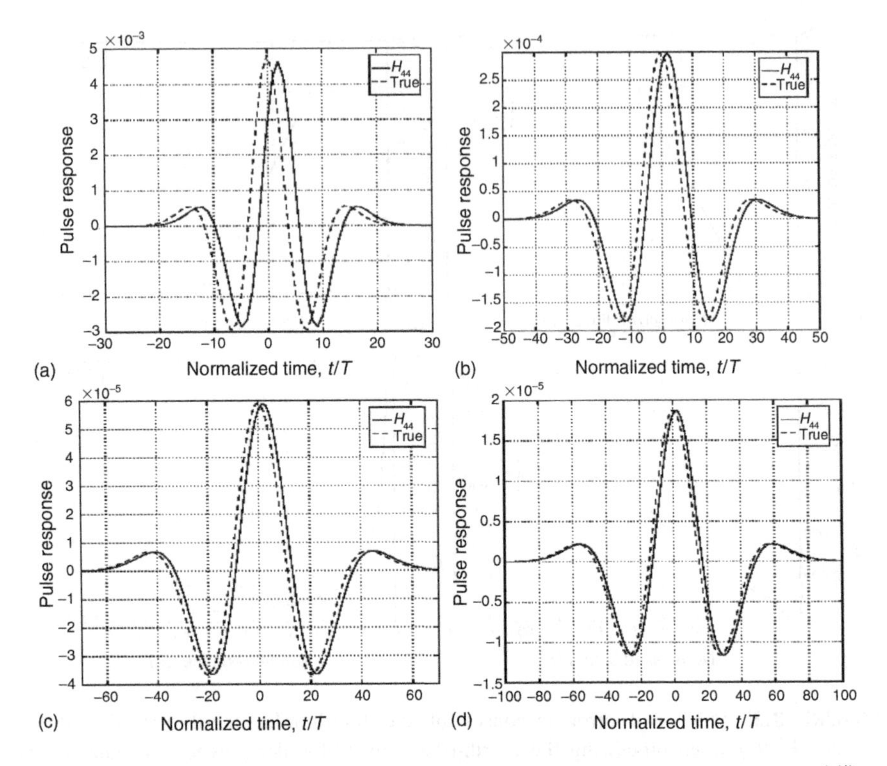

FIGURE 3.26 Pulse responses of the fourth-order fourth-derivative differentiator $\hat{H}_4^{(4)}(z)$ when processing the Gaussian input pulse with various pulse widths w. (a) $w = 5$, (b) $w = 10$, (c) $w = 15$, and (d) $w = 20$.

error over the initial time interval $0 \leq m < q$ but has zero processing error over the longer time interval $m \geq q$.

3. For a Gaussian input pulse (i.e., $x[m] = \exp[-m^2/(2w^2)]$), the qth-order pth-derivative differentiator, $\hat{H}_q^{(p)}(z;\, q = p)$, whose processing accuracy increases with increasing pulse width w, is the optimum filter when compared with the higher-order pth-derivative differentiator, $\hat{H}_q^{(p)}(z;\, q > p)$, because of its structural simplicity and impressive performance. The results for the Gaussian pulse are similar to those for the exponential pulse (i.e., $x[m] = \exp(-m/w)$, $m \geq 0$).
4. In general, regardless of the type of input pulse, the optical differentiators have large processing error over the initial time interval corresponding to q sampling periods but their processing errors reduce significantly over a longer time interval.
5. Optical differentiation is still a new research area with many potential applications. One example is to apply the first-order first-derivative FIR optical differentiator described here to the design of an optical dark-soliton detector, which is outlined in a later chapter.

Appendix A: Generalized Theory of the Newton–Cotes Digital Integrators

In this appendix, a classical numerical integration scheme together with the digital signal processing technique is employed to develop a generalized theory of the Newton–Cotes digital integrators, which is believed to be described for the first time. This theory has been used in the synthesis of the programmable incoherent Newton–Cotes optical integrator (INCOI) as described in Chapter 5. A definition of numerical integration is first given and is used as a basis in the derivation process. The Newton's interpolating polynomial is then described, based on which a general form of the Newton–Cotes closed integration formulas is derived. Finally, a generalized theory of the Newton–Cotes digital integrators is obtained.

A.1 DEFINITION OF NUMERICAL INTEGRATION

It is assumed that a continuous-time signal $x(t)$ is given and that its integral

$$y(t) = \int_{0}^{t} x(t)\,\mathrm{d}t \qquad \text{(A.3.1)}$$

is to be determined from a sequence of samples of the continuous-time signal $x(t)$ at the discrete time $t = t_n$ where

$$t_n = nT, \quad n = 0, 1, 2, \ldots \qquad \text{(A.3.2)}$$

with $T > 0$ being the period between successive samples. Intuitively, the integral $y(t)$ cannot be obtained for all t, but only for $t = t_n$. Thus, Equation A.3.1 can be written as

$$y_n = y(t_n) = \int_{0}^{t_n} x(t)\,\mathrm{d}t \qquad \text{(A.3.3)}$$

To simplify the numerical integration algorithm, the integration interval $[0,t_n]$ is divided into a number of equal segments, each with a step size of T. The underlying principle of the numerical integration algorithm is shown in Figure A.3.1.

From Figure A.3.1, the integral in Equation A.3.3 can be divided into two integrals as

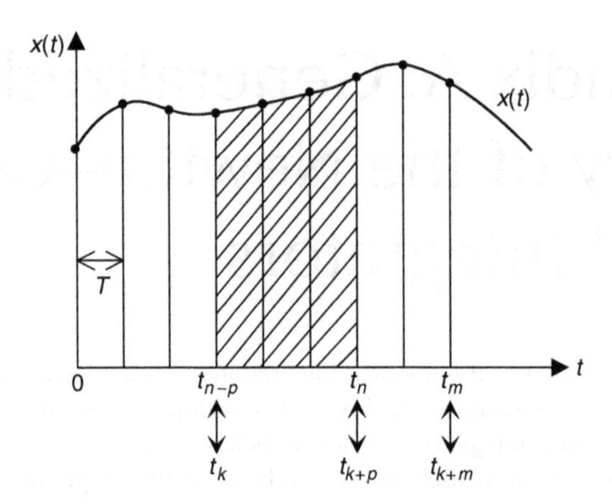

FIGURE A.3.1 Graphical illustration of the numerical integration technique.

$$y_n = \int_0^{t_{n-p}} x(t)\,\mathrm{d}t + \int_{t_{n-p}}^{t_n} x(t)\,\mathrm{d}t = y_{n-p} + i_p \tag{A.3.4}$$

where the partial integral i_p, which represents the area of the hatched region of Figure A.3.1, is given by [36]

$$i_p = \int_{t_{n-p}}^{t_n} x(t)\,\mathrm{d}t \tag{A.3.5}$$

The z-transform of Equation A.3.4 is given by [36]

$$Y(z) = z^{-p}Y(z) + I_p(z) \tag{A.3.6a}$$

or

$$Y(z) = \left[\frac{1}{1 - z^{-p}}\right] I_p(z) \tag{A.3.6b}$$

where $Y(z) = Z\{y_n\}$ and $I_p(z) = Z\{i_p\}$ with $Z\{.\}$ being the z-transform of $\{.\}$. In Equations A.3.6a and A.3.6b, the z-transform parameter is defined as $z = \exp(j\omega T)$ where $j = \sqrt{-1}$, ω is the angular frequency, and T is the sampling period of the integrator [67]. The z-transform of the partial integral $I_p(z)$ is to be determined in Section A.4.

A.2 NEWTON'S INTERPOLATING POLYNOMIAL

For analytical simplicity, the discrete-time variables in Figure A.3.1 are redefined as

$$t_k = t_{n-p}, \quad k = n - p \tag{A.3.7a}$$

$$t_{k+p} = t_n \tag{A.3.7b}$$

$$t_{k+m} = t_m \tag{A.3.7c}$$

Using Equations A.3.7a and A.3.7b, Equation A.3.5 becomes

$$i_p = \int_{t_k}^{t_{k+p}} x(t)\,\mathrm{d}t \tag{A.3.8}$$

For the time interval $[t_k, t_{k+m}]$ as shown in Figure A.3.1, the curve $x(t)$ can be approximated by the mth-order Newton's interpolating polynomial, which passes through $m + 1$ data points, as [40]

$$x(t) = x(t_k) + \frac{\Delta x(t_k)}{T}(t - t_k) + \frac{\Delta^2 x(t_k)}{2!T^2}(t - t_k)(t - t_{k+1}) + \cdots + \frac{\Delta^m x(t_k)}{m!T^m}(t - t_k)(t - t_{k+1})\cdots(t - t_{k+m-1}) \tag{A.3.9}$$

where the ith discrete-time variable is given by

$$t_{k+i} = t_k + iT, \quad i = 0, 1, \ldots, m - 1 \tag{A.3.10}$$

and the qth forward difference equation is given by [38]

$$\Delta^q x(t_k) = \sum_{i=0}^{q} (-1)^i \binom{q}{i} x(t_{k+q-i}), \quad q = 0, 1, \ldots, m \tag{A.3.11}$$

The binomial coefficient in Equation A.3.11 is defined as [38]

$$\binom{q}{i} = \frac{q(q-1)\cdots(q-(i-1))}{i!} = \frac{q!}{(q-i)!i!} \tag{A.3.12a}$$

with

$$\binom{q}{0} = \binom{q}{q} = 1 \tag{A.3.12b}$$

Equation A.3.11 can be written in a recursive form as [38]

$$\Delta^0 x(t_k) = x(t_k) \tag{A.3.13a}$$

$$\Delta^1 x(t_k) = \Delta x(t_k) = x(t_{k+1}) - x(t_k) \tag{A.3.13b}$$

$$\Delta^q x(t_k) = \Delta^{q-1} x(t_{k+1}) - \Delta^{q-1} x(t_k), \quad q = 2, \ldots, m \tag{A.3.13c}$$

Equation A.3.9 can be simply expressed as

$$x(t) = x(t_k) + \sum_{q=1}^{m} \frac{\Delta^q x(t_k)}{q!T^q} \left\{ \prod_{i=0}^{q-1} (t - t_{k+i}) \right\} \quad \text{(A.3.14)}$$

which can be further simplified by defining a new quantity

$$\eta = \frac{t - t_k}{T} \quad \text{(A.3.15)}$$

Substituting Equation A.3.15 into Equation A.3.10, the following equation is obtained:

$$t - t_{k+i} = T(\eta - i), \quad i = 0, 1, \ldots, m - 1 \quad \text{(A.3.16)}$$

Substituting Equation A.3.16 into Equation A.3.14 results in

$$\begin{aligned} x(t) &= x(t_k) + \sum_{q=1}^{m} \frac{\Delta^q x(t_k)}{q!T^q} \left\{ \prod_{i=0}^{q-1} [T(\eta - i)] \right\} \\ &= x(t_k) + \sum_{q=1}^{m} \frac{\Delta^q x(t_k)}{q!} \left\{ \prod_{i=0}^{q-1} (\eta - i) \right\} \\ &= x(t_k) + \Delta x(t_k)\eta + \Delta^2 x(t_k)\frac{\eta(\eta - 1)}{2!} + \cdots + \Delta^m x(t_k)\frac{\eta(\eta - 1)\cdots(\eta - (m - 1))}{m!} \end{aligned} \quad \text{(A.3.17)}$$

which can further be simplified to

$$\begin{aligned} x(t) &= x(t_k) + \sum_{q=1}^{m} \binom{\eta}{q} \Delta^q x(t_k) \\ &= \sum_{q=0}^{m} \binom{\eta}{q} \Delta^q x(t_k). \end{aligned} \quad \text{(A.3.18)}$$

Thus, for the time interval $[t_k, t_{k+m}]$, the mth-order Newton's interpolating polynomial of the curve $x(t)$ can be simply described by Equation A.3.18.

A.3 GENERAL FORM OF THE NEWTON–COTES CLOSED INTEGRATION FORMULAS

Substituting Equation A.3.18 into Equation A.3.8 results in

$$i_p = \int_{t_k}^{t_{k+p}} \left[\sum_{q=0}^{m} \binom{\eta}{q} \Delta^q x(t_k) \right] \mathrm{d}t \quad \text{(A.3.19)}$$

From Equation A.3.15, $dt = Td\eta$ and the limits of integration are changed from $t = t_k$ to $\eta = 0$ and from $t = t_{k+p}$ to $\eta = p$. Substituting these parameters into Equation A.3.19 results in

$$i_p = \int_0^p \left[\sum_{q=0}^{m} \binom{\eta}{q} \Delta^q x(t_k) \right] \cdot T\,d\eta = T \sum_{q=0}^{m} \left[\int_0^p \binom{\eta}{q} d\eta \cdot \Delta^q x(t_k) \right] \quad \text{(A.3.20)}$$

which can be rearranged to give

$$i_p = T \sum_{q=0}^{m} C_q(p) \Delta^q x(t_k) \quad \text{(A.3.21)}$$

where the qth coefficient $C_q(p)$ is given by

$$C_q(p) = \int_0^p \binom{\eta}{q} d\eta, \quad q = 0, 1, \ldots, m \quad \text{(A.3.22)}$$

The qth forward difference equation, as described by Equation A.3.11, can be further simplified by substituting $u = q - i$ or $i = q - u$ into Equation A.3.11 to give

$$\begin{aligned} \Delta^q x(t_k) &= (-1)^q \sum_{u=0}^{q} (-1)^{-u} \binom{q}{q-u} x(t_{k+u}) \\ &= (-1)^q \sum_{u=0}^{q} (-1)^{u} \binom{q}{u} x(t_{k+u}) \end{aligned} \quad \text{(A.3.23)}$$

where

$$(-1)^{-u} = (-1)^u, \quad u \in \text{integer} \quad \text{(A.3.24a)}$$

$$\binom{q}{q-u} = \binom{q}{u} \quad \text{(A.3.24b)}$$

have been used. Thus, the general form of the Newton–Cotes closed integration formulas can be simply described by three closed-form formulas, as given in Equations A.3.21 through A.3.23.

A.4 GENERALIZED THEORY OF THE NEWTON–COTES DIGITAL INTEGRATORS

Taking the z-transform of Equation A.3.21 results in

$$I_p(z) = T \sum_{q=0}^{m} C_q(p) \Delta^q X(z) \quad \text{(A.3.25)}$$

where $\Delta^q X(z) = Z\{\Delta^q x(t_k)\}$ is the z-transform of Equation A.3.23, which is given by

$$\Delta^q X(z) = (-1)^q \sum_{u=0}^{q} (-1)^u \binom{q}{u} \cdot X(z) z^{-u} \quad \text{(A.3.26)}$$

where $X(z)z^{-u} = Z\{x(t_{k+u})\}$. Equation A.3.26 can be rearranged to give

$$\begin{aligned} \frac{\Delta^q X(z)}{[X(z)(-1)^q]} &= \sum_{u=0}^{q} (-1)^u \binom{q}{u} z^{-u} \\ &= 1 - qz^{-1} + \frac{q(q-1)}{2!} z^{-2} + \cdots + (-1)^r \frac{q!}{(q-r)!r!} z^{-r} \\ &\quad + \cdots + (-1)^q z^{-q} \end{aligned} \quad \text{(A.3.27)}$$

Note that Equation A.3.27 can be recognized as [68]

$$\frac{\Delta^q X(z)}{[X(z)(-1)^q]} = (1 - z^{-1})^q \quad \text{(A.3.28)}$$

or

$$\Delta^q X(z) = \Delta^q D(z) \cdot X(z) \quad \text{(A.3.29)}$$

where

$$\Delta^q D(z) = (-1)^q (1 - z^{-1})^q, \quad q = 0, 1, \ldots, m \quad \text{(A.3.30)}$$

Substituting Equation A.3.29 into Equation A.3.25 gives

$$I_p(z) = X(z) T \sum_{q=0}^{m} C_q(p) \Delta^q D(z) \quad \text{(A.3.31)}$$

Substituting Equation A.3.31 into Equation A.3.6b, the pth-order transfer function of the Newton–Cotes digital integrators can be generally described by

$$\begin{aligned} H_{mp}(z) &= \frac{Y(z)}{X(z)} = \frac{T}{1 - z^{-p}} \sum_{q=0}^{m} C_q(p) \Delta^q D(z) \\ &= \frac{T}{1 - z^{-p}} \left[C_0(p) + C_1(p) \Delta D(z) + C_2(p) \Delta^2 D(z) + \cdots + C_m(p) \Delta^m D(z) \right] \end{aligned} \quad \text{(A.3.32)}$$

where the qth coefficient is given, from Equation A.3.22, as

$$C_q(p) = \int_0^p \binom{\eta}{q} \mathrm{d}\eta \quad \text{(A.3.33)}$$

and the qth difference equation is given, from Equation A.3.30, as

$$\Delta^q D(z) = (-1)^q(1 - z^{-1})^q, \quad \text{for } q = 0, 1, \ldots, m \text{ and } 1 \leq p \leq m \qquad \text{(A.3.34)}$$

In summary, a generalized theory of the Newton–Cotes digital integrators has been derived, and Equations A.3.32 through A.3.34 correspond to Equations 3.37 and 3.38 in Section 3.2.

REFERENCES

1. H.J. Caufield, W.T. Rhodes, M.J. Foster, and S. Horvitz, Optical implementation of systolic array processing, *Opt. Commun.*, 40, 86–90, 1981.
2. D. Casasent, Acoustooptic linear algebra processors: Architectures, algorithms, and applications, *Proc. IEEE*, 72, 831–849, 1984.
3. W.T. Rhodes and P.S. Guilfoyle, Acoustooptic algebraic processing architectures, *Proc. IEEE*, 72, 820–830, 1984.
4. R.A. Athale and J.N. Lee, Optical processing using outer-product concepts, *Proc. IEEE*, 72, 931–941, 1984.
5. B. Moslehi, J.W. Goodman, M. Tur, and H.J. Shaw, Fiber-optic lattice signal processing, *Proc. IEEE*, 72, 909–930, 1984.
6. K.P. Jackson, S.A. Newton, B. Moslehi, M. Tur, C.C. Cutler, J.W. Goodman, and H.J. Shaw, Optical fiber delay-line signal processing, *IEEE Trans. Microwave Theory Tech.*, MTT-33, 193–210, 1985.
7. A. Goutzoulis, E. Malarkey, D.K. Davies, J. Bradley, and P.R. Beaudet, Optical processing with residue LED/LD lookup tables, *Appl. Opt.*, 25, 3097–3112, 1986.
8. B. Drake, R. Bocker, M. Lasher, R. Patterson, and W. Miceli, Photonic computing using the modified signed-digit number representation, *Opt. Eng.*, 25, 38–43, 1986.
9. R.A. Athale, Highly redundant number representation for medium accuracy optical computing, *Appl. Opt.*, 25, 3122–3127, 1986.
10. K.H. Brenner, Programmable optical processor based on symbolic substitution, *Appl. Opt.*, 27, 1687–1691, 1988.
11. H.J. Whitehouse and J.M. Speiser, Linear signal processing architectures, in *Aspects of Signal Processing. Part 2*, G. Tacconi, Ed., Boston, MA: NATO Advanced Study Institute, 1976, pp. 669–702.
12. E.E. Swartzlander, The quasi-serial multiplier, *IEEE Trans. Comp.*, C-22, 317–321, 1973.
13. D. Psaltis, D. Casasent, D. Neft, and M. Carlotto, Accurate numerical computation by optical convolution, in *1980 International Optical Computing Conference II*, W.T. Rhodes, Ed., *Proc. Soc. Photo-Opt. Instrum. Eng.*, 232, 151–156, 1980.
14. R.A. Athale and W.C. Collins, Optical matrix–matrix multiplier based on outer product decomposition, *Appl. Opt.*, 21, 2089–2090, 1982.
15. R.P. Bocker, Optical digital RUBIC (rapid unbiased bipolar incoherent calculator) cube processor, *Opt. Eng.*, 23, 26–32, 1984.
16. P.S. Guilfoyle, Systolic acousto-optic binary convolver, *Opt. Eng.*, 23, 20–25, 1984.
17. S. Cartwright, New optical matrix–vector multiplier, *Appl. Opt.*, 23, 1683–1684, 1984.
18. A.P. Goutzoulis, Systolic time-integrating acoustooptic binary processor, *Appl. Opt.*, 23, 4095–4099, 1984.
19. D. Casasent and B.K. Taylor, Banded-matrix high-performance algorithm and architecture, *Appl. Opt.*, 24, 1476–1480, 1985.

20. D. Psaltis and R.A. Athale, High accuracy computation with linear analog optical systems: A critical study, *Appl. Opt.*, 25, 3071–3077, 1986.
21. D. Casasent and S. Riedl, Direct finite element solution on an optical laboratory matrix–vector processor, *Opt. Commun.*, 65, 329–333, 1988.
22. E.J. Baranoski and D.P. Casasent, High accuracy optical processors: A new performance comparison, *Appl. Opt.*, 28, 5351–5357, 1989.
23. F.T.S. Yu and M.F. Cao, Digital optical matrix multiplication based on a systolic outer-product method, *Opt. Eng.*, 26, 1229–1233, 1987.
24. G. Eichmann, Y. Li, P.P. Ho, and R.R. Alfano, Digital optical isochronous array processing, *Appl. Opt.*, 26, 2726–2733, 1987.
25. Y. Li, G. Eichmann, and R.R. Alfano, Fast parallel optical digital multiplication, *Opt. Commun.*, 64, 99–104, 1987.
26. Y. Li, B. Ha, and G. Eichmann, Fast digital optical multiplication using an array of binary symmetric logic counters, *Appl. Opt.*, 30, 531–539, 1991.
27. N.Q. Ngo and L.N. Binh, Fibre-optic array processors for algebra computations, *Proc. IREE, 18th Australian Conf. Opt. Fibre Technol.*, Wollongong, pp. 356–359, 1993.
28. N.Q. Ngo and L.N. Binh, Fiber-optic array algebraic processing architectures, *Appl. Opt.*, 34, 803–815, 1995.
29. R.P. Bocker, S.R. Clayton, and K. Bromley, Electrooptical matrix multiplication using the twos complement arithmetic for improved accuracy, *Appl. Opt.*, 22, 2019–2021, 1983.
30. S. Cartwright, Optical matrix multiplication, in *Optical Computing: Digital and Symbolic*, R. Arrathoon, Ed., New York: Marcel Dekker, 1989, pp. 185–219.
31. P.R. Prucnal, M.A. Santoro, and S.K. Sehgal, Ultrafast all-optical synchronous multiple access fiber networks, *IEEE J. Selected Areas Commun.*, SAC-4, 1484–1493, 1986.
32. C.A. Liechti, High speed transistors: Directions for the 1990s, *Microwave J.*, 30, 165–177, 1989.
33. R.A. Becker, C.E. Woodward, F.J. Leonberger, and R.C. Williamson, Wide-band electrooptic guided-wave analog-to-digital converters, *Proc. IEEE*, 72, 802–819, 1984.
34. G.F. Franklin, J.D. Powell, and M.L. Workman, *Digital Control of Dynamic Systems*, 2nd ed., Reading, MA: Addison-Wesley, 1990.
35. W.J. Tompkins and J.G. Webster (Eds.), *Design of Microcomputers-Based Medical Instrumentation*, Englewood Cliffs, NJ: Prentice-Hall, 1981.
36. R. Vich, *Z Transform Theory and Applications*, Norwell, MA: Kluwer Academic Publishers, 1987.
37. R. Pintelon and J. Schoukens, Real-time integration and differentiation of analog signals by means of digital filtering, *IEEE Trans. Instrum. Meas.*, 39, 923–927, 1990.
38. M. Abramowitz and I.A. Segun, *Handbook of Mathematical Function*, New York: Dover Publications, 1964.
39. S.C. Chapra and R.P. Canale, *Numerical Methods for Engineers*, 2nd ed., Singapore: McGraw-Hill, 1989.
40. N.Q. Ngo and L.N. Binh, Programmable incoherent Newton–Cotes optical integrator, *Opt. Commun.*, 119, 390–402, 1995.
41. A. Ehrhardt, M. Eiselt, G. Groákoptf, L. Küller, R. Ludwig, W. Pieper, R. Schnabel, and G. Weber, Semiconductor laser amplifier as optical switching data, *J. Lightwave Technol.* 11, 1287–1295, 1993.
42. S. Shimada and H. Ishio (Eds.), *Optical Amplifiers and Their Applications*, London: John Wiley & Sons, 1994, chapters 1–5.
43. R. Ludwig, K. Magari, and Y. Suzuki, Properties of 1.55 μm high gain polarisation insensitive MQW semiconductor laser amplifiers, *OSA Tech. Dig. Ser.*, 17, 148–151, 1992.

44. K. Okamoto, K. Takiguchi, and Y. Ohmori, 16-channel optical add/drop multiplexer using silica-based array-waveguide gratings, *Electron. Lett.*, 31, 723–724, 1995.
45. D.A.T.A. Business Publishing (Ed.), Digest: Microprocessors (handbook, edn. 22), 1993.
46. A.P. Goutzoulis, D.K. Davies, and J.M. Zomp, Hybrid electronic fiber optic wavelength-multiplexed system for true time-delay steering of phased array antennas, *Opt. Eng.*, 31, 2312–2322, 1992.
47. S. Usui and I. Amidror, Digital low-pass differentiation for biological signal processing, *IEEE Trans. Biomed. Eng.*, BME-29, 686–693, 1982.
48. M.I. Skolnik, *Introduction to Radar Systems*, 2nd ed., New York: McGraw-Hill, 1980.
49. B.V.K. Vijaya Kumar and C.A. Rahenkamp, Calculation of geometric moments using Fourier plane intensities, *Appl. Opt.*, 25, 997–1007, 1986.
50. L.R. Rabiner and R.W. Schafer, On the behavior of minimax relative error FIR digital differentiators, *Bell Syst. Tech. J.*, 53, 333–360, 1974.
51. C.A. Rahenkamp and B.V.K. Vijaya Kumar, Modifications to the McClellan, Parks, and Rabiner computer program for designing higher order differentiating FIR filters, *IEEE Trans. Acoust. Speech Signal Process.*, ASSP-34, 1671–1674, 1986.
52. G.P. Agrawal, *Fiber-Optic Communication Systems*, London: John Wiley & Sons, 1992.
53. A. Antoniou, Design of digital differentiators satisfying prescribed specifications, *Proc. IEE*, 127, pt. E, 24–30, 1980.
54. A. Antoniou and C. Charalambous, Improved design method for Kaiser differentiators and comparison with equiripple method, *Proc. IEE*, 128, pt. E, 190–196, 1981.
55. B. Kumar and S.C. Dutta Roy, Design of digital differentiators for low frequencies, *Proc. IEEE*, 76, 287–289, 1988.
56. R.R.R. Reddy, B. Kumar, and S.C. Dutta Roy, Design of efficient second and higher order FIR digital differentiators for low frequencies, *Signal Process.*, 20, 219–225, 1990.
57. B. Kumar and S.C. Dutta Roy, Maximally linear FIR digital differentiators for midband frequencies, *Intern. J. Circuit Theory Appl.*, 17, 21–27, 1989.
58. B. Kumar and S.C. Dutta Roy, Design of efficient FIR digital differentiators and Hilbert transformers for midband frequency ranges, *Intern. J. Circuit Theory Appl.*, 17, 483–488, 1989.
59. B. Kumar and S.C. Dutta Roy, Maximally linear FIR digital differentiators for high frequencies, *IEEE Trans. Circuits Syst.*, CAS-36, 890–893, 1989.
60. S.C. Pei and J.J. Shyu, Design of FIR Hilbert transformers and differentiators by eigenfilter, *IEEE Trans. Circuits Syst.*, 35, 1457–1461, 1988.
61. S.C. Pei and J.J. Shyu, Eigenfilter design of higher order digital differentiators, *IEEE Trans. Acoust. Speech Signal Process.*, 37, 505–511, 1989.
62. S. Sunder and R.P. Ramachandran, Least-squares design of higher order nonrecursive differentiators, *IEEE Trans. Signal Process.*, 42, 956–961, 1994.
63. K. Sasayama, M. Okuno, and K. Habara, Coherent optical transversal filter using silica-based waveguides for high-speed signal processing, *J. Lightwave Technol.*, 9, 1225–1230, 1991.
64. S. Sunder, W.S. Lu, A. Antoniou, and Y. Su, Design of digital differentiators satisfying prescribed specifications using optimisation techniques, *Proc. IEE*, 138, pt. G, 315–320, 1991.
65. G.S. Pandian and F.E. Seraji, Optical pulse response of a fiber ring resonator, *Proc. IEE*, 42, part J, 235–239, 1991.
66. N.Q. Ngo and L.N. Binh, Theory of a FIR optical digital differentiator, *Fiber Integrated Opt.*, 14, 359–385, 1995.
67. A.V. Oppenheim and R.W. Schafer, *Discrete-Time Signal Processing*, Englewood Cliffs, NJ: Prentice-Hall, 1989.
68. H.B. Dwight, *Tables of Integrals and Other Mathematical Data*, 4th ed., Toronto, Canada: The Macmillan Company, 1961.

4 Ultrashort Pulse Photonic Generators

Ultrashort pulse generators are the lightwave sources that emit short pulses of high intensity. That means the energy has been concentrated to a very short time and periodically distributed along the pulses of the sequence. The peak power of these pulses would reach the nonlinear threshold of several materials that are used as the interaction and guided media for photonic signal processing.

This chapter thus presents a number of important and emerging light sources of very narrow pulse width sequence, which are considered to be application in soliton communications and soliton logics. Photonic generators for bright and dark solitons as well as bound solitons are given.

Mode-locked fiber lasers are also well known over the last decades and are emerging as the important sources for advanced lightwave communications technology. Practical implementations of these types of lasers are subsequently treated after the sections on solitons.

4.1 OPTICAL DARK-SOLITON GENERATOR AND DETECTORS

In this section, the trapezoidal optical integrator described in Chapter 3 is proposed as an optical dark-soliton generator and the first-order first-derivative optical differentiator outlined in Chapter 6 and a first-order Butterworth lowpass optical filter are proposed as optical dark-soliton detectors. A brief review of solitons in optical fibers is presented in Section 4.1. The nonlinear Schroedinger equation describing soliton propagation in a lossless optical fiber and the parameters of the dispersion-shifted fiber and laser source used are given in Section 4.1.2. The design and performance of the optical dark-soliton detectors are first investigated in Section 4.1.3 so that they can be characterized. The design (Section 4.1.4.2) and performance (Section 4.1.5.1) of the optical dark-soliton generator are then outlined. The performances of the combined dark-soliton generator and detectors are also described in Section 4.1.5.2. The theory of coherent integrated-optic signal processing described in Section 2.3 is employed in this chapter where electric-field amplitude signals are considered.

4.1.1 INTRODUCTION

Optical solitons are seen to be promising candidates as information carriers in ultralong-distance or ultrahigh-speed repeaterless optically amplified communication systems in the near future. The history of solitons in optical fibers began almost a quarter century ago when Hasegawa and Tappert [1,2] proposed that optical solitons can propagate without distortion over an infinitely long distance in a lossless single-mode optical fiber through the exact balancing of the inherent effects of group velocity dispersion (GVD) and self-phase modulation: bright solitons exist in the negative (or anomalous) GVD regime, while dark solitons appear in the positive (or normal) GVD regime. The pioneer work of Hasegawa and Tappert has led to the field of optical soliton communications, which has been under intensive worldwide investigation in the last decade.

The practical importance of bright solitons in optical fibers was not realized until 1980 when Mollenauer et al. [3] reported the first experimental verification of the existence of a bright soliton over a 700 m standard single-mode optical fiber. Several years after the prediction made by Hasegawa [4] that practical propagation of bright solitons over long distances could be made possible by periodically compensating for the fiber loss through the use of Raman amplifier gain, Mollenauer and Smith [5] exploited this idea and presented the first bright-soliton transmission experiment ever carried out over 4000 km. One year later, Nakazawa et al. [6] reported the first soliton transmission experiment that used erbium-doped fiber amplifier (EDFA) as an optical repeater for the 50 km fiber link. Recent technological advances have generated many successful ultrahigh-speed or ultralong-reach transmission experiments, in which EDFAs were used as optical repeaters; see, for example, the recent papers by Nakazawa [7] and Aubin et al. [8].

Dark solitons have been predicted to offer better stability than bright solitons against fiber loss [9], interactions between neighboring solitons [10] and amplified noise-induced timing jitter [11,12]. Dark-soliton propagation experiments, however, are much more difficult to implement than bright-soliton propagation experiments because of the difficulty of generating and detecting dark solitons. In 1987, Emplit et al. [13] experimentally studied the propagation of odd-symmetry optical dark pulses that were generated using amplitude and phase filtering techniques. However, dark pulse propagation was not convincingly observed in their experiment because the fiber length was shorter than the soliton period and because of the low resolution in their pulse-shaping and pulse-measurement apparatus. Nevertheless, it was the first experimental propagation of dark pulses in optical fiber. One year later, Krökel et al. [14] experimentally demonstrated the evolution of an even-symmetry dark pulse into a pair of low contrast dark pulses. In the same year, Weiner et al. [15] successfully developed a technique for synthesizing femtosecond optical pulses with almost arbitrary shape and phase, and presented a more convincing experimental observation of soliton-like propagation of odd-symmetry dark pulse superimposed upon a broader Gaussian background pulse.

Unlike previous schemes for generating dark pulses with a finite background [16,17], techniques for generating dark pulses with a CW (continuous wave) background (or dark solitons) have also been proposed and experimentally demonstrated [18,19]. Richardson et al. [16] reported the first experimental demonstration of the generation of a 100 GHz dark-soliton train by means of nonlinear conversion of a high-intensity beat signal in a positive GVD decreasing fiber. They also confirmed the stability of the generated dark-soliton train by propagating it through a 2 km length of positive GVD shifted fiber. Zhao and Bourkoff [17] proposed the use of an electro-optic intensity modulator driven by square pulses to generate dark solitons. Recently, Nakazawa and Suzuki [18] modified this scheme by using an AND circuit and a T-flip-flop circuit to perform data conversion to obtain a pseudorandom dark-soliton train. They also developed the first dark-soliton detection scheme, based on a one-bit-shifting technique with a Mach–Zehnder interferometer, which converts a dark-soliton signal into a modified return-to-zero (RZ) signal and into an inverted non-return-to-zero (NRZ) signal. Using the developed dark-soliton generation and detection schemes, Nakazawa and Suzuki [19] conducted a dark-soliton transmission experiment over 1200 km for the first time.

In this chapter, the trapezoidal optical integrator is described as an optical dark-soliton generator and the first-order first-derivative optical differentiator and the first-order Butterworth lowpass optical filter are proposed as optical dark-soliton detectors. Most of the work presented here has been described by Ngo et al. [20].

The effect of fiber loss on dark-soliton propagation is not considered here so that the underlying principles of dark-soliton generation and detection can be demonstrated.

4.1.2 Optical Fiber Propagation Model

To demonstrate the effectiveness of the proposed optical dark-soliton generator and detectors, the nonlinear Schroedinger equation describing pulse propagation in a lossless single-mode optical fiber

$$j\frac{\partial A}{\partial Z} - \frac{\beta_2}{2}\frac{\partial^2 A}{\partial t^2} + \xi|A|^2 A = 0 \tag{4.1}$$

is numerically solved using the split-step Fourier method [21]. In Equation 4.1, $j = \sqrt{-1}$, $A(Z, t) = \sqrt{P_0}\, U(Z, t)$, P_0 is the peak power of the incident pulse with normalized amplitude $U(Z, t)$, Z the distance of propagation, t the retarded time,* $\beta_2 = -D\lambda^2/(2\pi c)$ the GVD parameter, D the fiber dispersion parameter, λ the operating wavelength, c the speed of light in vacuum, $\xi = 2\pi n_2/(A_{\text{eff}}\lambda)$ the fiber nonlinear coefficient, n_2 the nonlinear refractive index, and A_{eff} is the effective core area. It is useful to write the peak power and soliton period as $P_0 = N^2|\beta_2|/(\xi T_0^2)$ and $Z_0 = \pi T_0{}^2/(2|\beta_2|)$, respectively, where the integer N is the soliton order and T_0 the initial pulse width. The dispersion-shifted fiber, with zero-dispersion wavelength at 1550 nm, and soliton source are assumed to have the following typical parameters: $\beta_2 = +1.27\ \text{ps}^2/\text{km}$ (for $D = -1.0$ ps/nm/km), $\xi = 3.2\ \text{W}^{-1}\text{km}^{-1}$ (for $A_{\text{eff}} = 40\ \mu\text{m}^2$ and $n_2 = 3.2 \times 10^{-20}\ \text{m}^2/\text{W}$), $P_0 = 0.494$ mW (for $N = 1$ and $T_0 = 28.4$ ps), and $Z_0 = 995$ km.

* The retarded time is defined as the normalized time measured in a frame of reference moving with the pulse at the group velocity, i.e., $t = \bar{t} - Z/v_g$ where $\bar{t}$ is the actual time and v_g the group velocity.

4.1.3 Design and Performance of Optical Dark-Soliton Detectors

The optical differentiator and Butterworth lowpass optical filter are now designed as optical dark-soliton detectors. The fundamental dark-soliton normalized pulse at the fiber input is given by [10]

$$U(0,t) = \begin{cases} \tanh(t/T_0 + q_0), & -\infty < t/T_0 < 0 \\ -\tanh(t/T_0 - q_0), & 0 \le t/T_0 < \infty \end{cases} \tag{4.2}$$

where

q_0 is a constant
$T_b = 2q_0T_0$ is the bit time
$T_{w1} = 1.76T_0 = 50$ ps is the FWHM of the soliton pulse

It is considered that $T_b = 200$ ps (for $q_0 = 3.52$), which corresponds to a bit rate of $1/T_b = 5$ Gb/s.

4.1.3.1 Design of Optical Dark-Soliton Detectors

By definition [22], the derivative of a fundamental dark-soliton pulse with amplitude tanh(t) is given by a bright-squared* soliton pulse with a temporal amplitude function sech^2(t) given by

$$\frac{\mathrm{d}[\tanh(t)]}{\mathrm{d}t} = \mathrm{sech}^2(t) \tag{4.3}$$

The design of an optical differentiator, which can perform the above derivative function, and a first-order Butterworth lowpass optical filter are now described as optical dark-soliton detectors. Figure 4.1 shows the schematic diagram of the asymmetrical Mach–Zehnder interferometer (AMZI) used for dark-soliton detection. The AMZI, which can be implemented using the power line communication (PLC) technology, namely, silica-based waveguides embedded on a silicon substrate as described in Section 2.3, consists of two directional couplers, DC1 and DC2, with

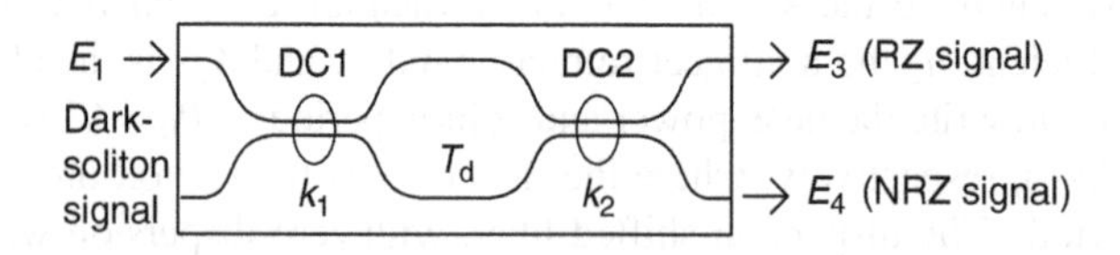

FIGURE 4.1 Schematic diagram of the asymmetrical Mach–Zehnder interferometer (AMZI) used for dark-soliton detection. All lines are integrated optical waveguides using planar lightwave technology.

* It is well known that the electric-field amplitude of a bright-soliton pulse is given by a secant pulse sech(t). Here, the electric-field amplitude of a secant-squared pulse sech^2(t) is referred to as a bright-squared soliton pulse. Thus, the electric-field amplitude of the bright-squared soliton pulse is the square of that of the bright-soliton pulse.

cross-coupled intensity coefficients k_1 and k_2, which are interconnected by two unequal waveguide arms with a differential time delay of T_d.

Neglecting the propagation delay, insertion loss and waveguide birefringence,* the transfer functions of the AMZI are given by[†]

$$H_{31}(z) = \frac{E_3}{E_1} = \exp(-j\omega T_u)[(1-k_1)(1-k_2)]^{1/2}\{1 - z_{31}z^{-1}\} \tag{4.4}$$

$$H_{41}(z) = \frac{E_4}{E_1} = \exp[-j(\omega T_u + \pi/2)][(1-k_1)k_2]^{1/2}\{1 - z_{41}z^{-1}\} \tag{4.5}$$

whose zeros in the z-plane are given by

$$z_{31} = \left[\frac{k_1 k_2}{(1-k_1)(1-k_2)}\right]^{1/2} \tag{4.6}$$

$$z_{41} = -\left[\frac{k_1(1-k_2)}{(1-k_1)k_2}\right]^{1/2} \tag{4.7}$$

where E_1 and (E_3, E_4) are the electric-field amplitudes at the input and output ports, respectively, $z = \exp(j\omega T_d)$ is the z-transform parameter, ω the angular optical frequency, and $T_d = T_\ell - T_u$, the sampling period of the AMZI, is the differential time delay between the lower arm (with delay T_ℓ) and the upper arm (with delay T_u). For $k_1 = k_2 = 0.5$ and hence $z_{31} = 1$, $H_{31}(z)$ corresponds to the transfer function of the first-order first-derivative optical differentiator as described in Chapter 6. Note that $H_{31}(z)$ is also the transfer function of the first-order Butterworth highpass optical filter with a 3 dB cutoff frequency at $\omega T_d = \pi/2$. While, for $k_1 = k_2 = 0.5$ and hence $z_{41} = -1$, $H_{41}(z)$ corresponds to the transfer function of the first-order Butterworth lowpass optical filter with a 3 dB cutoff frequency also at $\omega T_d = \pi/2$.

4.1.3.2 Performance of the Optical Differentiator

The sampling period of the AMZI is chosen to be equal to the bit period, i.e., $T_d = T_b$, for reasons to be given below. Figure 4.2 shows the 5 Gb/s 4-bit fundamental dark-soliton signals at $Z = 0$ and $Z = 100Z_0$. Compared with the input dark-soliton signal at $Z = 0$, the initial separation of the dark-soliton signal at $Z = 100Z_0$ is increased by 2.5% as a result of the repulsive force between neighboring dark solitons at a long distance [10].

The propagated dark-soliton signal at $Z = 100Z_0$ is detected by the optical differentiator and the resulting RZ signals are shown in Figure 4.2 for $0.8 \le z_{31} \le 1.2$. The solid curve corresponds to the performance of the ideal differentiator which requires $k_1 = k_2 = 0.5$ and hence $z_{31} = 1$. Such a requirement is often difficult to

* The waveguide birefringence of the AMZI can be eliminated by a fiber polarization controller or by inserting polyimide half waveplates into the gap of the waveguide paths.

† These transfer functions have been obtained by using the waveguide directional coupler defined in Equation 2.2 and the signal-flow graph technique described in Chapter 3.

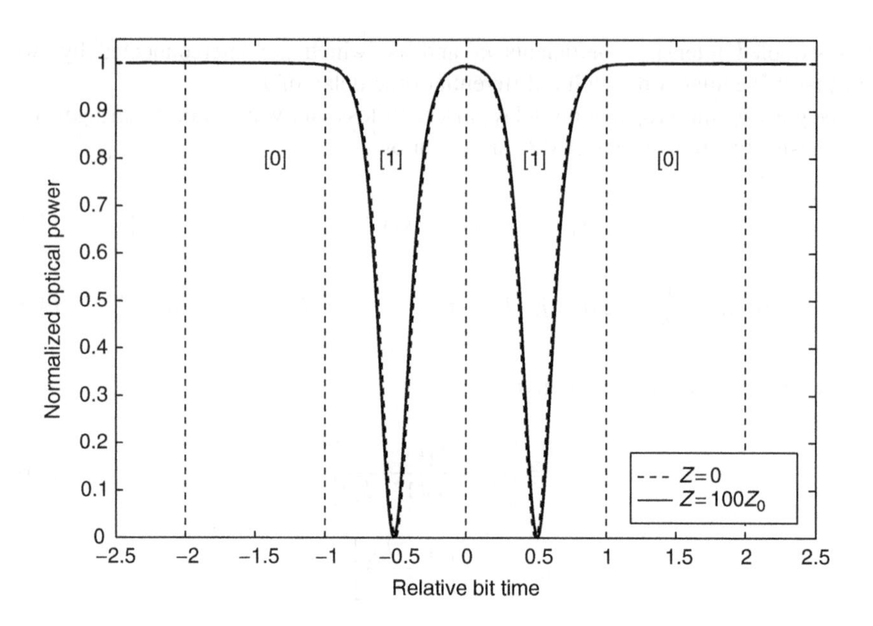

FIGURE 4.2 5 Gb/s 4-bit fundamental dark-soliton signals at $Z = 0$ and $Z = 100Z_0$.

achieve in practice because of the difficulty of fabricating directional couplers with the very precise values of the coupling coefficients. From Figure 4.2, as z_{31} deviates from its nominal value, the power levels of the space signals are raised from the otherwise zero values to $P \leq 0.0125$, and this results in decreasing the otherwise infinite mark-to-space ratio (MTSR) to MTSR ≥ 80. Surprisingly, the middle portion of the adjacent mark signals is not affected by this variation in z_{31}. For MTSR ≥ 80, the allowable values of the coupling coefficients of the AMZI lie in the range of $0.45 \leq k_1, k_2 \leq 0.55$, which is obtained by imposing the condition $0.8 \leq z_{31} \leq 1.2$ on Equation 4.6. Such variation of the coupling coefficients can be easily tolerated using the PLC technology.

Note that the detected RZ signals are of the square-type intensity pulse shape rather than the $\text{sech}^4(t)$ shape expected from the derivative of a $\tanh(t)$ amplitude pulse shape. This is because the chosen sampling period is large when compared with the dark-soliton pulse width (i.e., $T_d = 4T_{w1}$). When $T_d = 0.05T_{w1} = 2.5$ ps, the RZ signals have the expected $\text{sech}^4(t)$ intensity pulse shape, showing the high processing accuracy of the differentiator. However, the drawbacks of using $T_d \ll 4T_{w1}$ are that the performance of the differentiator significantly deteriorates through the large variation of z_{31} and that a small differential length of interferometer arms is required, which then requires very high fabrication accuracy. It is clear from Figure 4.3, where the RZ signals are detectable, that high processing accuracy of the differentiator is not necessary as far as dark-soliton detection is concerned.

The bit time of the dark-soliton signal must be sufficiently large to minimize the effect of dark-soliton interactions especially at very long distances. Thus, the sampling period of the AMZI, T_d, the bit period of the dark-soliton signal, T_b, and the dark-soliton pulse width, T_{w1}, must be related to each other in such a way that the

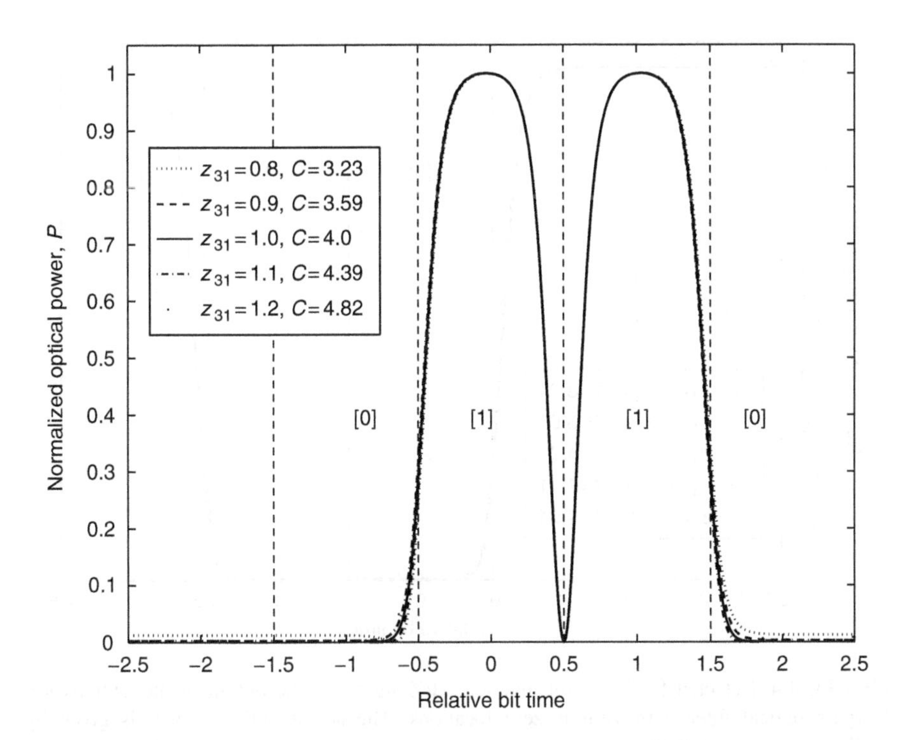

FIGURE 4.3 Detected 5 Gb/s 4-bit RZ signals at the output of the optical differentiator with various zero locations. The actual optical power is given by $|A(Z,t)|^2 = (1 - k_1)(1 - k_2)PCP_0$ where P and C are constant factors.

optical bandwidth of the AMZI (i.e., $1/(2T_d)$) is sufficiently large to accommodate the optical spectrum of the input dark-soliton signal. It was found that $T_d = T_b = 4T_{w1}$ satisfies this bandwidth requirement. In this study, for example, the 2.5 GHz bandwidth of the AMZI is wide enough to accommodate the 1 GHz spectral width of the dark-soliton spectrum at $Z = 100Z_0$.

4.1.3.3 Performance of the Butterworth Lowpass Optical Filter

The propagated dark-soliton signal at $Z = 100Z_0$, as shown in Figure 4.2, is detected by the Butterworth lowpass optical filter, resulting in the inverted NRZ signals as shown in Figure 4.4 for $-1.2 \leq z_{41} \leq -0.8$. The solid curve corresponds to the performance of the ideal lowpass filter which requires $k_1 = k_2 = 0.5$ and $z_{41} = -1$. As z_{41} deviates from its nominal value, the power levels of the space signals are increased from the otherwise zero values to $P \leq 0.0125$, which gives MTSR ≥ 80. For MTSR ≥ 80, the allowable values of the coupling coefficients of the AMZI are also in the range of $0.45 \leq k_1, k_2 \leq 0.55$.

For both cases of the differentiator and lowpass filter, the variations in z_{31} and z_{41} have more pronounced effects on the space signals than the mark signals. Compared with the propagated dark-soliton signal at $Z = 100Z_0$ (see Figure 4.2), Figures 4.3 and 4.4 show that the detected signals are shifted by a 1/2-bit to the right of the time

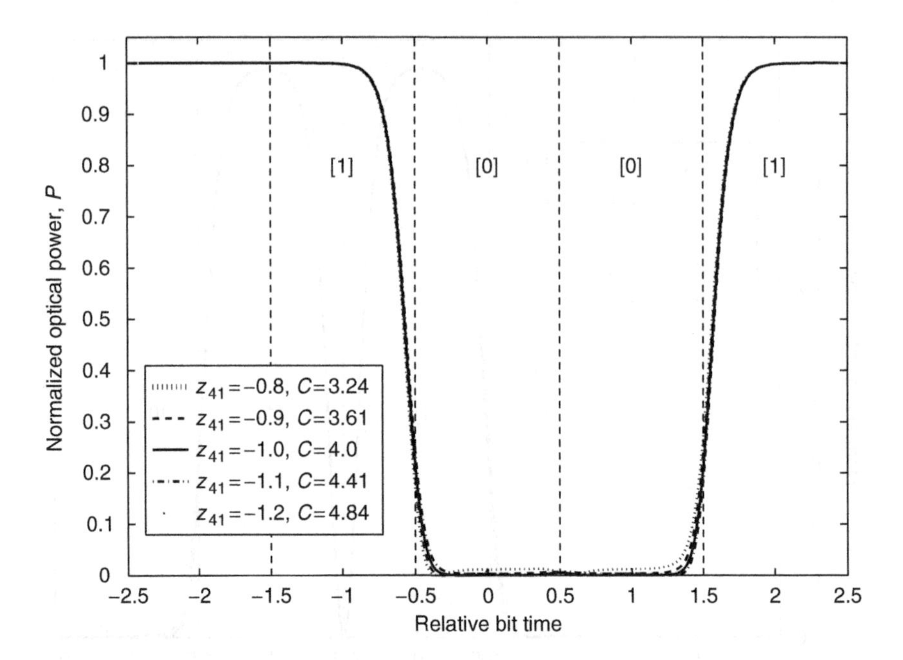

FIGURE 4.4 Detected 5 Gb/s 4-bit inverted NRZ signals at the output of the Butterworth lowpass optical filter with various zero locations. The actual optical power is given by $|A(Z,t)|^2 = (1 - k_1)k_2PCP_0$.

axis. The simulation results for both cases are also consistent with the experimental results of Nakazawa and Suzuki [19].

The proposed optical differentiator is based on the concept of optical differentiation whereas the AMZI described by Nakazawa and Suzuki [18] was based on the one-bit-shifting technique. However, these two completely different concepts can be implemented by the AMZI. The lowpass filter is based on the experimental investigations of Nakazawa and Suzuki [18]. Thus, the simulation results presented here provide the groundwork for the design of optical dark-soliton detectors since no theoretical results are given in Ref. [18].

4.1.4 Design of the Optical Dark-Soliton Generator

4.1.4.1 Design of the Optical Integrator

By definition [21], the integral of a bright-squared soliton pulse with amplitude $\text{sech}^2(t)$ is given by a fundamental dark-soliton pulse with its amplitude following the shape of $\tanh(t)$:

$$\int_{-\infty}^{t} \text{sech}^2(t')\mathrm{d}t' = \tanh(t) \tag{4.8}$$

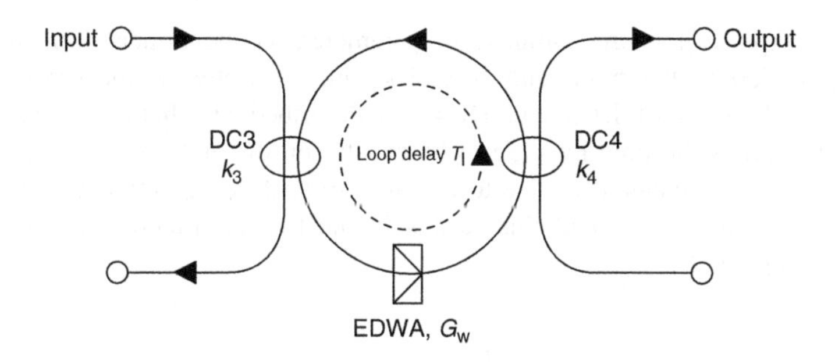

FIGURE 4.5 Schematic diagram of the all-pole optical filter using the PLC technology, which is used as the optical integrator.

The design of an optical integrator that can perform the above integral function is now described. Figure 4.5 shows the schematic diagram of the all-pole optical filter using the PLC technology, which is designed to be used as the optical integrator as described in Chapter 3. Note that the physical structure is exactly the same as that of the incoherent recursive fiber-optic signal processor in which fiber-optic components have been used (see Figure 3.6). The filter consists of an active optical waveguide loop formed by two directional couplers, DC3 and DC4, which are interconnected by waveguides.

A portion of the input signal is coupled by DC3 to the lower active waveguide path where it is amplified by an erbium-doped waveguide amplifier (EDWA) of intensity gain G_w, as described in Section 2.3.4. A portion of the amplified signal is coupled to the output port by DC4, while the rest is coupled to the upper path. A portion of the signal in the upper path is then coupled to the lower path where it is further amplified by the EDWA and the process continues. Optical signals can be maintained circulating in the loop for as long as the losses are compensated by the EDWA. Following the EDWA, an integrated waveguide isolator [21] and a narrow-band optical filter may be required to suppress the stimulated Brillouin scattering (SBS) and to minimize the amplified spontaneous emission (ASE), respectively.

Neglecting the propagation delay, insertion loss, and waveguide birefringence, the electric-field transfer function of the all-pole optical integrator is given by*

$$H(z) = \frac{-[k_3 k_4 G_w]^{1/2}}{1 - [(1-k_3)(1-k_4)G_w]^{1/2} z^{-1}} \tag{4.9}$$

which has one zero at the origin and one pole in the z-plane given by

$$\text{Pole} = [(1-k_3)(1-k_4)G_w]^{1/2} \tag{4.10}$$

* This transfer function has been derived by using the waveguide directional coupler defined in Equation 2.2 and the signal-flow graph method described in Chapter 3.

where $z = \exp(j\omega T_I)$ is the z-transform parameter, T_I the optical loop delay or sampling period of the filter, and k_3 and k_4 are, respectively, the cross-coupled intensity coefficients of DC3 and DC4. As described in Chapter 5, the optical integrator requires the pole location to be exactly on the unit circle in the z-plane, i.e., pole = 1. A convenient choice to accomplish such a requirement is to choose $k_3 = k_4 = 0.5$ and $G_w = 4$, and this results in the transfer function of the optical integrator as given by

$$-H(z) = \frac{1}{1 - z^{-1}} \tag{4.11}$$

Note that the excess losses of DC3 and DC4 and the losses of the straight and bend waveguides of the optical loop, which are not taken into account here, can also be compensated by the EDWA. Note also that the transfer function of the optical integrator is the reciprocal of that of the optical differentiator as described by Equation 3.4 when $k_1 = k_2 = 0.5$ is used. This is as expected from the inverse operation between the integral and derivative functions.

The unmodulated (or CW) signal of the soliton source is assumed to be externally modulated by an ideal optical intensity modulator. The modulated signal to be processed by the optical integrator is assumed to be of the normalized form

$$x(t) = \sum_n (-1)^n a_n S^{1/2}(t/T_0 - 2nq_0) \tag{4.12}$$

where $a_n \in (0,1)$ is the digital sequence and the intensity pulse shape

$$S(t) = \text{sech}^4(t) \tag{4.13}$$

The processed signal at the output of the optical integrator is given, by the convolution of the modulated signal with the impulse response of the optical integrator $(-h(t))$, as

$$y(t) = x(t)*(-h(t)) \cong \sum_n (-1)^n a_n \tanh(t/T_0 - 2nq_0) + 1 \tag{4.14}$$

where $*$ denotes the convolution operation and $(-h(t))$ is given by the inverse z-transform of $-H(z)$. The integrated signal in Equation 4.14 is a positive function because of the positive nature of the input pulse $S(t)$ and the positive impulse response $(-h(t))$, described in Chapter 5. Thus, the amplitude level of this integrated signal must be lowered by an amount equal to a CW signal of unity amplitude to obtain the required fundamental dark-soliton signal. As a result, the fundamental dark-soliton signal $y_I(t)$ can be generated according to

$$y_I(t) \cong y(t) - y_{max}/2 \cong \sum_n (-1)^n a_n \tanh(t/T_0 - 2nq_0) \tag{4.15}$$

where $y_{max} = \max[y(t)] = 2$ is the maximum amplitude of $y(t)$.

4.1.4.2 Design of an Optical Dark-Soliton Generator

The optical dark-soliton generator is now designed based on the mathematical formulations given in Equations 4.12 through 4.15. Figure 4.6 shows the schematic diagram of the proposed optical dark-soliton generation scheme using the PLC technology. The unmodulated light signal of the laser source is equally split by a 3 dB waveguide directional coupler into the upper and lower waveguide paths, which are of equal length. The processed modulated signal at the end of the upper path and the unmodulated signal at the end of the lower path are synchronously combined by another waveguide 3 dB directional coupler and then amplified by the EDFA, resulting in the required fundamental dark-soliton signal at the fiber input. The EDFA of intensity gain G_f is used as a booster amplifier to provide sufficient power to the fiber input to enable dark-soliton propagation. The requirement that the signals in the upper and lower paths be synchronously combined can be achieved by having both paths to be of the same length. In the upper waveguide arm, the optical intensity modulator can be of a $LiNbO_3$ Mach–Zehnder type and the optical integrator is implemented using the PLC technology as described in Section 4.1.4.1. In the lower waveguide arm, the optical tunable coupler, having a transfer function* $\sqrt{K}\exp(j\theta_{32})$ with K being the cross-coupled intensity coefficient and θ_{32} the phase shift, and the optical phase shifter with a phase shift ϕ, can also be implemented using the PLC technology. The characteristics of the phase shifter and tunable coupler have been described in Sections 2.3.2 and 2.3.3. Thus, the optical integrator, tunable coupler, phase shifter, 3 dB directional couplers, and the waveguide paths can all be implemented using the same PLC technology.

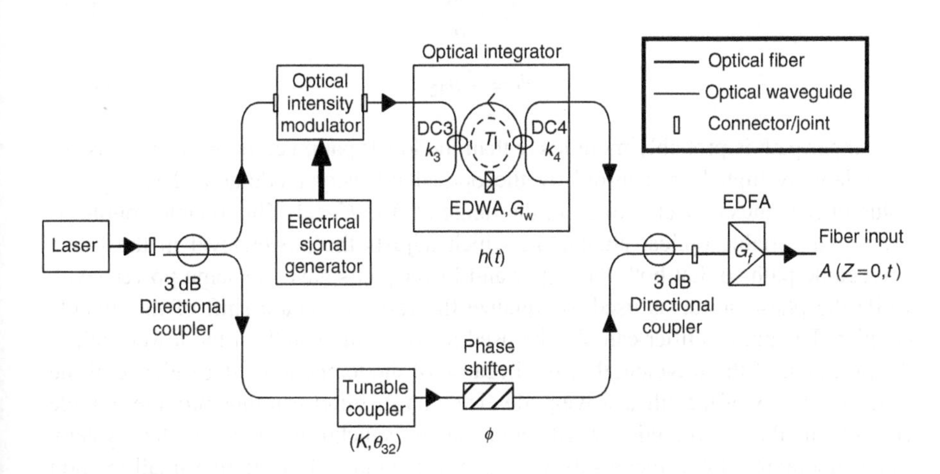

FIGURE 4.6 Schematic diagram of the proposed optical dark-soliton generator using the PLC technology.

* This transfer function corresponds to the transfer function E_4/E_1 in Equation 2.7.

The fundamental dark-soliton signal at the fiber input is given by

$$A(Z=0,t) = -j\sqrt{G_f P_s/4}\left[\sqrt{\alpha_u}\ x(t)*(-h(t)) - \sqrt{\alpha_\ell K}\exp(j(\theta_{32}+\phi))\right] \quad (4.16)$$

where P_s is the peak power of the laser source and α_u and α_ℓ are, respectively, the intensity losses of the upper and lower waveguide paths, which include all possible optical losses such as the connector loss between the fiber and waveguide, the straight and bend waveguide losses, and the insertion losses of the optical intensity modulator, the optical integrator, and the tunable coupler. The first and second terms in Equation 4.16 represent the processed modulated signal and the unmodulated signal in the upper and lower paths, respectively. Generation of the fundamental dark-soliton signal requires the terms inside the square brackets in Equation 4.16. Figures 4.12 and 4.13 show the composition of a mode-locked regenerative laser (MLRL) without and with feedback loop used in this study, respectively. It consists principally, for a nonfeedback ring, an optical close loop with an optical gain medium, an optical modulator (intensity or phase type), an optical fibers coupler, and associated optics. An optoelectronically (O/E) feedback loop detecting the repetition rate signal, and generating radiofrequency (RF) sinusoidal waves to electro-optically drive the intensity modulator, is necessary for the regenerative configuration as shown in Figure 4.13. Equation 4.16 is equal to the respective terms in Equation 4.15 such that

$$\sqrt{\alpha_\ell K}\exp(j(\theta_{32}+\phi)) = \max\left[\sqrt{\alpha_u}x(t)*(-h(t))\right]/2 \quad (4.17)$$

from which

$$K = \frac{\alpha_u}{\alpha_\ell} \quad (4.18)$$

$$\phi = -\theta_{32} \quad (4.19)$$

The upper path is probably more lossy than the lower path, i.e., $\alpha_u < \alpha_\ell$, because of the relatively high loss incurred by the optical intensity modulator. The required value of K is thus expected to be in the range of $0 < K < 1$. The tunable coupler is used as a variable optical attenuator, which adjusts the power level in the lower waveguide path so that both the upper and lower paths have the same power level, while the phase shifter is used to equalize the resulting phase shift of the tunable coupler. The phase shifter can also be used to adjust the length of the lower path to the precision of the wavelength order because of the temperature dependence of the refractive index of the silica waveguide. The phase shifter can thus provide flexible control of the waveguide length and hence synchronization of the system. The temperature of the silicon substrates can be maintained to within a small fraction of a degree to stabilize the refractive index of the waveguides, and this stability enables the optical dark-soliton generation scheme to efficiently and stably generate high-speed dark-soliton signals.

The following practical measures must be considered for stable operation. The CW light of the laser source must be in one polarization state; easily achieved by means of a fiber polarization controller (PC) placed at the laser output. A fiber PC is also required at the fiber input because of the presence of the waveguide birefringence.

4.1.5 Performance of the Optical Dark-Soliton Generator and Detectors

It is considered that the input amplitude pulse to the optical integrator is a bright-squared soliton pulse pair (normalized):

$$U(0,t) = -\text{sech}^2(t/T_0 - q_0) + \text{sech}^2(t/T_0 + q_0) \tag{4.20}$$

which has an FWHM of $T_{w2} = 1.21T_0 = 34.4$ ps. The bit time $T_b = 2q_0T_0$ is chosen to be $T_b = 200$ ps (for $q_0 = 3.52$), the same as that used for the dark-soliton detectors in Section 4.1.3. The sampling period of the optical integrator is considered to be $T_I = 25$ ps, which corresponds to an optical bandwidth of 20 GHz.

4.1.5.1 Performance of the Optical Dark-Soliton Generator

Figure 4.7 shows the ideal, generated, and propagated 5 Gb/s 4-bit dark-soliton signals. The solid curve corresponds to the ideal dark-soliton signal whose amplitude

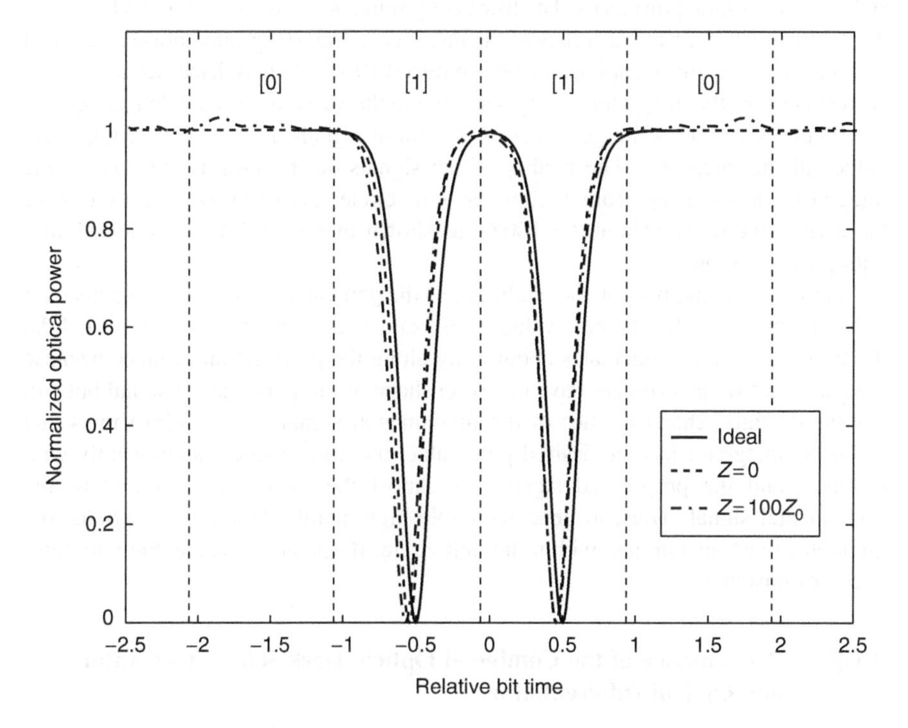

FIGURE 4.7 5 Gb/s 4-bit dark-soliton signals, showing the ideal signal, the signal at the output of the dark-soliton generator, and the signal at a large distance of $Z = 100Z_0$.

pulse is described by Equation 4.2. The bright-squared soliton pulse, as described by Equation 4.20, is processed by the ideal optical dark-soliton generator, in which the optical integrator has the desired pole location pole = 1, and this results in the generated dark-soliton signal at $Z = 0$ (dashed curve), which is slightly shifted to the left of the time axis compared with the ideal dark-soliton signal. This is because of the low processing accuracy of the integrator as a result of its relatively large sampling period, which occupies a large fraction of the pulse width, i.e., $T_I = 0.73T_{w2}$. It was found that when the sampling period is reduced by 10-fold to $T_I = 0.073T_{w2}$, the generated dark-soliton signal has almost the same shape as the ideal dark-soliton signal, showing the high processing accuracy of the integrator. As discussed in Chapter 3 for the case of the differentiator, the disadvantages associated with using a small sampling period are that the performance of the integrator significantly deteriorates when the pole position deviates from its nominal value and that high fabrication accuracy of the loop length of the integrator is required. The sampling period of the integrator is therefore chosen so that its optical bandwidth is large enough to accommodate the optical spectrum of the input bright-squared soliton signal. In this study, for example, the 20 GHz bandwidth of the integrator is sufficient to cover the 20 GHz spectral width of the input bright-squared soliton signal.

Figure 4.7 shows that the signal at a very large distance of $Z = 100Z_0$ still preserves the characteristics of the generated signal. This shows the effectiveness of the proposed scheme to generate high-quality dark solitons, which can propagate very stably over very long distances. The low background noise of the propagated signal is due to the low processing accuracy of the integrator. The background noise is reduced significantly when the sampling period is reduced 10-fold but, as discussed above, the performance of the integrator is very sensitive to the variation of its pole location.

Figure 4.8a shows the generated dark-soliton signals at $Z = 0$ for various pole values of the integrator. The quality of the signals deteriorates as the pole of the integrator moves away from within the unit circle. As expected, the quality of the corresponding signals at $Z = 100Z_0$, as shown in Figure 4.8b, also deteriorates with pole variation.

Figures 4.9a and b show the evolution of the generated dark-soliton signals over 100 soliton periods for the pole values 0.994 and 1, respectively. From Figure 4.9a, the background noise fluctuates about unity along the propagation distance because the generated signal does not have the exact shape of the fundamental signal but still has the essential characteristics of the fundamental signal. Figure 4.9b shows that when the integrator has the desired pole value, the noise reduces significantly with distance, and the propagated signal has almost the same shape as that of the fundamental signal. Thus, to generate stable high-quality dark-soliton signals, the integrator must operate just within the unit circle. If operated outside, the integrator becomes unstable.

4.1.5.2 Performance of the Combined Optical Dark-Soliton Generator and Optical Differentiator

The individual performances of the optical dark-soliton generator and optical differentiator have been separately analyzed in Sections 4.1.5.1 and 4.1.3.3, respectively.

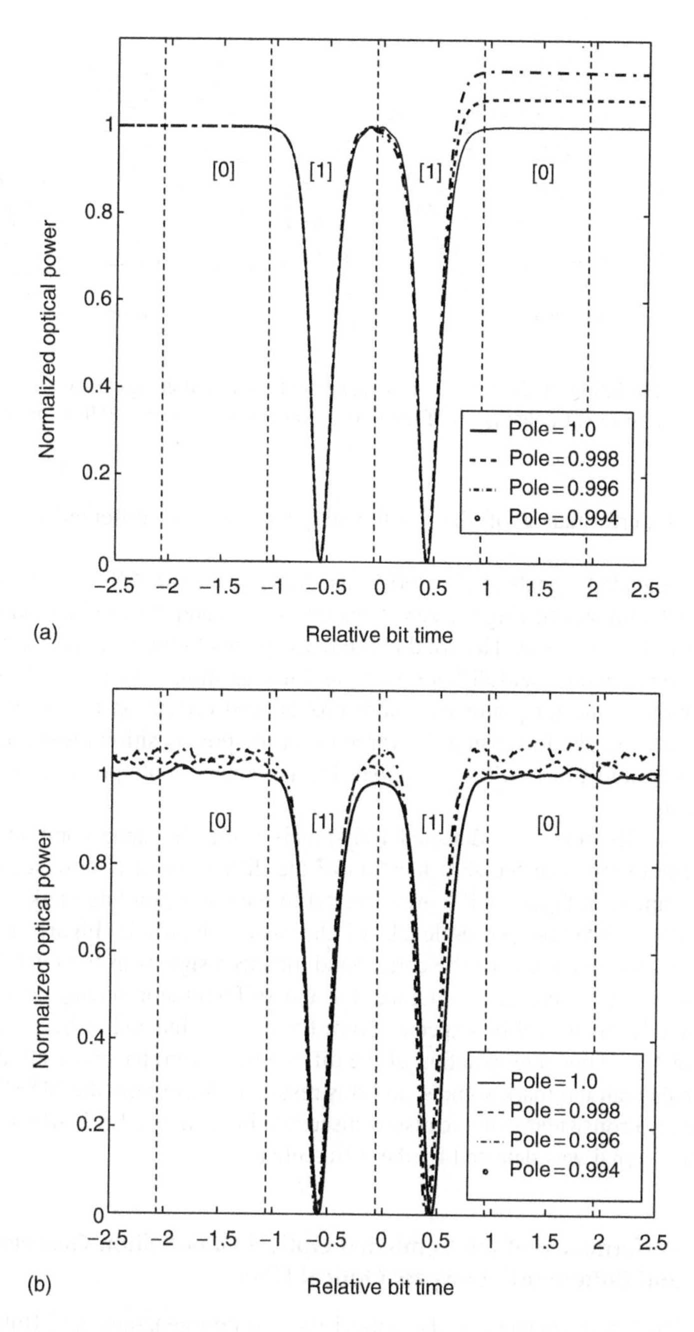

FIGURE 4.8 5 Gb/s 4-bit generated and propagated dark-soliton signals for various pole values of the optical integrator. (a) Generated dark-soliton signals at $Z = 0$. (b) The corresponding dark-soliton signals of Figure 4.8a at $Z = 100Z_0$.

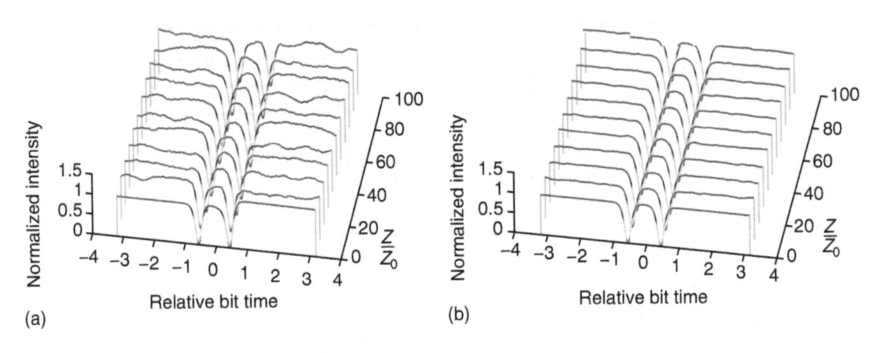

FIGURE 4.9 Evolution of the 5 Gb/s 4-bit generated dark-soliton signals over 100 soliton periods for two different pole values of the optical integrator. (a) Pole = 0.994. (b) Pole = 1.

The overall performance of the combined generator and differentiator is now described.

The propagated signals at $Z = 100Z_0$, as shown in Figure 4.8b, are detected by the ideal differentiator having the zero location $z_{31} = 1$, and the resulting RZ signals are shown in Figure 4.10a. The solid curve corresponds to the ideal performance of the combined generator and differentiator, and has an infinite value of MTSR. The pole variation of the integrator has more pronounced effects on the space signals than the mark signals. Increasing the variation of the pole position results in raising the pulse amplitude, otherwise zero power levels to some finite power levels of the space signals.

Figure 4.10b shows the detected RZ signals using the same conditions as in Figure 4.10a except that the zero location of the differentiator is now at $z_{31} = 0.8$. The mark signals in Figure 4.10b are identical to the corresponding mark signals in Figure 4.10a, while the power levels of the space signals in Figure 4.10b are relatively higher than those of the corresponding space signals in Figure 4.10a.

Similar results were also obtained for the differentiator having zero values $z_{31} = 0.9, 1.1, 1.2$ as would be expected from the results obtained in Section 4.1.3.2 (see Figure 4.3). Thus, the variation of the differentiator zero has more effect on the space signals than the mark signals, and this results in decreasing the MTSR. These results are also consistent with the results discussed in Section 4.1.3.2, where an ideal dark-soliton signal was detected by the differentiator.

4.1.5.3 Performance of the Combined Optical Dark-Soliton Generator and Butterworth Lowpass Optical Filter

The individual performances of the optical dark-soliton generator and Butterworth lowpass optical filter have been separately analyzed in Sections 4.1.5.1 and 4.1.3.3, respectively. The overall performance of the combined generator and lowpass filter is now described.

The propagated signals at $Z = 100Z_0$, as shown in Figure 4.8b, are detected by the ideal lowpass filter having the zero location $z_{41} = -1$, and the resulting inverted

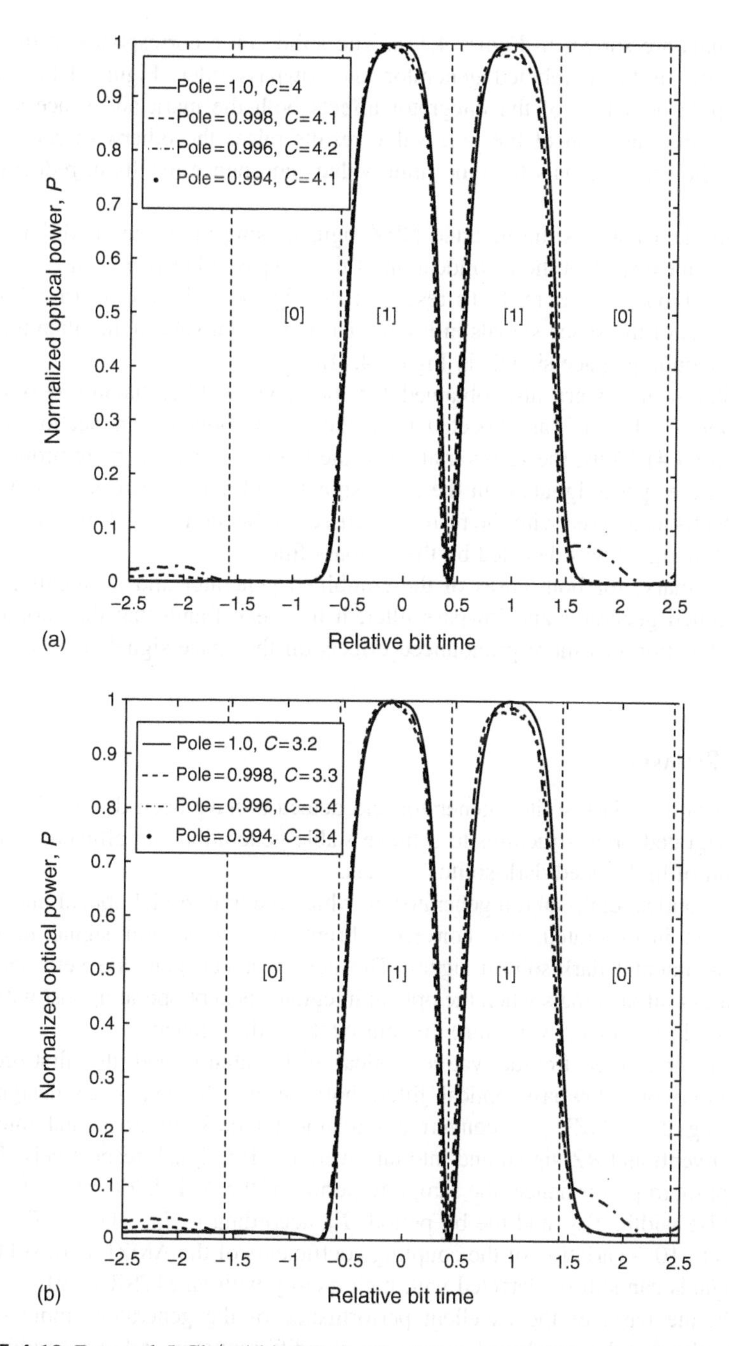

FIGURE 4.10 Detected 5 Gb/s 4-bit RZ signals at the output of the optical differentiator, having two different zero locations, as a result of processing the corresponding propagated dark-soliton signals as shown in Figure 4.8b. The actual optical power is given by $|A(Z,t)|^2 = (1 - k_1)(1 - k_2)PCP_0$. (a) Zero location at $z_{31} = 1$. (b) Zero location at $z_{31} = 0.8$.

NRZ signals are shown in Figure 4.11a. The solid curve corresponds to the ideal performance of the combined generator and lowpass filter. Figure 4.11a shows that the pole deviation of the integrator affects both the mark and space signals. Increasing the variation of the pole value results raises the otherwise zero power levels of the space signals to some finite values and hence results in reducing the MTSRs.

Figure 4.11b shows the inverted NRZ signals using the same conditions as in Figure 4.11a except that the zero location of the lowpass filter is now at $z_{41} = -0.8$. The mark signals in Figure 4.11b resemble the signals in Figure 4.11a, while the power levels of the space signals in Figure 4.11b are relatively higher than those of the corresponding space signals in Figure 4.11a.

Similar results were also obtained for the lowpass filter having zero values $z_{41} = -0.9, -1.1, -1.2$ as expected from the results obtained in Section 4.1.3.3 (see Figure 4.4). Thus, the zero variation of the lowpass filter has more pronounced effects on the space signals than the mark signals and hence decreases the MTSR. These results also agree with the results discussed in Section 4.1.3.3, where an ideal dark-soliton signal was detected by the lowpass filter.

In summary, for both cases of the combined generator and differentiator and the combined generator and lowpass filter, it has been found that the variation of the zero location has more pronounced effects on the space signals than the mark signals.

4.1.6 Remarks

- An optical dark-soliton generator and detectors have been designed using integrated-optic structures to achieve stable generation and efficient detection of high-speed dark-soliton signals.
- The optical dark-soliton generator, in which the trapezoidal optical integrator is incorporated, can convert a bright-squared soliton signal into a fundamental dark-soliton signal. The generator can generate very stable dark-soliton signals when the optical integrator is kept operating just within (i.e., less than 1% variation) the unit circle in the z-plane.
- The first-order first-derivative optical differentiator and the first-order Butterworth lowpass optical filter, both of which have been designed using the AMZI, can convert a fundamental dark-soliton signal into a conventional RZ signal and into an inverted NRZ signal, respectively. For optimum performance, the sampling period of the AMZI, T_d, relates to the pulse width, T_{w1}, and the bit period, T_b, according to $T_d = 4T_{w1} = T_b$. For up to 10% variation of the coupling coefficients of the AMZI, dark-soliton signals can still be detected with high quality with an MTSR $\geq$ 80.
- On the basis of the excellent performance of the generator, impressive results for the combined generator and differentiator and the combined generator and lowpass filter are obtained.

The AMZI and the all-pole optical filter described here will be used in the design of tunable optical filters in a later chapter.

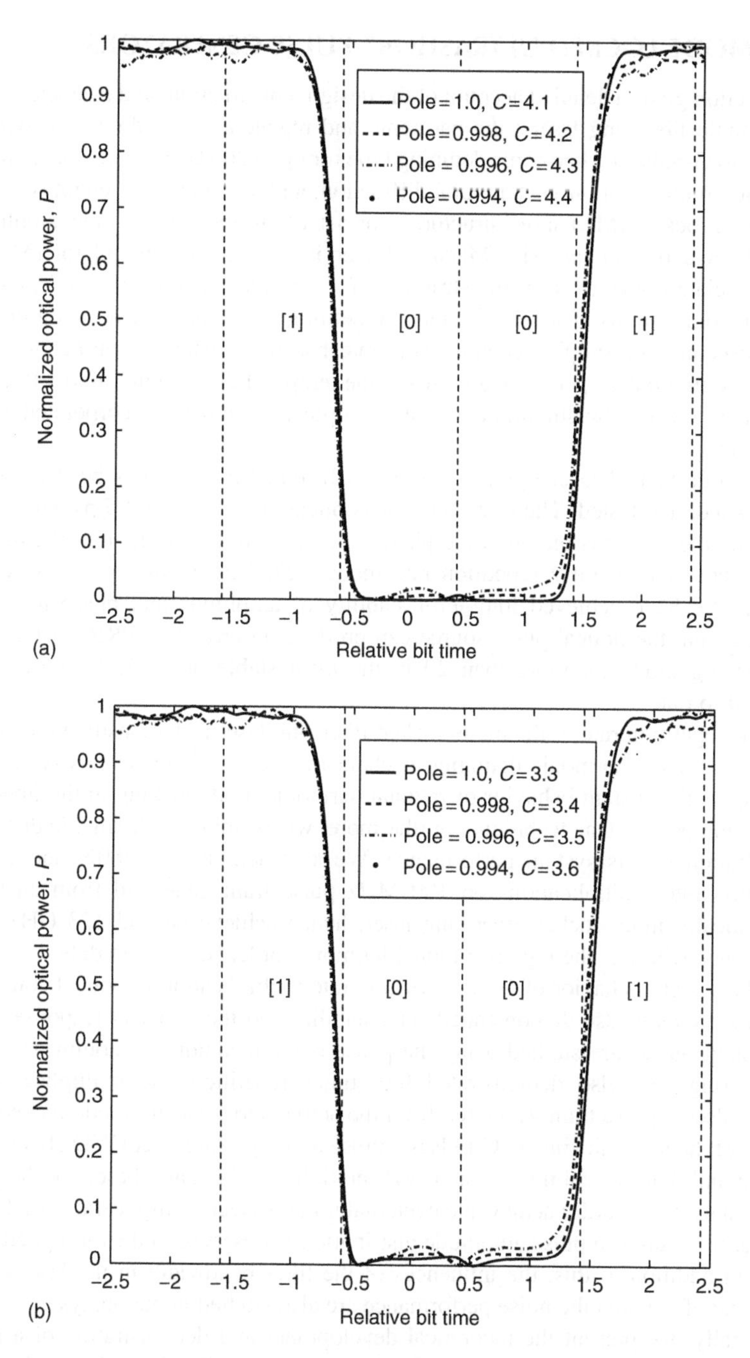

FIGURE 4.11 Detected 5 Gb/s 4-bit inverted NRZ signals at the output of the Butterworth lowpass optical filter, having two different zero locations, as a result of processing the corresponding propagated dark-soliton signals as shown in Figure 5.8b. The actual optical power is given by $|A(Z,t)|^2 = (1 - k_1)k_2PCP_0$. (a) Zero location $z_{41} = -1$. (b) Zero location $z_{41} = -0.8$.

4.2 MODE-LOCKED ULTRASHORT PULSE GENERATORS

This section gives a detailed account of the design, construction, and characterization of photonic fiber ring lasers: the harmonic and regenerative mode-locked type for 10 and 40 G-pulses/s, harmonic detuning type for up to 200 G-pulses/s, the harmonic repetition multiplication via temporal diffraction, and the multiwavelength type.

Mode-locked (ML) laser structures consist of in-line optical fiber amplifiers, guided-wave optical intensity Mach–Zehnder interferometric modulator (MZIM), and associate optics to form the structure of a ring resonator for the generation of photonic pulse trains of several gigahertz repetition rate with pulse duration in order of picoseconds or subpicoseconds. This rate has been extended further to a few hundreds giga-pulses per second by using the temporal diffraction Talbot effect that eliminates the bandwidth limitation of the optical modulator incorporated in the laser ring.

A mode-locked laser operating at 10 GHz repetition rate has been designed, constructed, and tested. The laser generates optical pulse train of 4.5 ps pulse width when the modulator is biased at the phase quadrature quiescent region. Preliminary experiment of a 40 GHz repetition rate mode-locked laser has also been demonstrated. We have achieved long-term stability of amplitude and phase noise that indicates that the optical pulse source can produce an error-free PRBS pattern in a self-locking mode for more than 20 h, the most stable photonic fiber ring laser reported to date.

The repetition rate of the mode-locked fiber ring laser is demonstrated up to 200 G-pulses/s using harmonic detuning mechanism in a ring laser. In this system, we investigate the system behavior of rational harmonic mode locking in the fiber ring laser using phase plane technique. Furthermore, we examine the harmonic distortion contribution to this system performance. We also demonstrate 660× and 1230× repetition rate multiplications on 100 MHz pulse train generated from an active harmonically mode-locked fiber ring laser, hence achieve 66 and 123 GHz pulse operations, which is the highest rational harmonic order reported to date.

The system behavior of GVD repetition rate multiplication in optical communication systems is also demonstrated. The stability and the transient response of the multiplied pulses are studied using the phase plane technique of nonlinear control engineering. We also demonstrated four times repetition rate multiplication on 10 G-pulses/s pulse train generated from the active harmonically mode-locked fiber ring laser, hence achieving 40 G-pulses/s pulse train by using fiber GVD effect. It has been found that the stability of the GVD multiplied pulse train, based on the phase plane analysis, is hardly achievable even under the perfect multiplication conditions. Furthermore, uneven pulse amplitude distribution is observed in the multiplied pulse train. In addition to this, the influences of the filter bandwidth in the laser cavity, nonlinear effect, and the noise performance are also studied in our analyses.

Finally, we present the theoretical development and demonstration of a multiwavelength optically amplifier fiber ring laser using an all-polarization-maintaining fiber (PMF) Sagnac loop. The Sagnac loop simply consists of a PMF coupler and a segment of stress-induced PMF, with a single-polarization coupling point in the loop. The Sagnac loop is shown to be a stable comb filter with equal frequency period, which determines the possible output power spectrum of the fiber ring laser. The number of output lasing wavelengths is obtained by adjusting the polarization state of the light in the unidirectional ring cavity by means of a polarization controller.

Multiwavelength ultrashort-width ultrahigh repetition rate photonic generators will be proposed and reported in the near future, by combining the principles of the above photonic fiber ring lasers.

4.2.1 Regenerative Mode-Locked Fiber Lasers

Generation of ultrashort optical pulses with multiple gigabits repetition rate is critical for ultrahigh bit rate optical communications, particularly for the next generation of terabits per second optical fiber systems. As the demand for the bandwidth of the optical communication systems increases, the generation of short pulses with ultrahigh repetition rate becomes increasingly important in the coming decades.

The mode-locked fibers laser offers a potential source of such pulse train. Although the generation of ultrashort pulses by mode locking of a multimodal ring laser is well known, the applications of such short pulse trains in multigigabits per second optical communications challenge its designers on its stability and spectral properties. Recent reports on the generation of short pulse trains at repetition rates in order of 40 Gb/s, possibly higher in the near future [23], motivate us to design and experiment with these sources to evaluate whether they can be employed in practical optical communication systems.

Further, the interest of multiplexed transmission at 160 Gb/s and higher in the foreseeable future requires us to experiment with optical pulse source having a short pulse-duration and high repetition rates. This report describes laboratory experiments of a mode-locked fibers ring laser (MLFRL), initially with a repetition rate of 10 GHz and preliminary results of higher multiple repetition rates up to 40 GHz. The mode-locked ring lasers reported hereunder adopt an active mode-locking scheme whereby partial optical power of the output optical waves is detected, filtered, and a clock signal is recovered at the desired repetition rate. It is then used as an RF drive signal to the intensity modulator incorporated in the ring laser. A brief description on the principle of operation of the MLFRL is given in the next section followed by a description of the mode-locked laser experimental setup and characterization.

Active mode-locked fiber lasers remain as a potential candidate for the generation of such pulse trains. However, the pulse repetition rate is often limited by the bandwidth of the modulator used or the RF oscillator that generates the modulation signal. Hence, some techniques have been proposed to increase the repetition frequency of the generated pulse trains. Rational harmonic mode locking is widely used to increase the system repetition frequency [23–25]. 40 GHz repetition frequency has been obtained with fourth-order rational harmonic mode locking at 10 GHz baseband modulation frequency [26]. Wu and Datta [25] have reported 22nd-order rational harmonic detuning in the active mode-locked fiber laser, with 1 GHz base frequency, leading to 22 GHz pulse operation. This technique is simple and achieved by applying a slight deviated frequency from the multiple of fundamental cavity frequency. Nevertheless, it is well known that it suffers from inherent pulse amplitude instability as well as poor long-term stability. Therefore, pulse amplitude equalization techniques are often applied to achieve better system performance [25,27,28].

Other than this rational harmonic detuning, some other techniques have been reported and used to achieve the same objective. Fractional temporal Talbot-based repetition rate multiplication technique [29,30] uses the interference effect between the dispersed pulses to achieve the repetition rate multiplication. The essential element of this technique is the dispersive medium, such as linearly chirped fiber grating (LCFG) [29,31] and dispersive fiber [32–34]. Intracavity optical filtering [26,35] uses modulators and a high finesse Fabry–Perot filter (FFP) within the laser cavity to achieve higher repetition rate by filtering out certain lasing modes in the mode-locked laser. Other techniques used in repetition rate multiplication include higher-order FM mode locking [28], optical time-domain multiplexing [36], etc.

The stability of high repetition rate pulse train generated is one of the main concerns for practical multigigabits per second optical communication systems. Qualitatively, a laser pulse source is considered as stable if it is operating at a state where any perturbations or deviations from this operating point are not increased but suppressed. Conventionally, the stability analyses of such laser systems are based on the linear behavior of the laser in which we can analytically analyze the system behavior in both time and frequency domains. However, when the mode-locked fiber laser is operating under nonlinear regime, none of these standard approaches can be used, since direct solution of nonlinear different equation is generally impossible, hence frequency-domain transformation is not applicable. Some inherent nonlinearities in the fiber laser may affect its stability and performance, such as the saturation of the embedded gain medium, nonquadrature biasing of the modulator, nonlinearities in the fiber, etc., hence, nonlinear stability approach should be used in any laser stability analysis.

In Section 4.2.3, we focus on the stability and transient analyses of the rational harmonic mode locking in the fiber ring laser system using phase plane method, which is commonly used in nonlinear control system. This technique has been previously used in Ref. [34] to study the system performance of the fractional temporal Talbot repetition rate multiplication systems. It has been shown that it is an attractive tool in system behavior analysis. However, it has not been used in the rational harmonic mode-locking fiber laser system. In Section 4.2.3.1, the rational harmonic detuning technique is briefly discussed. Section 4.2.3.2 describes the experimental setup for the repetition rate multiplication used. Section 4.2.3.3 investigates the dynamic behavior of the phase plane of the fiber laser system, followed by some simulation results. Section 4.2.3.4 presents and discusses the results obtained from the experiment and simulation. Finally, some concluding remarks and possible future developments for this type of ring laser are given.

Rational harmonic detuning [25,37] is achieved by applying a slight deviated frequency from the multiple of fundamental cavity frequency. This technique is simple in nature. However, this technique suffers from inherent pulse amplitude instability, which includes both amplitude noise and inequality in pulse amplitude. Furthermore, it gives poor long-term stability. Hence, pulse amplitude equalization techniques are often applied to achieve better system performance. Fractional temporal Talbot-based repetition rate multiplication technique [8] uses the interference effect between the dispersed pulses to achieve the repetition rate multiplication. The essential element of this technique is the dispersive medium, such as LCFG [31,37] and single-mode fiber [31,32]. Intracavity optical filtering [28,35] uses modulators and a high finesse FFP within the laser cavity to achieve higher repetition rate by filtering out certain lasing modes in the mode-locked laser. Other techniques used in repetition rate multiplication include higher-order FM mode locking [35], optical time-domain multiplexing, etc.

Although Talbot-based repetition rate multiplication systems are based on the linear behavior of the laser, there are still some inherent nonlinearities affecting its stability, such as the saturation of the embedded gain medium, nonquadrature biasing of the modulator, nonlinearities in the fiber, etc., hence, nonlinear stability approach must be adopted. In Section 4.2.4, we focus on the stability and transient analyses of the GVD multiplied pulse train using the phase plane analysis of nonlinear control analytical technique [24]. This is the first time, to the best of our knowledge, that the phase plane analysis is being used to study the stability and transient performances of the GVD repetition rate multiplication systems. In Section 4.2.4.1, the GVD repetition rate multiplication technique is briefly given. Section 4.2.4.2 describes the experimental setup for the repetition rate multiplication. Section 4.2.4.3 investigates the dynamic behavior of the phase plane of GVD multiplication system, followed by some simulation results. Section 4.2.4.4 presents and discusses the results obtained from the experiment. Finally, some concluding remarks and possible future developments for this type of laser are given.

Owing to the enormous practical applications in components testing and dense wavelength-division multiplexing (DWDM) optical networks, and considering that the complexity and cost of optical sources will increase as the number of wavelength channels increases, multiwavelength optical sources capable of generating a large number of wavelengths have been an area of strong and continuing worldwide research interest. Erbium-doped fiber (EDF) lasers operating in the 1550 nm window are widely used in wavelength-division multiplexing (WDM) systems, fiber-optic sensor systems, and photonic true-time delay beam forming systems [38,39]. Among these different types of fiber lasers, unidirectional traveling-wave ring lasers have been studied extensively in recent years because of several advantages such as elimination of back scattering and spatial hole-burning effects. Several types of optical filters such as fiber Bragg gratings (FBGs) [40,41], FPF [42], multimode filters [43], and high-birefringence fibers [44] have been used in the construction of multiwavelength fiber ring lasers (MWFRLs). Multiwavelength fiber lasers using FBGs as wavelength-selective filters will require a large number of individual FBGs to be written on one single segment of fiber depending on the number of lasing wavelengths to be generated. This type of fiber laser has high insertion loss and is costly due to a large number of phase masks required to fabricate the FBGs.

The Sagnac loop simply consists of a PMF coupler and a segment of stress-induced PMF, with a single-polarization coupling point in the loop. The Sagnac loop is shown to be a stable comb filter with equal frequency period, which determines the possible output power spectrum of the fiber ring laser. The number of output lasing wavelengths is obtained by adjusting the polarization state of the light in the unidirectional ring cavity by means of a polarization controller.

Sagnac fiber interferometers have several applications, such as gyroscopes, magnetic field sensors, and secure optical communications. A Sagnac loop filter simply consists of a fiber coupler, the two ends of which are spliced to the two ends of a segment of optical fiber. The Sagnac loop can function as a reflector or periodic filter depending on the birefringence with the fiber loop. The Sagnac loop is simply a reflector when there is no birefringence in the fiber loop. Owing to the frequency dependence of the coupler, the coupler acts as a reflector when its coupling coefficient is 50% and as a partial reflector for other values of coupling coefficient. The Sagnac loop as a reflector has been used in the fabrication of fiber laser with a linear cavity [45]. Also the Sagnac loop can function as a periodic filter when there is some birefringence induced in the loop [46,47]. Compared with a Mach–Zehnder interferometer, the Sagnac loop filter is more robust against environmental changes because it is a two-beam interferometer with one common path. However, the Sagnac filter may not operate stably when only standard single-mode fibers are used to form the coupler and the loop. This is because that the small fiber core imperfection, external stress, bending, twisting, and temperature variation may cause the spliced sections of the coupler and the loop to behave as a birefringent media, the birefringence of which changes randomly with time. It should be noted that Refs. [46,47] do not provide a complete theoretical basis of the effect of the linear birefringence on the performance of the Sagnac filter, which is presented in this section. This kind of birefringent loop filter has been placed inside the linear cavity of a fiber laser with a PC inside the loop because the coupler was made of standard single-mode fiber [48].

In Section 4.2.5, we present a theoretical analysis and implementation of an MWFRL with a Sagnac PMF loop filter. It is shown that when a single-polarization coupling point is formed by the stress-induced PMF loop, the fiber laser can provide stable generation of multiple lasing wavelengths with equal wavelength spacing. Section 4.2.5.1 presents the theory of the Sagnac PMF loop filter, which consists of a PMF coupler (instead of a standard single-mode fiber coupler used in previous works as described above) and a segment of PMF. Section 4.2.5.2 presents the experimental results and discussion.

4.2.2 Ultrahigh Repetition Rate Fiber Mode-Locked Lasers

This section gives a detailed account of the design, construction, and characterization of a mode-locked (ML) fibers ring laser. The ML laser structure employs inline optical fiber amplifiers, a guided-wave optical intensity MZIM, and associate optics to form a ring resonator structure generating optical pulse trains of several gigahertz repetition rate with pulse duration in order of picoseconds. Long-term stability of amplitude and phase noise has been achieved, which indicates that the optical pulse source can produce an error-free PRBS pattern in a self-locking mode for more than 20 h. A mode-locked laser operating at 10 GHz repetition rate has been designed, constructed, tested, and packaged. The laser generates optical pulse train of 4.5 ps pulse width when the modulator is biased at the phase quadrature quiescent region. Preliminary experiment of a 40 GHz repetition rate mode-locked laser has also been demonstrated. Although it is still unstable in long term, without an O/E feedback loop, optical pulse trains have been observed.

4.2.2.1 Mode-Locking Techniques and Conditions for Generation of Transform Limited Pulses from a Mode-Locked Laser

4.2.2.1.1 Schematic Structure of MLRL

Figures 4.12 and 4.13, respectively, show the composition of an MLRL without and with feedback loop used in this study. It consists principally, for a nonfeedback ring, an optical close loop with an optical gain medium, an optical modulator (intensity or phase type), an optical fibers coupler, and associated optics. An O/E feedback loop detecting the repetition-rate signal and generating RF sinusoidal waves to electro-optically drive the intensity modulator is necessary for the regenerative configuration as shown in Figure 4.13.

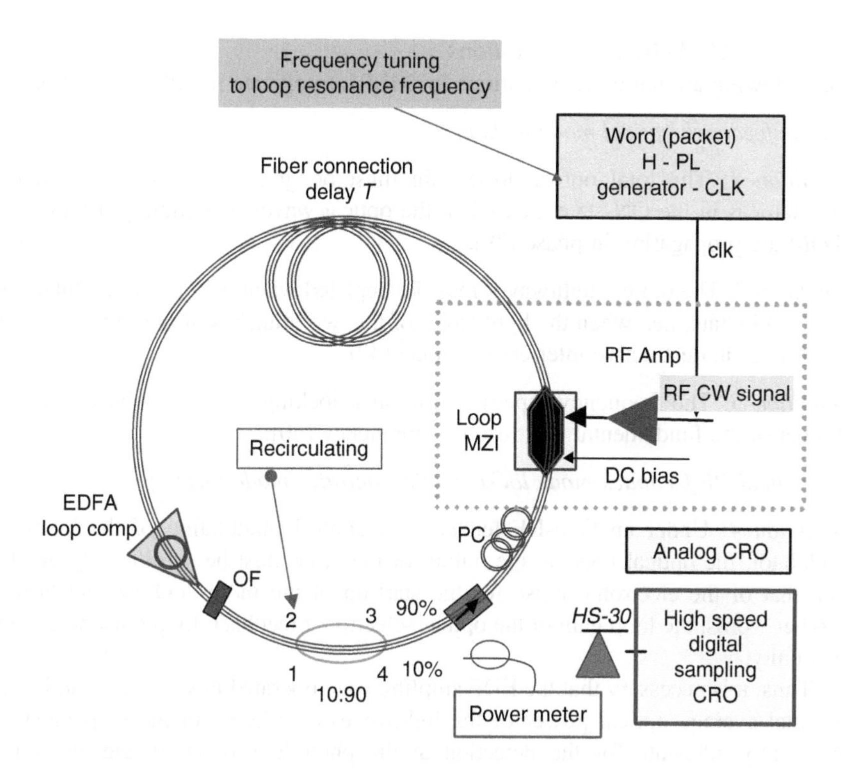

FIGURE 4.12 Schematic arrangement of a mode-locked ring laser without the active feedback control.

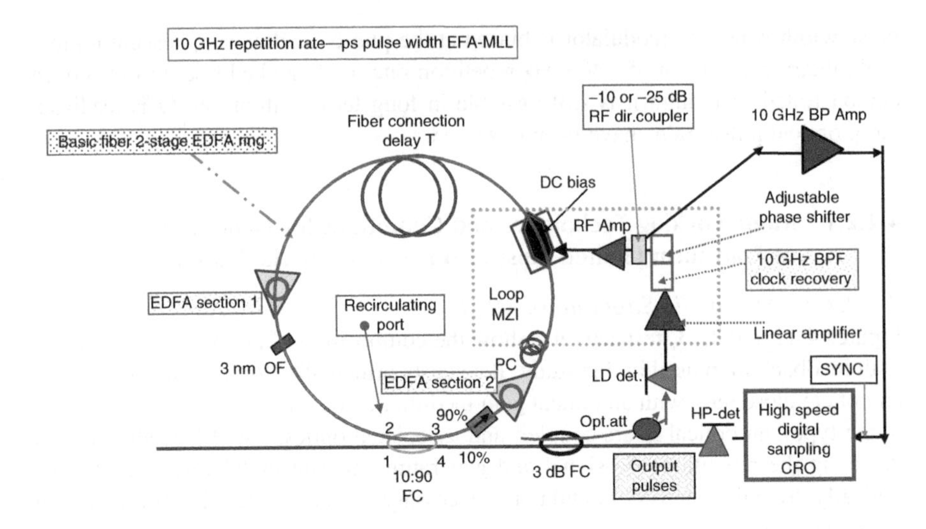

FIGURE 4.13 Schematic arrangement of a mode-locked ring laser with an O/E-RF electronic active feedback loop.

4.2.2.1.2 Mode-Locking Conditions

The following are the basic conditions for MLRL to operate in pulse oscillation:

For nonfeedback optical mode-locking

Condition 1: The total optical loop gain must be greater than unity when the modulator is in the ON-state, i.e., when the optical waves transmitting through the MZIM are propagating in phase [29].

Condition 2: The optical lightwaves must be depleted when the optical modulator is in the OFF-state, i.e., when the lightwaves of the two branches of the MZIM are out of phase or in destructive interference mode [49].

Condition 3: The frequency repetition rate at a locking state must be a multiple number of the fundamental ring resonant frequency [50].

For optical-RF feedback mode locking—Regenerative mode-locking

Condition 4: Under an O/E-RF feedback to control modulation of the intensity modulator, the optical noise at the output of the laser must be significantly greater than that of the electronic noise for the start-up of the mode locking and lasing. In other words, the loop gain of the optical–electronic feedback loop must be greater than unity.

Thus, it is necessary that the EDF amplifiers are operated in saturation mode and the total average optical power of the lightwaves circulating in the loop must be sufficiently adequate for the detection at the photodetector (PD) and electronic preamplifier. Under this condition, the optical quantum shot noise dominates the electronic shot noise.

4.2.2.1.3 *Factors Influencing the Design and Performance of Mode Locking and Generation of Optical Pulse Train*

The locking frequency is a multiple of the fundamental harmonic frequency of the ring defined as the inverse of the traveling time around the loop and is given by

$$f_{\mathrm{RF}} = \frac{Nc}{n_{\mathrm{eff}}L} \tag{4.21}$$

where

f_{RF} is the RF frequency required for locking and the required generation rate
N is an integer and indicates mode number order
c is the velocity of light in vacuum
n_{eff} is effective index of the guided propagating mode
L is loop length including that of the optical amplifiers

Under the requirement of the OC-192 standard bit rate, the locking frequency must be in the region of 9.95 G-pulses/s. That is, the laser must be locked to a very high order of the fundamental loop frequency that is in the region of 1 MHz to 10 MHz depending on the total ring length. For an optical ring of length about 30 m and a pulse repetition rate of 10 GHz, the locking occurs on approximately the 1400th harmonic mode.

It is also noted that the effective refractive index n can be varied in different sections of the optical components forming the laser ring. Furthermore, the two polarized states of propagating lightwaves in the ring, if the fibers are not of polarization maintaining (PM) type, would form two simultaneously propagating rings, and they could interfere or hop between these dual polarized rings.

The pulse width, denoted as $\Delta\tau$, of the generated optical pulse trains can be found to be given by [51]

$$\Delta\tau = 0.45\left(\frac{\alpha_{\mathrm{t}}G_{\mathrm{t}}}{\Delta_m}\right)^{1/4}\frac{1}{(f_{\mathrm{RF}}\Delta\nu)^{1/2}} \tag{4.22}$$

with $\alpha_{\mathrm{t}}G_{\mathrm{t}}$ is the round trip gain coefficient as a product of all the loss and gain coefficients of all optical components including their corresponding fluctuation factor, Δ_m is the modulation index, and $\Delta\nu$ is the overall optical bandwidth (in units of hertz) of the laser.

Hence, the modulation index and the bandwidth of the optical filter influence the generated pulse width of the pulse train. However, the optical characteristics of the optical filters and optical gain must be flattened over the optical bandwidth of the transform limit for which a transform limited pulse must satisfy, for a sech^2 pulse intensity profile, the relationship

$$\Delta\tau\Delta\nu = 0.315 \tag{4.23}$$

Similarly, for Gaussian pulse shape the constant becomes 0.441. The fluctuation of the gain or loss coefficients over the optical flattened region can also influence the generated optical pulse width and mode-locking condition.

In the case for regenerative mode-locking case as illustrated in Figure 4.13, the optical output intensity is split and O/E detected, we must consider the sensitivity and noises generated at the PD. Two major sources of noises are generated at the input of the PD, firstly the optical quantum shot noises generated by the detection of the optical pulse trains and secondly, the random thermal electronic noises of the small signal electronic amplifier following the detector. Usually the electronic amplifier would have a 50 Ω equivalent input resistance R referred to the input of the optical preamplifier as evaluated at the operating repetition frequency, this gives a thermal noise spectral density of

$$S_R = \frac{4kT}{R} \tag{4.24}$$

in A^2/Hz with k being the Boltzmann's constant. This equals to 3.312×10^{-22} A^2/Hz at 300 K. Depending on the electronic bandwidth B_e of the electronic preamplifier, i.e., wideband or narrowband type, the total equivalent electronic noise (square of noise current) is given by $i_{NT}^2 = S_R B_e$. Under the worst case, when a wideband amplifier of a 3 dB, electrical bandwidth of 10 GHz, the equivalent electronic noise at the input of the electronic amplifier is 3.312×10^{-11} A^2, i.e., an equivalent noise current of 5.755 μA is present at the input of the clock recovery circuit. If a narrowband-pass amplifier of 50 MHz 3 dB bandwidth centered at 10 GHz is employed, this equivalent electronic noise current is 0.181 μA.

Now considering the total quantum shot noise generated at the input of the clock recovery circuit, suppose that a 1.0 mW (or 0 dBm) average optical power is generated at the output of the MLRL, then a quantum shot noise* of approximately 2.56×10^{-22} A^2/Hz (i.e., an equivalent electronic noise current of 16 nA) is present at the input of the clock recovery circuit. This quantum shot noise current is substantially smaller than that of the electronic noise.

For the detected signal at the optical receiver incorporated in the clock recovery circuit to generate a high signal-to-noise ratio, the optical average power of the generated pulse trains must be high, at least at a ratio of 10. We estimate that this optical power must be at least 0 dBm at the PD for the MLRL to lock efficiently to generate a stable pulse train.

Given that a 10% fibers coupler is used at the optical output and an estimate optical loss of about 12 dB due to coupling, connector loss and attenuation of all optical components employed in the ring, the total optical power generated by the amplifiers must be about 30 dBm. To achieve this, we employ two EDF amplifiers of 16.5 dBm output power, each positioned before and after the optical coupler, one is used to compensate for the optical losses and one for generating sufficient optical gain and power to dominate the electronic noise in the regenerative loop.

4.2.2.2 Experimental Setup and Results

The experimental setups for MLL and RMLL are as shown, again, in Figures 4.12 and 4.13. Associate equipment used for monitoring of the mode locking and measurement of the lasers is also included. However, we note the following:

---

* By using the relationship of the quantum noise spectral density of $2qRP_{av}$ with P_{av} the average optical power, q the electronic charge, and R the responsivity of the detector.

To lock the lasing mode of the MLL to a certain repetition rate or multiple harmonic of the fundamental ring frequency, a synthesizer is required to generate the required sinusoidal waves for modulating the optical intensity modulator and tuned to a harmonic of the cavity fundamental frequency.

A signal must be created for the purpose of triggering the digital oscilloscope to observe the locking of the detected optical pulse train. For the HP-54118A, amplitude of this signal must be >200 mV. This is also critical for the RMLL setup as the RF signal detected and phase locked via the clock recovery circuitry must be spit to generate this triggering signal.

The following are the typical experimental procedures: (1) After the connection of all optical components with the ring path broken, ideally at the output of the fibers coupler, a CW optical source can be used to inject optical waves at a specific wavelength to monitor the optical loss of the ring; (2) close the optical ring and monitor the average optical power at the output of the 90:10 fibers coupler and hence estimate the optical power available at the PD is about −3 dBm after a 50:50 fibers coupler; (3) determine whether an optical amplifier is required for detecting the optical pulse train or whether this optical power is sufficient for O/E RF feedback condition as stated above; (4) set the biasing condition and hence the bias voltage of the optical modulator; and (5) tune the synthesizer or the electrical phase to synchronize the generation and locking of the optical pulse train.

One could observe the following: (1) The optical pulse train generated at the output of the MLL or RMML. Experimental setup is shown in Figure 4.14. (2) Synthesized modulating sinusoidal waveforms can be monitored as shown in

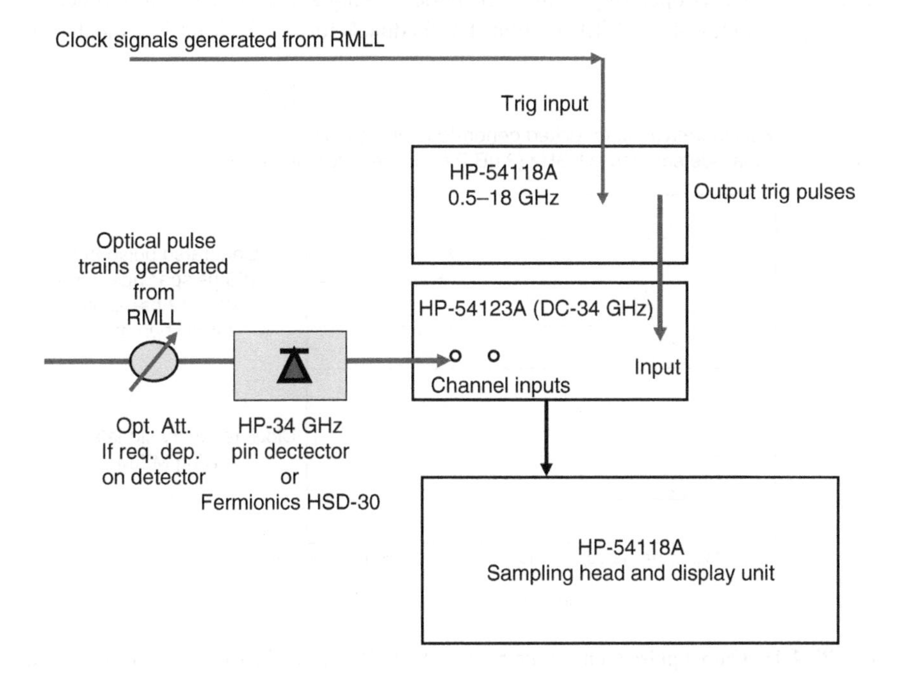

FIGURE 4.14 Experimental setup for monitoring the locking of the photonic pulse train.

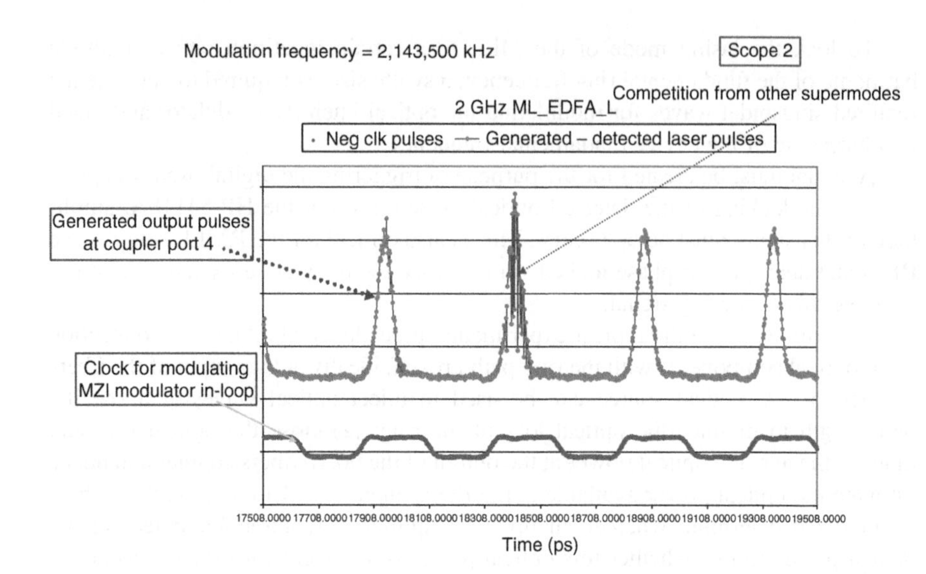

FIGURE 4.15 Detected pulse train at the MLRL output tested at a multiple frequency in the range of 2 GHz repetition frequency.

Figures 4.15 through 4.17. Figure 4.15 illustrates the mode locking of an MLL operating at around 2 GHz repetition rate with the modulator driven from a pattern generator while Figures 4.16 and 4.17 show the sinusoidal waveforms generating when the MLRL is operating at the self-mode-locking state. (3) The interference of other super-modes of the MLL without RF feedback for self-locking is indicated in

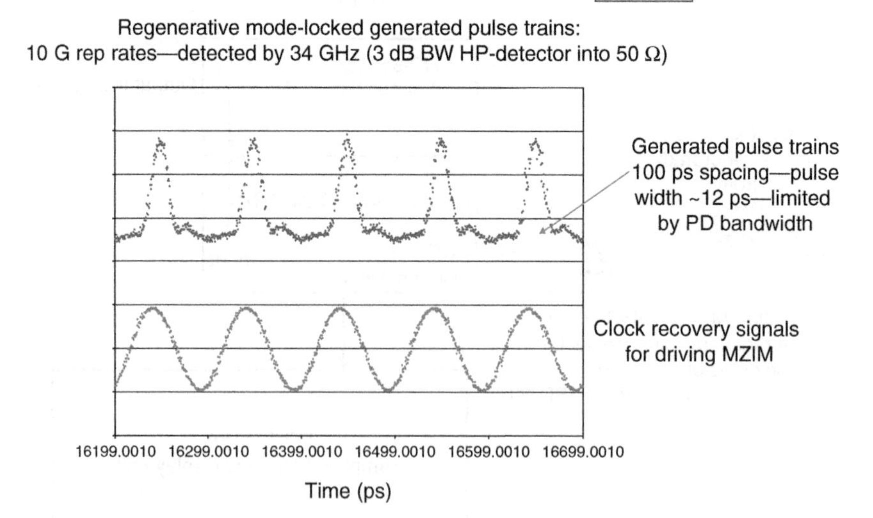

FIGURE 4.16 Output pulse trains of the regenerative MLRL and the RF signals as recovered for modulating the MZIM for self-locking.

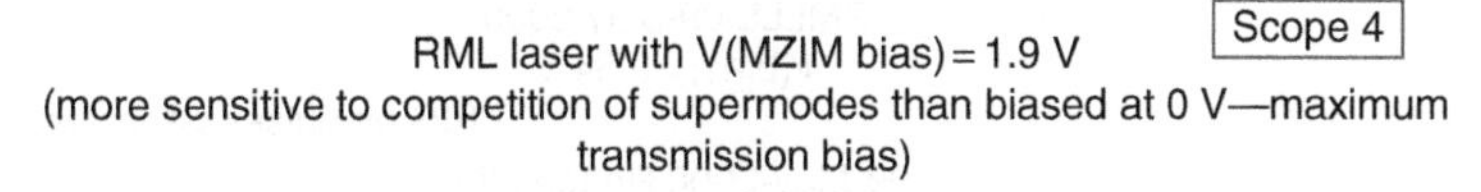

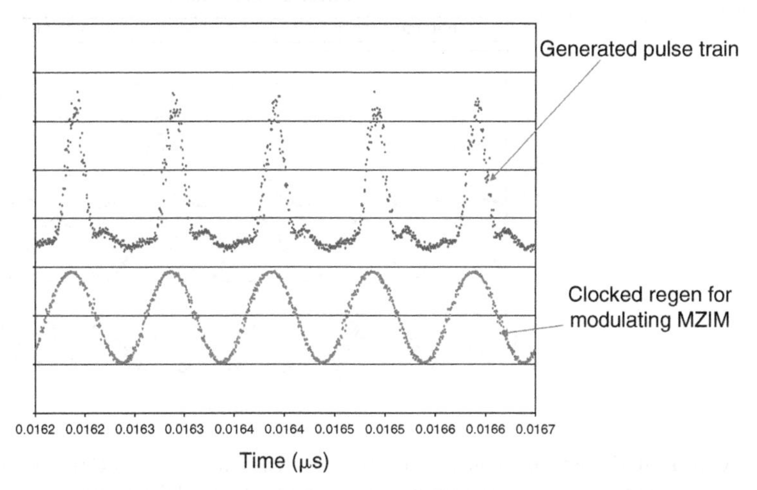

FIGURE 4.17 Detected output pulse trains of the regenerative MLRL and recovered clock signal when the MZIM is biased at a negative going slope of the operating characteristics of the modulator.

Figure 4.15. (4) Observed optical spectrum (not available in electronic form). (5) Electrical spectrum of the generated pulse trains was observed showing a −70 dB super-mode suppression under the locked state of the regenerative MLRL. (6) Figures 4.16 and 4.18 show that the regenerative MLRL can be operating under the cases when the modulator is biased either at the positive or at the negative going slope of the optical transfer characteristics of the Mach–Zehnder modulator.

Optical pulse width is measured using an optical autocorrelator (slow- or fast-scan mode). Typical pulse width obtained with the slow-scan autocorrelator is shown in Figure 4.19. Minimum pulse duration obtained was 4.5 ps with a time-bandwidth product of about 3.8 showing that the generated pulse is near transform limited.

The BER measurement can be used to monitor the stability of the regenerative MLRL. The BER error detector was then programmed to detect all "1" at the decision level at a tuned amplitude level and phase delay. The clock source used is that produced by the laser itself. This setup is shown in Figure 4.20 and an error-free measurement has been achieved for over 20 h. The O/E detected waveform of the output pulse train for testing the BER is shown in Figure 4.21 after 20 h operation, the recorded waveform is obtained under infinite persistence mode of the digital oscilloscope.

A drift of clock frequency of about 20 kHz over 1 h in open laboratory environment is observed. This is acceptable for a 10 GHz repetition rate.

The clock recovered waveforms were also monitored at the initial locked state and after the long-term test as shown in Figures 4.16 and 4.17, respectively. Figure 4.17 is obtained under the infinite persistence mode of the digital oscilloscope.

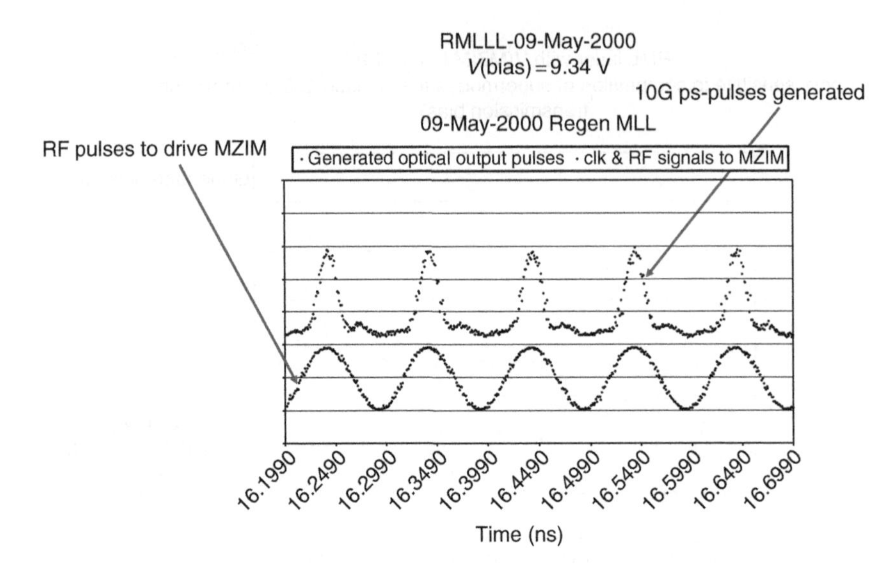

FIGURE 4.18 Output pulse trains and clock recovered signals of the 10G regenerative MLRL when the modulator is biased at the positive going slope of the modulator operating transfer curve.

We note the following factors, which are related to the above measurements (Figures 4.15 through 4.17):

All the above measurements have been conducted with two distributed optical amplifiers (GTi EDF optical amplifiers) driven at 180 mA and a specified output optical power of 16.5 dBm.

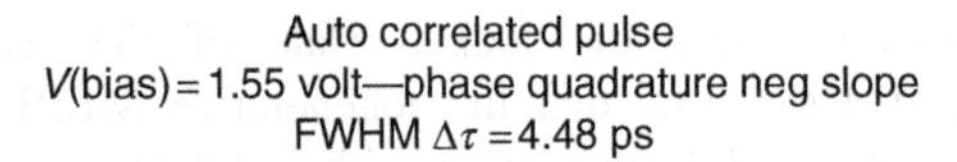

Autocorrelator set thumbwheel 6 and 0 (100 ps range at 10 ps/s delay rate)

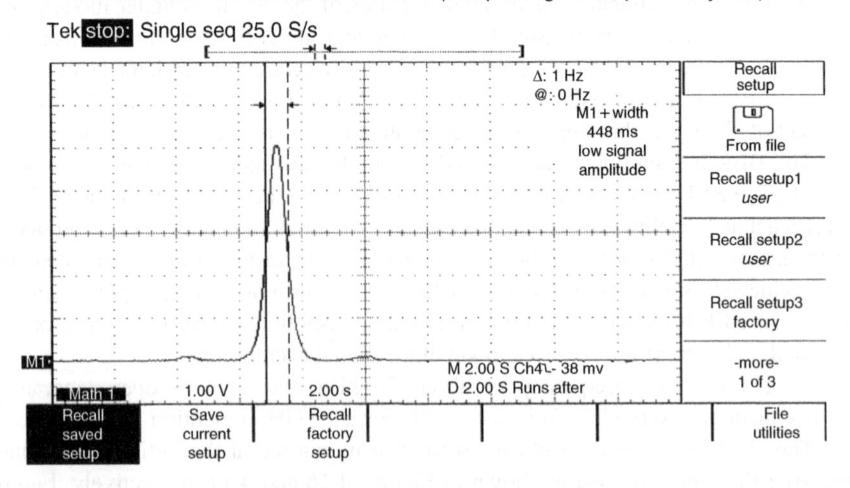

FIGURE 4.19 Autocorrelation trace of output pulse trains of 9.95 GHz regenerative MLRL.

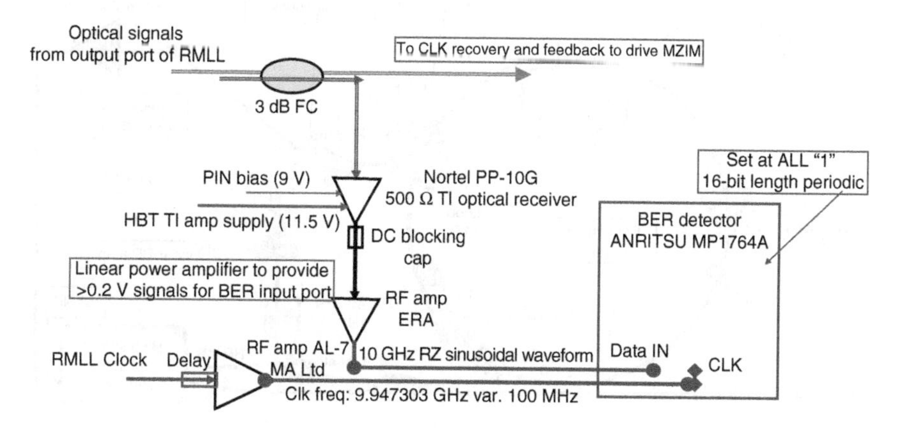

FIGURE 4.20 Experimental setup for monitoring the BER of the photonic pulse train.

Optical pulse trains are detected with 34 GHz 3 dB bandwidth HP pin detector directly coupled to the digital oscilloscope without using any optical preamplifier.

4.2.2.2.1 40 GHz Regenerative Mode-Locked Laser

Following our initial success of the construction and testing of a regenerative MLRL, a 40 GHz repetition rate regenerative mode-locked laser was constructed.

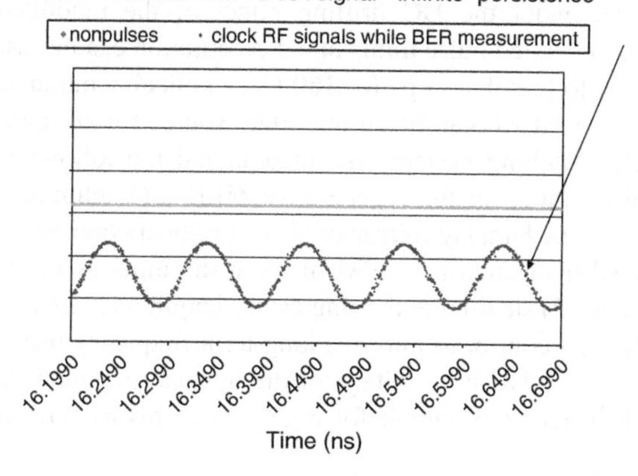

FIGURE 4.21 BER measurement—O/E detected signals from the generated output pulse trains for BER test set measurement. The waveform is obtained after 20 h persistence.

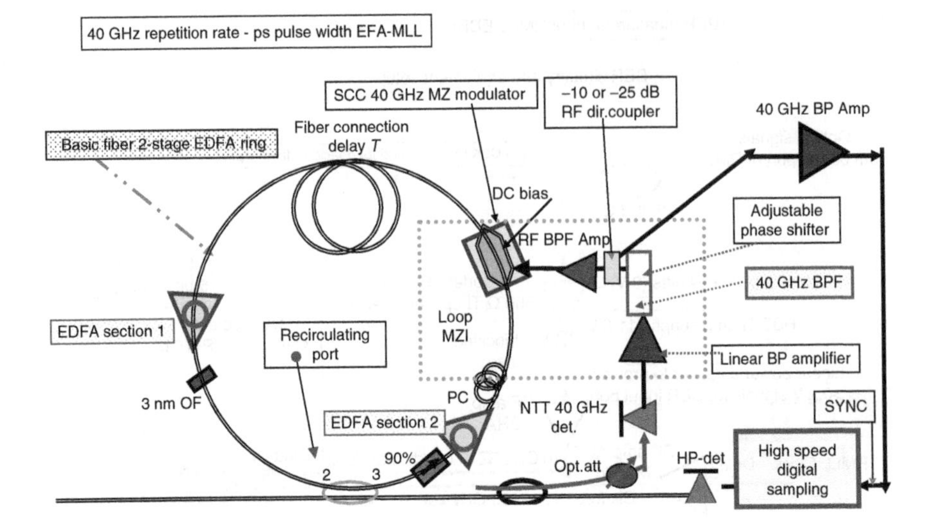

FIGURE 4.22 Setup of a 40 G regenerative MLRL.

The schematic arrangement of the 40 G regenerative MLRL is shown in Figure 4.22. Initial observation of the locking and generation of the laser has been observed and progress of this laser design and experiments will be reported in the near future.

4.2.2.3 Remarks

We have successfully constructed a mode-locked laser operating under an open loop condition and with O/E RF feedback providing regenerative mode locking. The O/E feedback can certainly provide a self-locking mechanism under the condition that the polarization characteristics of the ring laser are manageable. This is done by ensuring that all fibers path is under constant operating condition. The regenerative MLRL can self-lock even under the DC drifting effect of the modulator bias voltage (over 20 h).* The generated pulse trains of 4.5 ps duration can be, without difficulty, compressed further to less than 3 ps for 160 Gb/s optical communication systems.

The regenerative MLRL can be an important source for all-optical switching of an optical packet switching system. We recommend the following for successful construction and operation of the regenerative MLRL: (1) eliminating polarization drift through the use of Faraday mirror or all PM optical components, for example, polarized Er-doped fiber amplifiers, PM fibers at the input and output ports of the intensity modulator; (2) stabilizing the ring cavity length with appropriate packaging and via piezo/thermal control to improve long-term frequency drift; (3) control and automatic tuning of the DC bias voltage of the intensity modulator; (4) developing electronic RF clock recovery circuit for regenerative MLRL operating at 40 GHz

* Typically, the DC bias voltage of a $LiNbO_3$ intensity modulator is drifted by 1.5 V after 15 h of continuous operation.

repetition rate together with appropriate polarization control strategy; (5) studying the dependence of the optical power circulating in the ring laser by varying the output average optical power of the optical amplifiers under different pump power conditions; (6) incorporating a phase modulator, in lieu of the intensity modulator, to reduce the complexity of polarization dependence of the optical waves propagating in the ring cavity, thus minimizing the bias drift problem of the intensity modulator.

4.2.3 Active Mode-Locked Fiber Ring Laser by Rational Harmonic Detuning

In this section, we investigate the system behavior of rational harmonic mode-locking in the fiber ring laser using phase plane technique of the nonlinear control engineering. Furthermore, we examine the harmonic distortion contribution to this system performance. We also demonstrate 660× and 1230× repetition rate multiplications on 100 MHz pulse train generated from an active harmonically mode-locked fiber ring laser, hence achieve 66 and 123 GHz pulse operations by using rational harmonic detuning, which is the highest rational harmonic order reported to date.

4.2.3.1 Rational Harmonic Mode Locking

In an active harmonically mode-lock fiber ring laser, the repetition frequency of the generated pulses is determined by the modulation frequency of the modulator, $f_m = qf_c$, where q is the qth harmonic of the fundamental cavity frequency, f_c, which is determined by the cavity length of the laser, $f_c = c/nL$, where c is the speed of light, n is the refractive index of the fiber, and L is the cavity length. Typically, f_c is in the range of kilohertz or megahertz. Hence, to generate gigahertz pulse train, mode locking is normally performed by modulation in the states of $q \gg 1$, i.e., q pulses circulating within the cavity, which is known as harmonic mode locking. By applying a slight deviation or a fraction of the fundamental cavity frequency, $\Delta f = f_c/m$, where m is the integer, the modulation frequency becomes

$$f_m = qf_c \pm \frac{f_c}{m} \tag{4.25}$$

This leads to m-times increase in the system repetition rate, $f_r = mf_m$, where f_r is the repetition frequency of the system [2]. When the modulation frequency is detuned by an m fraction, the contributions of the detuned neighboring modes are weakened, only every mth lasing mode oscillates in phase and the oscillation waveform maximums accumulate, hence achieving in m times higher repetition frequency. However, the small but not negligible detuned neighboring modes affect the resultant pulse train, which leads to uneven pulse amplitude distribution and poor long-term stability. This is considered as harmonic distortion in our modeling, and it depends on the laser line width and amount detuned, i.e., fraction m. The amount of the allowable detunes or rather the obtainable increase in the system repetition rate by this technique is very much limited by the amount of harmonic distortion. When the amount of frequency detuned is too small relative to the modulation frequency, i.e., large m, contributions of the neighboring lasing modes become prominent, thus reduce the repetition rate multiplication capability significantly. In another words, no repetition frequency multiplication is achieved when the detuned frequency is unnoticeably small. Often the case, it is considered as the system noise due to improper modulation frequency tuning. In addition, the pulse amplitude fluctuation is also determined by this harmonic distortion.

4.2.3.2 Experimental Setup

The experimental setup of the active harmonically mode-locked fiber ring laser is shown in Figure 4.23. The principal element of the laser is an optical close loop with an optical gain medium, a Mach–Zehnder amplitude modulator (MZM), an optical PC, an optical bandpass filter (BPF), optical couplers, and other associated optics.

The gain medium used in our fiber laser system is an EDFA with saturation power of 16 dBm. A polarization-independent optical isolator is used to ensure unidirectional lightwave propagation as well as to eliminate back reflections from the fiber splices and optical connectors. A free space filter with 3 dB bandwidth of 4 nm at 1555 nm is inserted into the cavity to select the operating wavelength of the generated signal and to reduce the noise in the system. In addition, it is responsible for the longitudinal modes selection in the mode-locking process. The birefringence of the fiber is compensated by a polarization controller, which is also used for the polarization alignment of the linearly polarized lightwave before entering the planar structure modulator for better output efficiency. Pulse operation is achieved by introducing an asymmetric coplanar traveling wave 10 Gb/s lithium niobate, $Ti:LiNbO_3$ Mach–Zehnder amplitude modulator into the cavity with half wave voltage, V_π of 5.8 V, and insertion loss of $\leq$7 dB. The modulator is DC biased near the quadrature point and not more than the V_π such that it operates around the linear region of its characteristic curve. The modulator is driven by a 100 MHz, 100 ps step recovery diode (SRD), which is in turn driven by an RF amplifier (RFA) an RF

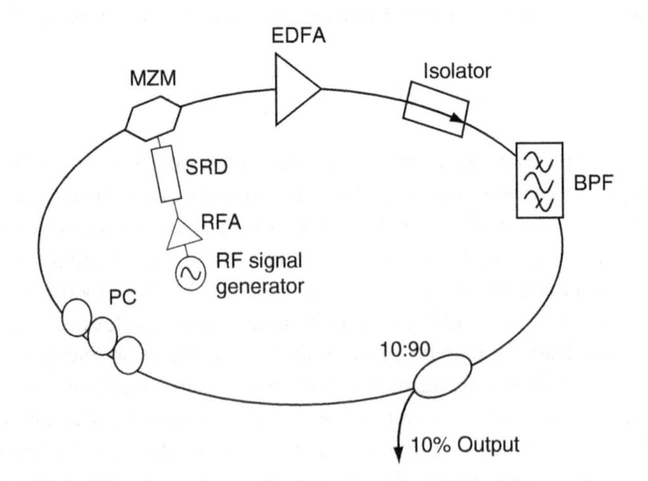

FIGURE 4.23 Schematic diagram of an active mode-locked fiber ring laser.

signal generator. The modulating signal generated by the SRD is a ~1% duty cycle Gaussian pulse train. The output coupling of the laser is optimized using a 10/90 coupler. Ninety percent of the optical field power is coupled back into the cavity ring loop, while the remaining portion is taken out as the output of the laser and analyzed.

4.2.3.3 Phase Plane Analysis

Nonlinear system frequently has more than one equilibrium point. It can also oscillate at fixed amplitude and fixed period without external excitation. This oscillation is called limit cycle. However, limit cycles in nonlinear systems are different from linear oscillations. First, the amplitude of self-sustained excitation is independent of the initial condition, whereas the oscillation of a marginally stable linear system has its amplitude determined by the initial conditions. Second, marginally stable linear systems are very sensitive to changes, whereas limit cycles are not easily affected by parameter changes [52].

Phase plane analysis is a graphical method of studying second-order nonlinear systems. The result is a family of system motion of trajectories on a two-dimensional plane, which allows us to visually observe the motion patterns of the system. Nonlinear systems can display more complicated patterns in the phase plane, such as multiple equilibrium points and limit cycles. In the phase plane, a limit cycle is defined as an isolated closed curve. The trajectory has to be both closed, indicating the periodic nature of the motion, and isolated, indicating the limiting nature of the cycle [52].

The system modeling of the rational harmonic mode-locked fiber ring laser system is done based on the following assumptions: (1) detuned frequency is perfectly adjusted according to the fraction number required, (2) small harmonic distortion, (3) no fiber nonlinearity is included in the analysis, (4) no other noise sources are involved in the system, and (5) Gaussian lasing mode amplitude distribution analysis.

The phase plane of a perfect 10 GHz mode-locked pulse train without any frequency detune is shown Figure 4.24 and the corresponding pulse train is shown in Figure 4.25a. The shape of the phase plane exposes the phase between the displacement and its derivative. From the phase plane obtained, one can easily observe that the origin is a stable node and the limit cycle around that vicinity is a stable limit cycle, hence leading to stable system trajectory. $4\times$ multiplication pulse trains, i.e., $m=4$, without and with 5% harmonic distortion are shown in Figures 4.25b and c. Their corresponding phase planes are shown in Figures 4.26a and b. For the case of zero harmonic distortion, which is the ideal case, the generated pulse train is perfectly multiplied with equal amplitude and the phase plane has stable symmetry periodic trajectories around the origin too. However, for the practical case, i.e., with 5% harmonic distortion, it is obvious that the pulse amplitude is unevenly distributed, which can be easily verified with the experimental results obtained in Ref. [25]. Its corresponding phase plane shows more complex asymmetry system trajectories.

One may naively think that the detuning fraction, m, could be increased to a very large number, so a very small frequency is deviated, Δf, so as to obtain a very high

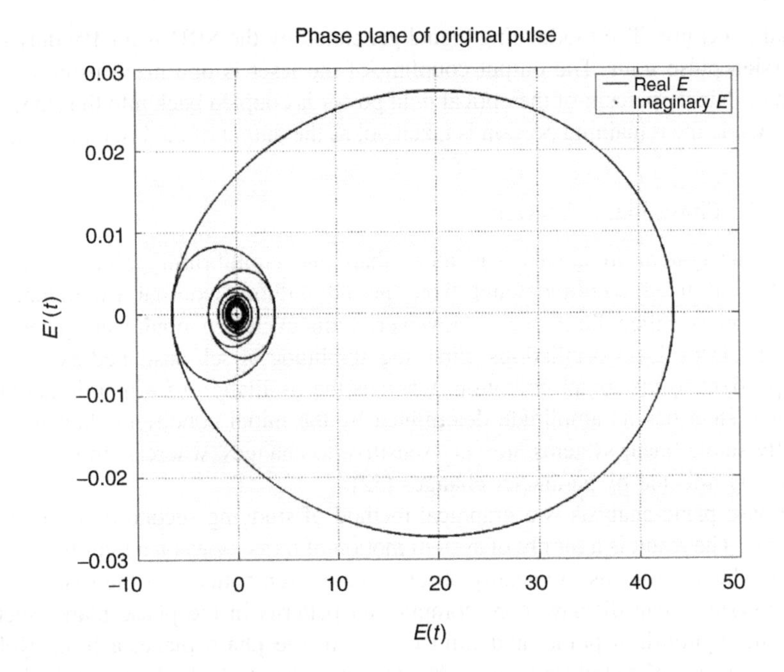

FIGURE 4.24 Phase plane of a 10 GHz mode-locked pulse train (solid line, real part of the energy; dotted line, imaginary part of the energy; *x*-axes, *E*(*t*) and *y*-axes, *E*′(*t*)).

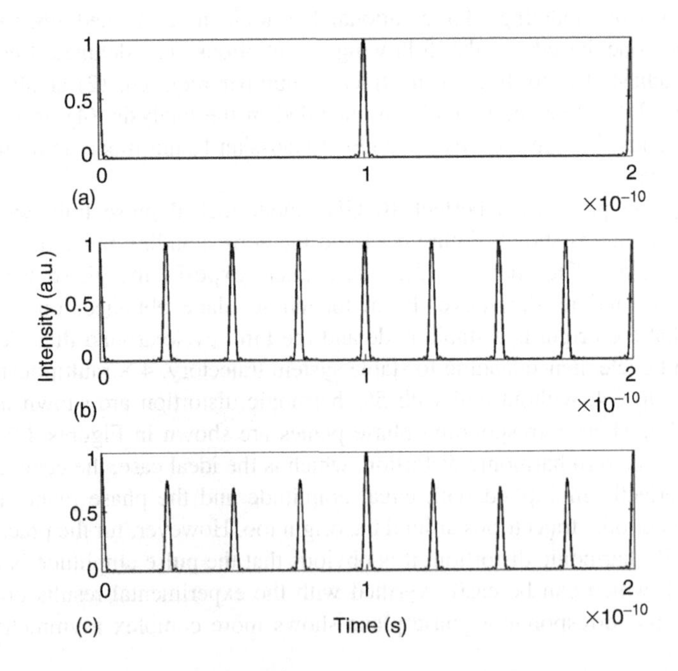

FIGURE 4.25 Normalized pulse propagation of original pulse (a); detuning fraction of 4, with 0% (b) 5% (c) harmonic distortion noise.

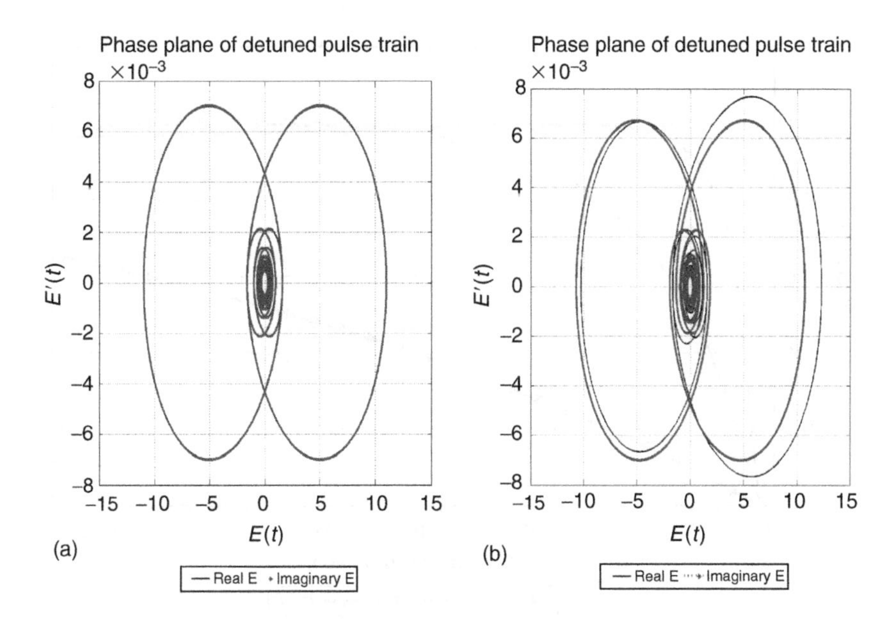

FIGURE 4.26 Phase plane of detuned pulse train, $m = 4$, 0% harmonic distortion (a), and (b) 5% harmonic distortion (solid line, real part of the energy; dotted line, imaginary part of the energy; x-axes, $E(t)$ and y-axes, $E'(t)$).

repetition frequency. This is only true in the ideal world, if no harmonic distortion is present in the system. However, this is unreasonable for a practical mode-locked laser system.

We define the percentage fluctuation, $\%F$ as follows:

$$\%F = \frac{E_{max} - E_{min}}{E_{max}} \times 100\% \tag{4.26}$$

where E_{max} and E_{min} are the maximum and minimum peak amplitude of the generated pulse train. For any practical mode-locked laser system, fluctuations above 50% should be considered as poor laser system design. Therefore, this is one of the limiting factors in a rational harmonic mode-locking fiber laser system. The relationships between the percentage fluctuation and harmonic distortion for three multipliers ($m = 2$, 4, and 8) are shown in Figure 4.27. Thus, the obtainable rational harmonic mode locking is very much limited by the harmonic distortion of the system. For 100% fluctuation, it means no repetition rate multiplication, but with additional noise components; a typical pulse train and its corresponding phase plane are shown in Figures 4.28 (lower plot) and 4.29 with $m = 8$ and 20% harmonic distortion. The asymmetric trajectories of the phase graph explain the amplitude unevenness of the pulse train. Furthermore, it shows a more complex pulse formation system. Thus, it is clear that for any harmonic mode-locked laser system, the small side pulses generated are largely due to improper or not exact tuning of the modulation frequency of the system. An experimental result is depicted in Figure 4.32 for a comparison.

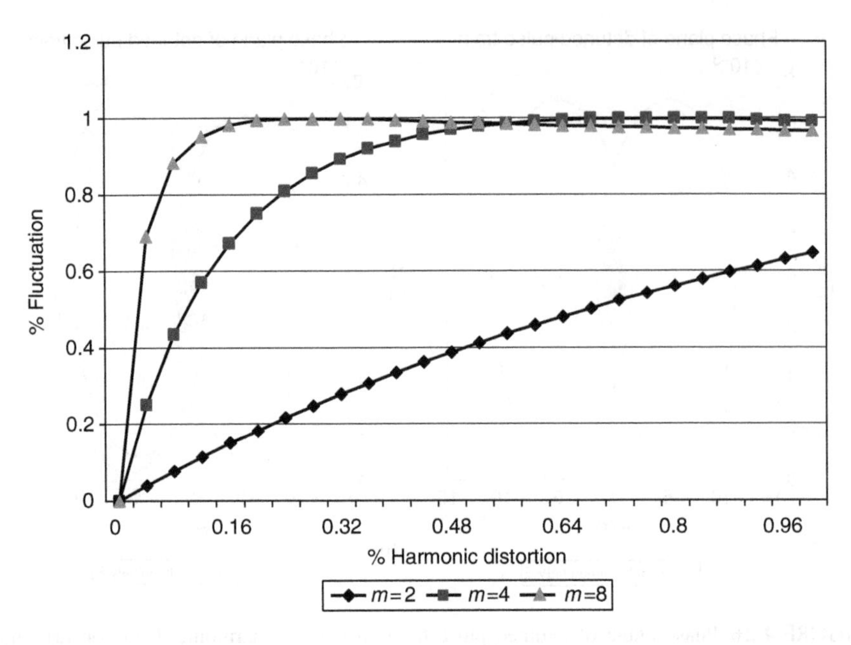

FIGURE 4.27 Relationship between the amplitude fluctuation and the percentage harmonic distortion (diamond, $m=2$; square, $m=4$; triangle, $m=8$).

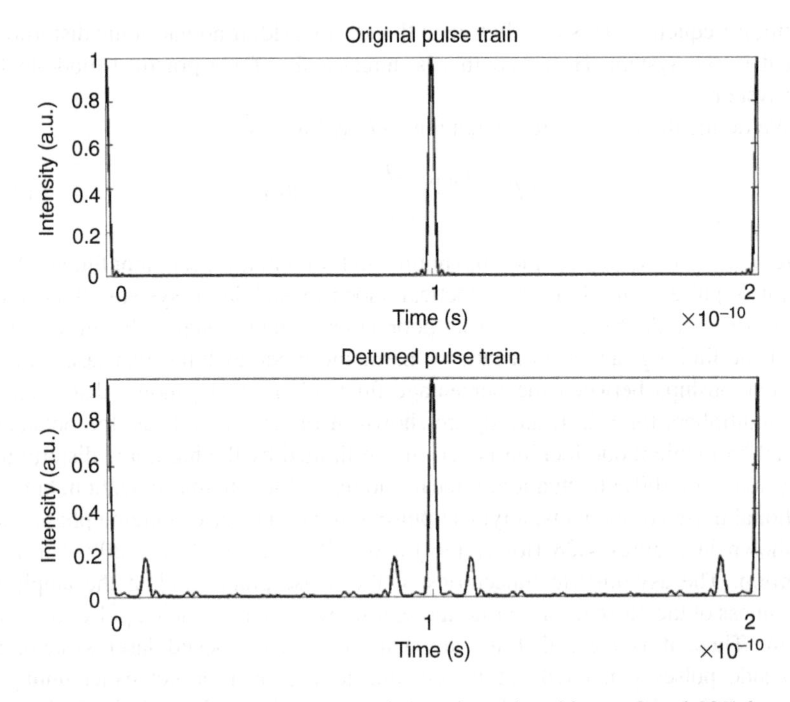

FIGURE 4.28 10 GHz pulse train (upper plot), pulse train with $m=8$ and 20% harmonic distortion (lower plot).

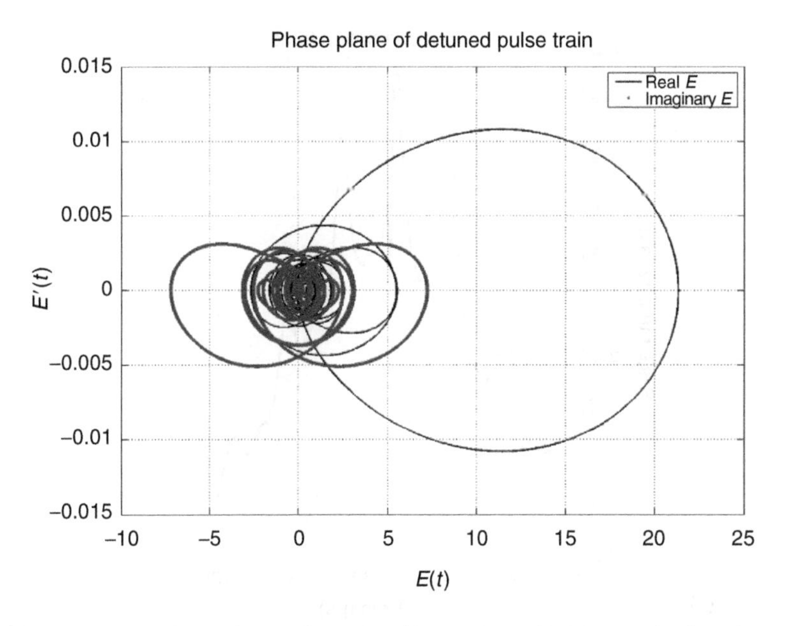

FIGURE 4.29 Phase plane of the pulse train with $m=8$ and 20% harmonic distortion.

4.2.3.4 Results and Discussion

By careful adjustment of the modulation frequency, polarization, gain level, and other parameters of the fiber ring laser, we have managed to obtain the 660th and 1230th order of rational harmonic detuning in the mode-locked fiber ring laser with base frequency of 100 MHz, hence achieving 66 and 123 GHz repetition frequency pulse operation. The autocorrelation traces and optical spectrums of the pulse operations are shown in Figure 4.30. With Gaussian pulse assumption, the obtained pulse widths of the operations are 2.5456 ps and 2.2853 ps, respectively. For the 100 MHz pulse operation, i.e., without any frequency detune, the generated pulse width is about 91 ps. Thus, we achieved not only an increase in the pulse repetition frequency but also a decrease in the generated pulse widths. This pulse narrowing effect is partly due to the self-phase modulation effect of the system, as observed in the optical spectrums. Another reason for this pulse shortening is stated by Haus et al. [49], where the pulse width is inversely proportional to the modulation frequency, as follows:

$$\tau^4 = \frac{2g}{M\omega_{\mathrm{m}}^2\omega_{\mathrm{g}}^2} \tag{4.27}$$

where

τ is the pulse width of the mode-locked pulse
ω_{m} is the modulation frequency
g is the gain coefficient
M is the modulation index
ω_{g} is the gain bandwidth of the system

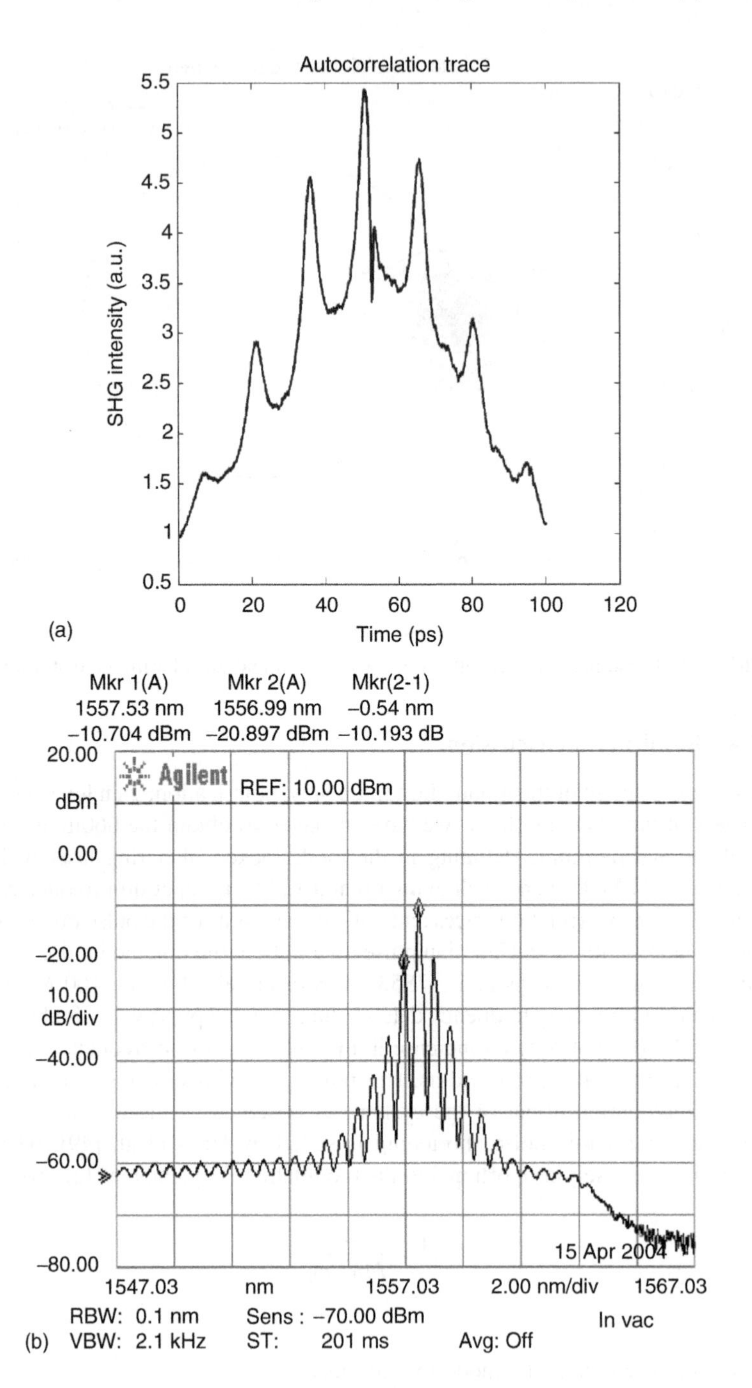

FIGURE 4.30 (a) Autocorrelation traces of 66 GHz and (c) 123 GHz pulse operation; (b) optical spectrums of 66 GHz and (d) 123 GHz.

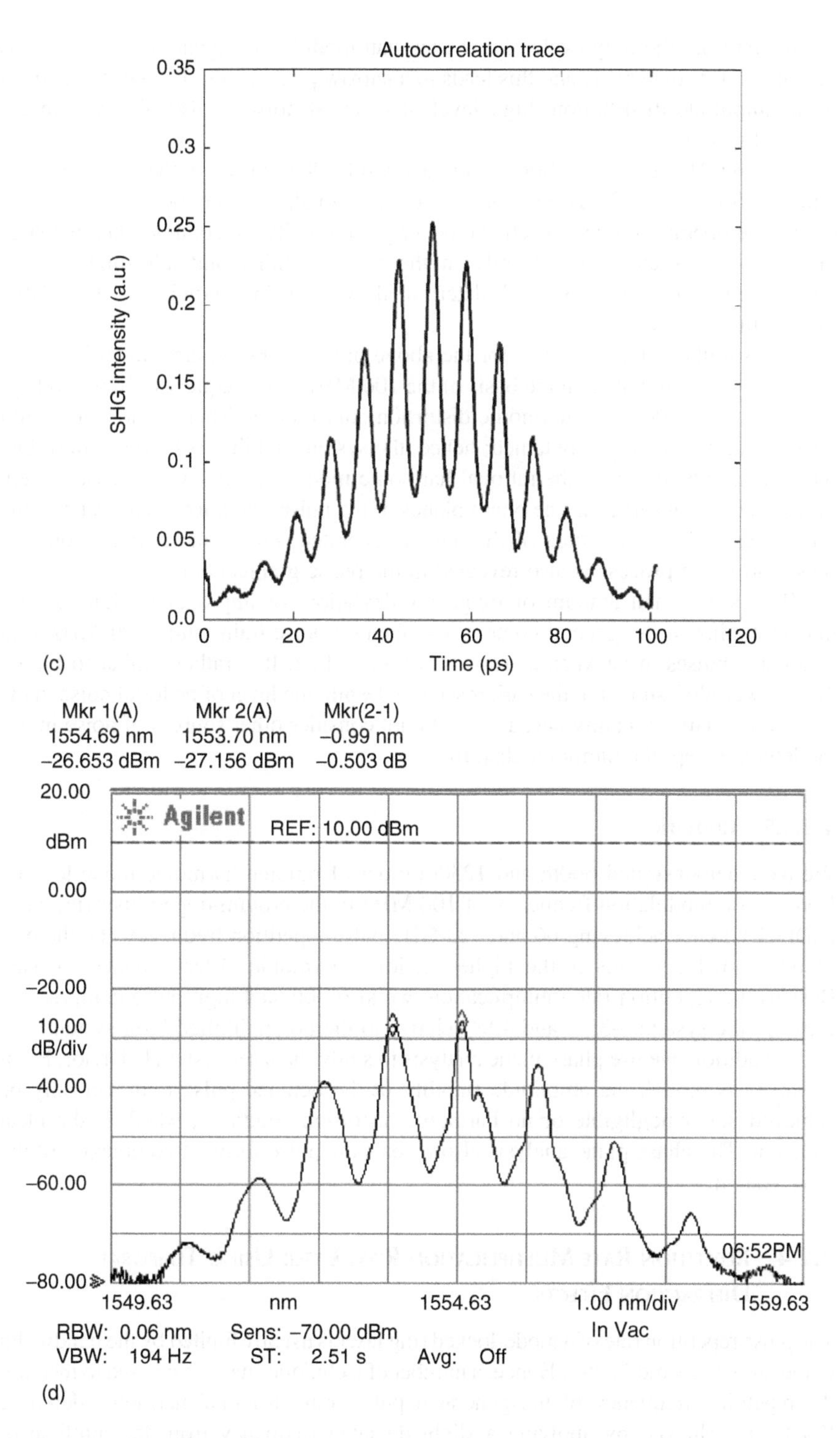

FIGURE 4.30 (continued)

In addition, the duty cycle of our Gaussian modulation signal is ~1%, which is very much less than 50%, and this leads to a narrow pulse width. Besides the uneven pulse amplitude distribution, high level of pedestal noise is also observed in the obtained results.

For 66 GHz pulse operation, 4 nm bandwidth filter is used in the setup, but it is removed for the 123 GHz operation. It is done so to allow more modes to be locked during the operation, thus, to achieve better pulse quality. In contrast, this increases the level of difficulty significantly in the system tuning and adjustment. As a result, the operation is very much determined by the gain bandwidth of the EDFA used in the laser setup.

The simulated phase planes for the above pulse operation are shown in Figure 4.31. They are simulated on the basis of the 100 MHz base frequency, 10 round trips condition, and 0.001% of harmonic distortion contribution. There is no stable limit cycle in the phase graphs obtained; hence, the system stability is hardly achievable, which is a known fact in the rational harmonic mode locking. Asymmetric system trajectories are observed in the phase planes of the pulse operations. This reflects the unevenness of the amplitude of the pulses generated. Furthermore, more complex pulse formation process is also revealed in the phase graphs obtained.

By a very small amount of frequency deviation, or improper modulation frequency tuning in the general context, we obtain a pulse train with ~100 MHz with small side pulses in between as shown in Figure 4.32. It is rather similar to Figure 4.28 (lower plot) shown in the earlier section despite the level of pedestal noise in the actual case. This is mainly because we do not consider other sources of noise in our modeling, except the harmonic distortion.

4.2.3.5 Remarks

We have demonstrated 660th and 1230th order of rational harmonic mode locking from a base modulation frequency of 100 MHz in the erbium-doped fiber ring laser (EDFRL), hence achieving 66 and 123 GHz pulse repetition frequency. To the best of our knowledge, this is the highest rational harmonic order obtained to date. Besides the repetition rate multiplication, we also obtained high pulse compression factor in the system, ~35× and 40× relative to the nonmultiplied laser system.

In addition, we use phase plane analysis to study the laser system behavior. From the analysis model, the amplitude stability of the detuned pulse train can only be achieved under negligible or no harmonic distortion condition, which is the ideal situation. The phase plane analysis also reveals the pulse forming complexity of the laser system.

4.2.4 Repetition Rate Multiplication Ring Laser Using Temporal Diffraction Effects

The pulse repetition rate of a mode-locked ring laser is usually limited by the bandwidth of the intracavity modulator. Hence, a number of techniques have to be used to increase the repetition frequency of the generated pulse train. Rational harmonic detuning [25,37] is achieved by applying a slight deviated frequency from the multiple of

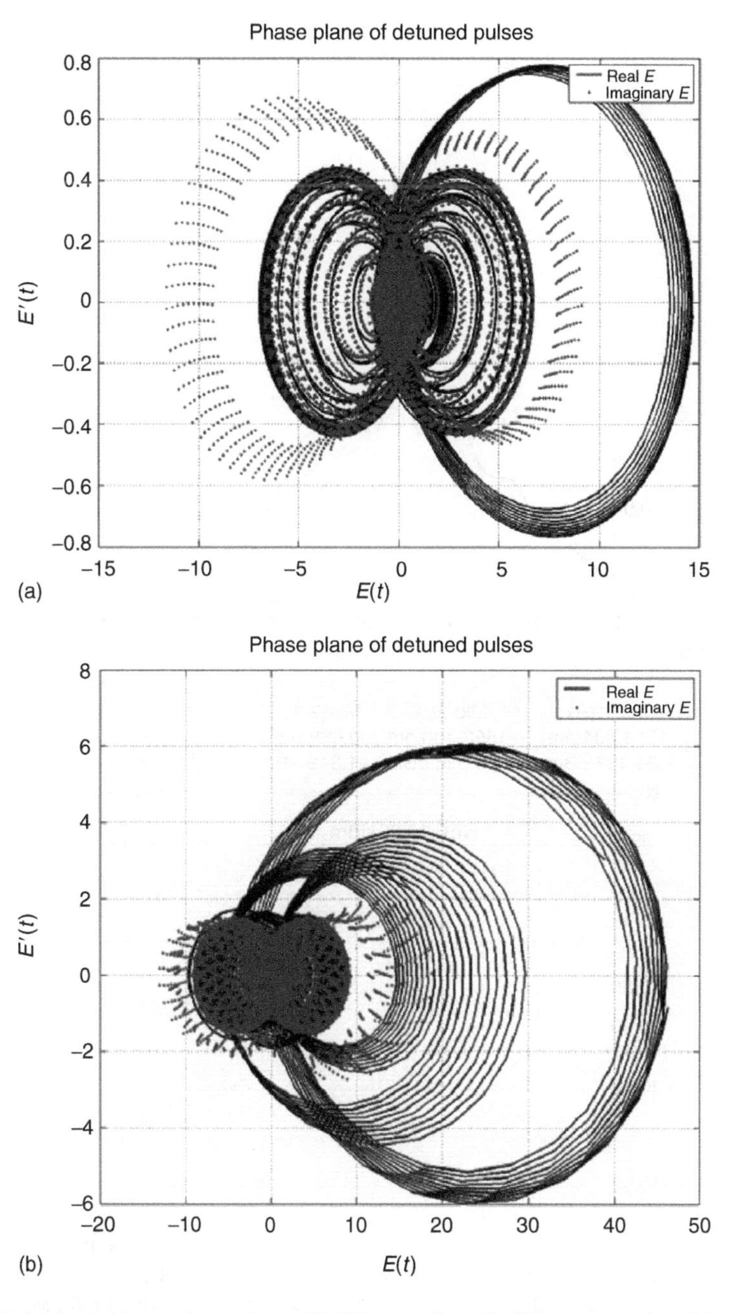

FIGURE 4.31 (a) Phase plane of the 66 GHz and (b) 123 GHz pulse train with 0.001% harmonic distortion noise.

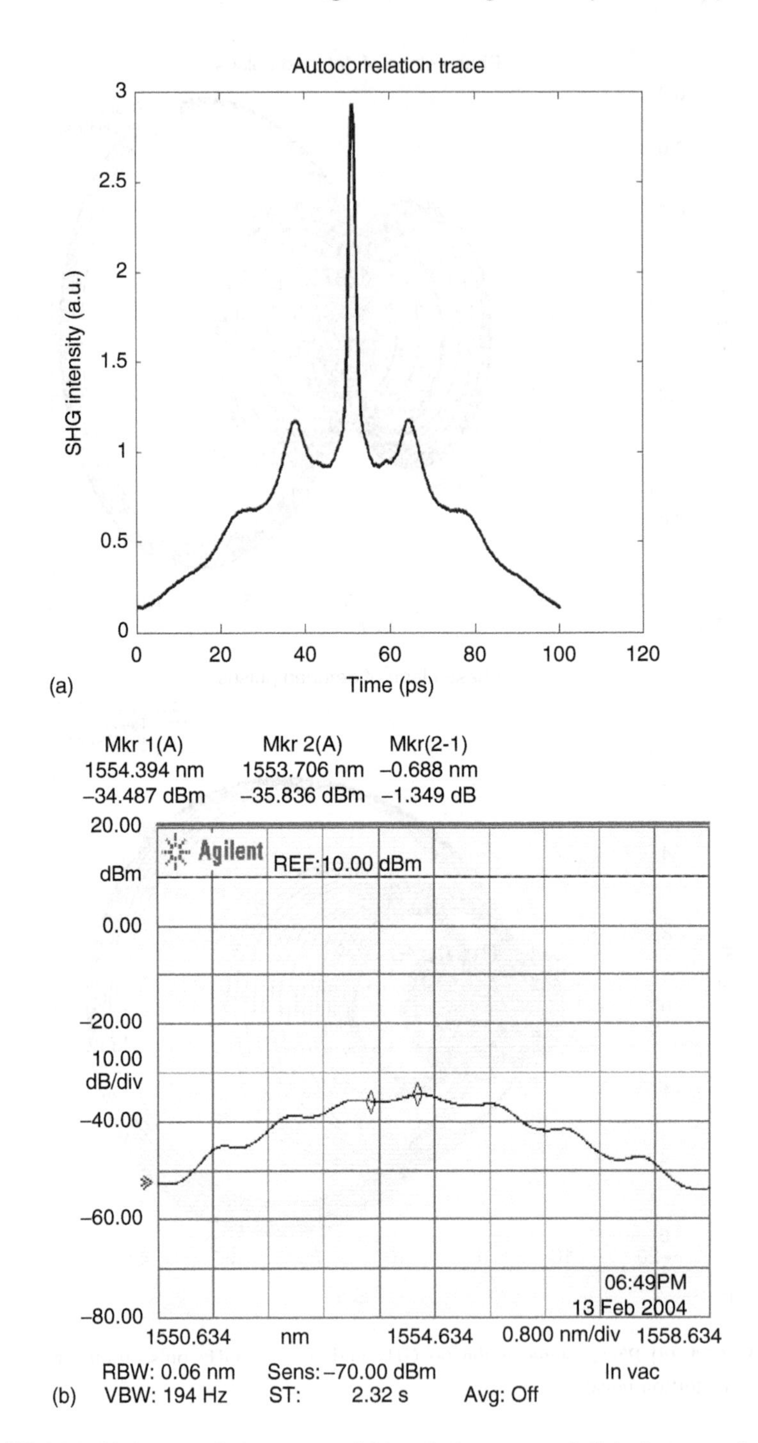

FIGURE 4.32 (a) Autocorrelation trace and (b) optical spectrum of slight frequency detune in the mode-locked fiber ring laser.

fundamental cavity frequency. 40 Ghz repetition frequency has been obtained by Wu and Dutta [25] using 10 GHz baseband modulation frequency with fourth-order rational harmonic mode locking. This technique is simple in nature. However, this technique suffers from inherent pulse amplitude instability, which includes both amplitude noise and inequality in pulse amplitude, furthermore, it gives poor long-term stability. Hence, pulse amplitude equalization techniques are often applied to achieve better system performance [24,27,53]. Fractional temporal Talbot-based repetition rate multiplication technique [27,29–31,53] uses the interference effect between the dispersed pulses to achieve the repetition rate multiplication. The essential element of this technique is the dispersive medium, such as LCFG [31,37] and single-mode fiber [31,32]. This technique will be discussed further in Section II. Intracavity optical filtering [28,35] uses modulators and a high finesse FFP within the laser cavity to achieve higher repetition rate by filtering out certain lasing modes in the mode-locked laser. Other techniques used in repetition rate multiplication include higher-order FM mode locking [35], optical time-domain multiplexing, etc.

The stability of high repetition rate pulse train generated is one of the main concerns for practical multigigabits per second optical communication system. Qualitatively, a laser pulse source is considered as stable if it is operating at a state where any perturbations or deviations from this operating point are not increased but suppressed. Conventionally, the stability analyses of such laser systems are based on the linear behavior of the laser in which we can analytically analyze the system behavior in both time and frequency domains. However, when the mode-locked fiber laser is operating under nonlinear regime, none of these standard approaches can be used, since direct solution of nonlinear different equation is generally impossible, hence frequency-domain transformation is not applicable. Although Talbot-based repetition rate multiplication systems are based on the linear behavior of the laser, there are still some inherent nonlinearities affecting its stability, such as the saturation of the embedded gain medium, nonquadrature biasing of the modulator, nonlinearities in the fiber, etc., hence, nonlinear stability approach must be adopted.

We investigate the stability and transient analyses of the GVD multiplied pulse train using the phase plane analysis of nonlinear control analytical technique [24]. This phase plane analysis is very common in modern control engineering for the study of stability and transient performances of the GVD repetition rate multiplication systems.

The stability and the transient response of the multiplied pulses are studied using the phase plane technique of nonlinear control engineering. We also demonstrated four times repetition rate multiplication on 10 Gb/s pulse train generated from the active harmonically mode-locked fiber ring laser, hence achieving 40 Gb/s pulse train by using fiber GVD effect. It has been found that the stability of the GVD multiplied pulse train, based on the phase plane analysis, is hardly achievable even under the perfect multiplication conditions. Furthermore, uneven pulse amplitude distribution is observed in the multiplied pulse train. In addition, the influences of the filter bandwidth in the laser cavity, nonlinear effect, and the noise performance are also studied in our analyses.

In Section 4.2.4.1, the GVD repetition rate multiplication technique is briefly given. Section 4.2.4.2 describes the experimental setup for the repetition rate multiplication. Section 4.2.4.4 investigates the dynamic behavior of the phase plane of GVD multiplication system, followed by simulation and experimental results. Finally, some concluding remarks and possible future developments are given.

4.2.4.1 GVD Repetition Rate Multiplication Technique

When a pulse train is transmitted through an optical fiber, the phase shift of kth individual lasing mode due to GVD is

$$\varphi_k = \frac{\pi\lambda^2 D z k^2 f_r^2}{c} \tag{4.28}$$

where

- λ is the center wavelength of the mode-locked pulses
- D is the fiber's GVD factor
- z is the fiber length
- f_r is the repetition frequency
- c is the speed of light in vacuum

This phase shift induces pulse broadening and distortion. At Talbot distance, $z_T = 2/\Delta\lambda f_r/D$ [29], the initial pulse shape is restored, where $\Delta\lambda = f_r\lambda^2/c$ is the spacing between Fourier-transformed spectrum of the pulse train. When the fiber length is equal to $z_T/(2m)$ (where $m = 2, 3, 4, \ldots$), every mth lasing mode oscillates in phase and the oscillation waveform maximums accumulate. However, when the phases of other modes become mismatched, this weakens their contributions to pulse waveform formation. This leads to the generation of a pulse train with a multiplied repetition frequency with m times. The pulse duration does not change significantly even after the multiplication, because every mth lasing mode dominates in pulse waveform formation of m times multiplied pulses. The pulse waveform therefore becomes identical to that generated from the mode-locked laser, with the same spectral property. Optical spectrum does not change after the multiplication process, because this technique utilizes only the change of phase relationship between lasing modes and does not use the fiber's nonlinearity.

The effect of higher-order dispersion might degrade the quality of the multiplied pulses, i.e., pulse broadening, appearance of pulse wings, and pulse-to-pulse intensity fluctuation. In this case, any dispersive media to compensate the fiber's higher-order dispersion would be required to complete the multiplication process. To achieve higher multiplications, the input pulses must have a broad spectrum and the fractional Talbot length must be very precise in order to receive high-quality pulses. If the average power of the pulse train induces the nonlinear suppression and experience, anomalous dispersion along the fiber, solitonic action would occur and prevent the linear Talbot effect from occurring.

The highest repetition rate obtainable is limited by the duration of the individual pulses, as pulses start to overlap when the pulse duration becomes comparable to the pulse train period, i.e., $m_{max} = \Delta T/\Delta t$, where ΔT is the pulse train period and Δt is the pulse duration.

4.2.4.2 Experimental Setup

Group velocity dispersion repetition rate multiplication is used to achieve 40 Gb/s operation. The input to the GVD multiplier is a 10.217993 Gb/s laser pulse source, obtained from active harmonically mode-locked fiber ring laser, operating at 1550.2 nm.

The principal element of the active harmonically mode-locked fiber ring laser is an optical closed loop with an optical gain medium (i.e., EDF with 980 nm pump source), an optical 10 GHz amplitude modulator, optical bandpass filter, optical fiber couplers, and other associated optics. The schematic construction of the active mode-locked fiber ring laser is shown in Figure 4.33. The active mode-locked fiber laser design is based on a fiber ring cavity where the 25 m EDF with Er^{3+} ion concentration of 4.7×10^{24} ions/m^3 is pumped by two diode lasers at 980 nm: SDLO-27–8000–300 and CosetK1116 with maximum forward pump power of 280 mW and backward pump power of 120 mW. The pump lights are coupled into the cavity by the 980/1550 nm WDM couplers; with insertion loss for 980 and 1550 nm signals are about 0.48 and 0.35 dB, respectively. A polarization-independent optical isolator ensures the unidirectional lasing. The birefringence of the fiber is compensated by a PC. A tunable FP filter with 3 dB bandwidth of 1 nm and wavelength tuning range from 1530 to 1560 nm is inserted into the cavity to select the center wavelength of the generated signal as well as to reduce the noise in the system. In addition, it is used for the longitudinal modes selection in the mode-locking process. Pulse operation is achieved by introducing a JDS Uniphase 10 Gb/s lithium niobate, Ti:$LiNbO_3$ Mach–Zehnder amplitude modulator into the cavity with half wave

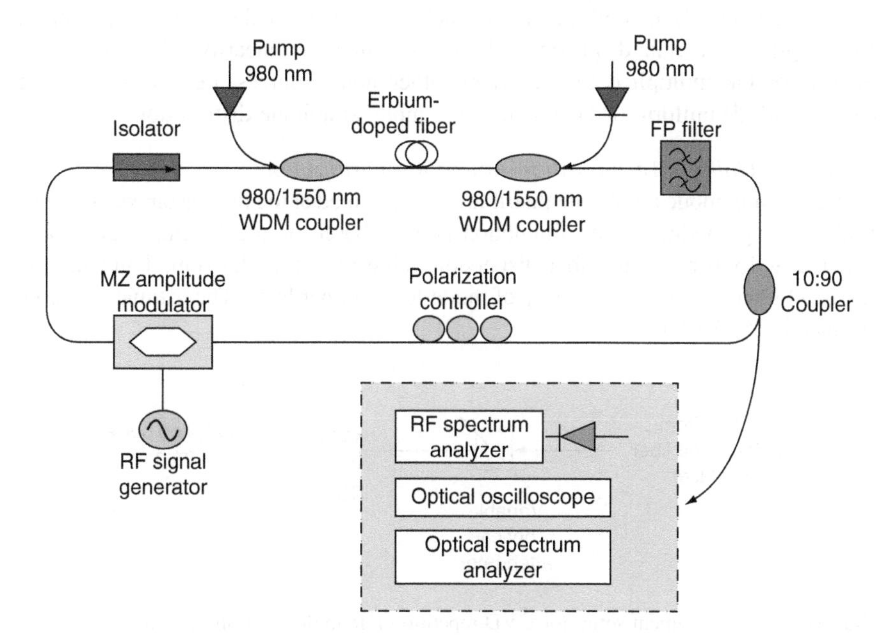

FIGURE 4.33 Schematic diagram for active mode-locked fiber ring laser.

voltage, V_π of 5.8 V. The modulator is DC biased near the quadrature point and not more than the V_π such that it operates on the linear region of its characteristic curve and driven by the sinusoidal signal derived from an Anritsu 68347C synthesized signal generator. The modulating depth should be less than unity to avoid signal distortion. The modulator has an insertion loss of $\leq$7 dB. The output coupling of the laser is optimized using a 10/90 coupler. Ninety percent of the optical field power is coupled back into the cavity ring loop, while the remaining portion is taken out as the output of the laser and is analyzed using a New Focus 1014B 40 GHz photodetector, Ando AQ6317B Optical Spectrum Analyzer, Textronix CSA 8000 80E01 50GHz Communications Signal Analyzer, or Agilent E4407B RF Spectrum Analyzer.

One fiber spool of about 3.042 km of dispersion compensating fiber (DCF), with a dispersion value of −98 ps/nm/km, was used in the experiment; the schematic of the experimental setup is shown in Figure 4.34. The variable optical attenuator used in the setup is to reduce the optical power of the pulse train generated by the mode-locked fiber ring laser, hence to remove the nonlinear effect of the pulse. A DCF length for 4× multiplication factor on the ~10 GHz signal is required and estimated to be 3.048173 km. The output of the multiplier (i.e., at the end of DCF) is then observed using Textronix CSA 8000 80E01 50GHz Communications Signal Analyzer.

4.2.4.3 Phase Plane Analysis

The system modeling for the GVD multiplier is done based on the following assumptions: (1) perfect output pulse from the mode-locked fiber ring laser without any timing jitter, (2) the multiplication is achieved under ideal conditions (i.e., exact fiber length for a certain dispersion value), (3) no fiber nonlinearity is included in the analysis of the multiplied pulse, (4) no other noise sources are involved in the system, and (5) uniform or Gaussian lasing mode amplitude distribution.

4.2.4.3.1 Uniform Lasing Mode Amplitude Distribution

Uniform lasing mode amplitude distribution is assumed at the first instance, i.e., ideal mode-locking condition. The simulation is done based on the 10 Gb/s pulse train, centered at 1550 nm, with fiber dispersion value of –98 ps/km/nm, 1 nm flat-top passband filter is used in the cavity of mode-locked fiber laser. The estimated Talbot distance is 25.484 km.

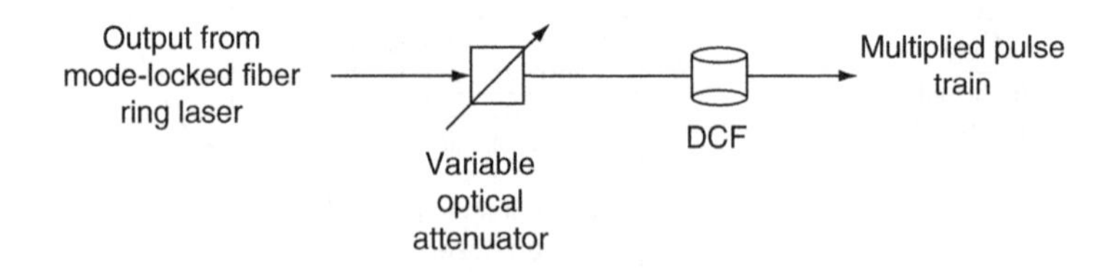

FIGURE 4.34 Experiment setup for GVD repetition rate multiplication system.

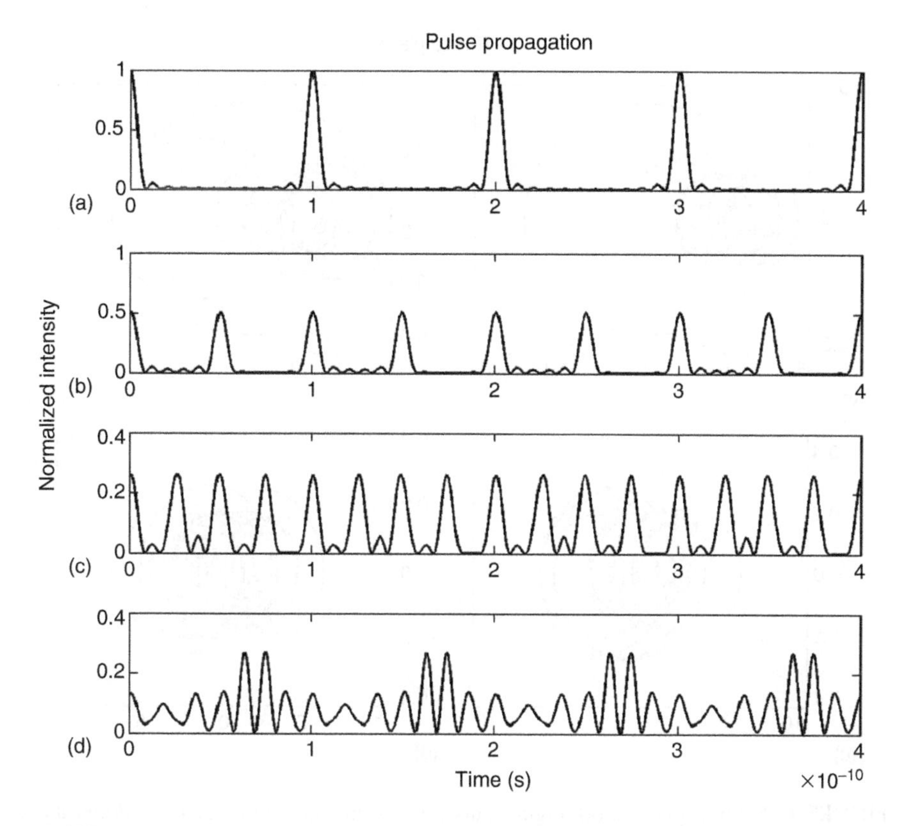

FIGURE 4.35 Pulse propagation of (a) original pulse, (b) 2× multiplication, (c) 4× multiplication, and (d) 8× multiplication with 1 nm filter bandwidth and equal lasing mode amplitude analysis.

The original pulse (direct from the mode-locked laser) propagation behavior and its phase plane are shown in Figures 4.35a and 4.36a. From the phase plane obtained, one can observe that the origin is a stable node and the limit cycle around that vicinity is a stable limit cycle. This agrees very well to our first assumption: ideal pulse train at the input of the multiplier. Also, we present the pulse propagation behavior and phase plane for two times, four times, and eight times GVD multiplication system in Figures 4.37a through d and 4.38. The shape of the phase graph exposes the phase between the displacement and its derivative.

As the multiplication factor increases, the system trajectories are moving away from the origin. As for the four times and eight times multiplications, there is neither stable limit cycle nor stable node on the phase planes even with the ideal multiplication parameters. Here, we see the system trajectories spiral out to an outer radius and back to inner radius again. The change in the radius of the spiral is the transient response of the system. Hence, with the increase in multiplication factor, the

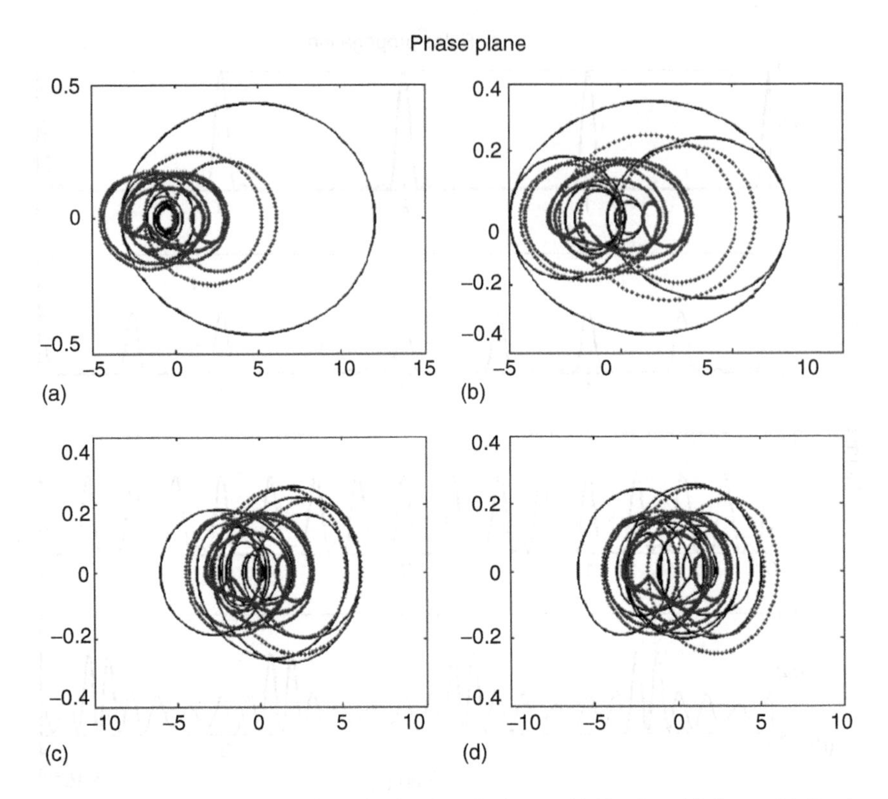

FIGURE 4.36 Phase plane of (a) original pulse, (b) 2× multiplication, (c) 4× multiplication, and (d) 8× multiplication with 1 nm filter bandwidth and equal lasing mode amplitude analysis (solid line, real part of the energy; dotted line, imaginary part of the energy; x-axes, $E(t)$ and y-axes, $E'(t)$).

system trajectories become more sophisticated. Although GVD repetition rate multiplication uses only the phase change effect in multiplication process, the inherent nonlinearities still affect its stability indirectly. Despite the reduction in the pulse amplitude, we observe uneven pulse amplitude distribution in the multiplied pulse train. The percentage of unevenness increases with the multiplication factor in the system.

4.2.4.3.2 Gaussian Lasing Mode Amplitude Distribution

This set of the simulation models the practical filter used in the system. It gives us a better insight on the GVD repetition rate multiplication system behavior. The parameters used in the simulation are exactly the same except the filter of the laser has been changed to 1 nm (125 GHz at 1550 nm) Gaussian-profile passband filter. The spirals of the system trajectories and uneven pulse amplitude distribution are more severe than those in the uniform lasing mode amplitude analysis as observed in Figures 4.38 through 4.41.

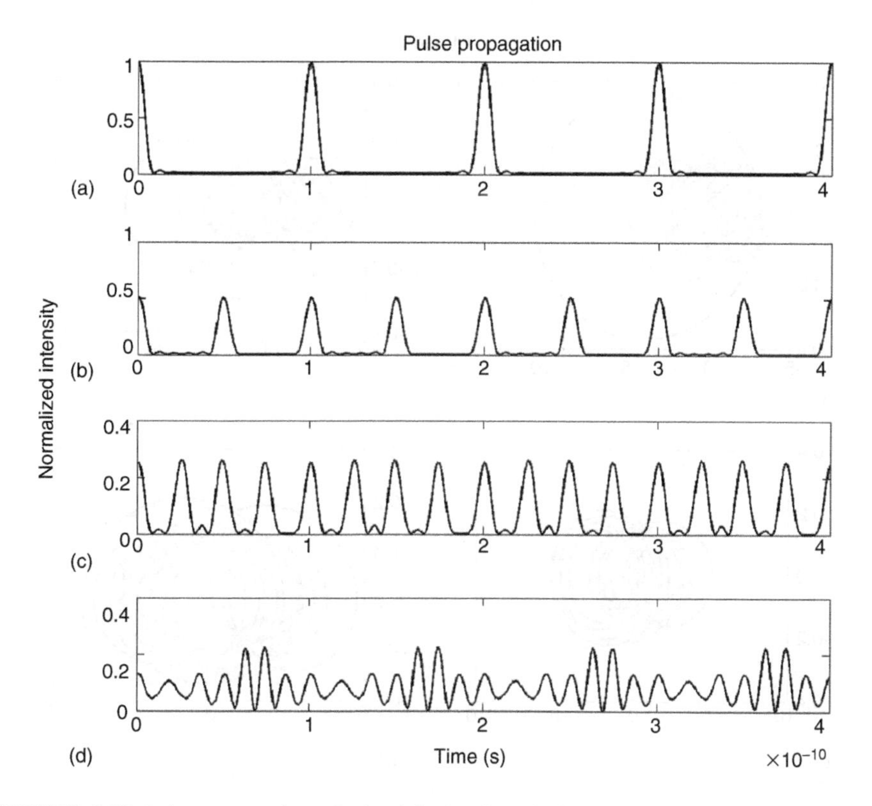

FIGURE 4.37 Pulse propagation of (a) original pulse, (b) 2× multiplication, (c) 4× multiplication, and (d) 8× multiplication with 1 nm filter bandwidth and Gaussian lasing mode amplitude analysis.

4.2.4.3.3 Effects of Filter Bandwidth

Filter bandwidth used in the mode-locked fiber ring laser will affect the system stability of the GVD repetition rate multiplication system as well. The analysis done above is based on 1 nm filter bandwidth. The number of modes locked in the laser system increases with the bandwidth of the filter used, which gives us a better quality of the mode-locked pulse train. The simulation results shown below are based on the Gaussian lasing mode amplitude distribution and 3 nm filter bandwidth used in the laser cavity; other parameters remain unchanged.

With wider filter bandwidth, the pulse width and the percentage pulse amplitude fluctuation decrease. This suggests a better stability condition. Instead of spiraling away from the origin, the system trajectories move inward to the stable node. However, this leads to a more complex pulse formation system.

4.2.4.3.4 Nonlinear Effects

When the input power of the pulse train enters the nonlinear region, the GVD multiplier loses its multiplication capability as predicted. The additional nonlinear

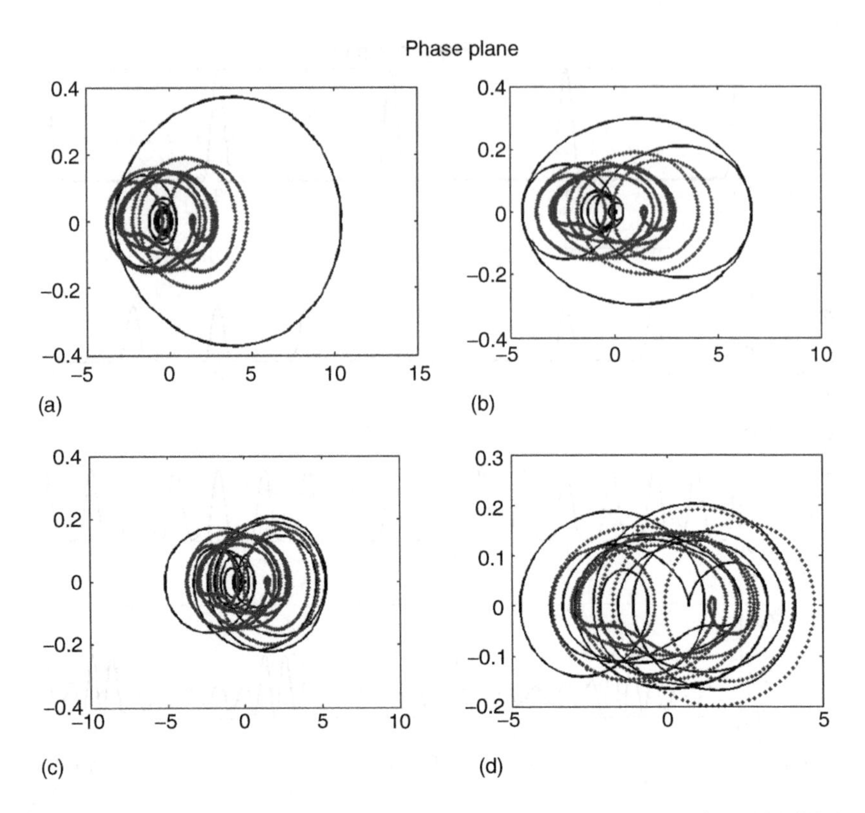

FIGURE 4.38 Phase plane of (a) original pulse, (b) 2× multiplication, (c) 4× multiplication, and (d) 8× multiplication with 1 nm filter bandwidth and Gaussian lasing mode amplitude analysis (solid line, real part of the energy; dotted line, imaginary part of the energy; x-axes, $E(t)$ and y-axes, $E'(t)$).

phase shift due to the high input power is added to the total pulse phase shift and destroys the phase change condition of the lasing modes required by the multiplication condition. Furthermore, this additional nonlinear phase shift also changes the pulse shape and the phase plane of the multiplied pulses.

4.2.4.3.5 Noise Effects

The above simulations are all based on the noiseless situation. However, in the practical optical communication systems, noises are always sources of nuisance, which can cause system instability, therefore it must be taken into consideration for the system stability studies.

Since the optical intensity of the m times multiplied pulse is m times less than the original pulse, it is more vulnerable to noise. The signal is difficult to differentiate from the noise within the system if the power of multiplied pulse is too small. The phase plane the multiplied pulse is distorted due to the presence of the noise, which leads to poor stability performance.

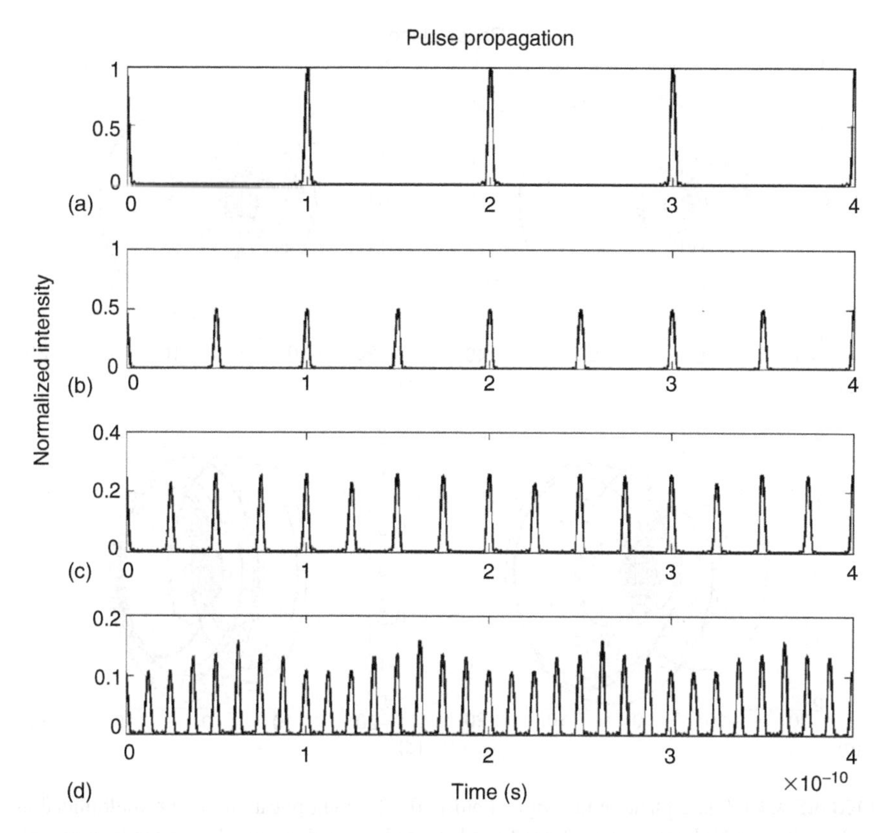

FIGURE 4.39 Pulse propagation of (a) original pulse, (b) 2× multiplication, (c) 4× multiplication, and (d) 8× multiplication with 3 nm filter bandwidth and Gaussian lasing mode amplitude analysis.

4.2.4.4 Demonstration

The obtained 10 GHz output pulse train from the mode-locked fiber ring laser is shown in Figure 4.42. Its spectrum is shown in Figure 4.43. This output was then used as the input to the dispersion compensating fiber, which acts as the GVD multiplier in our experiment. The obtained four times multiplication by the GVD effect and its spectrum are shown in Figures 4.44 and 4.45.

The spectrums for both cases (original and multiplied pulse) are exactly the same since this repetition rate multiplication technique utilizes only the change of phase relationship between lasing modes and does not use fiber's nonlinearity.

The multiplied pulse suffers an amplitude reduction in the output pulse train; however, the pulse characteristics should remain the same. The instability of the multiplied pulse train is mainly due to the slight deviation from the required DCF length (0.2% deviation). Another reason for the pulse instability, which derived from our analysis, is the divergence of the pulse energy variation in the vicinity around the origin, as the multiplication factor gets higher. The pulse amplitude decreases with

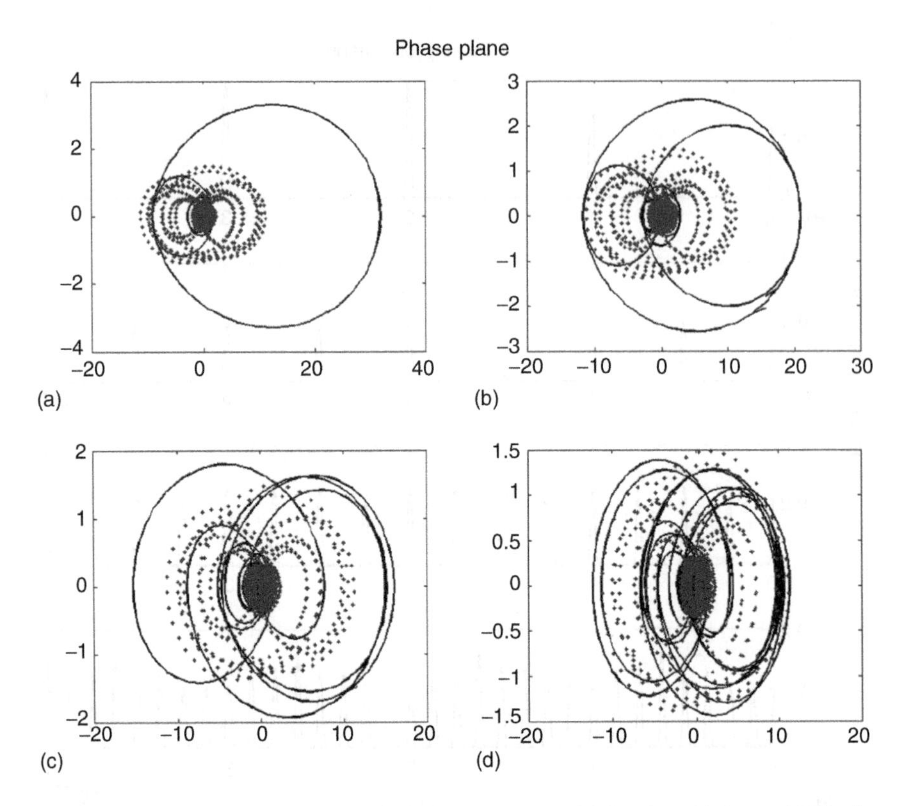

FIGURE 4.40 Phase plane of (a) original pulse, (b) 2× multiplication, (c) 4× multiplication, and (d) 8× multiplication with 3 nm filter bandwidth and Gaussian lasing mode amplitude analysis (solid line, real part of the energy; dotted line, imaginary part of the energy; *x*-axes, $E(t)$ and *y*-axes, $E'(t)$).

the increase in multiplication factor, as the fact of energy conservation, when it reaches certain energy level, which is indistinguishable from the noise level in the system, the whole system will become unstable and noisy.

4.2.4.5 Remarks

We have demonstrated four times repetition rate multiplication by using fiber GVD effect; hence, 40 GHz pulse train is obtained from 10 GHz mode-locked fiber laser source. However, its stability is of great concern for the practical use in the optical communication systems. Although the GVD repetition rate multiplication technique is linear in nature, the inherent nonlinear effects in such system may disturb the stability of the system. Hence, any linear approach may not be suitable in deriving the system stability. Stability analysis for this multiplied pulse train has been studied by using the nonlinear control stability theory, which is the first time, to the best of our knowledge, that phase plane analysis is being used to study the transient and stability performance of the GVD repetition rate multiplication system. Surprisingly, from the analysis model, the stability of the multiplied pulse train can hardly be

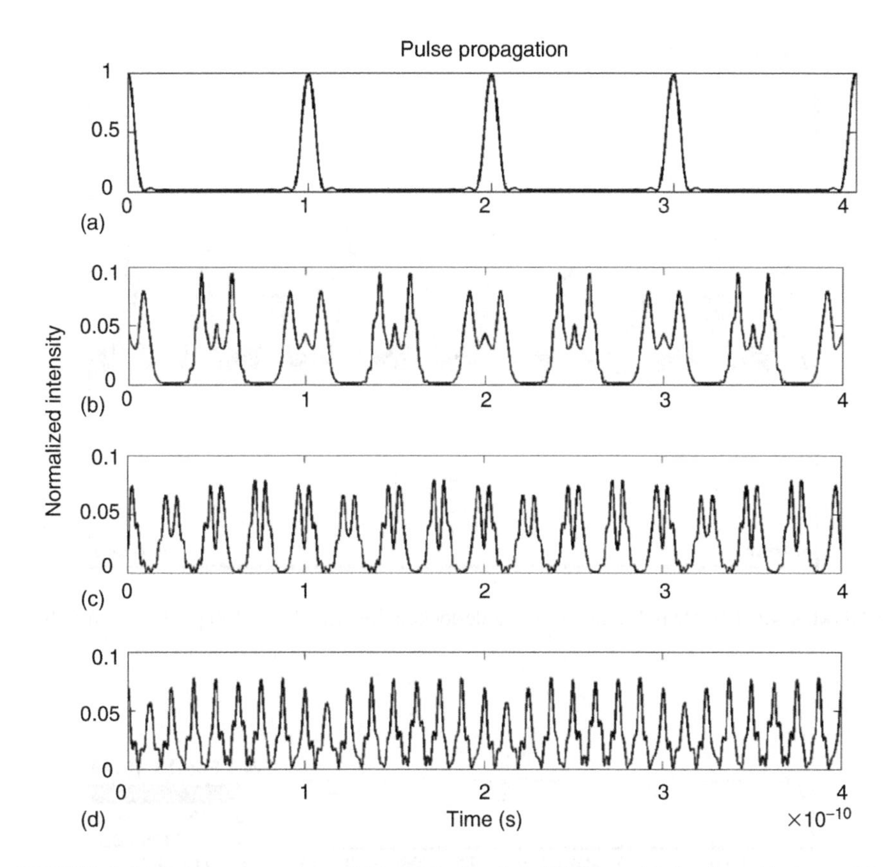

FIGURE 4.41 Pulse propagation of (a) original pulse, (b) 2× multiplication, (c) 4× multiplication, and (d) 8× multiplication with 3 nm filter bandwidth, Gaussian lasing mode amplitude analysis, and input power = 1 W.

achieved even under perfect multiplication conditions. Furthermore as shown in Figures 4.41 and 4.46, we observed uneven pulse amplitude distribution in the GVD multiplied pulse train, which is due to the energy variations between the pulses that cause some energy beating between them. Another possibility is the divergence of the pulse energy variation in the vicinity around the equilibrium point that leads to instability as seen in Figures 4.47 and 4.48.

The pulse amplitude fluctuation increases with the multiplication factor. Also, with wider filter bandwidth used in the laser cavity, better stability condition can be achieved. The nonlinear phase shift and noises in the system challenge the system stability of the multiplied pulses. These not only change the pulse shape of the multiplied pulses but also distort the phase plane of the system. Hence, the system stability is greatly affected by the self-phase modulation as well as the system noises.

This stability analysis model can further be extended to include some system nonlinearities, such as the gain saturation effect, nonquadrature biasing of the modulator, fiber nonlinearities, etc. The chaotic behavior of the system may also be studied by applying different initial phase and injected energy conditions to the model.

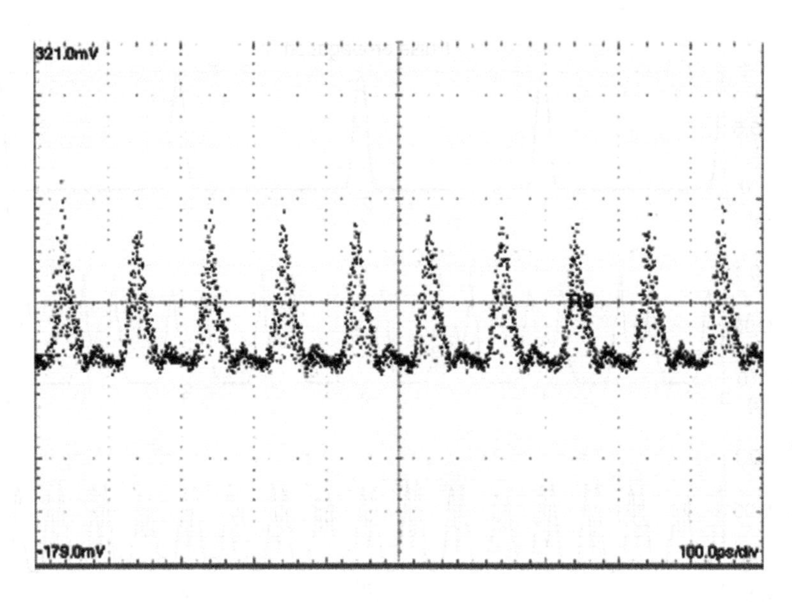

FIGURE 4.42 10 GHz pulse train from mode-locked fiber ring laser (100 ps/div, 50 mV/div).

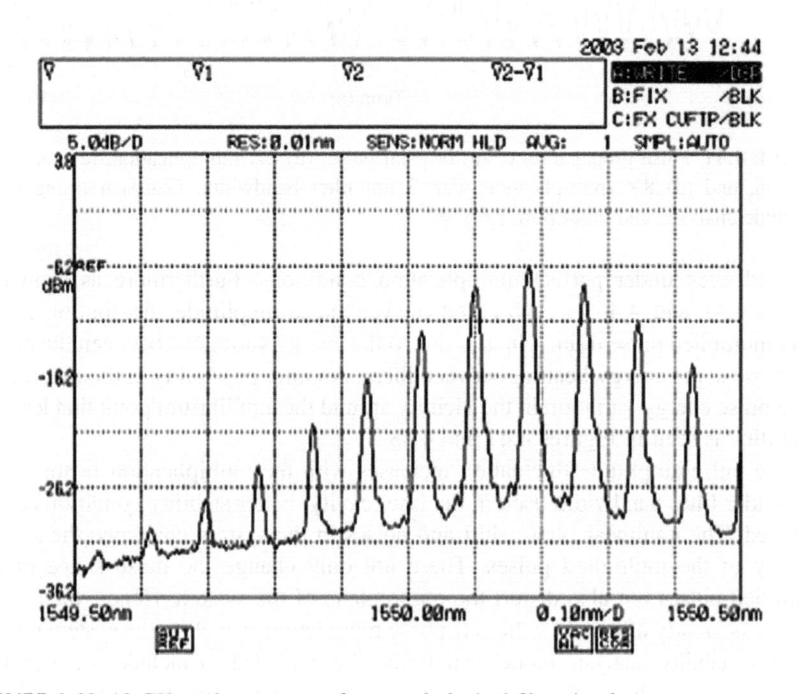

FIGURE 4.43 10 GHz pulse spectrum from mode-locked fiber ring laser.

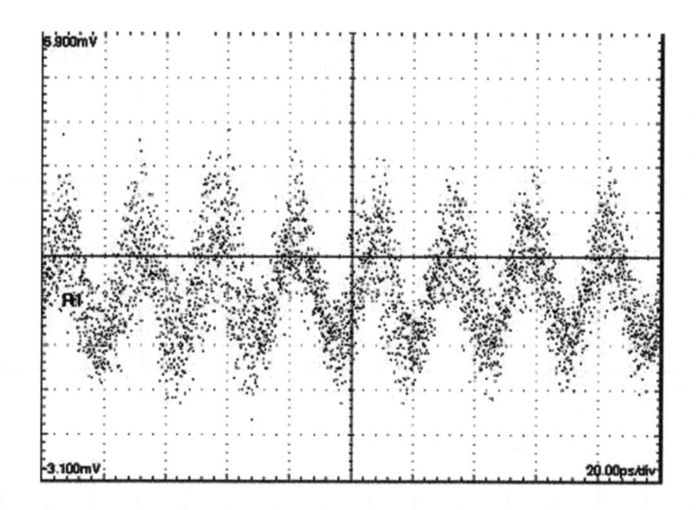

FIGURE 4.44 40 GHz multiplied pulse train (20 ps/div, 1 mV/div).

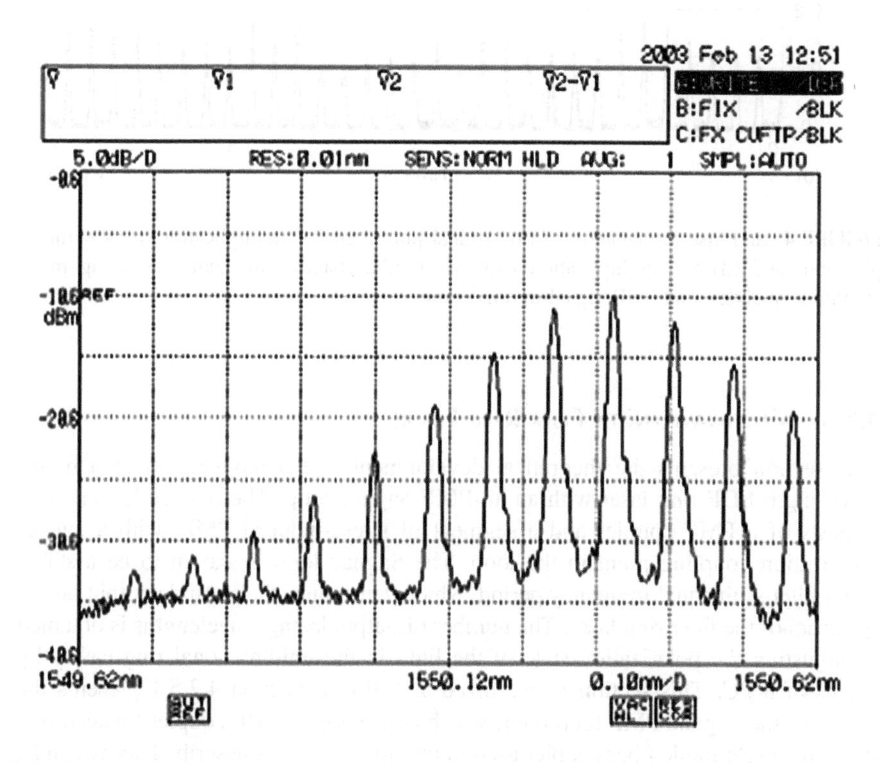

FIGURE 4.45 40 GHz pulse spectrum from GVD multiplier.

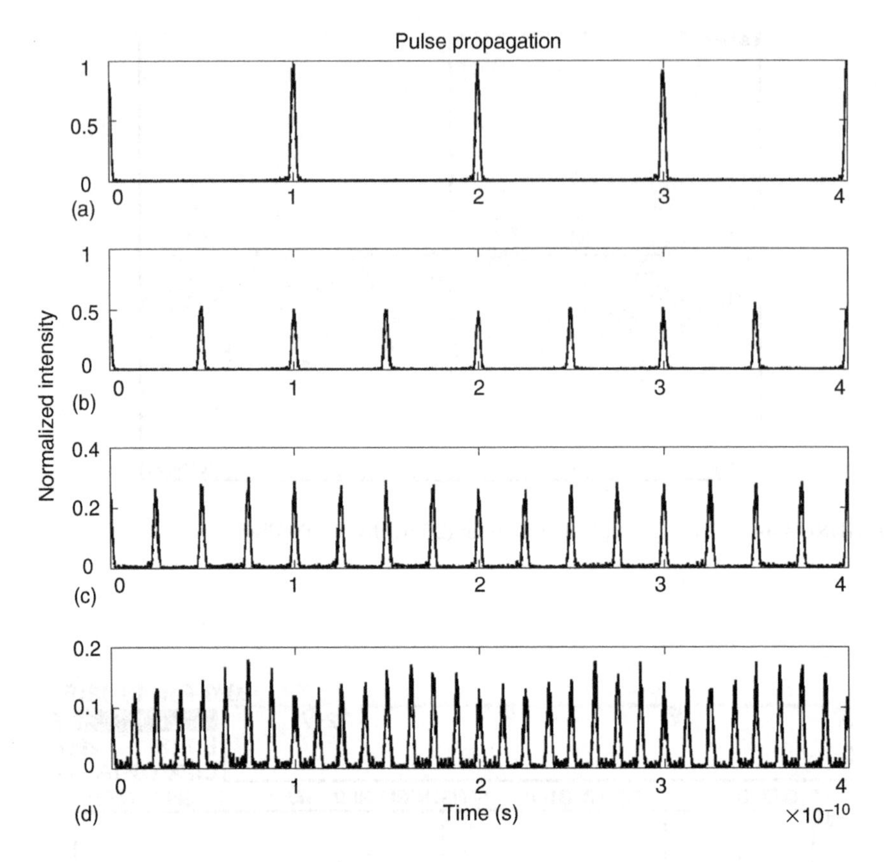

FIGURE 4.46 Pulse propagation of (a) original pulse, (b) 2× multiplication, (c) 4× multiplication, and (d) 8× multiplication with 3 nm filter bandwidth, Gaussian lasing mode amplitude analysis, and 0 dB signal-to-noise ratio.

4.2.5 Multiwavelength Fiber Ring Lasers

This section presents the theoretical development and demonstration of a multiwavelength EDF ring laser with an all-PMF Sagnac loop. The Sagnac loop simply consists of a PMF coupler and a segment of stress-induced PMF, with a single-polarization coupling point in the loop. The Sagnac loop is shown to be a stable comb filter with equal frequency period, which determines the possible output power spectrum of the fiber ring laser. The number of output lasing wavelengths is obtained by adjusting the polarization state of the light in the unidirectional ring cavity by means of a PC. This section is organized as follows: Section 4.2.5.1 presents the theory of the Sagnac PMF loop filter, which consists of a PMF coupler (instead of a standard single-mode fiber coupler used in previous works as described above) and a segment of PMF. Section 4.2.5.2 presents the experimental results and discussion. Concluding remarks are given.

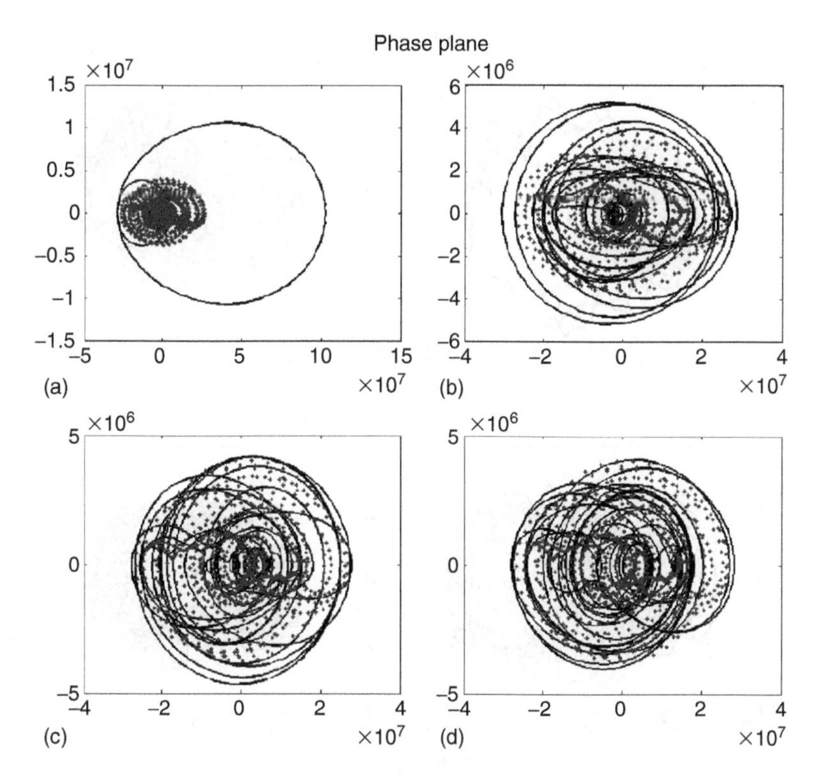

FIGURE 4.47 Phase plane of (a) original pulse, (b) 2× multiplication, (c) 4× multiplication, and (d) 8× multiplication with 3 nm filter bandwidth, Gaussian lasing mode amplitude analysis, and input power = 1 W (solid line, real part of the energy; dotted line, imaginary part of the energy; *x*-axes, $E(t)$ and *y*-axes, $E'(t)$).

4.2.5.1 Theory

In this section, we present a theoretical analysis of the all-PMF Sagnac loop, which is the key component in the fiber ring laser. We consider the simplest case (see Figure 4.49a), in which only one polarization mode-coupling point exists. That is, the loop filter is constructed by splicing the two pigtails (with lengths l_1 and l_2) of the PMF coupler with a phase difference θ along a certain principal axis at the spliced point. The input light is equally split into two waves by the 3 dB PMF coupler, and the two counter-propagating waves are recombined at the coupler output port after traveling through the loop. The electric components, $E_{ix}(\omega)$ and $E_{iy}(\omega)$, of the input light, $E_{in}(\omega)$, can be defined as

$$E_{in}(\omega) = \begin{pmatrix} E_{ix}(\omega) \\ E_{iy}(\omega) \end{pmatrix} \tag{4.29}$$

where ω is the angular optical frequency. The PMF with length l can be considered as an ideal waveguide with linear birefringence, which is described by the Jones propagation matrix as [16]

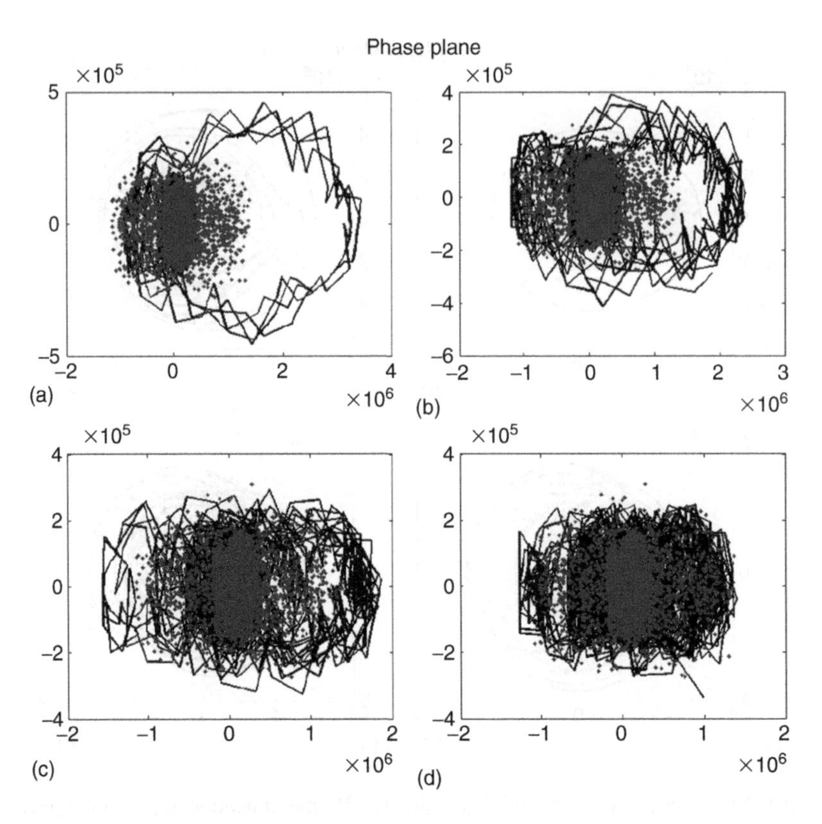

FIGURE 4.48 Phase plane of (a) original pulse, (b) 2× multiplication, (c) 4× multiplication, and (d) 8× multiplication with 3 nm filter bandwidth, Gaussian lasing mode amplitude analysis, and 0 dB signal-to-noise ratio (solid line, real part of the energy; dotted line, imaginary part of the energy; *x*-axes, $E(t)$; and *y*-axes, $E'(t)$).

$$J_{\text{PMF}}(\omega,l) = \begin{pmatrix} \exp(j\Delta\beta(\omega)l/2) & 0 \\ 0 & \exp(-j\Delta\beta(\omega)l/2) \end{pmatrix} \tag{4.30}$$

where $j = \sqrt{-1}$ and $\Delta\beta(\omega) = \beta_x(\omega) - \beta_y(\omega)$ is the difference between the two propagation constants of a high-birefringence fiber, which supports two linearly orthogonal fundamental modes (i.e., $HE^x{}_{11}$ and $HE^y{}_{11}$). It should be noted that a common average phase shift of $\exp(\overline{j\beta(\omega)l})$ is omitted in Equation 4.30, because the Sagnac interferometer cannot distinguish between the common phase term for the clockwise wave (CW) and counterclockwise wave (CCW). The transfer matrix of the coordinate (i.e., $\Theta(\theta)$) of the polarization mode-coupling point at the principal axes with a phase difference of θ (see Figure 4.49b) is given as

$$\Theta(\theta) = \begin{pmatrix} \cos\theta & \sin\theta \\ \sin\theta & -\cos\theta \end{pmatrix} \tag{4.31}$$

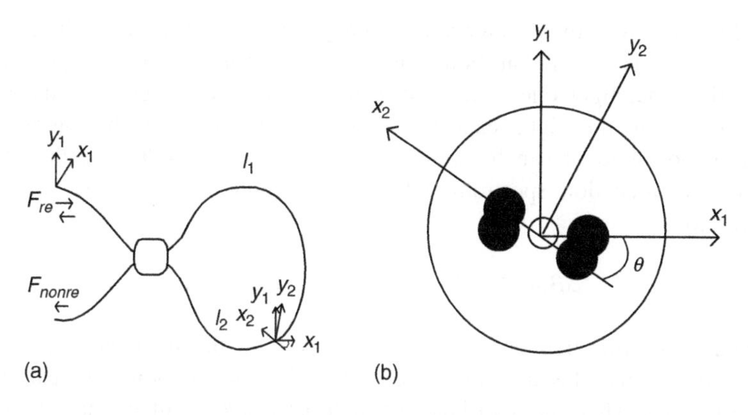

FIGURE 4.49 (a) Schematic diagram of the proposed all-PMF Sagnac loop filter with a single coupling point. (b) Representation of the coordinates of the single coupling point.

The 3 dB PMF coupler is assumed to be ideal so that polarization coupling, polarization-dependent loss, and frequency dependence of the coupler are negligible (i.e., the coupling ratio is 50% in the operating wavelength range). The CW and CCW will experience the same phase shift and a 3 dB loss through the coupler; thus, there is no phase difference at the reciprocal port. Hence, the CW's Jones matrix for the Sagnac loop is given by

$$G_{CW}(\omega) = \frac{1}{2} J_{PMF}(\omega, l_1)\Theta(\theta) J_{PMF}(\omega, l_2) \tag{4.32}$$

where $J_{PMF}(\omega,l_1)$ and $J_{PMF}(\omega,l_2)$ are defined in Equation 4.30. The CCW's Jones matrix is simply the transpose of the CW's Jones matrix and is given by

$$G_{CCW}(\omega) = G_{CW}^T(\omega) \tag{4.33}$$

The electric components of the light at the output ports are given by

$$E_{out}(\omega) = \begin{pmatrix} E_{ox}(\omega) \\ E_{oy}(\omega) \end{pmatrix} \tag{4.34}$$

The relationship between the electric components at the input and output ports is given by

$$E_{out}(\omega) = [G_{CCW}(\omega) + G_{CW}(\omega)]E_{in}(\omega) \tag{4.35}$$

Using Equations 4.21 through 4.26, the intensity transfer function, F_{re}, for the reciprocal port can be derived as

$$F_{re} = (1 - \sin^2\theta)\sin^2(\Delta\beta \cdot \Delta l/2) \tag{4.36}$$

where $\Delta l = l_2 - l_1$ is the difference between the length of the two PMF segments in the loop. It is noted that Equation 4.36 is independent of the polarization state of the input light due to the fact that the interference terms of the x- and y-components of the light cancel out with each other at the output port. From Equation 4.36, it can be shown that when $\theta \neq 0$ or $\theta \neq \pi$ the spectral peaks of the reflection spectrum will have maximum intensity at frequencies according to

$$\Delta\beta(\omega_{\rm m}) \cdot \Delta l = 2\pi m \quad (m = 1, 2, 3, \ldots) \tag{4.37}$$

Note that light with frequency ω_m will disappear at the nonreciprocal port for the case of $\theta = \pi/2$ because the transfer function of the nonreciprocal port is $F_{\rm nonre} = 1 - F_{\rm re}$. There are two kinds of high-birefringence fibers, namely, stress-induced birefringent fiber and geometry-induced birefringent fiber, where the former birefringent fiber has greater birefringence. Here, we only consider the stress-induced birefringent fiber, where the effective index of each polarization is influenced by stress alone. Thus, the modal birefringence B is independent of wavelength over a particular wavelength range, and is given by

$$\Delta\beta(\omega_m) = \frac{2\pi}{\lambda} B \quad (m = 0, 1, 2, \ldots) \tag{4.38}$$

Using Equations 4.37 and 4.38, the spectral peaks of the reflection spectrum will have maximum intensity at frequencies $f_{\rm m}$ given by

$$f_m = \frac{mc}{B\Delta l} \quad (m = 0, 1, 2, \ldots) \tag{4.39}$$

where c is the speed of the light in vacuum. From Equation 4.39, the Sagnac loop is a comb filter whose spectral peaks are separated by frequency spacing given by

$$f_{m+1} - f_m = \frac{c}{B\Delta l} \tag{4.40}$$

It is noted that although the frequency-dependent intensity transfer function of the loop filter is independent of the state of polarization of the input light, the polarization state of the output light generally depends on the polarization state and frequency of the input light.

4.2.5.2 Experimental Results and Discussion

This section presents the experimental verification of the theoretical analysis of the all-PM Sagnac loop filter described in Section 4.2.5 and the experimental results of the fiber ring laser. Figure 4.50 shows a typical reflective spectrum of the all-PMF Sagnac loop filter. The loop filter was constructed by splicing the two pigtails (with lengths l_1 and l_2) of the PMF coupler in 0° and 90° with respect to their principal axes to form a single coupling point in the loop (i.e., a phase difference of $\theta = 90°$).

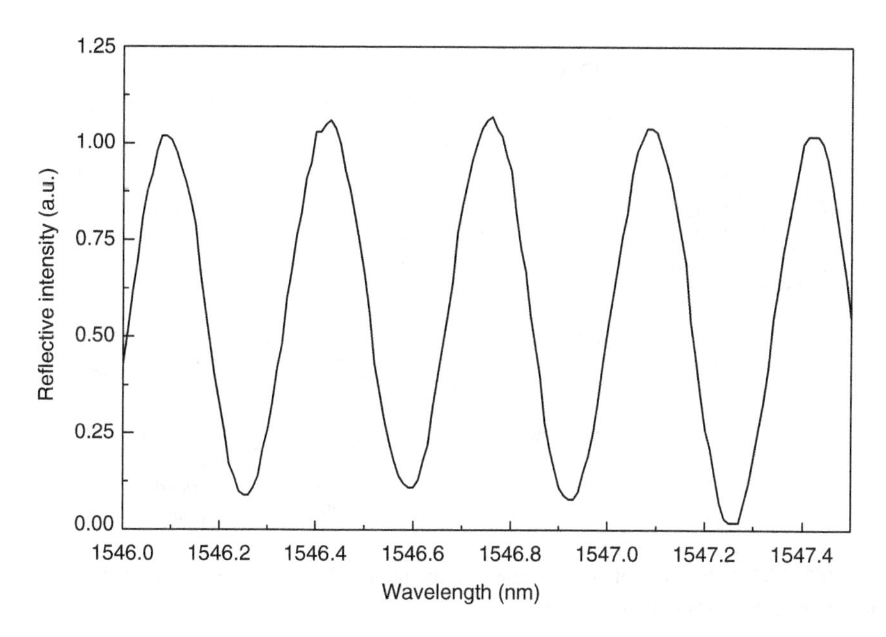

FIGURE 4.50 A typical reflective spectrum of the all-PMF Sagnac loop filter with a single coupling point.

The loop filter is highly stable as expected because all the components used are all-PM components. From Figure 4.50, the frequency period of the filter is 0.35 nm, which agrees well with the theoretical value as predicted by Equation 4.40 when the following parameter values, $B = 5.2e^{-4}$ and $\Delta l = 13.2$ m (in the 1550 nm window), are substituted into the equation.

Figure 4.51 shows the schematic diagram of the proposed unidirectional fiber ring laser with the all-PMF Sagnac loop filter. It consists of a 15 m long of Er^{3+} silica fiber doped with ~200 ppm of erbium. The EDF has a numerical aperture (NA) of 0.21, a cutoff wavelength of 920 nm, and an absorption coefficient of 12 dB/m at 980 nm. To increase the optical pump efficiency, the EDF is pumped by a 980 nm laser diode (LD), which generates 70 mW power in both directions in the ring cavity through the two 980/1550 nm WDM couplers. A polarization-independent fiber isolator is used to provide unidirectional ring oscillation so as to avoid spatial hole burning in the gain medium. The coupler-2 is used as the output coupler for the fiber laser and also to direct the light wave to the Sagnac loop filter. The periodic spectral peaks of the Sagnac filter will determine the lasing wavelengths. A PC is used in the cavity to adjust the polarization state to obtain several lasing wavelengths at the output port.

Figures 4.52a and b show the experimental results of the output lasing wavelengths of the fiber ring laser under different polarization conditions by adjusting the PC in the cavity. It can be seen that the wavelength spacing is 1.0 nm, which is defined by the 1.0 nm frequency period of the Sagnac loop filter with parameter values of $B = 5.2e^{-4}$ and $\Delta l = 4.5$ m. Figure 4.53 shows the output spectra of the

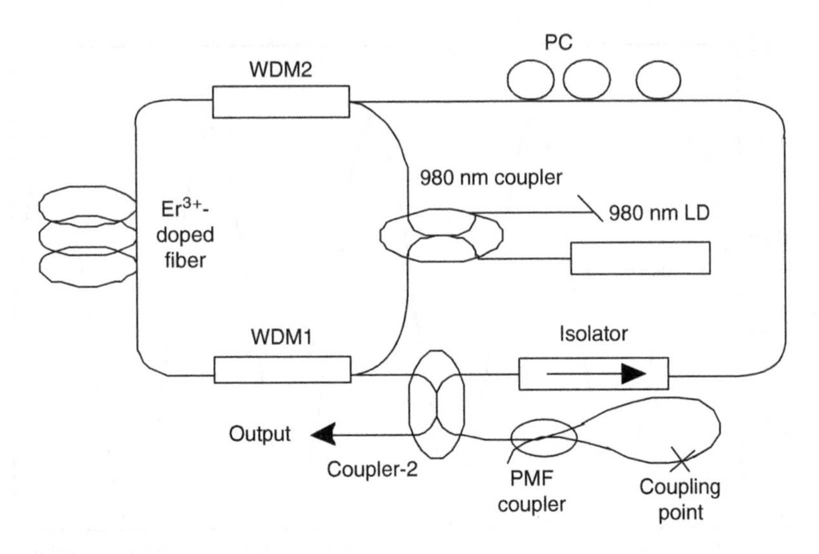

FIGURE 4.51 Schematic of the proposed unidirectional fiber ring laser using the all-PMF Sagnac loop as a stable periodic filter.

lasing wavelengths of the fiber ring laser under a particular polarization condition in the cavity, where the wavelength spacing is 0.50 nm, which is defined by the 0.50 nm frequency period of the Sagnac loop filter with parameter values of $B = 5.2e^{-4}$ and $\Delta l = 9.0$ m. It should be noted that the number of output lasing wavelengths of the proposed fiber ring laser could be greatly increased by overcoming the large homogeneous broadening of the gain medium of the EDF at room temperature [33]. This problem can be overcome by cooling the EDF to 77 K, but

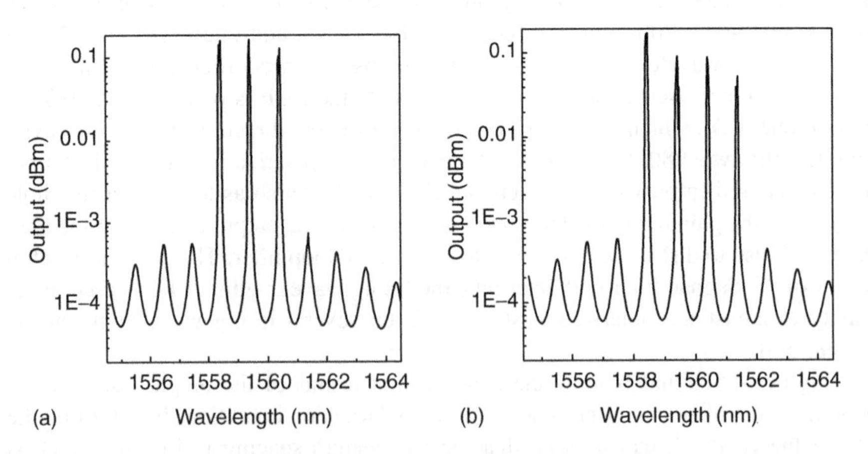

FIGURE 4.52 (a, b) Typical output lasing wavelengths of the fiber ring laser under different polarization conditions of the PC in the ring cavity. Wavelength spacing is 1.0 nm.

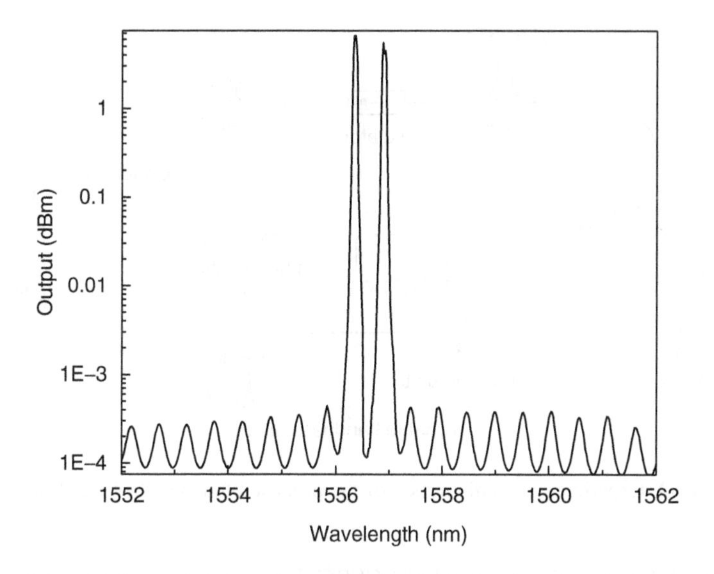

FIGURE 4.53 Typical output lasing wavelength of the fiber ring laser under a particular polarization condition of the PC in the ring cavity. Wavelength spacing is 0.50 nm.

this technique is probably not suitable for practical applications [34]. A more practical approach is to use an acousto-optic frequency shifter in the ring cavity to prevent the steady-state laser oscillation to generate a larger number of stable lasing wavelengths [26].

In this section, we have demonstrated an EDF ring laser using an all-PMF Sagnac loop filter for multiwavelength operation. The theoretical analysis and experimental results of the all-PMF Sagnac loop as a stable comb filter have been presented. The Sagnac loop filter is a simple and all-fiber device that consists of a PMF coupler and a segment of stress-induced PMF to form the loop. The number of output lasing wavelengths has been obtained by adjusting the polarization state of the light in the ring cavity using a polarization controller. The wavelength spacing is determined by the frequency period of the comb filter with equal frequency interval.

4.2.6 Multiwavelength Tunable Fiber Ring Lasers

Furthermore, tunable EDFR lasers [54–61] with a continuous tuning range of 15.5 nm (i.e., from 1546.8 to 1562.3 nm) can be demonstrated. The laser output power is ~9 dBm and the power variation is <2 dB over the tuning range. The laser is tuned using our recently developed tunable narrow-bandpass filter based on a phase-shifted linearly chirped FBG (LCFBG). Heating the LCFBG at a small contact point using a resistance wire would introduce a temporary phase-shift into the LCFBG, and hence creating a very narrow passband peak within the stopband of the LCFBG operating in the transmission mode. Scanning the resistance wire along the LCFBG, which is controlled by a linear travel stage, the narrow passband

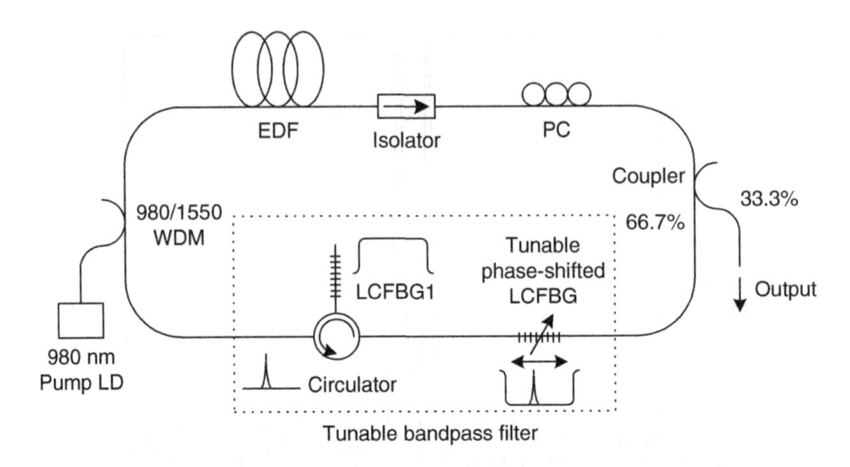

FIGURE 4.54 Experimental configuration of the proposed tunable fiber ring laser.

peak will shift across the stopband of LCFBG. This tuning technique overcomes the limited tuning range of FBG due to the failure feature of silica fiber.

Figure 4.54 shows the schematic of the proposed tunable EDFRL. The laser consists of 25 m of EDF pumped by a 980 nm LD, a tunable phase-shifted LCFBG, an optical circulator, an LCFBG1, a 2:1 fiber coupler, a PC, and an isolator. The mode field diameter of the EDF is 6.6 μm and its cladding diameter is 80 μm.

Our recently developed tunable phase-shifted LCFBG is used as a tunable bandpass filter (TBF) in the laser cavity. The length of the LCFBG is 6 cm and the chirp rate is 2.25 nm/cm. The stopband of the LCFBG is 21 nm (i.e., from 1544 to 1565 nm). The LCFBG was inscribed into a hydrogen-loaded germanium-doped cladding-mode suppression fiber using the UV scanning beam technique. The cladding-mode loss, which can cause an undesirable effect on the filter characteristics, is eliminated using this cladding-mode suppression fiber. A resistance wire is used as a thermal head to heat a small contact point of LCFBG. A thermally-induced temporary phase shift will be introduced into the LCFBG, and hence a narrow passband peak will be created in the stopband of the LCFBG. When the wire heater is scanned along the LCFBG using a linear travel stage, the center wavelength of the narrow passband peak will shift according to the position of the resistance wire. As a result, an all-fiber TBF can be achieved. To eliminate the broadened spectral disturbance due to thermal conduction, a heat sink beside the heating section is used to cool the LCFBG. It should be noted that, in this experiment, an identical LCFBG1 operating in the reflection mode and a circulator are needed in the cavity to suppress the transmission outside the stopband of the LCFBG. This is to eliminate the unwanted laser output outside the stopband of the LCFBG to achieve good laser performance. The measured spectrum of the TBF (with LCFBG1 and the circulator) is shown in Figure 4.55. The 3 dB bandwidth of the TBF is smaller than 0.01 nm, which is only limited by the 0.01 nm resolution of the optical spectrum analyzer. The total insertion loss is ~8 dB. However, the LCFBG1 and the optical circulator will not be needed when the stopband of

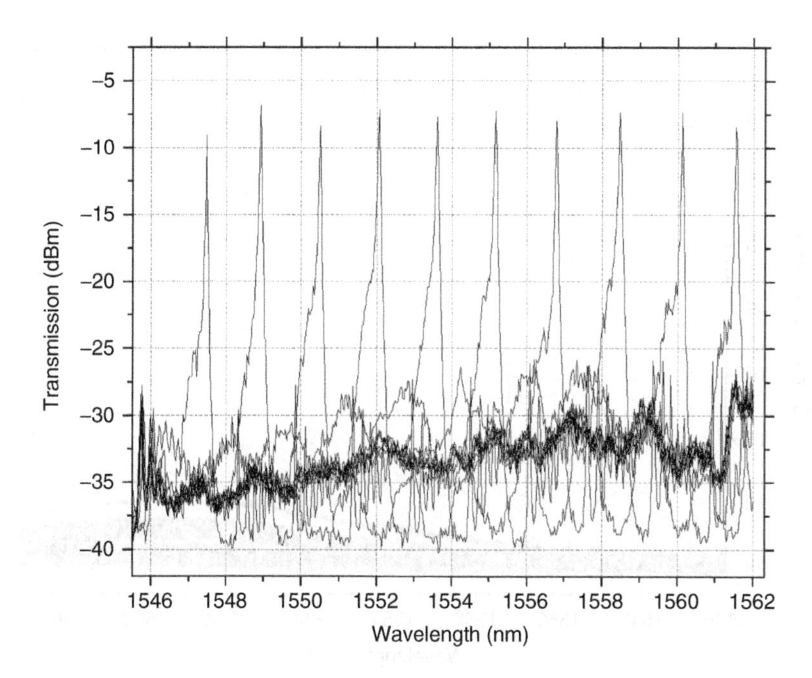

FIGURE 4.55 Measured transmission spectrum of TBF over the tuning range.

the LCFBG can cover the whole gain bandwidth of EDF, and this will reduce the insertion loss and cost of the laser.

The optical spectrum of the tunable fiber ring laser over the tuning range of 15.5 nm is shown in Figure 4.56. The power of the pump LD is ~98 mW. The laser output is ~9 dBm. When scanning the resistance wire along the LCFBG over 50 mm, the lasing wavelength was shifted from 1546.8 to 1562.3 nm (or 15.5 nm range). It is observed that there is no spectral distortion of the lasing wavelength over the tuning range and that the output power is relatively constant with a power variation of <2 dB. The side-mode suppression ratio is greater than 30 dB. Figure 4.57 shows the output spectrum of the fiber ring laser and its stability with time. The time interval between the scans is 10 min. It can be seen that the laser output is very stable with time. The laser linewidth is measured using a scanning Fabry–Perot interferometer with 10 GHz free spectral range and 50 MHz resolution. The measured linewidth is ~6.5 MHz, which is limited by the resolution of the scanning F–P interferometer.

The resolution of the travel stage to control the resistance wire is 0.01 mm, which corresponds to a tuning step of 3 pm of the laser. The tuning step can be made smaller using a travel stage with higher resolution or LCFBG with smaller chirp rate. It should be noted that the tunable range of the fiber ring laser is limited only by the stopband width of the LCFBG. Using LCFBG with longer length or larger chirp rate, the laser tuning range can be increased to cover the whole C-band. The fiber

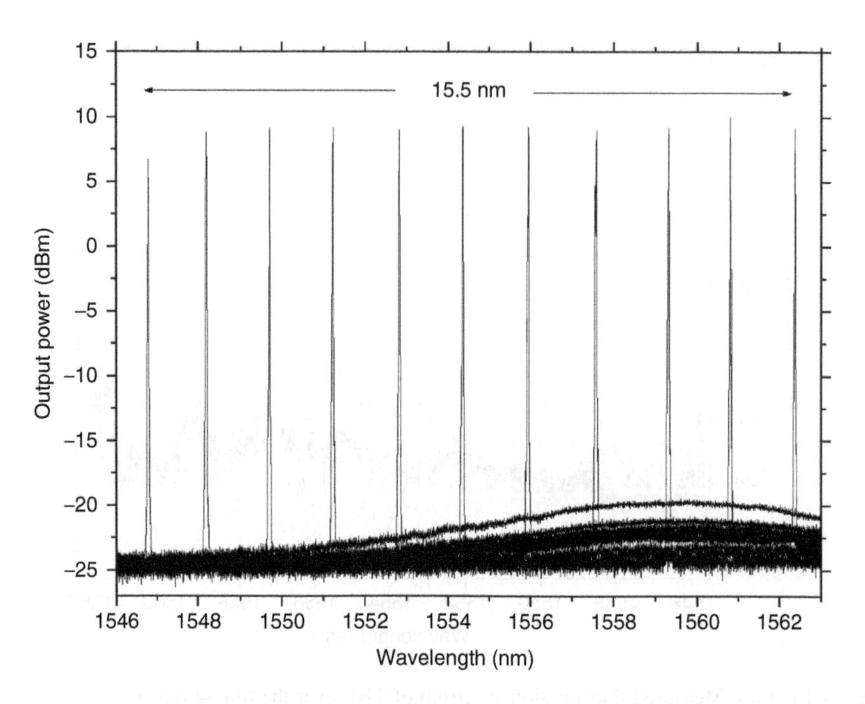

FIGURE 4.56 Output power of the tunable laser over the 15.5 nm tuning range.

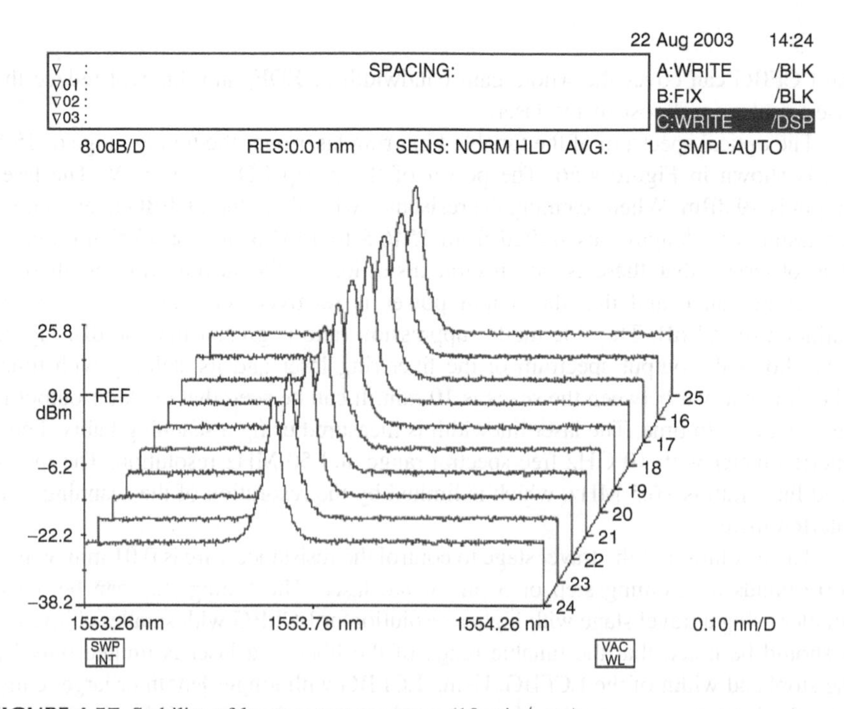

FIGURE 4.57 Stability of laser output spectrum (10 min/scan).

ring laser can also operate in the S- and L-bands using LCFBG with appropriate stopband.

4.2.6.1 Remarks

This chapter describes about a mode-locked laser operating under the open loop condition and with O/E RF feedback providing regenerative mode locking. The O/E feedback can certainly provide a self-locking mechanism under the condition that the polarization characteristics of the ring laser are manageable. The regenerative MLRL can self-lock even under the DC drifting effect of the modulator bias voltage (over 20 h). The generated pulse trains of 4.5 ps duration can be, with minimum difficulty, compressed further to less than 3 ps for 160 Gb/s optical communication systems.

The temporal Talbot phenomenon is also given by demonstrating the 660th and 1230th order of rational harmonic mode locking from a base modulation frequency of 100 MHz in the optically amplified fiber ring laser, hence achieving 66 and 123 GHz pulse repetition frequency. Besides the repetition rate multiplication, we also obtain high pulse compression factor in the system, ~35× and 40× relative to the nonmultiplied laser system. In addition, we use phase plane analysis to study the laser system behavior. From the analysis model, the amplitude stability of the detuned pulse train can only be achieved under negligible or no harmonic distortion condition, which is the ideal situation. The phase plane analysis also reveals the pulse forming complexity of the laser system.

We have demonstrated four times repetition rate multiplication by using fiber GVD effect; hence, 40 GHz pulse train is obtained from 10 GHz mode-locked fiber laser source. Stability analysis for this multiplied pulse train has been studied by using the nonlinear control stability theory that phase plane analysis is being used to study the transient and stability performance of the GVD repetition rate multiplication system. Surprisingly, from the analysis model, the stability of the multiplied pulse train can hardly be achieved even under perfect multiplication conditions. Furthermore, we observed uneven pulse amplitude distribution in the GVD multiplied pulse train, which is due to the energy variations between the pulses that cause some energy beating between them. Another possibility is the divergence of the pulse energy variation in the vicinity around the equilibrium point that leads to instability.

The pulse amplitude fluctuation increases with the multiplication factor. Also, with wider filter bandwidth used in the laser cavity, better stability condition can be achieved. The nonlinear phase shift and noises in the system challenge the system stability of the multiplied pulses. These not only change the pulse shape of the multiplied pulses but also distort the phase plane of the system. Hence, the system stability is greatly affected by the self-phase modulation as well as the system noises. This stability analysis model can further be extended to include some system nonlinearities, such as the gain saturation effect, nonquadrature biasing of the modulator, fiber nonlinearities, etc. The chaotic behavior of the system may also be studied by applying different initial phase and injected energy conditions to the model.

In Section 4.2.5, we demonstrated an EDF ring laser using an all-PMF Sagnac loop filter for multiwavelength operation. The theoretical analysis and experimental results of the all-PMF Sagnac loop as a stable comb filter were presented. The Sagnac loop filter is a simple and all-fiber device that consists of a PMF coupler and a segment of stress-induced PMF to form the loop. The number of output lasing wavelengths has been obtained by adjusting the polarization state of the light in the ring cavity using a polarization controller. The wavelength spacing is determined by the frequency period of the comb filter with equal frequency interval.

We are currently pursuing the design and demonstration of multiwavelength mode-locked lasers to generate ultrashort and ultrahigh rep-rate pulse sequences by employing a multispectral filter demuxes and muxes within the photonic fiber ring. These lasers will be reported in the near future.

REFERENCES

1. A. Hasegawa and F. Tappert, Transmission of stationary nonlinear optical pulses in dispersive dielectric fibers: I. Anomalous dispersion, *Appl. Phys. Lett.*, 23, 142–144, 1973.
2. A. Hasegawa and F. Tappert, Transmission of stationary nonlinear optical pulses in dispersive dielectric fibers: II. Normal dispersion, *Appl. Phys. Lett.*, 23, 171–173, 1973.
3. L.F. Mollenauer, R.H. Stolen, and J.P. Gordon, Experimental observation of picosecond pulse narrowing and solitons in optical fibers, *Phys. Rev. Lett.*, 45, 1095–1098, 1980.
4. A. Hasegawa, Numerical study of optical soliton transmission amplified periodically by the stimulated Raman process, *Appl. Opt.*, 23, 3302–3309, 1984.
5. L.F. Mollenauer and K. Smith, Demonstration of soliton transmission over more than 4000 km in fiber with loss periodically compensated by Raman gain, *Opt. Lett.*, 13, 675–677, 1988.
6. M. Nakazawa, Y. Kimura, and K. Suzuki, Soliton amplification and transmission with Er^{3+}-doped fibre repeater pumped by GainAsP laser diode, *Electron. Lett.*, 25, 199–200, 1989.
7. M. Nakazawa, Ultrahigh speed optical soliton communication and related technology, IOOC'95 Technical Digest Series, Hong Kong, vol. 4, pp. 96–98, 1995.
8. G. Aubin, E. Jeanney, T. Montalant, J. Moulu, F. Pirio, J.-B. Thomine, and F. Devaux, Electro absorption modulator for a 20 Gbit/s soliton transmission experiment over 1 million km with a 140 km amplifier span, IOOC'95 Technical Digest Series, Hong Kong, vol. 5, pp. 29–30, Post-deadline paper PD2–5, 1995.
9. W. Zhao and E. Bourkoff, Propagation properties of dark solitons, *Opt. Lett.*, 14, 703–705, 1989.
10. W. Zhao and E. Bourkoff, Interactions between dark solitons, *Opt. Lett.*, 14, 1371–1373, 1989.
11. J.P. Hamaide, P. Emplit, and M. Haelterman, Dark-soliton jitter in amplified optical transmission systems, *Opt. Lett.*, 16, 1578–1580, 1991.
12. Y.S. Kivshar, M. Haelterman, P. Emplit, and J.P. Hamaide, Gordon–Haus effect on dark solitons, *Opt. Lett.*, 19, 19–21, 1994.
13. P. Emplit, J.P. Hamaide, F. Reynaud, C. Froehly, and A. Barthelemy, Picosecond steps and dark pulses through nonlinear single mode fibers, *Opt. Commun.*, 62, 374–379, 1987.

14. D. Krökel, N.J. Halas, G. Giuliani, and D. Grischkowsky, Dark-pulse propagation in optical fibers, *Phys. Rev. Lett.*, 60, 29–32, 1988.
15. A.M. Weiner, J.P. Heritage, R.J. Hawkins, R.N. Thurston, E.M. Kirschner, D.E. Leaird, and W.J. Tomlinson, Experimental observation of the fundamental dark soliton in optical fibers, *Phys. Rev. Lett.*, 61, 2445–2448, 1988.
16. D.J. Richardson, R.P. Chamberlin, L. Dong, and D.N. Payne, Experimental demonstration of 100 GHz dark soliton generation and propagation using a dispersion decreasing fibre, *Electron. Lett.*, 30, 1326–1327, 1994.
17. W. Zhao and E. Bourkoff, Generation of dark solitons under a cw background using waveguide electro-optic modulators, *Opt. Lett.*, 15, 405–407, 1990.
18. M. Nakazawa and K. Suzuki, Generation of a pseudorandom dark soliton data train and its coherent detection by one-bit-shifting with a Mach–Zehnder interferometer, *Electron. Lett.*, 31, 1084–1085, 1995.
19. M. Nakazawa and K. Suzuki, 10 Gbit/s pseudorandom dark soliton data transmission over 1200 km, *Electron. Lett.*, 31, 1076–1077, 1995.
20. N.Q. Ngo, L.N. Binh, and X. Dai, Optical dark-soliton generators and detectors, *Opt. Commun.*, 132, 389–402, 1996.
21. G.P. Agrawal, *Nonlinear Fiber Optics*, Boston, MA: Academic Press, 1989.
22. M. Abramowitz and I.A. Segun, *Handbook of Mathematical Function*, New York: Dover Publications, 1964.
23. K. Kuroda and H. Takakura, Mode-locked ring laser with output pulse width of 0.4 ps, *IEEE Trans. Inst. Meas.*, 48, 1018–1022, 1999.
24. G. Zhu, H. Chen, and N. Dutta, Time domain analysis of a rational harmonic mode-locked ring fiber laser, *J. Appl. Phys.*, 90, 2143–2147, 2001.
25. C. Wu and N.K. Dutta, High repetition rate optical pulse generation using a rational harmonic mode-locked fiber laser, *IEEE J. Quantum Electron.*, 36, 145–150, 2000.
26. K.K. Gupta, N. Onodera, K.S. Abedin, and M. Hyodo, Pulse repetition frequency multiplication via intracavity optical filtering in AM mode-locked fiber ring lasers, *IEEE Photon. Technol. Lett.*, 14, 284–286, 2002.
27. Y. Shiquan, L. Zhaohui, Z. Chunliu, D. Xiaoyi, Y. Shuzhong, K. Guiyun, and Z. Qida, Pulse-amplitude equalization in a raional harmonic mode-locked fiber ring laser by using modulator as both mode locker and equalizer, *IEEE Photon. Technol. Lett.*, 15, 389–391, 2003.
28. K.S. Abedin, N. Onodera, and M. Hyodo, Higher order FM mode-locking for pulse-repetition-rate enhancement in actively mode-locked lasers: Theory and experiment, *IEEE J. Quantum Electron.*, 35, 875–890, 1999.
29. J. Azana and M.A. Muriel, Technique for multiplying the repetition rates of periodic trains of pulses by means of a temporal self-imaging effect in chirped fiber gratings, *Opt. Lett.*, 24, 1672–1674, 1999.
30. S. Atkins and B. Fischer, All optical pulse rate multiplication using fractional Talbot effect and field-to-intensity conversion with cross gain modulation, *IEEE Photon. Technol. Lett.*, 15, 132–134, 2003.
31. J. Azana and M.A. Muriel, Temporal self-imaging effects: theory and application for multiplying pulse repetition rates, *IEEE J. Quantum Electron.*, 7, 728–744, 2001.
32. D.A. Chestnut, C.J.S. de Matos, and J.R. Taylor, 4× Repetition rate multiplication and Raman compression of pulses in the same optical fiber, *Opt. Lett.*, 27, 1262–1264, 2002.
33. S. Arahira, S. Kutsuzawa, Y. Matsui, D. Kunimatsu, and Y. Ogawa, Repetition frequency multiplication of mode-locked using fiber dispersion, *J. Lightwave Technol.*, 16, 405–410, 1998.

34. W.J. Lai, P. Shum, and L.N. Binh, Stability and transient analyses of temporal Talbot-effect-based repetition-rate multiplication mode-locked laser systems, *IEEE Photon. Technol. Lett.*, 16, 437–439, 2004.
35. K.S. Abedin, N. Onodera, and M. Hyodo, Repetition-rate multiplication in actively mode-locked fiber lasers by higher-order FM mode-locking using a high finesse Fabry–Perot filter, *Appl. Phys. Lett.*, 73, 1311–1313, 1998.
36. W. Daoping, Z. Yucheng, L. Tangjun, and J. Shuisheng, 20 Gb/s optical time division multiplexing signal generation by fiber coupler loop-connecting configuration, presented at 4th Optoelctronics and Communications Conference, 1999.
37. D.L.A. Seixasn and M.C.R. Carvalho, 50 GHz fiber ring laser using rational harmonic mode-locking, presented at Microwave and Optoelectronics Conference, 2001.
38. R.Y. Kim, Fiber lasers and their applications, presented at Laser and Electro-Optics, CLEO/Pacific Rim'95, 1995.
39. H. Zmuda, R.A. Soref, P. Payson, S. Johns, and E.N. Toughlian, Photonic beamformer for phased array antennas using a fiber grating prism, *IEEE Photon. Technol. Lett.*, 9, 241–243, 1997.
40. G.A. Ball, W.W. Morey, and W.H. Glenn, Standing-wave monomode erbium fiber laser, *IEEE Photon. Technol. Lett.*, 3, 613–615, 1991.
41. D. Wei, T. Li, Y. Zhao, et al., Multiwavelength erbium-doped fiber ring laser with overlap-written fiber Bragg gratings, *Opt. Lett.*, 25, 1150–1152, 2000.
42. S.K. Kim, M.J. Chu, and J.H. Lee, Wideband multiwavelength erbium-doped fiber ring laser with frequency shifted feedback, *Opt. Commun.*, 190, 291–302, 2001.
43. Z. Li, L. Caiyun, and G.Yizhi, A polarization controlled multiwavelength Er-doped fiber laser, presented at APCC/OECC99, 1999.
44. R.M. Sova, C.S. Kim, and J.U. Kang, Tunable dual-wavelength all-PM fiber ring laser, *IEEE Photon. Technol. Lett.*, 14, 287–289, 2002.
45. I.D. Miller, D.B. Mortimore, P. Urquhart, et al., A Nd3+-doped cw fiber laser using all-fiber reflectors, *Appl. Opt.*, 26, 2197–2201, 1987.
46. X. Fang and R.O. Claus, Polarization-independent all-fiber wavelength-division multiplexer based on a Sagnac interferometer, *Opt. Lett.*, 20, 2146–2148, 1995.
47. X. Fang, H. Ji, C.T. Aleen, et al., A compound high-order polarization-independent birefringence filter, *IEEE Photon. Technol. Lett.*, 19, 458–460, 1997.
48. X.P. Dong, S. Li, K.S. Chiang, et al., Multiwavelength erbium-doped fiber laser based on a high-birefringence fiber loop, *Electron. Lett.*, 36, 1609–1610, 2000.
49. D. Jones, H. Haus, and E. Ippen, Subpicosecond solitons in an actively mode locked fiber laser, *Opt. Lett.*, 1818–1820, 1996.
50. X. Zhang, M. Karlson, and P. Andrekson, Design guideline for actively mode locked fiber ring lasers, *IEEE Photon. Tech. Lett.*, 1103–1105, 1998.
51. A.E. Siegman, *Laser*. Mill Valley, CA: University Press, 1986.
52. J.J.E. Slotine and W. Li, *Applied Nonlinear Control*. Englewood Cliffs, NJ: Prentice-Hall, 1991.
53. K.K. Gupta, N. Onodera, and M. Hyodo, Technique to generate equal amplitude, higher-order optical pulses in rational harmonically modelocked fiber ring lasers, *Electron. Lett.*, 37, 948–950, 2001.
54. E.L. Goldstein, L. Eskilden, V. da Silva, M. Andrejco, and Y. Silberberg, Inhomogeneously broadened fiber-amplifier cascades for transparent multiwavelength lightwave network, *IEEE J. Lightwave Technol.*, LT-13, 782–790, 1995.
55. S. Yamashita and K. Hotate, Multiwavelength erbium-doped fiber laser using intracavity etalon and cooled by liquid nitrogen, *Electron. Lett.*, 2, 1298–1299, 1996.

56. A. Bellemare, M. Karasek, M. Rochette, S. Lrochelle, and M. Tetu, Room temperature multifrequency erbium-doped fiber lasers anchored on the ITU frequency grid, *IEEE J. Lightwave Technol.*, 18, 825–831, 2000.
57. S. Yamashita and M. Nishihara, Widely tunable erbium-doped fiber ring laser covering both C-band and L-band, *IEEE J. Select Topics Quantum Electron.*, 7, 41–43, 2001.
58. M.Y. Jeon, H.K. Lee, K.H. Kim. E.H. Lee, S.H. Yun, B.Y. Kim, and Y.W. Koh, An electronically wavelength-tunable mode-locked fiber laser using an all-fiber acoustooptic tunable filter, *IEEE Photon. Technol. Lett.*, 8, 1618–1620, 1996.
59. Y.T. Chieng and R.A. Minasian, Tunable erbium-doped fiber laser with a reflection Mach–Zehnder interferometer, *IEEE Photon. Technol. Lett.*, 6, 153–155, 1994.
60. S.K. Kim. G. Stewart, and B. Culshaw, Mode-hop-free single-longitudinal-mode erbium-doped fiber laser frequency scanned with a fiber ring resonator, *Appl. Opt.*, 38, 5154–5157, 1999.
61. Y.W. Song, S.A. Havstad, D. Starodubov, Y. Xie, A.E. Willner, and J. Feinberg, 40-nm-wide tunable fiber ring laser with single-mode operation using a highly stretchable FBG, *IEEE Photon. Technol. Lett.*, 13, 1167–1169, 2001.

5 Dispersion Compensation Using Photonic Filters

Compensating of dispersion effects in the transmission of lightwave-modulated signals is a very critical task in long-haul ultrahigh-speed optical communications. Several compensation and equalization techniques have been reported. However, this chapter describes only the techniques, employing optical filters and resonators that can be designed and implemented using photonic signal processing methodology. The generation of highly dispersive effects in resonators that are operated under resonance and eigenfiltering is given.

5.1 DISPERSION COMPENSATION USING OPTICAL RESONATORS

In recent years, there were arising interests in studying the characteristics of the all-fiber photonic circuit components, especially the recirculating delay line [1–3] and an optical resonator [4–8]. It was found that the photonic circuits have great potential of applications in many areas, such as communications [9,10], sensing devices [11,12], and signal processing devices [13]. The progress in fiber-optic technology also makes the implementation of these circuits more easily. For instance, because of the development in optical amplifier technology [14], active optical devices [2,3] can be possibly realized. As the trend of investigation of photonic circuits proceeds, it is expected that the circuits analyzed may consist of more and more elements within them. This surely causes difficulties in the examination of the circuits. Thus, an efficient method is needed to facilitate these problems.

As far as the examination of the circuit behavior and its design is concerned, the analysis of the circuit involves the determination of the circuit's transfer function. This is usually achieved by simultaneously solving a set of linear field or intensity equations. However, this method becomes error-prone and time consuming as the complexities of the circuit increase due to the increasing number of components and interconnections in the optical circuit network. It makes the formulation of the transfer functions for the circuit and the analysis on the circuits' characteristics more difficult and tedious. Although scattering matrix has been used in describing the circuit characteristics [15,16] and is shown to be a more systematic approach to the problem, a simple method of determining the circuit's transfer functions from the set of governing equations has not been proposed, to the best of our knowledge. This restricts our analysis to simple circuits only.

The new method employs the use of signal-flow graph (SFG) theory [9] in optical circuits. The signal-flow graph theory was first introduced by Mason [1,2] some 40 years ago. The theory has already been applied in electrical and electronic circuits for a long time. It is the first time, to the best of our knowledge, to apply the SFG theory for the analysis of optical circuits. The unique feature of our work is that the optical circuits can be represented in form of SFG and their circuits' transfer functions can then be determined systematically using the corresponding manipulation rules. The analysis of optical circuits which employs the new method has been published recently by our group [17–20]. The new method is found to be much more efficient than the conventional method. The main advantage of our method over the conventional one is that more simple and systematic procedures are used in deriving the circuits' transfer functions. This certainly lowers the possibility of error occurred. Moreover, the graphical representation of the optical circuits also allows easier examination of recirculating or resonating loops in the circuit. Thus, the locations of loops in the circuits can be identified with lesser effort and then the mechanism of resonance can be studied easily. Furthermore, our new method incorporates a stability test, which is useful in the assessment of the photonic circuit operation. The stability of the photonic circuit should be an important part in the design. However, very few studies have been conducted on aspects of stable and unstable states of the photonic processing unit. The stability test should prove to be an important tool in photonic circuit design in the future. The result is significant as a systematic method is presented to the analysis of photonic circuit. The new method should lead to a better understanding, regarding the performance of various photonic circuits and may initiate studies on complex photonic circuits on a large scale.

Passive resonators have been studied for some simple photonic circuit configurations [4–8]. However, optical amplifiers can be inserted in the fiber paths of the circuit to enhance the circuit's performance. It has been shown [3] that optical amplifiers can provide flexibility in the design of the photonic circuit, not only just as compensators for losses in fiber paths. An amplified double coupler double ring (DCDR) photonic circuit is studied in detail in this work. The passive operation of this circuit has been examined in the case of a resonator [21,22] but its use as recirculating delay line and the circuit's temporal response has never been studied. The development of the SFG theory in optical circuits allows us to carry out the generalized study of the DCDR circuit easily and efficiently. We can identify therecirculating or resonant loops in the circuit immediately, once its SFG has been drawn. This would significantly increase our understanding on the circuit performance, particularly, the effects from the physical structure of the circuit. The special feature of the DCDR circuit is that it consists of two loops, which share one common fiber path in the circuit. This special feature produces some useful results, like the application as an adder, which are not realizable in other simple configurations.

The DCDR circuit, like most of the optical circuits, is discrete in nature and can be treated as a discrete-time signal system. The discrete behavior of the all-fiber optical circuit is due to the delays in the fiber paths of the circuit. Hence, z-transform for the analysis of discrete-time signal processing is used in our work. The DCDR circuit system will be represented by transfer functions in the z-domain. The transfer function would contain poles and zeros on the z-plane, which are the roots of the denominator and numerator of the transfer function, respectively. The report would show that the pole and zero locations or distributions of the transfer functions determine the characteristics of the circuit, for instance, the filtering properties. Thus, the circuit responses would be explained in terms of the pole and zero values of the transfer functions. In this work, the circuit is examined under two conditions—illuminated by a temporal incoherent source and a coherent source with finite linewidth. In the former case, the analysis is carried out on an intensity basis and the circuit is operating as a recirculating delay line. In the latter case, the use of the circuit as a resonator is investigated together with the transient response of the circuit. The computation of transient response is generalized for optical circuits.

Investigation is also carried out on the possibility of employing the DCDR as a fiber dispersion equalizer. The use of optical resonators as dispersion equalizers was studied recently [9,10], but they were confined to simple resonators only which may limit the design flexibility. In this work, the DCDR circuit is shown to give more freedom in the design as an equalizer.

The section is organized as follows. In Section 5.1.1, the graphical technique is used to derive the transfer function of the resonator in association with Mason's gain rules. The SFG theory and Mason's rule [23,24] are found to be applicable in the analysis of photonic circuits. The unique feature of this method is shown and the advantages of our method over the conventional ones are presented.

Section 5.1.4 discusses the DCDR circuit operating under temporally incoherent source. The circuit is considered on an intensity basis and it is treated as recirculating delay lines. Several operation modes of the circuit are displayed, including passive, active and active with negative optical gain. Both the frequency and temporal responses are studied. Possible applications of the circuit in signal processing are suggested. Procedures in design of the circuit are proposed for the case with negative optical gain; this acts as an illustration of a general design procedure. The DCDR circuit under coherent source is studied with emphasis on the resonance effects and the transient response of the circuit. The resonance of the DCDR circuit is displayed and its uses in certain signal processing applications are mentioned. The effects of the source coherence and the input pulse shape on the transient response of the circuit are studied in detail. Also, algorithms for computations of transient responses for general photonic circuits are given.

In Section 5.1.6, one of the applications of the DCDR circuit, a fiber dispersion equalizer, is examined specifically. The equalization achieved by the circuit under different circuit's parameters is demonstrated. The capability of the DCDR circuit as a dispersion equalizer is shown with different operating points.

5.1.1 Signal-Flow Graph Application in Optical Resonators

The mathematical tools that are used in our analysis are introduced here. They include the z-transform and SFG theory. Z-transform of discrete-time signal processing is useful for analysis of photonic circuits, especially, for those consisting of delay elements. The SFG theory enables us to examine the efficiency in the investigation of the characteristics of photonic circuits. The graph reduction rule and Mason's rule [1,2], which are associated with the SFG theory is also included. Stability of the circuit is usually an important criterion for evaluating the circuit performance. Therefore, the stability test for the system is also introduced.

For the sake of simplicity, a simple photonic circuit, a single loop resonator circuit, is used as an example to illustrate the manipulation of the SFG theory. The z-transform technique, commonly known in the digital signal processing [3] (or can be called discrete-time signal processing) is employed for the analysis of the photonic circuit in this work. The discrete-time signal can be considered as equally spaced sampling of a continuous-time signal. The z-transform of a sequence $x[n]$ is defined as

$$X(z) = \sum_{n=-\infty}^{\infty} x[n]z^{-n} \tag{5.1}$$

$x[n]$ can be interpreted as the nth term of the sequence of numbers, which describe the discrete-time signal in the time domain. The ratio between the output transform $Y(z)$ and the input transform $U(z)$ of a system is called the transfer function $H(z)$ which is given as

$$H(z) = \frac{Y(z)}{U(z)} \tag{5.2}$$

Rearranging Equation 5.2 gives

$$Y(z) = H(z) \cdot U(z) \tag{5.3}$$

This shows that in the z-domain multiplication of the input with the transfer function produces the output. The transfer function given in Equation 5.2 is related to the corresponding time-domain sequence $h[n]$; indeed, it is usually defined as the impulse response of the system. The corresponding operation in the time domain of Equation 5.3 is

$$y[n] = \sum_{r=-\infty}^{\infty} u[r]h[n-r] \tag{5.4}$$

which is the convolution sum that can be expressed as

$$y[n] = u[n]*h[n] = h[n]*u[n] \tag{5.5}$$

where the symbol $*$ denotes the convolution. It is to be noted that the convolution operation is commutative. There are two important properties of convolution. The convolution of two functions in the time domain corresponds to the multiplication of their z-transforms. On the other hand, the multiplication of two functions in the time domain corresponds to their convolution in the z-domain.

Now writing the numerator and denominator, NUM(z) as $\sum_{i=0}^{B} b_i z^{-i}$ and DEN(z) as $\sum_{i=0}^{A} a_i z^{-i}$, respectively, the $H(z)$, the z-domain transfer function of the photonic processor, can be written as

$$H(z) = \frac{\mathrm{NUM}(z)}{\mathrm{DEN}(z)} = \frac{\sum_{i=0}^{B} b_i z^{-i}}{\sum_{i=0}^{A} a_i z^{-i}} \tag{5.6}$$

which can also be rewritten into product form as

$$H(z) = \frac{b_0}{a_0}\left(z^{A-B}\right)\frac{\prod_{k=1}^{B}(z - q_k)}{\prod_{j=1}^{A}(z - p_j)} \tag{5.7}$$

It is assumed that $H(z)$ has been expressed in a irreducible form. The values p_j are called the poles of $H(z)$ such that $H(p_j) = \infty$. The values q_k are called the zeros of $H(z)$ for which $H(q_k) = 0$. Also, there is a pole at $z = 0$ of the multiplicity of $(B - A)$, if $B > A$. If $A > B$, there would be a zero at $z = 0$ of the multiplicity of $(A - B)$. As it has been found that the transfer functions can be expressed in z-domain, thus the transfer characteristics are dependent on the zero–pole patterns [4] of these transfer functions in the z-plane. The magnitude–frequency response at a particular frequency as the operating point moving on the unit circle, $|z| = 1$, of the z-plane is given by

$$|H(z)| = |H|(e^{j\omega\tau}) = \frac{|b_0|}{|a_0|}\frac{\prod_{k=1}^{B} l_{zk}}{\prod_{j=1}^{A} l_{pj}} \tag{5.8}$$

where

$z = e^{j\omega\tau}$

ω is the angular frequency in rad/s

τ is the sampling period of signals in seconds

l_{zk} and l_{pj} are the lengths from the operating point to the position of the kth zero and the jth pole of the transfer function, respectively

The corresponding phase–frequency response at a particular operating frequency (wavelength) is given by

$$\arg(H(z)) = \sum_{k=1}^{B} \phi_{zk} - \sum_{j=1}^{A} \phi_{pj} + (A - B)\omega\tau \tag{5.9}$$

where ϕ_{zk} and ϕ_{pj} are the phase angles of the zeros and poles, respectively, formed by the horizontal real axis and the lines connecting the poles and zeros to the operating point in the z-plane.

Thus from Equations 5.8 and 5.9, we can design the magnitude–frequency response by adjusting the pole and zero patterns of the transfer functions. To obtain a maximum magnitude at a particular operating wavelength (frequency), we require a pole or a very small value of $\prod_{j=1}^{A} l_{pj}$ at that wavelength. Similarly, to obtain a minimum at a particular wavelength, a zero or an infinitesimal value of $\prod_{k=1}^{B} l_{zk}$ is required at that wavelength.

There are some relationships between the positions of poles in the z-plane to those correspondingly in the s-plane (the continuous frequency domain). Recall that $z = e^{sT}$ where $s = j\omega$ is one basic property of this relationship: when a pole position moves on the imaginary axis of the s-plane, and it would move along the unit circle of the z-plane. In this case, we would have marginal stability and a lossless system. When a pole moves on the imaginary axis towards the left-half of the s-plane, its equivalent z-plane pole would move inside the unit circle in the z-plane. The system would thus become lossy and stable. If one of the system poles lies outside the unit circle in the z-plane, the system becomes unstable. Its temporal response would increase with time. In general, the system would be stable if all the system poles lie inside or on the unit circle in the z-plane. Stability plays an important role in design of photonic circuits. Stability test for the photonic circuit is introduced in Section 5.1.2.

The SFG is well known in the analyses and formulation of massive electrical and electronic circuits. We employ this technique extensively in the analysis of optical circuits here. As mentioned in Chapter 2, an SFG is a graphical diagram that comprises of directed branches and nodes, in other words, a graph with nodes linked up by directed branches in some way. A branch represents the relationships between two nodes and it is usually associated with a transmittance or transfer function. The signal flows through the branch in a direction indicated by the arrowhead. The transmittance denotes the functional operation that the signal would undergo as it travels through the branch. In general, that functional operation can be linear or nonlinear. The rules of engagement of the transmission path are described in this chapter.

The nodes in the SFG represent the circuit variables. In photonic circuits, they are usually optical fields and intensities in the case for coherence and incoherence, respectively. The value of a node variable is taken to be the sum of all incoming signals entering the node. This node value or signal travels along each outgoing branch connected to it. There are two special kinds of nodes in SFG, source, and sink. A source is a node at which branches connected to it are all outgoing. Likewise, a sink is a node with incoming branches only.

There are several terms which are frequently used for describing SFG. For instance, a feed forward path from nodes a to b is a sequence of nodes and branches that the signal passes through from nodes a to b, in which all nodes are traversed only once. A feedback loop is a path, which starts and terminates at the same node such that no node is traversed more than once. A feedback loop, which contains only a single node is called a self-loop.

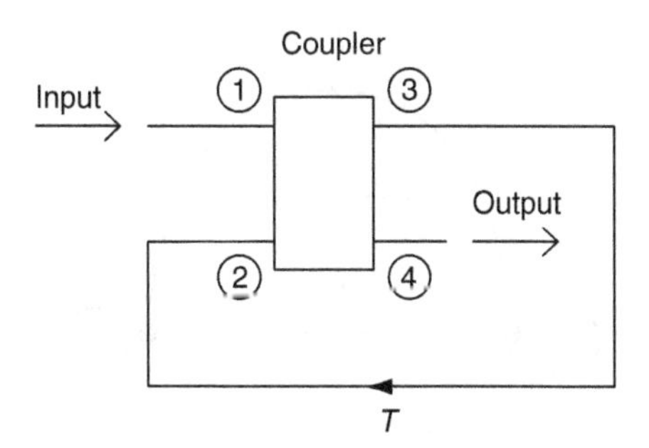

FIGURE 5.1 The schematic diagram of the single loop resonator.

Thus, the SFG is a graphical illustration of the relationships among the variables in a photonic circuit. In the following presentation, a single loop resonator circuit is used as a simple example to illustrate different aspects of the SFG theory and associated rules. A schematic diagram of the single loop resonator is shown in Figure 5.1.

A single loop resonator comprises a 2×2 directional coupler and a fiber path as shown in Figure 5.1. The fiber path is connected in such a way that one of the output ports of the coupler is joined to one of the coupler's input ports through the path. The excess power loss of the coupler is assumed to be zero, i.e., the coupler is lossless. This assumption of the coupler is valid for the whole scope of our analysis. The power coupling coefficient of the coupler is given as k. T represents the transmittance of the fiber path and is given as

$$T = t_a G z^{-1} \tag{5.10}$$

t_a is the passive transmission coefficient of the fiber path (3)–(2), related to the fiber's loss. If t_a is equal to its maximum value of 1, the fiber is lossless. G is the optical intensity gain factor of the fiber-optical amplifiers (if any) inserted in the path. G equals to one corresponds to the case where there is no amplifier in the feedback fiber path, i.e., passive operation of the circuit. By using the z-transform representation, the delay element in the photonic circuit with a unit delay can be represented by z^{-1}. The unit delay or the basic time delay is defined as the time required for the optical waves to travel along a length l of optical fiber such that it is equal to the sampling period of the input optical signal. In this case, z^{-1} is used to show that the signal at port 2 of the circuit is a unit-delayed version of the signal at port 3, apart from the scaling of the signal's magnitude. This represents the discrete nature of the signals in the circuit.

As the circuit is taken to be an example of the SFG illustration only, we further assume that the source is temporally incoherent so as to simplify the situation. By this assumption, it means that the source coherence length is much shorter than the shortest delay path in the circuit so that the phase change can be ignored. Intensity rather than field amplitude is thus considered. The directional coupler has a direct coupling of $(1 - k)$ and cross coupling of k of the intensity at the input port. Further

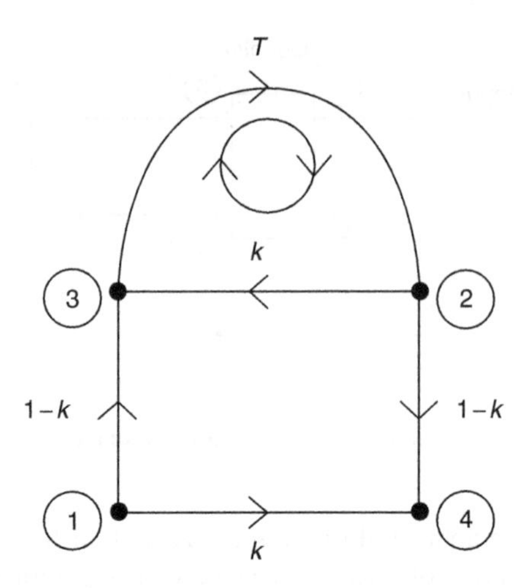

FIGURE 5.2 The signal-flow graph (SFG) of the single loop resonator circuit.

discussion of the temporal incoherent source is given in Chapter 2. The SFG in terms of intensity for the single loop resonator circuit is shown in Figure 5.2.

It can be seen that the single loop in the circuit and the direction of signal flow in the loop can be identified easily in the SFG in Figure 5.2. In all-fiber photonic circuits, the directional couplers and the delay lines usually form the basic building blocks. From Figure 5.2, it can be seen that the SFGs of these two elements are simple.

After obtaining the SFG of the photonic circuit, the derivation of the transfer functions between node variables is very straightforward. There are generally two methods to achieve this objective, the graphical reduction of the SFG and Mason's rule.

If the SFG reduction rules (3), (4), and (1) are applied in succession to the SFG in Figure 5.2, the output–input intensity transfer function of the single loop circuit is obtained as

$$H_{14} = \frac{k + (1 - 2k)T}{1 - kT} \tag{5.11}$$

It is not necessary to reduce the SFG to find the desired transfer functions of the circuit. Mason's rule [1,2] can also be applied to give the transfer functions directly. Considering the independent node a and dependent node b of an arbitrary photonic graphical network, which are its input and any port. Let the transfer function H_{ab} for the SFG defined as the overall transmittance between the two nodes a and b, then Mason's rule can be stated for the circuit as [1,2]

$$H_{ab} = \frac{\sum_{q=1}^{N} (F_{ab})_q (\Delta_{ab})_q}{\Delta} \tag{5.12}$$

where N is the total number of optical transmission paths from nodes a to b, $(F_{ab})_q$ is the optical transmittance of the qth path from nodes a to b in the SFG, and Δ is given by

$$\Delta = 1 - \sum_{u} T_{lu} + \sum_{u,v} T_{lu}T_{lv} - \sum_{u,v,w} T_{lu}T_{lv}T_{lw} + \cdots \tag{5.13}$$

where T_{lu} is the loop transmittance and only products of nontouching loops are included in the Δ expression. Thus, Equation 5.13 can be written in plain words as

> $\Delta = 1$ − (sum of all loop transmittances) + (sum of products of all loop transmittances of two nontouching optical loops) − (sum of products of all loop transmittances of three nontouching loops) $+ \cdots$

where Δ is defined as the graph determinant. Therefore, it follows that $(\Delta_{ab})_q$ is the cofactor of the qth forward transmission path which is equal to Δ after all loops touching the qth path have been excluded. Two optical loops are considered as nontouching, if they do not share any common node. However, one criterion of using Mason's rule is needed to be satisfied. The SFG should be able to be represented in the planar form in which there are no crossings between the signal-flow paths.

From the SFG of the single loop photonic circuit, it can be observed that there is only one optical feedback loop in the circuit which is (3)(2)(3). Thereafter, the loops or the paths in the circuit are given by sequences of the node numbers the loop or path follows. Its output–input intensity transfer function can be determined using Mason's rule of Equation 5.13. Applying Mason's rule, H_{14} of the single loop resonator circuit can then be determined by

$$H_{14} = \frac{\sum_{q=1}^{2} (F_{14})_q (\Delta_{14})_q}{\Delta} \tag{5.14}$$

Firstly, the loop transmittance of the single loop circuit is given as

$$T_{l1} = kT \tag{5.15}$$

Thus, Δ can be obtained by

$$\Delta = 1T_{l1} \Rightarrow \Delta = 1 - kT \tag{5.16}$$

There are two forward paths going from nodes 1 to 4; the paths are as follows: (i) the direct path (1)(4) that has a transmittance of k, that is $(F_{14})_1 = k$, and $(\Delta_{14})_1 = 1 - kT$ and (ii) path (1)(3)(2)(4) with the transmittance of $(F_{14})_2 = (1-k)^2 T$ and $(\Delta_{14})_2 = 1$. Substituting Equations 5.5 through 5.8 into Equation 5.4 yields

$$H_{14} = \frac{k + (1 - 2k)T}{1 - kT} \tag{5.17}$$

which is the same expression as obtained in Equation 5.11 that used the SFG reduction rules.

5.1.2 Stability

Stability consideration of the photonic circuit is critical as it is directly related to the performance of the circuit especially in the time-domain applications. As the photonic circuit can be modeled under the discrete signaling, Jury's stability test, which is well known in digital control system engineering [7], can be employed.

Consider a characteristic equation in the form of

$$F(z) = a_n z^n + a_{n-1} z^{n-1} + \cdots + a_2 z^2 + a_1 z + a_0 = 0 \tag{5.18}$$

The characteristic equation is defined as the equation obtained by equating the denominator of the transfer function to zero. For $n = 2$, which is a second-order system, the necessary and sufficient conditions for a stable operation of the circuit require

$$F(1) > 0, \quad \text{and} \quad F(-1) > 0 \quad \text{and} \quad |a_0| < a_2 \tag{5.19}$$

That is, all the roots of the equation must lie inside the unit circle in the z-plane. In this example, $n = 1$, the stability criterion becomes $F(1) > 0$ and $F(-1) < 0$. Now consider Equation 5.10. The characteristic equation of the single loop circuit is

$$F(z) = z - kt_a G = 0 \tag{5.20}$$

Thus, the stability criterion is

$$1 - kt_a G > 0 \Rightarrow kt_a G < 1 \tag{5.21}$$

The expression on the LHS of the inequality in Equation 5.21 is, in fact, the pole value of the circuit. If the pole value is less than one then the pole will lie inside the unit circle in the z-plane and stability is achieved. Therefore, it is clearly shown that Jury's stability test gives the condition for which the pole of the system will lie inside the unit circle in the z-plane.

5.1.3 Frequency and Impulse Responses

5.1.3.1 Frequency Response

The magnitude and phase responses of the photonic circuit can be obtained by repeating computations at different values of frequency along the unit circle, i.e., $|z| = 1$. The frequency responses of H_{14} of the single loop resonator circuit that is plotted against $\omega\tau$ (ωT in the graphs should read $\omega\tau$) is shown in Figure 5.3 together with a plot showing the pole and zero positions. ω is the angular frequency of the source and τ is the unit time delay of the fiber path in the circuit. The parameters used in this example are $k = 0.4142$, $t_a = 1$, and $G = 2.414$, which give a pole at $+1$ and a

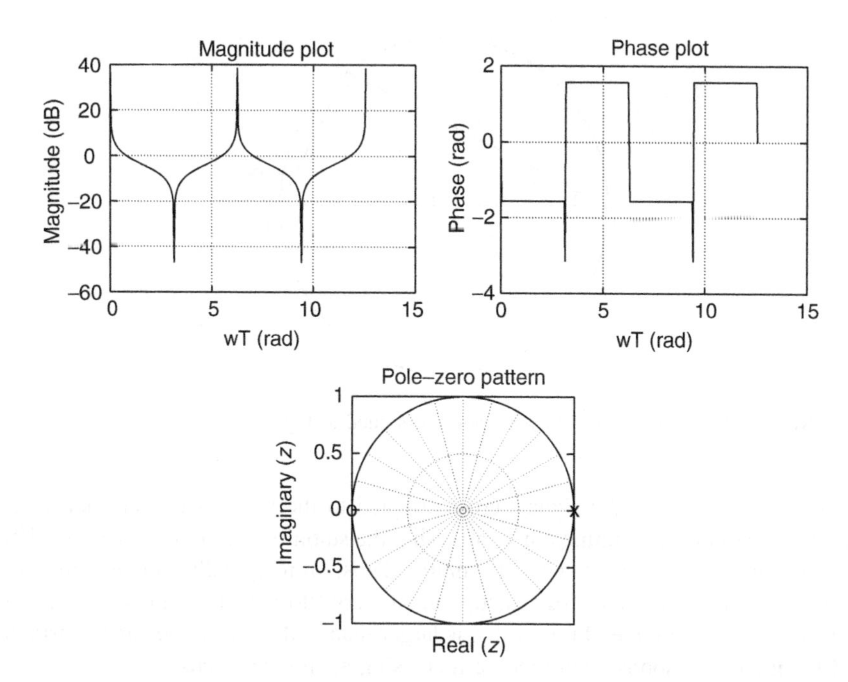

FIGURE 5.3 Frequency response for the single loop circuit together with the pole–zero positions with pole at +1 and zero at −1 (pole: *x*, zero: 0).

zero at −1. From the magnitude plot, we can see that the response has minima, when the frequency passing a pole and maxima for zeros. When passing through a pole, the phase changes from −1.57 rad (−90°) to 1.57 rad (90°), likewise; when traveling past a zero, the phase changes from 1.57 rad (90°) to −1.57 rad (−90°). At other frequencies, the phase remains constant at either −1.57 rad or 1.57 rad. This can be explained by examining the phase response expression for the circuit. The phase response of the single loop circuit is given by

$$\arg(H_{14}(z)) = \phi_{z1} - \phi_{p1} \tag{5.22}$$

Consider geometrically in the z-plane, from Figure 5.4, the difference between ϕ_{z1} and ϕ_{p1} is always 90°. Thus $(\phi_{z1} - \phi_{p1})$, depending on the position of the operating point, takes the value of either +90° or −90°.

5.1.3.2 Impulse and Pulse Responses

The impulse and pulse responses of the feedback resonator given for Figure 5.1 are shown in Figure 5.5. The pulse input can be obtained by launching a sequence of "1 1 1" into the circuit, the "1's" are delayed from each other by a unit time delay. It can be seen that the impulse response has a steady-state value. The system is thus undamped or lossless. From Equation 5.12, the magnitude of the loop transmittance has a pole at +1. Therefore, the signal maintains its magnitude in each circulation

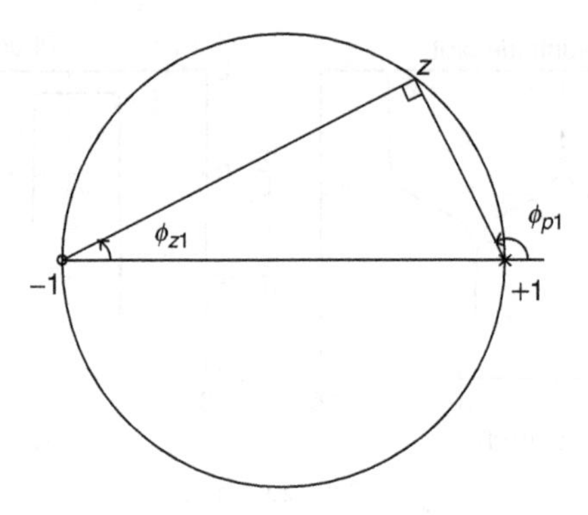

FIGURE 5.4 Representing the phase response obtained in Figure 5.3.

inside the circuit loop; the signals cross coupled to the circuit's output after every circulation are thus remaining constant. Using the stability criterion (Equation 5.19), kt_aG or the pole value equates to unity. Thus, a marginally stable condition is achieved, i.e., the pole lies exactly on or very close to the unit circle in the z-plane. If the magnitude of pole value is larger than 1, the system becomes unstable and the impulse response will then be increasing as time proceeds.

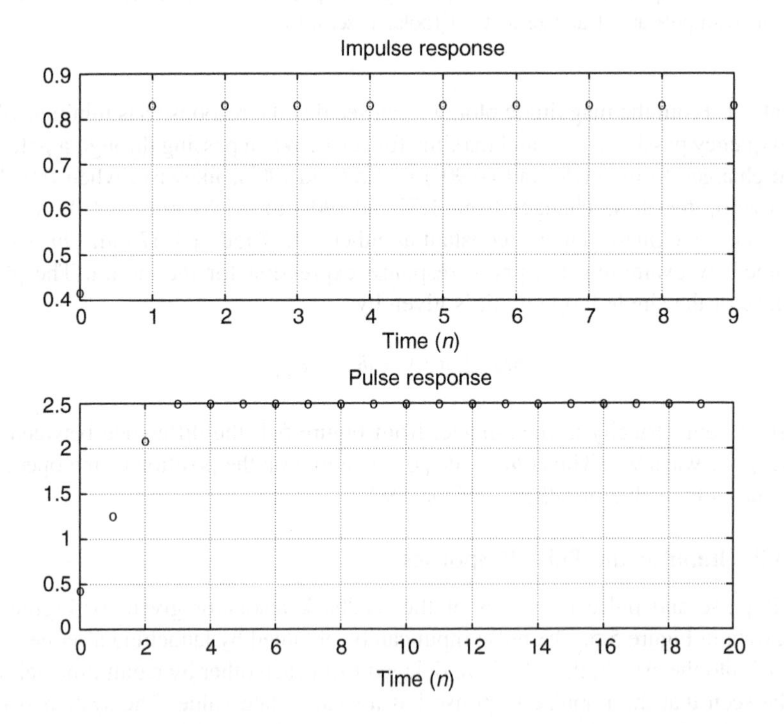

FIGURE 5.5 Impulse and pulse responses for the single loop circuit with the same circuit parameters as in Figure 5.3.

5.1.3.3 Cascade Networks

This photonic network theory can now be extended to a network of interconnected basic photonic circuits. In fact, we have so far treated the photonic circuits as elements of a signal processing system; in other words, we have systematized the circuit. Thus, the system theory can be applied in the manipulation of the circuit. For instance, there are two photonic circuits with output–input transfer functions H_a and H_b, respectively. If the output of the first circuit is connected to the input of the second circuit, i.e., the two circuits are connected in cascade, then the overall output–input transfer function of the system will then be equal to H_aH_b, the product of the two transfer functions. In general, if we connect n photonic circuits in cascade, the overall transfer function of the system can be easily determined as the product of all individual transfer functions.

5.1.3.4 Circuits with Bidirectional Flow Path

If in a photonic circuit, the signal travels in both directions through a fiber path, for instance one signal is transmitted and the other is reflected, the circuit can be represented by two SFGs. If the circuit is linear, the circuit transfer function can be obtained by the superposition of the two transfer functions as obtained individually from each SFG [8], the single ring resonator with two-wave mixing, or the z-shaped double-coupler resonator [9].

5.1.3.5 Remarks

The tools for analyzing general photonic circuit are introduced here. The application of z-transform in the analysis makes both manipulating and understanding easy. We have developed the application of the powerful SFG theory in the study of photonic circuits. This provides an examination of the circuit characteristics visually and it effectively reduces the effort of determining the transfer functions of the circuit. The techniques of handling the SFG are demonstrated, which includes Mason's rule and the graphical reduction rule. Together with the stability test provided, the circuit behavior can be examined thoroughly.

5.1.4 DCDR Circuit under Temporal Incoherent Condition

Resonator circuits with double or multiple couplers have been studied recently [1–4]. In this section, a DCDR circuit is analyzed. The use of this circuit as a resonator is studied [2,3], but its use as a recirculating delay line had not been examined. Here, the signal is assumed to be temporally incoherent [5,6]. Therefore, with this assumption, the circuit is operated as recirculating delay line instead of resonator. Incoherent processing is considered by using the intensities. A detailed study of the circuit is carried out for different sets of circuit parameters, including the case where negative optical gain is provided by the optical amplifier in the circuit. The frequency and impulse responses of the circuit are given. From the studies, several applications of the DCDR circuit are illustrated.

5.1.4.1 Transfer Function of the DCDR Circuit

The schematic diagram of the DCDR circuit is shown in Figure 5.6 as consisting of two 2×2 directional couplers interconnected with three optical fiber forward and feedback paths. The fiber paths (3)(6) and (4)(5) are referred to as the forward paths of the circuit while path (7)(2) is called the feedback path of the circuit. The circuit can be considered as a two-port device which incorporates one input port and one output port as visualized in Figure 5.6b. In this latter arrangement, the circuit can be viewed as a cascade form of two couplers with an overall feedback loop. k_1 and k_2 are the power coupling coefficients of the two couplers 1 and 2, respectively. T_1, T_2, and T_3 denotes the transmission functions of the forward paths (3)(6) and (4)(5), and the feedback path (7)(2), respectively.

The optical transmittances can be represented by

$$T_i = t_{ai} G_i z^{-m_i} \quad \text{for} \quad i = 1, 2, 3 \tag{5.23}$$

where

$i = 1, 2, 3$ corresponds to the three paths as shown in Figure 5.6
t_{ai} is the transmission coefficient of the ith fiber path as the same parameter t_a defined in Chapter 2
G_i is the optical intensity gain factor which is provided by the fiber-optical amplifiers incorporated in the paths
m_i is the order of the delay path

The SFG of the DCDR circuit is shown in Figure 5.7. The SFG of the DCDR circuit is represented in planar form, i.e., there is no crossing of optical paths in the graph. The node variable in the SFG represents the optical intensity at that point of the DCDR circuit. From the SFG of the DCDR circuit, it can be easily seen that there are two optical closed loops in the circuit which can be called the feedback loops or the recirculating loops of the circuit. One of the loops is the loop connecting nodes numbered (2), (3), (6), and (7), and the other through nodes numbered

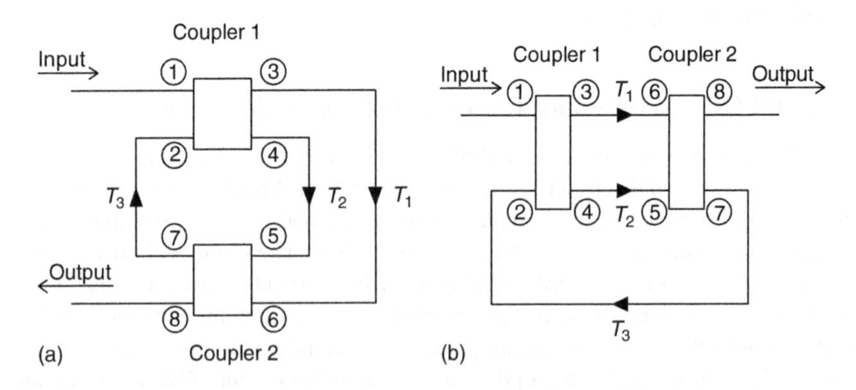

FIGURE 5.6 (a) The schematic diagram of the DCDR circuit and (b) the same circuit in a different arrangement.

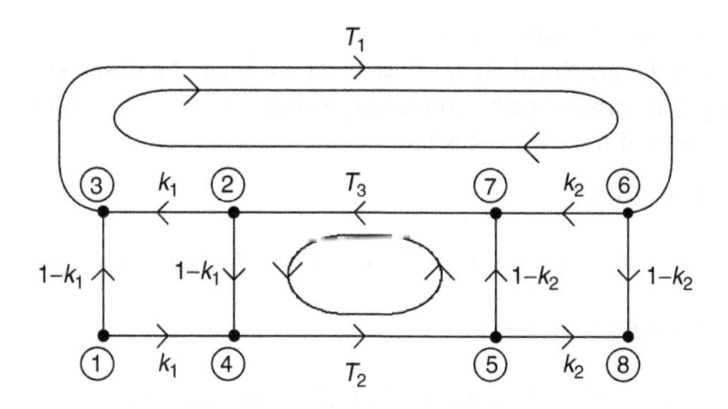

FIGURE 5.7 The SFG of the DCDR circuit.

(2), (4), (5), and (7). Furthermore, these two loops share one common path, the path (7)(2). This demonstrates one of the advantages of representing the photonic circuit with an SFG as we can identify the loops in the circuit without any difficulty. The SFG gives us an insight into the circuit properties. The next step in the analysis is to derive the optical transfer functions of the circuit from the SFG.

The output–input intensity transfer function of the DCDR circuit can be obtained by using either the graph reduction rules or Mason's rule as introduced in Ref. [7]. In deriving the optical transfer functions between nodes a and b, the subscript a or b is now omitted for the sake of clarity. By applying Mason's rule for the SFG of the DCDR photonic circuit, the output–input intensity transfer function is given by

$$H_{18} = \frac{I_8}{I_1} = \frac{\sum_{q=1}^{4} F_q \Delta_q}{\Delta} \tag{5.24}$$

where I_8 and I_1 are the output and input intensities, respectively, of the DCDR circuit. The path and loop transmittances can be obtained as follows:

(a) *Two loop transmittances*

(i) Loop 1: Loop 1 is formed by nodes (2), (3), (6), and (7), thus loop 1 can be written as (2)(3)(6)(7)(2). The loop optical transmittance of loop 1 is

$$T_{l1} = k_1 k_2 T_1 T_3 \tag{5.25}$$

(ii) Loop 2: Similarly, the optical transmittance of loop 2 that is (2)(4)(5)(7)(2) is given by

$$T_{l2} = (1 - k_1)(1 - k_2) T_2 T_3 \tag{5.26}$$

(b) *Forward path transmittances*

From the graphical signal-flow diagram, it can be observed that there are four optical forward paths connecting nodes 1 and 8. The four forward paths and its related transmittances are

Path 1: (1)(3)(6)(8)

$$F_1 = (1 - k_1)(1 - k_2)T_1 \quad \text{and} \quad \Delta_1 = 1 - T_{l2} \tag{5.27}$$

Path 2: (1)(3)(6)(7)(2)(4)(5)(8)

$$F_2 = (1 - k_1)^2 k_2{}^2 T_1 T_2 T_3 \quad \text{and} \quad \Delta_2 = 1 \tag{5.28}$$

Δ_2 is equal to unity because of its forward path touching both optical loops.

Path 3: (1)(4)(5)(7)(2)(3)(6)(8)

$$F_3 = k_1{}^2 (1 - k_2)^2 k_2{}^2 T_1 T_2 T_3 \quad \text{and} \quad \Delta_3 = 1 \tag{5.29}$$

Similar to Δ_2, Δ_3 is equal to unity because of the touching of two loops of the forward path.

Path 4: (1)(4)(5)(8)

$$F_4 = k_1 k_2 T_2 T_2 \quad \text{and} \quad \Delta_4 = 1 - T_{l1} \tag{5.30}$$

The loop determinant Δ in the denominator is then given by

$$\Delta = 1 - T_{l1} - T_{l2} \tag{5.31}$$

Therefore, the optical transfer function I_8/I_1 as H_{18} can be obtained as

$$H_{18} = \frac{(1 - k_1)(1 - k_2)T_1 + k_1 k_2 T_2 - (1 - 2k_1)(1 - 2k_2)T_1 T_2 T_3}{\text{DEN}} \tag{5.32}$$

where

$$\text{DEN} = 1 - k_1 k_2 T_1 T_3 - (1 - k_1)(1 - k_2)T_2 T_3 \tag{5.33}$$

Alternatively, the transfer functions of the DCDR circuit can be found by performing a graph reduction of the SFG in Figures 5.7 and 5.8a,b. After reduction to a certain stage, we see that the transfer function H_{18} obtained from Figure 5.8b is the same as that given in Equation 5.32. In Figure 5.8b, this can be performed by summing up the transmittances going to port 8, from path (1)(3)(8) and path (1)(4)(8).

5.1.4.2 Circulating-Input Intensity Transfer Functions

The same procedure as given in the above section can be used to get the circulating-input intensity transfer functions of the DCDR circuit. To obtain the transfer function

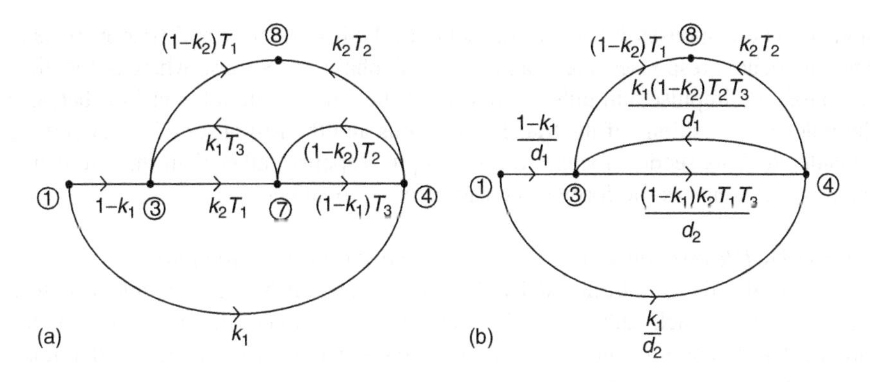

FIGURE 5.8 The reduced signal-flow graphs for the SFG (Figure 5.7) of the DCDR circuit. (a) First reduction and (b) further graphical reduction $d_1 = 1 - k_1k_2T_1T_3$ and $d_2 = 1 - (1 - k_1)(1 - k_2)T_2T_3$.

H_{13}, the circulating intensity I_3 with respect to the input intensity, we consider the SFG of the DCDR circuit again. There are two optical forward paths traveling from nodes 1 to 3; they are the direct path (1)(3) and path (1)(4)(5)(7)(2)(3). Their transmittances are given by:

$$\text{Path 1: (1)(3): } F_1 = 1 - k_1 \quad \text{and} \quad \Delta_1 = 1 - T_{l2} \tag{5.34}$$

$$\text{Path 2: (1)(4)(5)(7)(2)(3): } F_2 = k_1{}^2(1 - k_2)T_2T_3 \quad \text{and} \quad \Delta_2 = 1 \tag{5.35}$$

H_{13} is then given as

$$H_{13} = \frac{\sum_{i=1}^{2} F_i\Delta_i}{\Delta} \tag{5.36}$$

By substituting Equations 5.26 and 5.27 into Equation 5.28, we have

$$H_{13} = \frac{(1 - k_1) - (1 - 2k_1)(1 - k_2)T_2T_3}{\text{DEN}} \tag{5.37}$$

Similarly, the transfer functions H_{14} and H_{17} can be obtained as

$$H_{14} = \frac{I_1}{I_4} = \frac{k_1 + (1 - 2k_1)k_2T_1T_3}{\text{DEN}} \tag{5.38}$$

$$H_{17} = \frac{I_7}{I_1} = \frac{(1 - k_1)k_2T_1 + k_1(1 - k_2)T_2}{\text{DEN}} \tag{5.39}$$

5.1.4.3 Analysis

In this section, the behavior of the DCDR circuit under incoherent source condition is examined. The responses of the circuit with different circuit parameters are studied. The responses investigated, including both frequency and time responses. The frequency responses are magnitude and phase responses, whereas the time responses are impulse and pulse responses of the circuit. The relationships between the pole–zero positions of the transfer functions and the responses of the circuit are considered. This section focuses on the output response rather than the circulating intensity response as the former one can be easily obtained.

Case 1: DCDR resonance with unity delay in each transmission path
In case 1, the delay introduced by the interconnection between couplers in feed forward and feedback paths is equal to the sampling period of the signal entering the circuit, that means $m_1 = m_2 = m_3 = 1$. The denominator DEN of the transfer functions derived in Section 5.1.4.1 becomes

$$\text{DEN} = 1 - k_1k_2t_{a1}t_{a3}G_1G_3z^{-2} - (1 - k_1)(1 - k_2)t_{a2}t_{a3}G_2G_3z^{-2} \quad (5.40)$$

For simplicity, assuming that the fiber transmission paths are lossless, i.e., $t_{a1} = t_{a2} = t_{a3} = 1$, we take this assumption throughout this section unless otherwise specified. Thus, we have

$$\text{DEN} = 1 - k_1k_2G_1G_3z^{-2} - (1 - k_1)(1 - k_2)G_2G_3z^{-2} \quad (5.41)$$

Rearranging Equation 5.33 in the form

$$F(z) = z^2 - k_1k_2G_1G_3 - (1 - k_1)(1 - k_2)G_2G_3 = 0 \Leftrightarrow F(z) = a_2z^2 + a_0 = 0 \quad (5.42)$$

So with $a_2 = 1$ and $a_0 = -k_1k_2G_1G_3 - (1 - k_1)(1 - k_2)G_2G_3$.

Applying Jury's stability test, Equations 5.11 and 5.12, a stable operation of the circuit requires

$$F(1) > 0 \Rightarrow 1 - k_1k_2G_1G_3 - (1 - k_1)(1 - k_2)G_2G_3 > 0 \quad (5.43a)$$

$$F(-1) > 0 \Rightarrow 1 - k_1k_2G_1G_3 - (1 - k_1)(1 - k_2)G_2G_3 > 0 \quad (5.43b)$$

and

$$|a_0| < a_2 \Rightarrow |-k_1k_2G_1G_3 - (1 - k_1)(1 - k_2)G_2G_3| < 1 \quad (5.44)$$

If we consider the system stability in terms of the system poles, the necessary and sufficient condition for the optical networks to be stable is that all the poles of the system must lie on or inside the unit circle of the z-plane. In other words, the magnitude of these poles must be less than or equal to one. Thus, the pole magnitudes given by the characteristics of Equation 5.44 would result in

$$|k_1k_2G_1G_3 + (1 - k_1)(1 - k_2)G_2G_3| < 1 \quad (5.45)$$

It is shown once again that the stability test result is strongly linked to the pole positions in the z-plane. The system poles can be made imaginary when the expression inside the absolute sign on the left-hand side (LHS) of Equation 5.37 is set negative. This case would be treated later in the analysis.

Several scenarios of case 1 would now be studied depending on several parameters of the DCDR circuit.

(a) Case 1(a): Passive DCDR circuit

Operating the DCDR circuit under passive condition, i.e., when there is no optical amplification in the circuit implying $G_1 = G_2 = G_3 = 1$. The stability condition in Equation 5.37 becomes

$$|k_1 k_2 + (1 - k_1)(1 - k_2)| < 1 \tag{5.46}$$

The transfer function H_{18}, Equation 5.24, becomes

$$H_{18} = \frac{[(1 - k_1)(1 - k_2) + k_1 k_2]z^{-1} - (1 - 2k_1)(1 - 2k_2)z^{-3}}{1 - k_1 k_2 z^{-2} - (1 - k_1)(1 - k_2)z^{-2}} \tag{5.47}$$

where the zeros are at

$$z_z(1,2) = \pm\sqrt{\frac{(1 - 2k_1)(1 - 2k_2)}{(1 - k_1)(1 - k_2) + k_1 k_2}} \tag{5.48}$$

Let y be the expression of the LHS of Equation 5.39, the plot of y as a function of the coupling coefficients k_1 and k_2 is shown in Figure 5.9. As the ranges of k_1 and k_2 fall within "1" only, the stability inequality (Equation 5.7) is always satisfied. This can also be observed from Figure 5.9. It implies that the passive circuit is always stable. Recall that y is also the square of the system poles' magnitude, y attains its maximum value of 1 when $k_1 = k_2 = 0$ or $k_1 = k_2 = 1$. In either of these two circumstances, the circuit reduces to a single straight through optical path and contains no feedback loop. The system poles can never be positioned on the unit circle of the z-plane.

Examining the transfer function H_{18}, it is noted that the zeros of the output–input intensity transfer function for the passive DCDR circuit can be either purely real or purely imaginary depending on the values of k_1 and k_2.

The output–input frequency responses and impulse responses of the passive DCDR circuit have been computed using different values of k_1 and k_2. The corresponding responses are given in Figures 5.10 through 5.12. The data for the three figures are listed in Table 5.1.

In Figure 5.10, it can be seen that the variation in the magnitude plot is minute and less than 1 dB. This can be clearly observed that the zero and pole pairs are very close to each other and their effects counteract each other. The circuit thus resembles a single pole system, which has the pole at the center of the unit circle in the z-plane; the magnitude plot is therefore quite uniform with respect to the frequency. The output–input impulse response is mainly an impulse delayed by one unit delay time

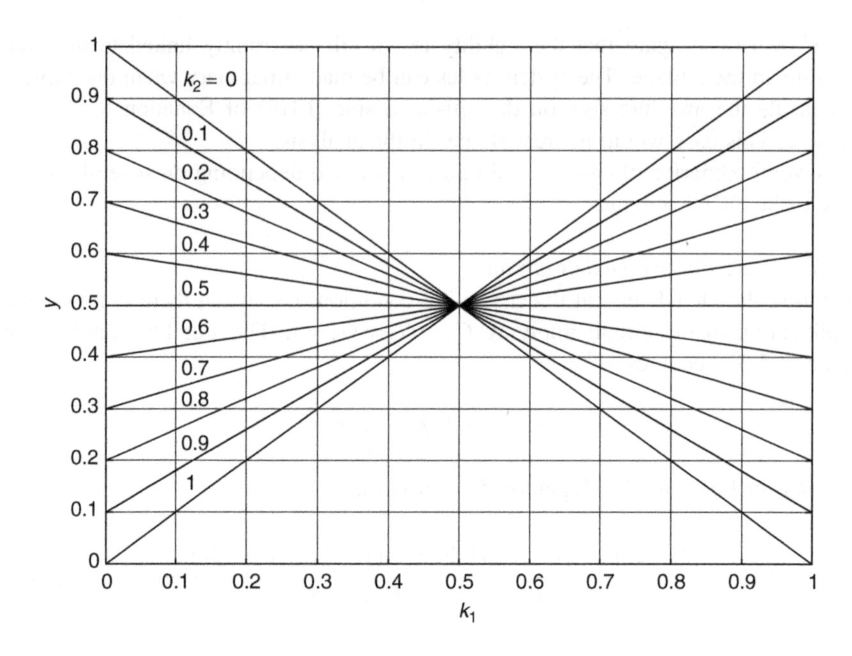

FIGURE 5.9 Variation of y (LHS of Equation 5.39) against the coupling coefficient k_1 with k_2 as a variable.

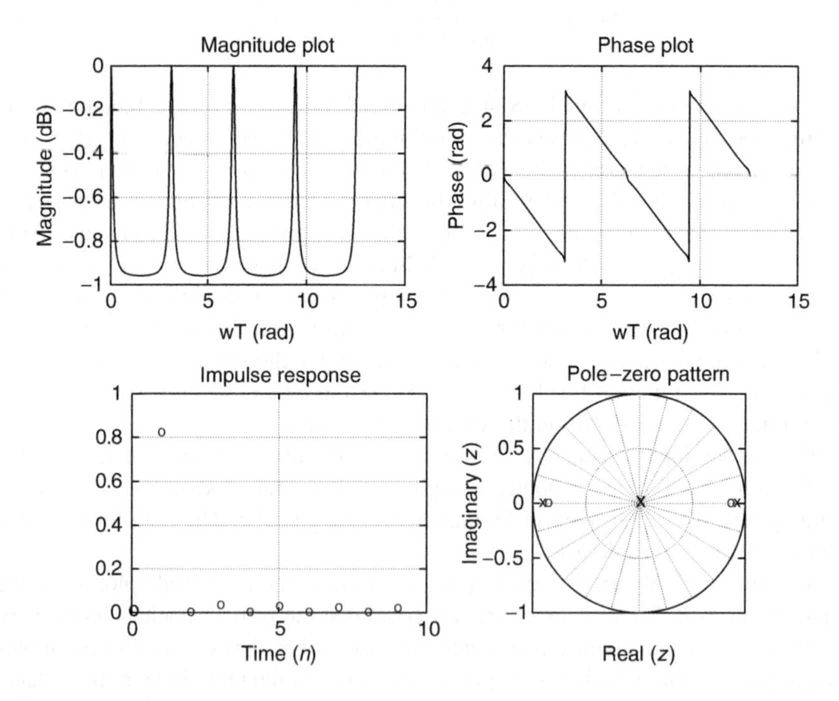

FIGURE 5.10 Frequency response, impulse response, and pole–zero plot (pole: x, zero: 0) for the passive DCDR circuit with circuit parameters given in Table 5.1.

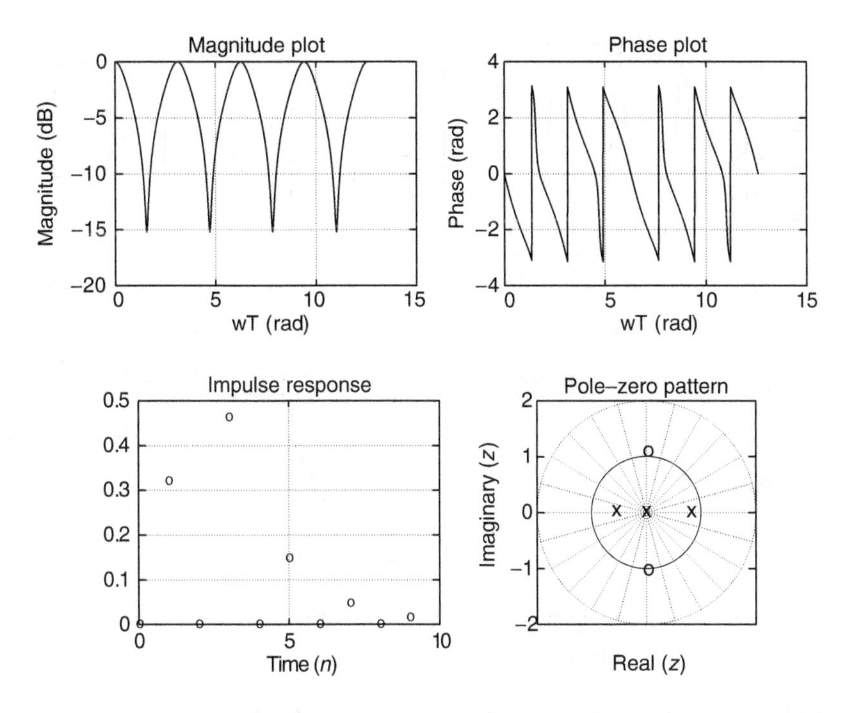

FIGURE 5.11 Frequency response, impulse response, and pole–zero plot for the passive DCDR circuit with circuit parameters given in Table 5.1.

from the input, which is similar to the response of a unit delay line with loss. It is noted that there are quasi-linear portions in the phase response which can be useful.

The output–input frequency response in Figure 5.11 shows quite sharp dips. Its impulse response shows a maximum after every circulation in the loop and then decays. In this case, the zeros are purely imaginary and the poles are purely real. Thus, they would not cancel each other as in the case in Figure 5.10.

Figure 5.12, where $k_1 = k_2 = 0.5$, gives the zeros at 0 which is at the center of the unit circle in the z-plane. As seen in Equation 5.39, if any one of k_1 and k_2 is equal to 0.5 then the zeros would be 0. Also, it is shown from Figure 5.9 that y (hence the square of system poles' magnitude) remains constant with different k_1 for $k_2 = 0.5$. It is also true when $k_1 = 0.5$, then y is a constant when we change k_2. In other words, if we set k_1 or k_2 equal to 0.5, the pole and zero patterns of the output–input transfer function stay the same when we change the other k. As the output response of the DCDR circuit depends solely on the pole and zero patterns of the output–input transfer function, the response would stay the same in this case. Moreover, it can be seen from the impulse response in Figure 5.12 that it is a lossy system. It is also found that in the passive circuit, the two couplers exhibit symmetrical behavior. For instance, the response of $k_1 = k_2 = K$ is equal to that of $k_1 = k_2 = 1 - K$.

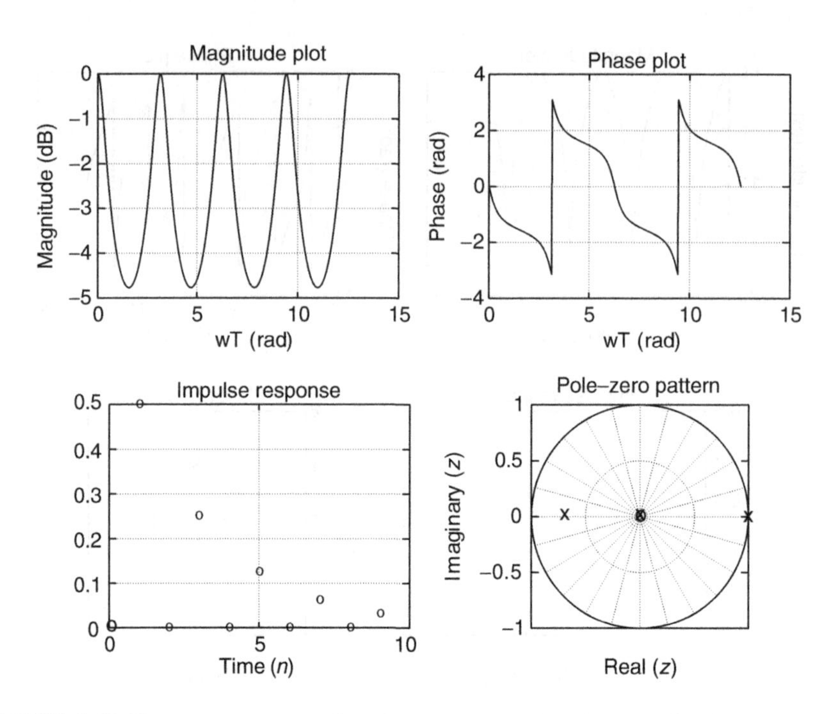

FIGURE 5.12 Frequency response, impulse response, and pole–zero plot for the passive DCDR circuit with circuit parameters given in Table 5.1.

(b) Case 1(b): Active DCDR circuit with unit delay in each path
(i) $G_1 > 1$. One optical amplifier in forward path
We investigate here the case, where only one optical amplifier is inserted in the DCDR circuit and it is inserted in the path connecting between ports 3 and 6, one of the two forward paths. The output–input intensity transfer function H_{18} simplifies to

$$H_{18} = \frac{[(1-k_1)(1-k_2)G_1 + k_1k_2]z^{-1} - (1-2k_1)(1-2k_2)G_1z^{-3}}{1-[k_1k_2G_1 + (1-k_1)(1-k_2)]z^{-2}} \tag{5.49}$$

TABLE 5.1
Parameters Used in the Analysis of the Passive DCDR Response with the Corresponding Poles and Zeros of the Output–Input Intensity Transfer Function H_{18}

	k_1	k_2	Poles	Zeros
Figure 5.10	0.9	0.9	0, ±0.905539	±0.883452
Figure 5.11	0.2	0.8	0, ±0.565685	±j1.06066
Figure 5.12	0.5	0.5	0, ±0.707107	0, 0

The zeros of H_{18} are thus located at

$$z_z(1,2) = \pm\sqrt{\frac{(1-2k_1)(1-2k_2)G_1}{(1-k_1)(1-k_2)G_1+k_1k_2}} \tag{5.50}$$

The stability condition now becomes

$$|k_1k_2G_1 + (1-k_1)(1-k_2)| < 1 \tag{5.51}$$

The pole values of the output intensity transfer function H_{18} and its zero values are plotted against the optical gain G_1 with k_1 and k_2 as parameters are shown in Figures 5.13a through c with different values of k_1 and k_2 as indicated.

To get a pole value close to "1" and just inside the unit circle, the value of G_1 needs to be 1.2, 5.2, and 3, respectively, in the cases shown in Figures 5.13a through c. These give us fast rolling off frequency response than the passive counterpart as are displayed in Figures 5.14 through 5.16. The data for the three figures are listed in Table 5.2.

Comparing Figures 5.13 and 5.14 with Figure 5.10, it is observed that the introduction of optical amplifier in the DCDR circuit enhances the performance, especially the frequency response. As the pole pairs are pushed very close to the unit circle in the

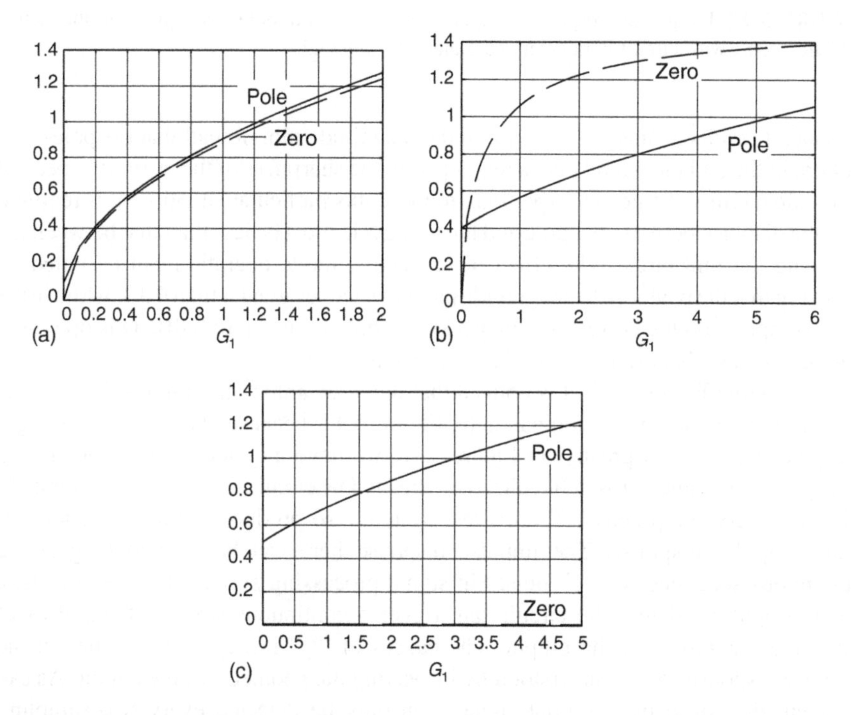

FIGURE 5.13 Plot of absolute magnitudes of the pole and zero of H_{18} against G_1 in case 1(b)(i). (a) $k_1 = 0.9$ and $k_2 = 0.9$, (b) $k_1 = 0.2$ and $k_2 = 0.8$, (c) $k_1 = 0.5$ and $k_2 = 0.5$.

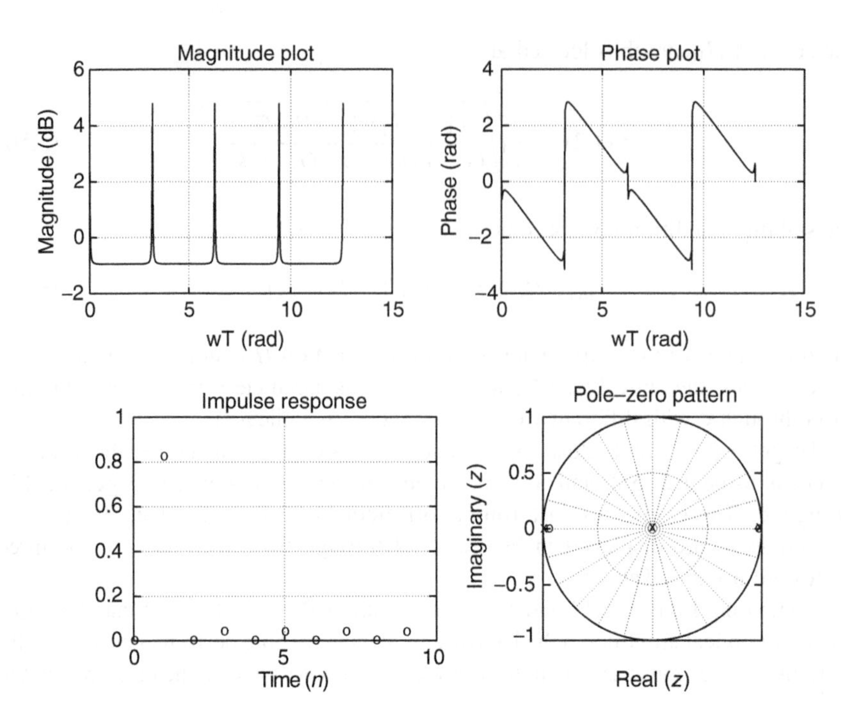

FIGURE 5.14 Frequency response, impulse response, and pole–zero plot for the active DCDR circuit with circuit parameters tabulated in Table 5.2.

z-plane, it results in sharper response in the magnitude plot. Recall that the pole–zero pattern in the z-plane would play a major part in the sharpness of the resonance peak of photonic circuits. To get a sharper maximum in this particular situation, it is required that the distance between the pole and the unit circle must be less than that between the zero and the unit circle. The effect of this active mode operation can be noted by realizing that the peaks in the magnitude plot have values greater than 0 dB, while in the passive operation the corresponding peaks can only come up to 0 dB. This operation can be used as a band-pass filter with narrow passband.

Inspecting Figure 5.15, it is noticed that the large amplifier gain has changed the response dramatically as compared to Figure 5.11 because its presence changes the pole–zero pattern greatly. In Figure 5.16, we have a system with a pole at 0, a pole pair at ± 1, and zeros at 0 realized by the DCDR circuit. This yields a magnitude plot with very sharp peaks. The lossless system also produces an infinite constant-value impulse response. The impulse response here can be applied to generate continuous sequences of "1" pulses in signal processing system. In other words, a single impulse fed into the circuit can trigger a continuous stream of impulses of the same magnitude at the output. The causes of "0" in between the ones in the impulse response can be understood by inspecting the geometry of the circuit. As can be seen, the consecutive output signals can only be detected every two sampling times, which is the round-loop time in this case. Thus, if we want to get a stream of

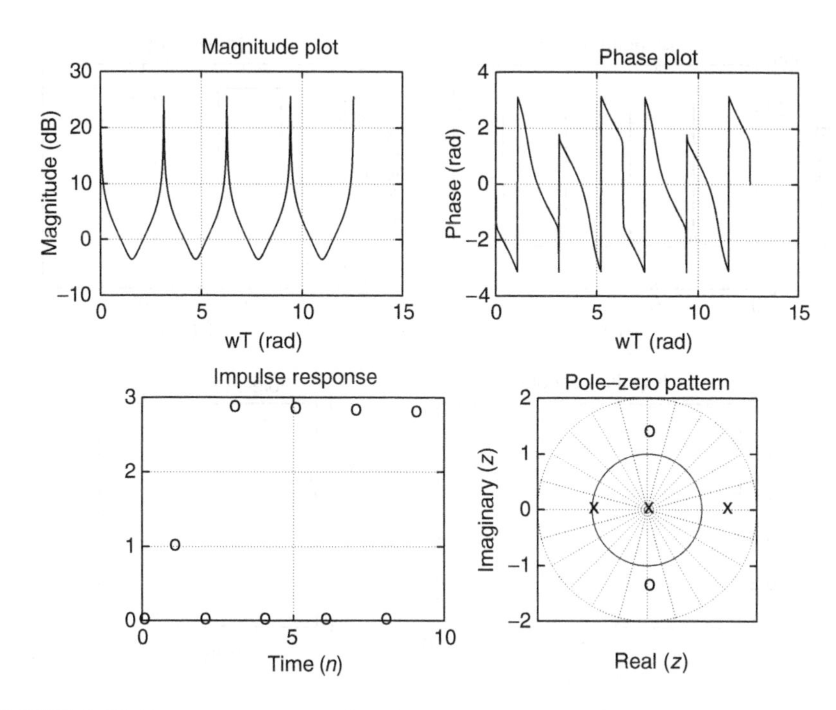

FIGURE 5.15 Frequency response, impulse response, and pole–zero plot for the active DCDR circuit with circuit parameters given in Table 5.2.

"1" signals, we need to acquire the output every two sampling times. If not, we would get "0 1 0 1 0 1 . . . ."

Another application is found with the impulse response generated by the DCDR circuit in the situation of Figure 5.16. We examine the pulse response as shown in Figures 5.17a through f for input sequences consisting of 1 and 0. For Figure 5.17a, the steady-state value of the pulse response is oscillating between 1 and 0 for an input stream with one "1." For Figure 5.17b, the steady-state value of the pulse response is oscillating between 1 and 1 for an input stream with two "1." In Figure 5.17c, the steady-state value of the pulse response is oscillating between 2 and 1 for an input stream with three "1." In Figure 5.17d, the steady-state value of the pulse response is oscillating between 1 and 2 for an input stream with three "1." Following the above pattern, we can observe that from the steady-state magnitude of the pulse response, the number of "1" in the input stream can be detected. It is shown in Figures 5.17e and f for longer input streams. If we look at the stream as sequence of two-digit binary numbers, from the response, we can indeed determine the occurrence of "1" in a certain digit position. For instance, in Figure 5.17d, the input streams are "1 1" and "0 1." The output stream in this case shows one in the first digit position and two in the second digit position that corresponds to the occurrence of "1" at those positions. The response, in fact, counts the numbers of "1" at the two digit positions and this may be used as an adder. This ability of counting arises from the geometry of the circuit and the orders of delay in each path.

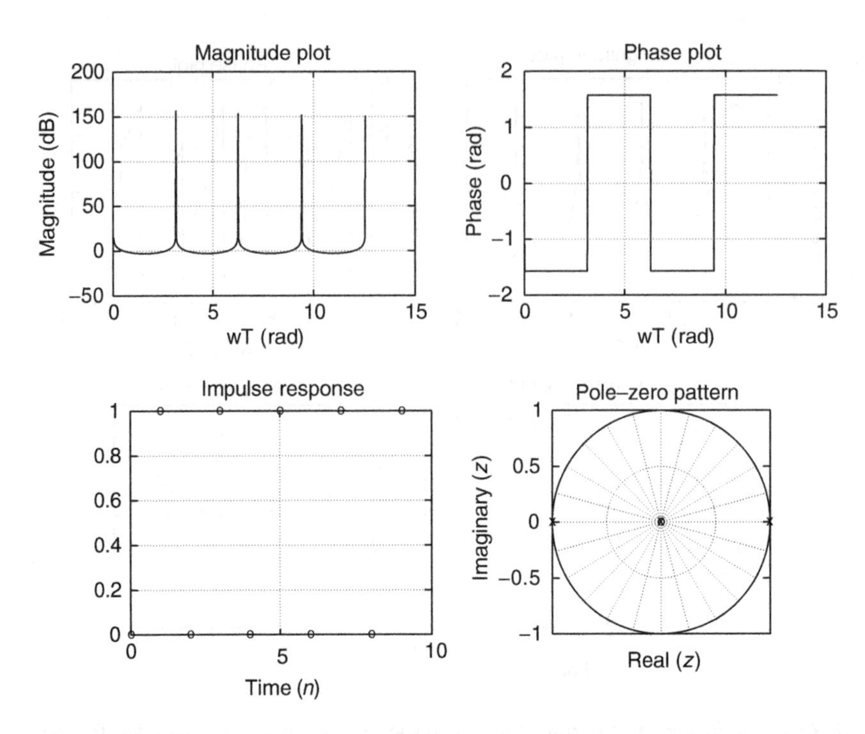

FIGURE 5.16 Frequency response, impulse response, and pole–zero plot for the active DCDR circuit with circuit parameters given in Table 5.2.

In Figure 5.17g, $m_1 = m_2 = 3$, the circuit can count numbers of "1" in four-digit numbers as expected.

(ii) $G_2 > 1$. Optical amplifier in the other feed forward path

This case would be similar to case 1(b)(i) where only one optical amplifier is placed in the other feed forward path. The output–input intensity H_{18} becomes

$$H_{18} = \frac{[(1-k_1)(1-k_2) + k_1k_2G_2]z^{-1} - (1-2k_1)(1-2k_2)G_2z^{-3}}{1 - [k_1k_2 + (1-k_1)(1-k_2)G_2]z^{-2}} \tag{5.52}$$

TABLE 5.2

Circuit Parameters Used in the Analysis of the DCDR Response in Case 1(b)(i) with the Corresponding Poles and Zeros of H_{18}

	k_1	k_2	G_1	Poles	Zeros
Figure 5.9	0.9	0.9	1.2	0, ±0.990959	±0.966595
Figure 5.10	0.2	0.8	5.2	0, ±0.995992	±j1.373716
Figure 5.11	0.5	0.5	3.0	0, ±1	0, 0

with the zeros at

$$z_z(1,2) = \pm\sqrt{\frac{(1-2k_1)(1-2k_2)G_2}{(1-k_1)(1-k_2)+k_1k_2G_2}} \tag{5.53}$$

The characteristic equation remains similar to that of Equations 5.7 and 5.11, except that the coupling coefficients are interchanged and G_2 is placed appropriately. Applying Jury's stability test again, we obtain the stability condition as

$$|k_1k_2 + (1-k_1)(1-k_2)G_2| < 1 \tag{5.54}$$

The responses of the DCDR circuit in this case are similar to case 1(b)(i), which has the amplifier inserted in the other forward path of the circuit. To illustrate the duality of the two cases, the following values can be used: $k_1 = k_2 = 0.1$, $G_1 = 1$, $G_2 = 1.2$, and $G_3 = 1$. The circuit responses are shown to be very closely similar to those

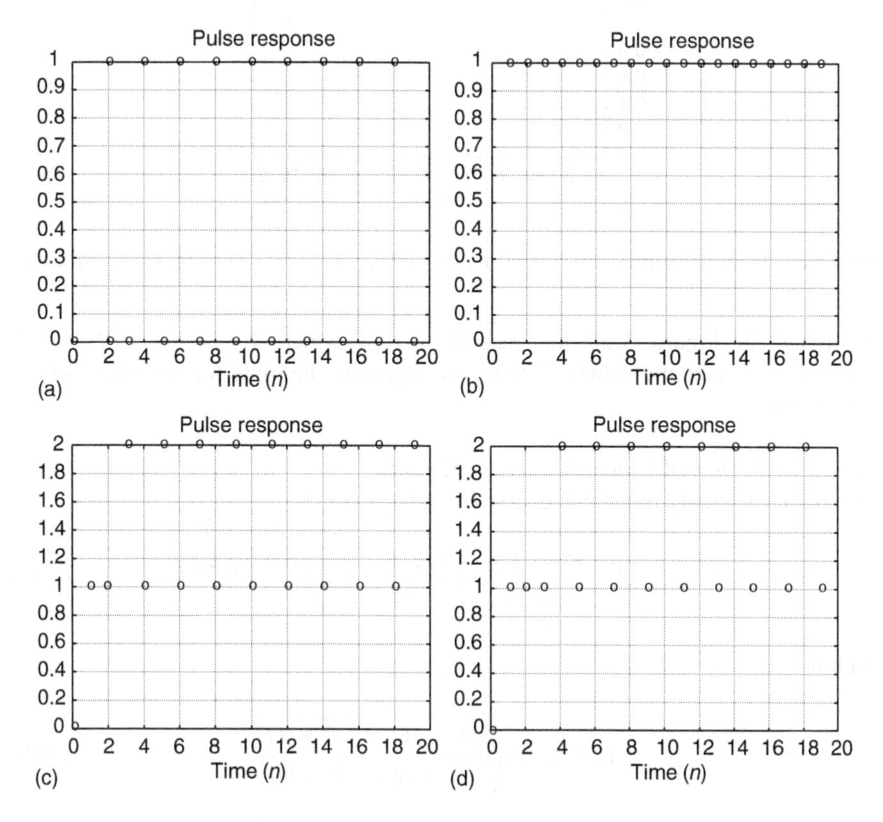

FIGURE 5.17 Pulse response of the DCDR circuit for different input sequence with $k_1 = k_2 = 0.5$, $G_1 = 3$, $G_2 = G_3 = 1$, and $m_1 = m_2 = m_3 = 1$ except (g) which has $m_1 = m_2 = 3$ and $m_3 = 1$. Input sequences for each figure: (a) [0 1], (b) [1 1], (c) [1 1 1], (d) [1 1 0 1], *(continued)*

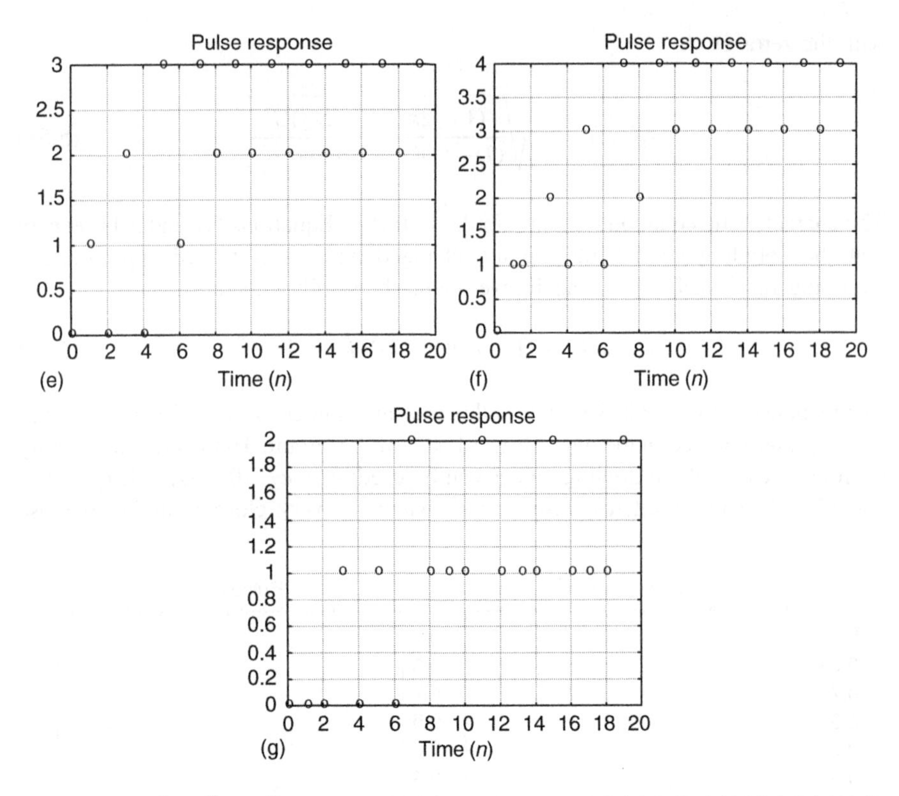

FIGURE 5.17 (continued) (e) [1 0 1 0 1 1 0 1], (f) [1 1 1 0 1 0 1 1 0 1], (g) [1 0 1 0 1 1 0 1].

illustrated in Figure 5.14 for case 1(b)(i) in which $k_1 = k_2 = 0.9$, $G_1 = 1.2$, $G_2 = G_3 = 1$, since the values of poles and zeros are the same for these two sets of parameters.

(iii) $G_3 > 1$. Optically amplified feedback path

The output intensity transfer function H_{18} now becomes

$$H_{18} = \frac{[(1-k_1)(1-k_2) + k_1k_2]z^{-1} - (1-2k_1)(1-2k_2)G_3z^{-3}}{1 - [k_1k_2 + (1-k_1)(1-k_2)]G_3z^{-2}} \tag{5.55}$$

with the zeros at

$$z_z(1,2) = \pm\sqrt{\frac{(1-2k_1)(1-2k_2)G_3}{(1-k_1)(1-k_2)+k_1k_2}} \tag{5.56}$$

The stability condition follows immediately that

$$|k_1k_2G_3 + (1-k_1)(1-k_2)G_3| < 1 \tag{5.57}$$

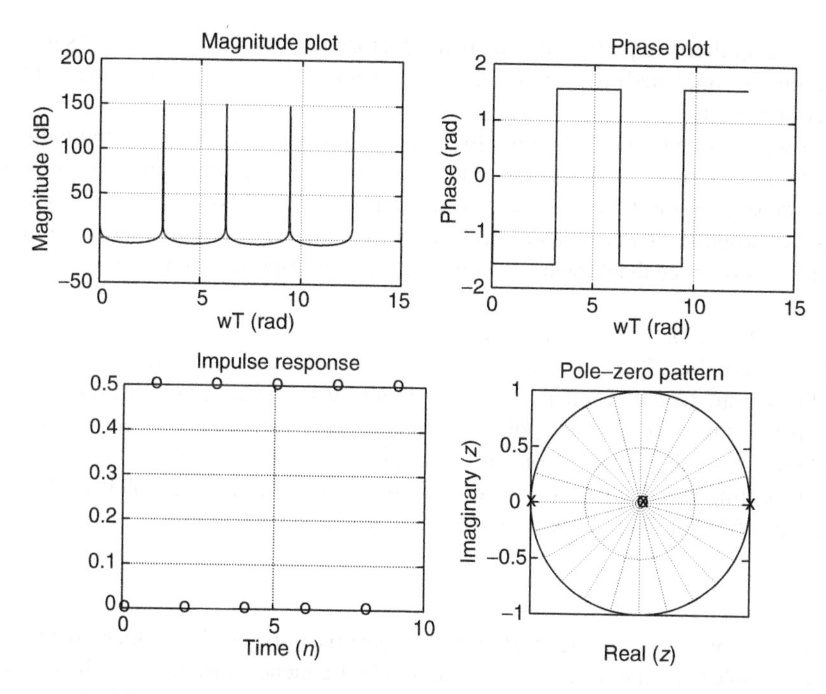

FIGURE 5.18 Frequency response, impulse response, and pole–zero plot for the active DCDR circuit (case 1(b)(iii)) with $k_1 = k_2 = 0.5$, $G_1 = G_2 = 1$, and $G_3 = 2$.

In the special case of $k_1 = k_2 = 0.5$, $G_1 = G_2 = 1$, and $G_3 = 2$, the output intensity responses are plotted in Figure 5.18. The frequency response of the DCDR circuit is the same as that in Figure 5.16, where the same k values are used there. However, with the optical amplifier situated at the feedback path rather than the feed forward path of the circuit, the optical gain required in the former case (to have the same pole and zero patterns and satisfying the stability condition) is less than that in the latter one. This makes one think to put the amplifier in the feedback path instead of putting it in the feed forward path. Nevertheless, the amplitude of the impulse response is smaller in this case, where we put the amplifier in the feedback path. This is because the gain we used in Figure 5.18 is smaller than that in Figure 5.16 and the output is obtained immediately after the amplifier for the case in Figure 5.16.

The three cases in case 1(b) have been studied, where only one optical amplifier is available for the DCDR circuit. In fact, this is very important in practice when several DCDR circuits are integrated to form a network then the least number of optical amplifiers in the loop is required. Therefore, the most important question remained to be addressed is where should we place the optical amplifier in the DCDR circuit? Should it be—in the feed forward or feedback path? The answer depends on the specific applications.

Considering the case when the DCDR circuit is applicable as a filter then the optical amplifier should be placed in the feedback path. As stated above, the amplifier in the feedback path requires less optical gain than that in the feed forward path to achieve the same performance. From another point of view, we can say that with an amplifier of a particular gain, better performance is achieved (as far as the use as a filter is concerned) if we put it in the feedback path (providing that the stability criterion is met).

If we are interested in the impulse response for applications in photonic digital processing systems, the position of the optical amplifier is very important [25–28]. Also, depending on the output ports where signals are tapped, the optical amplifier must be closed to that port. Thus, both cases where the optical amplifier is placed in feed forward or feedback paths can be used with appropriate applications.

Case 2: DCDR circuit with multiple delays
Different combinations of delay orders in the two feed forward paths and the feedback path would result in different circuit performances. One useful point to note here is that we can obtain poles, which are evenly and equally spaced around the circle in z-plane by certain choice of delay orders m_1, m_2, and m_3.

Inspecting the denominator of the circuit's transfer function, which is DEN in Equations 5.37 through 5.39, if $m_1 = m_2$, DEN will always follow

$$1 - az^{-b} \tag{5.58}$$

where a is any real number and b is any positive integer. Thus, there would be multiple poles for the system. The magnitude–frequency responses with $m_1 = m_2$ which equals to 2 and 3, respectively, are shown in Figures 5.19 and 5.20. The same k and G values used in Figure 5.16 are used here. In Figure 5.18, the pole values are located at 0, 0, $-0.5 \pm j0.866$, and 1. In Figure 5.20, the poles are located at 0, 0, 0, $\pm j$, and ± 1. An application of this feature is that optical frequencies or wavelengths of input optical signal can be filtered at equal interval, e.g., as a group of optical carriers at equal intervals.

It is expected that if the delays in the two forward paths are different, interesting responses would result. In this case, the intensities in the two forward paths do not always add up at coupler 2 at the same time. A result with $m_1 = 2$, $m_2 = 1$, $m_3 = 1$ is shown in Figure 5.21. The conjugate pole pairs are now located well inside the unit circle, leading to a lossy system and an oscillating time response. The appearance of the impulse response confirms this point.

Case 3: Special case with negative optical gain
(a) Purely real or purely imaginary poles, $m_1 = m_2 = m_3 = 1$
In this section, an optical amplifier with a negative optical small-signal gain is used; this would allow much greater flexibility in tuning the peaks of the magnitude plot corresponding to adjusting the pole–zero pattern of the optical transfer function of the DCDR circuit. The negative optical gain factor can be achieved by using an optical transistor as described in Ref. [5]. Alternatively, an optical amplifier incorporated with a π-phase shifter such as a LiNbO3-integrated optic phase modulator would allow a negative optical gain.

If $m_1 = m_2 = m_3 = 1$, the denominator DEN will be of the following form:

$$\text{DEN} = 1 - az^{-2} \tag{5.59}$$

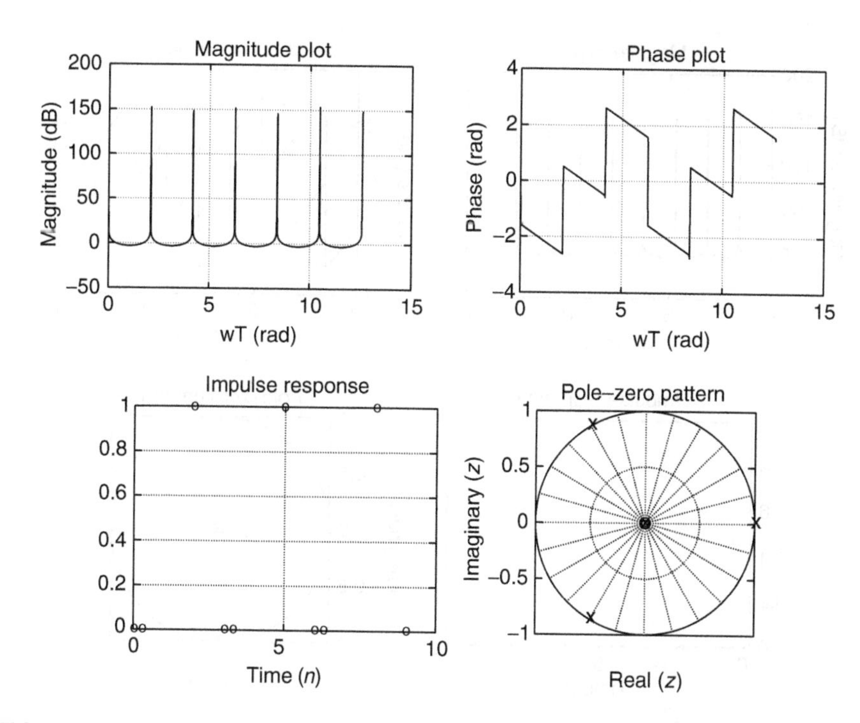

FIGURE 5.19 Frequency response, impulse response, and pole–zero plot for the active DCDR circuit (case 2) with $k_1 = k_2 = 0.5$, $G_1 = 3$, $G_2 = G_3 = 1$, and $m_1 = m_2 = 2$, $m_3 = 1$.

with $a = k_1k_2G_1G_3 + (1 - k_1)(1 - k_2)G_2G_3$. The optical system poles would be located at 0, π, $2\pi, \ldots,$ if $a > 0$ and at $\pi/2, 3\pi/2, \ldots$ if $a < 0$. As it can be easily seen, the characteristic equation of the photonic DCDR circuit would result in a quadratic equation with roots as purely real or purely complex conjugates in this situation. To achieve the latter case, it requires amplifiers in the DCDR circuit with negative gain.

Firstly, we examine a typical example in the above case where $k_1 = k_2 = 0.1$ and the gain factors for the three optical paths are $G_3 = 1$ and $G_2 = -1.2$, -1.25, and -1.3, respectively, with the gain G_1 as a variable. Figure 5.17a shows the magnitudes of the poles (solid line) and that of the zeros (dashed line) of H_{18} as a function of gain G_1 for $G_3 = 1$ and $G_2 = -1.2$. For stability, the values of the poles and zeros would never reach unity for this particular circumstance. In fact, it can reach close to unity, however, both zero and pole approach unity leading to the cancellation of their effects.

When the value of the negative gain G_2 is increased to -1.25 and -1.3, the values of the pole and zero are plotted in Figures 5.22b and c, respectively. Figure 5.22b exhibits a similar characteristic as in Figure 5.22a except that the pole and zero can now reach unity.

When $G_2 = -1.3$ the pole and zero at unity are separated as it can be observed in Figure 5.22c. Thus, we can select $G_1 = 5.3$ and $G_3 = 1$, $G_2 = -1.3$ for the pole equals

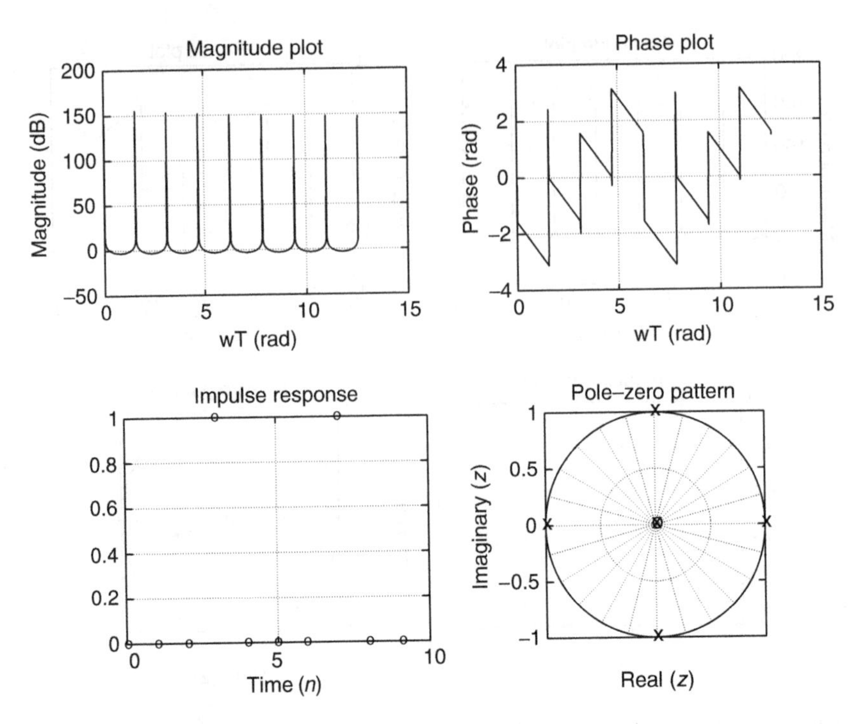

FIGURE 5.20 Frequency response, impulse response, and pole–zero plot for the active DCDR circuit (case 2) with $k_1 = k_2 = 0.5$, $G_1 = 3$, $G_2 = G_3 = 1$, and $m_1 = m_2 = 3$, $m_3 = 1$.

to unity corresponding to a zero value of 1.015. It is also noted here that for two identical couplers, the value of zero is always greater than that of the pole. Thus, the amplitudes of the frequency response always have minima according to the analysis of transfer function in the z-domain. Hence, the DCDR with negative gain in one of the forward or feedback paths can only be used as an optical notch filter at output port 8.

For the case, where $G_2 = -1.25$ and $G_1 = 2$, the DCDR circuit has altogether three system poles: one at the origin, $z_p(1) = 0$ and one complex pole pair at $z_p(2,3) = \pm j0.996243$. The transfer function H_{18} has two zeros at the location $z_z(1,2) = \pm j0.997663$. Clearly, the zeros closely follow the complex pole pair. Figure 5.23 shows the magnitude response of this optical transfer function.

The relation in Equation 5.8 allows us to design photonic circuit with adjustable response by appropriately positioning the poles and zeros of the optical transfer function. This fact is indeed the novel feature of our analysis using the SFG technique and z-transform. It is also determined that when G_1 is increased from 2 to 5 in the case for Figure 5.23, the magnitude $|H_{18}|$ changes from a ratio of 2336/3757 to 4655/18929.

Figure 5.24 shows the magnitude and phase responses together with the pole–zero position in the z-plane of the circulating transfer function H_{14} with the same circuit parameters as used in Figure 5.23. This H_{14} has maxima at an optical frequency corresponding to $\omega\tau$ that equals to odd multiple order of $\pi/2$. At these

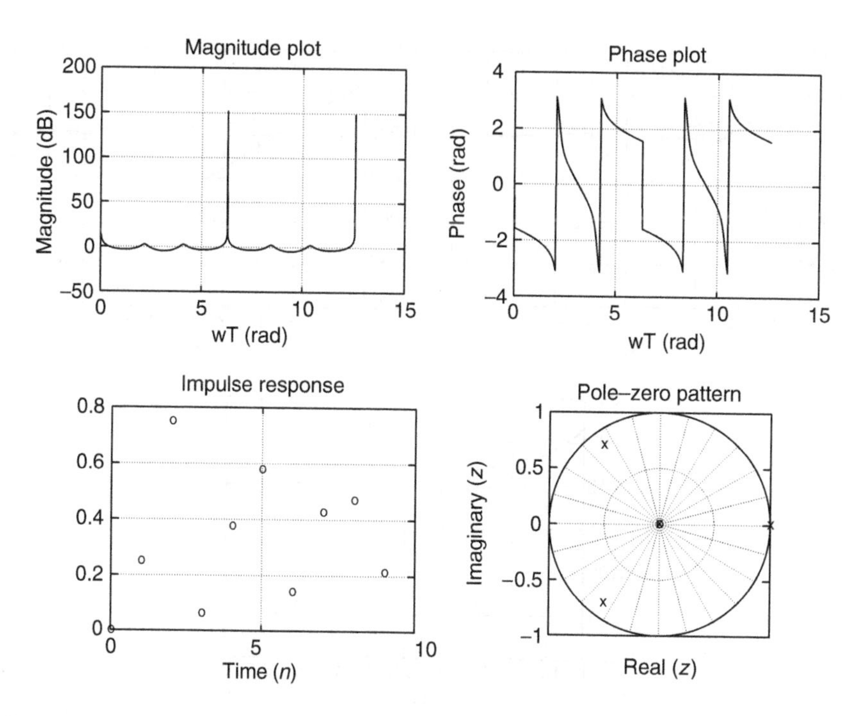

FIGURE 5.21 Frequency response, impulse response, and pole–zero plot for the active DCDR circuit (case 2) with $k_1 = k_2 = 0.5$, $G_1 = 3$, $G_2 = G_3 = 1$, and $m_1 = 2$, $m_2 = 1$, $m_3 = 1$.

values, the transfer functions H_{18} and H_{13} have minima as can be observed from Equations 5.37, 5.52, and 5.55. Thus, the circuit performs a kind of quasi-resonance.

Since it is assumed that the source is temporal incoherent here, thus the resonance effect should not occur in the circuit. This is because the beams traveling inside the circuit always add up constructively. However, with the application of negative optical gain in one of the paths, destructive interference can occur, which results in a behavior similar to the resonance in the coherent case. The resonance condition for the DCDR circuit under coherent source can be determined. H_{13} denotes the circulating transfer function in one of the two optical loops. Thus, we can conclude that at this resonance the energy is stored in one of the optical loops only for circuits with two optical loops, which share one common path. In addition, with this case of $G_2 = -1.25$, $G_3 = 1$, and $G_1 = 2$, the loop (3)(6)(7)(2)(3) has a positive transmittance value whereas the other loop (4)(5)(7)(2)(4) would have a negative loop transmittance. The negative gain in the forward path (4)(5) would interfere with the optical waves in the forward paths (3)(6) in a destructive manner and hence can generate a depletion of the output at a destructive interference. Hence, the optical energy is stored only in the loop (4)(5)(7)(2)(4). This helps explaining the quasi-resonant effect mentioned earlier. This finding is a significant development by applying the SFG technique in optical resonators in particular and photonic circuits in general.

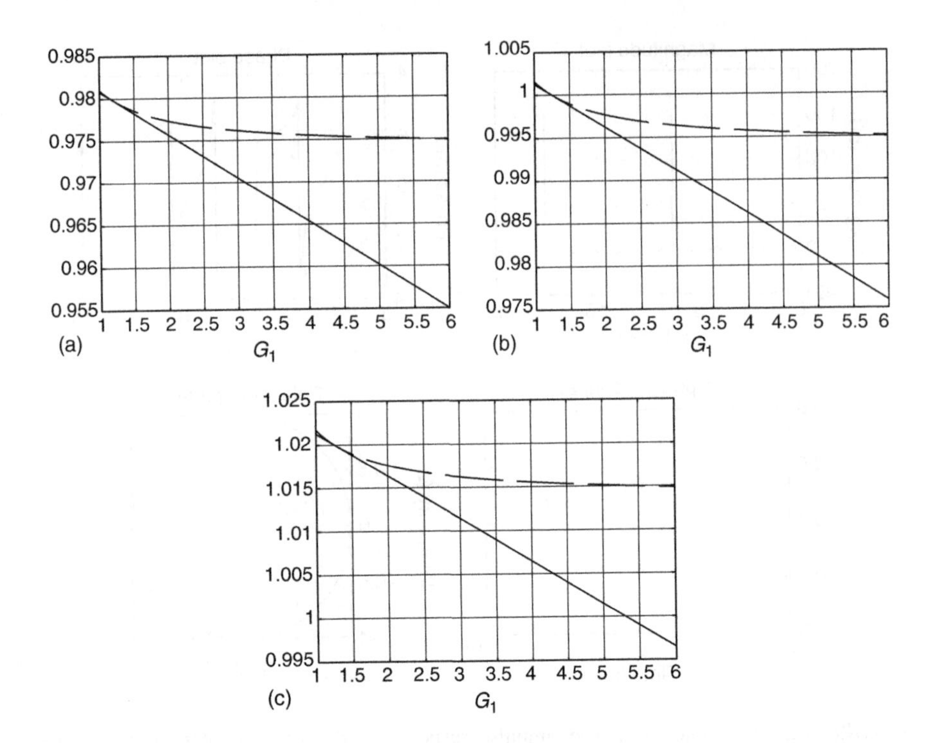

FIGURE 5.22 Values of poles (solid line) and zeros (dashed line) versus gain G_1 for DCDR circuit with $k_1 = k_2 = 0.1$, $G_3 = 1$ and (a) $G_2 = -1.2$, (b) $G_2 = -1.25$, (c) $G_2 = -1.3$.

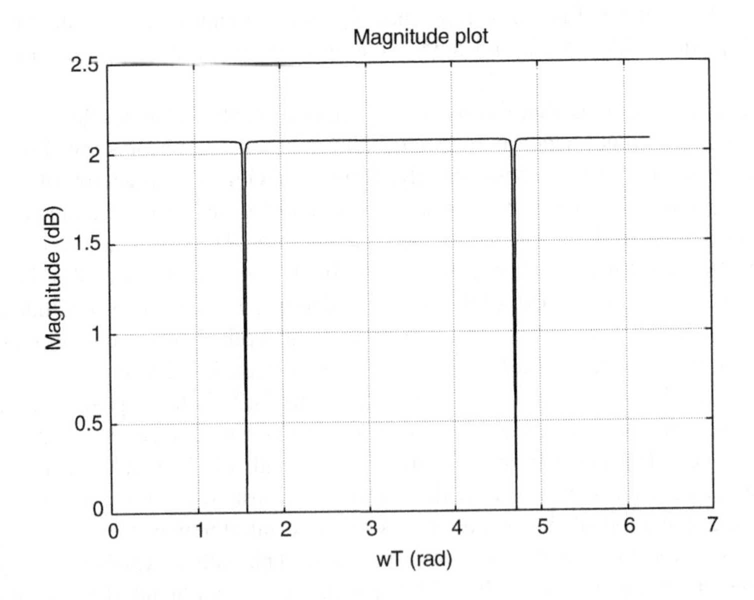

FIGURE 5.23 Magnitude–frequency response of H_{18} with $k_1 = k_2 = 0.1$, $G_1 = 2$, $G_2 = -1.25$, $G_3 = 1$, and $m_1 = m_2 = m_3 = 1$.

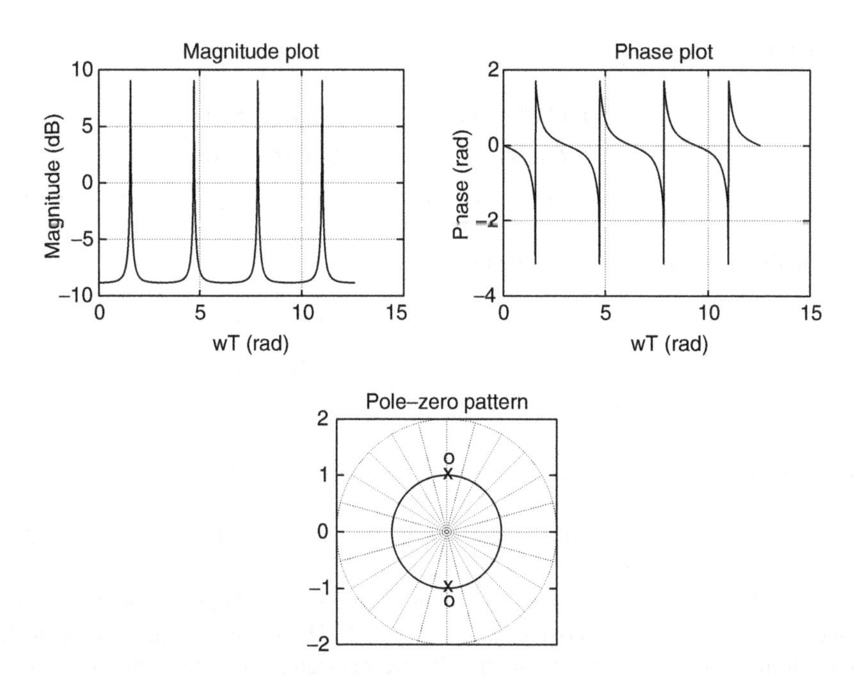

FIGURE 5.24 The responses of H_{14} with $k_1 = k_2 = 0.1$, $G_1 = 2$, $G_2 = -1.25$, $G_3 = 1$, and $m_1 = m_2 = m_3 = 1$.

(b) $m_3 = 1$ and either $m_1 = 0$ or $m_2 = 0$

Next, we consider another situation where we have direct connection in either of the forward paths (3)(6) or (4)(5) $\cdot\, m_3 = 1$. The direct connection would convert the DEN to a quadratic equation of the following form:

$$\text{DEN} = a + bz^{-1} + cz^{-2} \tag{5.60}$$

The values of the poles are then given by

$$z_p(1,2) = \frac{-b \pm \sqrt{b^2 - 4ac}}{2a} \tag{5.61}$$

where *a, b,* and *c* are three constants whose values depend on the photonic components. We consider here the case of $m_1 = 0$, $m_2 = 1$, $m_3 = 1$. The expression of DEN in terms of the circuit parameters is given below:

$$1 - k_1 k_2 G_1 G_3 z^{-1} - (1 - k_1)(1 - k_2) G_2 G_3 z^{-2} = 0 \tag{5.62}$$

We apply Jury's stability test for digital systems again, with

$$a_2 = 1, \quad a_1 = -k_1 k_2 G_1 G_3, \quad a_0 = (1 - k_1)(1 - k_2) G_2 G_3 \tag{5.63}$$

The stability criteria requires

$$
\begin{aligned}
&1 - k_1k_2G_1G_3 - (1 - k_1)(1 - k_2)G_2G_3 > 0 \\
&1 + k_1k_2G_1G_3 - (1 - k_1)(1 - k_2)G_2G_3 > 0 \\
&|(1 - k_1)(1 - k_2)G_2G_3| < 1
\end{aligned}
\tag{5.64}
$$

Rearranging Equation 5.64, we obtain

$$
\begin{aligned}
&k_1k_2G_1G_3 + (1 - k_1)(1 - k_2)G_2G_3 < 1 \\
&(1 - k_1)(1 - k_2)G_2G_3 - k_1k_2G_1G_3 < 1 \\
&|(1 - k_1)(1 - k_2)G_2G_3| < 1
\end{aligned}
\tag{5.65}
$$

To obtain a complex conjugate pole pair, it then requires that

$$
(k_1k_2G_1G_3)^2+4(1 - k_1)(1 - k_2)G_2G_3 < 0 \tag{5.66}
$$

To satisfy Equation 5.64, the amplifier gain G_2 or G_3 must take sufficient negative values if the couplers are passive, i.e., k_1, $k_2 < 1$. There are several combinations of photonic circuit parameters of the DCDR resonators to satisfy this equation. A procedure can be established as follows: select a combination of four of five of k_1, k_2, G_1, G_2, and G_3, plot Equation 5.66 against the fifth parameter as a variable; then find a range of operation so that this inequality is satisfied, and finally choose values in this range to design the photonic circuit and recheck the stability condition.

An example is given here to illustrate the above procedure where $k_1 = k_2 = 0.3$, $G_1 = 2$, $G_2 = -1$ are chosen. Figure 5.25 shows the LHS of Equation 5.65 as solid line and the other three lines represent the LHS of Equation 5.66 which are the stability expressions. The optical gain G_3 must be chosen so that it satisfies Equation 5.66 as well as the stability conditions. It follows that the maximum value of G_3 is approximately 2 for which the system remains stable.

This analytical examination of the stability of the photonic circuits is unique when a z-transform method is employed. Therefore, flexibility in design is achieved. The frequency response of the intensity transfer function H_{18} is shown in Figure 5.26 with the circuit parameters selected as given above. From the pole–zero pattern plotted, we can see that the poles and zeros of the optical transfer function H_{18} are complex in nature. It is noted here that the poles and zeros are complex conjugates and are not purely real or purely imaginary. This would allow the peaks of the magnitude plot designed to be closely spaced together. An application of this feature is that two or several wavelengths can be demultiplexed as we can observe from Figure 5.26. A phase change of π occurs at each pole position. The periodic variation of the output impulse response of the DCDR shows the circulation of the impulse delay in the recirculating loops.

Another example of the operation of the DCDR circuit is to choose the poles closely spaced. The pole and zero patterns of a DCDR with $m_1 = 0$, $m_2 = m_3 = 1$, and

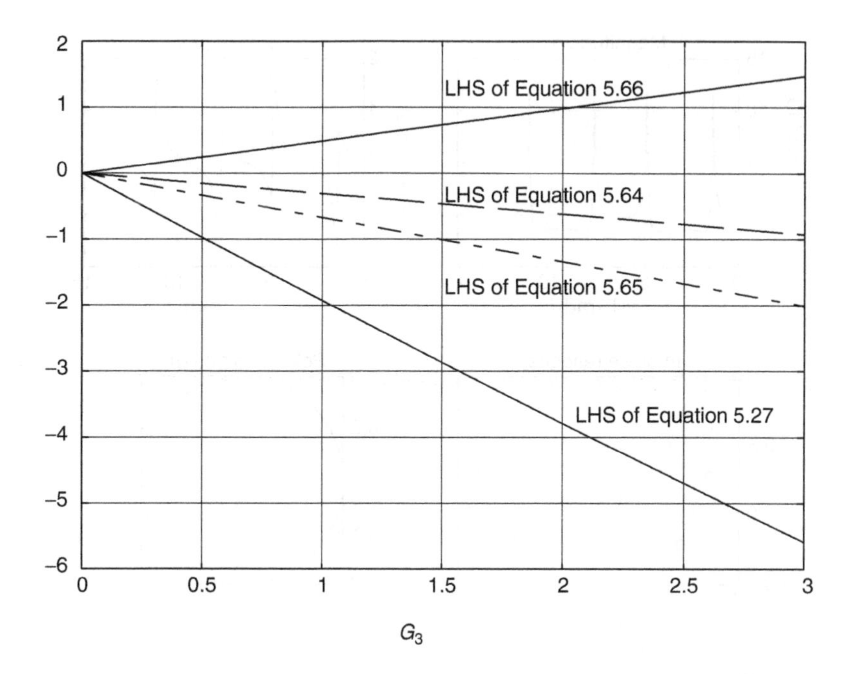

FIGURE 5.25 Selecting optical gain parameters for the design of a DCDR circuit when $m_1 = 0$ and $m_2 = m_3 = 1$.

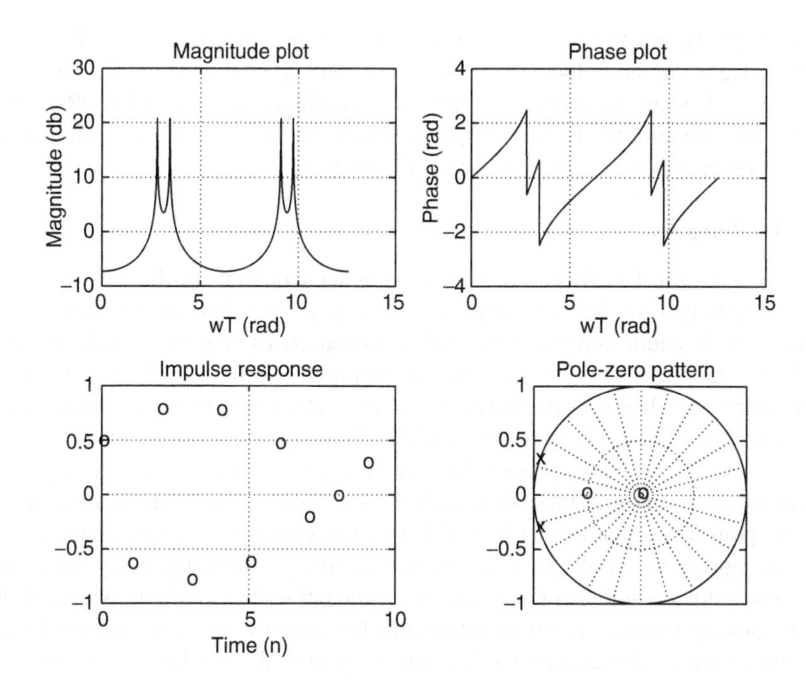

FIGURE 5.26 Frequency and discrete-time responses for H_{18} with $k_1 = k_2 = 0.3$, $G_1 = 2$, $G_2 = -1$, $G_3 = 2$ and $m_1 = 0$, $m_2 = m_3 = 1$. Poles: $0.1800 \pm j0.9734$, zeros: $0.0459 \pm j0.8068$.

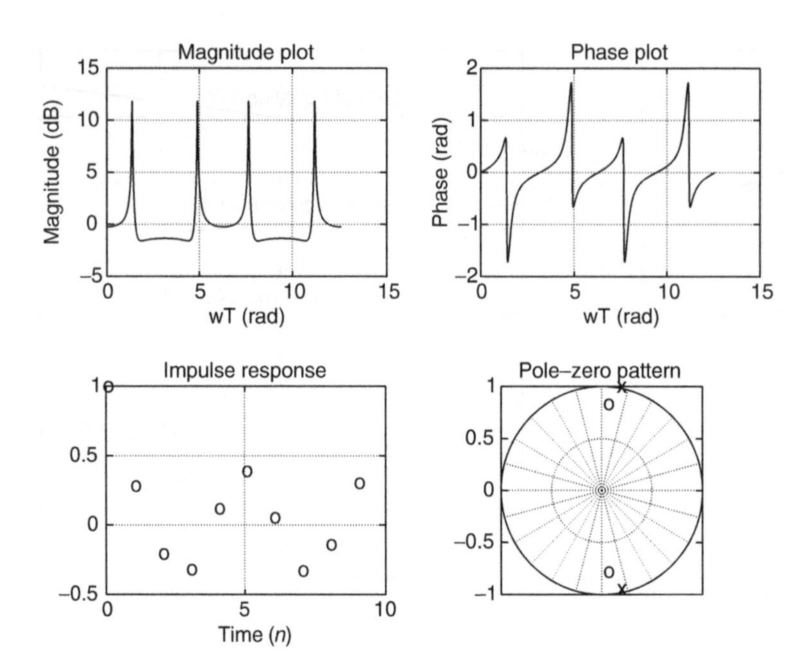

FIGURE 5.27 Frequency and discrete-time responses for H_{18} with $k_1 = k_2 = 0.5$, $G_1 = 1.9$, $G_2 = 1$, $G_3 = -4$ and $m_1 = 0$, $m_2 = m_3 = 1$. Poles: $0.9500 \pm j0.3123$, zeros: 0, 0.5263.

$k_1 = k_2 = 0.5$, $G_1 = 1.9$, $G_2 = 1$, and $G_3 = -4$ is shown in Figure 5.27. The zeros of the system are now located at the origin and a real negative of −0.5263. The magnitude response shows the feature of demultiplexing of this DCDR configuration. Furthermore, thc impulse response shows the variation of the output pulses with certain pattern different from that in Figure 5.26.

5.1.4.4 Remarks

In this section, the DCDR circuit is studied under temporal incoherent source for different operation modes. It is found that for active operation, the locations of the amplifiers in the circuit depend on the application required. Greater design flexibility is found with the presence of negative optical gain in the circuit. It is discovered that a quasi-resonance behavior is possible with the insertion of negative optical gain in the delay line. Procedures in design of the DCDR circuit are shown in the special case of negative optical gain but in general, this can be applied to other situations. Applications of the circuit in signal processing such as filters, counters, and adders are realized.

One point is needed to be clarified before proceeding to the next chapter. It is more appropriate to call the DCDR photonic circuit a recirculating delay line rather than a resonator when it is operating at the incoherent source situation. Although the output intensity response can have minimums for certain circuit's parameters (negative optical gain), the circulating intensity responses do not have maximums at the same condition. It would be shown in the next chapter that for a resonator to have resonance, there are generally two constraints it need to be met. They are the constraints on the circuit's parameters and the other one is the constraint on the operating point of the resonator. The latter one is usually related to the phase change the signal experiences when it travels along a loop in the circuit. For the source incoherent condition, we have ignored the effect of phase that could have on the signal. The parameter constraint can be easily met but due to the ignorance of the phase in the circuit, it cannot be met. The circuit is also very stable due to the ignorance of the phase effect on the circuit's performance.

5.1.5 DCDR under Coherence Operation

In the previous sections of this chapter, the DCDR circuit is studied under incoherent source condition. To study the resonance effect of the DCDR circuit, the analysis is required to be carried out on the field basis. Therefore, in this chapter, the circuit excited by coherent source with finite linewidth will be considered. The DCDR circuit is considered as a resonator here whereas in the previous chapter it is treated as a recirculating delay line.

Previous studies on the effect of source coherence on the performance of resonator circuit include Refs. [1–6], but no one had ever performed the analysis on a DCDR circuit or more complicated photonic circuits. This is mainly due to the complicated manipulations involved. However, with the help of our newly developed SFG theory in optical circuits, these difficulties can be overcome easily. In this chapter, an algorithm is derived to compute the response systematically. The temporal transient response of the circuit is shown with source coherence.

5.1.5.1 Field Analysis of the DCDR Circuit

Before considering the case for source with finite linewidth, we first derive the transfer functions of the DCDR circuit in terms of field rather than intensity as in the previous chapter. In this section, we derive the field version of the transfer functions for the DCDR circuit. The analysis in this section is different from the preceding intensity-basis analysis in the sense that the phase change encountered by the signal in the circuit would be taken into consideration; thus, interferometric effects can take place. Therefore, the resonance effect of the DCDR circuit can be investigated.

Firstly, we define the DCDR circuit parameters with lightwaves represented in the field amplitude. The subscript f distinguishes the nature of the variables as field variables from those in terms of intensity. The circuit parameters are defined as follows: $k_{pcf} = -j\sqrt{k_p}$ where $p = 1, 2$ as the field cross-coupling coefficient for the two couplers 1 and 2, $k_{pdf} = \sqrt{1 - k_p}$ where $p = 1, 2$ as the field direct-coupling coefficient for the two couplers 1 and 2, and $T_{if} = \sqrt{t_{ai}G_i}z^{-m_i}$ where $i = 1, 2$, and 3. The $-j$ term in the k_{pcf} expression above accounts for the $-\pi/2$ phase shift induced by the coupler during the cross coupling.

5.1.5.2 Output–Input Field Transfer Function

The output–input field transfer function is obtained by carrying out the same procedure as introduced in Section 5.1.1. The output–input field transfer function of the DCDR circuit, using Mason's rule given in Ref. [7], is given by

$$H_{18f} = \frac{E_8}{E_1} = \frac{\sum_{q=1}^{4} F_{qf}\Delta_{qf}}{\Delta_f} \tag{5.67}$$

E_8 and E_1 are the output and input field amplitudes, respectively, of the DCDR circuit. The loop transmittances and the forward path transmittances are stated as follows:

(a) Loop transmittances

(i) Loop 1: (2)(3)(6)(7)(2): The loop optical transmittance of loop 1 is

$$T_{l1f} = k_{1cf}k_{2cf}T_{1f}T_{3f}$$

(ii) Loop 2: (2)(4)(5)(7)(2): Optical transmittance of loop 2 is

$$T_{l2f} = k_{1df}k_{2df}T_{2f}T_{3f}$$

(b) Forward path transmittances

The four forward paths and their related transmittances are

Path 1: (1)(3)(6)(8)

$$F_{1f} = k_{1df}\ k_{2df}\ T_{1f} \quad \text{and} \quad \Delta_{1f}+ = 1 - T_{l2f}$$

Path 2: (1)(3)(6)(7)(2)(4)(5)(8)

$$F_{2f}+ = k_{1df}^2\ k_{2cf}^2\ T_{1f}2f\ T_{3f} \quad \text{and} \quad \Delta_{2f} = 1$$

where Δ_{2f} is equal to unity due to its forward path touching both optical loops.

Path 3: (1)(4)(5)(7)(2)(3)(6)(8)

$$F_{3f} = k_{1cf}^2\ k_{2df}^2\ T_{1f}\ T_{2f}\ T_{3f} \quad \text{and} \quad \Delta_{3f} = 1.$$

Δ_{3f} is equal to unity due to the touching of two loops of the forward path.

Path 4: (1)(4)(5)(8)

$$F_{4f} = k_{1c}\ k_{2cf}2f \quad \text{and} \quad \Delta_{4f} = 1 - T_{l1f}$$

Furthermore, the loop determinant Δ_f is given by $\Delta_f = 1 - T_{l1f} - T_{l2f}$

Hence, the output–input field transfer function, in terms of k_1, k_2, T_{1f}, T_{2f}, and T_{3f}, can be expressed as

$$H_{18f} = \frac{\sqrt{(1-k_1)(1-k_2)}T_{1f} - \sqrt{k_1k_2}T_{2f} - T_{1f}T_{2f}T_{3f}}{1 + \sqrt{k_1k_2}T_{1f}T_{3f} - \sqrt{(1-k_1)(1-k_2)}T_{2f}T_{3f}} \tag{5.68}$$

5.1.5.3 Circulating-Input Field Transfer Functions

Similarly, the circulating-input field transfer functions can be derived and are given as follows:

$$H_{13f} = \frac{\sqrt{(1-k_1)} - \sqrt{(1-k_2)}T_{2f}T_{3f}}{\text{DEN}_f} \quad (5.69)$$

$$H_{14f} = \frac{-j\sqrt{k_1} - j\sqrt{k_2}T_{1f}T_{3f}}{\text{DEN}_f} \quad (5.70)$$

$$H_{17f} = \frac{-j\sqrt{k_1(1-k_2)}T_{2f} - j\sqrt{(1-k_1)k_2}T_{1f}}{\text{DEN}_f} \quad (5.71)$$

where

$$\text{DEN}_f = 1 + \sqrt{k_1k_2}T_{1f}T_{3f} - \sqrt{(1-k_1)(1-k_2)}T_{2f}T_{3f}$$

5.1.5.4 Resonance of the DCDR Circuit

The usual definition of optical resonance in a photonic circuit is that at a particular frequency or wavelength, the optical output takes a minimum value while the optical energy is circulating in the loops of the photonic circuit. Thus, the resonant condition can be found by setting the output transfer function to zero or effectively finding the zeros of the output–input optical transfer function. In this case, the DCDR circuit can be referred to as a resonator. Considering Equation 5.66, if $t_{al} = t_{a2} = t_{a3} = t_a$, $k_1 = k_2 = k$, $G_1 = G_2 = G_3 = 1$, and $m_1 = m_2 = m_3 = 1$, the equation simplifies to

$$H_{18f} = \frac{(1-2k)\sqrt{t_a}\, z^{-1} - t_a\sqrt{t_a}\, z^{-3}}{1-(1-2k)t_a z^{-2}} \quad (5.72)$$

Rearranging, Equation 5.72 gives

$$H_{18f} = \frac{(1-2k)\sqrt{t_a}\, z^2 - t_a\sqrt{t_a}}{z^3-(1-2k)t_a z} \quad (5.73)$$

Setting the numerator in Equation 5.72 to zero results in two resonance conditions: (1) if $z^2 = -1$, $k = (1+t_a)/2$ and (2) if $z^2 = 1$, $k = (1-t_a)/2$. $z^2 = -1$ means $z = \pm j$, which can be interpreted as $\omega\tau = n\pi - \pi/2$, $n = 1, 2\ldots$. Recall that $\omega\tau$ is the phase change through a fiber path in the circuit with unit delay time. Similarly, $z^2 = 1$ which means $z = \pm 1$ and it can be interpreted as $\omega\tau = n\pi$, $n = 1, 2\ldots$. As t_a is close to 1 for low-loss fibers, k for resonance would be close to one and zero, respectively, for the two resonant conditions.

By inspecting the SFG of the DCDR circuit in Figure 5.28, it can be seen that there are two touching loops. They are loop 1, which is (2)(3)(6)(7)(2) and loop 2, which is (2)(4)(5)(7)(2) with their loop transmittances given in Equation 4.2–2 and (4.2–3), respectively. Equations (4.2–16) and (4.2–17) are the resonant conditions of the loop 1 and loop 2, respectively. The frequency response of the output–input intensity transfer function and the circulating-input intensity transfer functions under

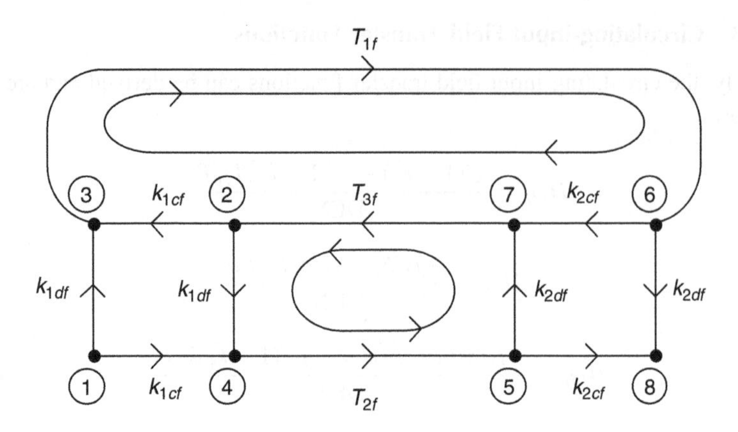

FIGURE 5.28 The SFG of the DCDR circuit in terms of the field variables.

the resonant conditions given in Equation 4.2–16 and Equation 4.2–17 are plotted in Figures 5.29 and 5.30, respectively. The instantaneous optical intensities are obtained by taking the square of the modulus of the corresponding field amplitudes. The vertical scale of the magnitude plot is shown with the absolute magnitude rather than dB as used in the previous chapters. It is found that this scale would give a better illustration of the resonance effect of the circuit.

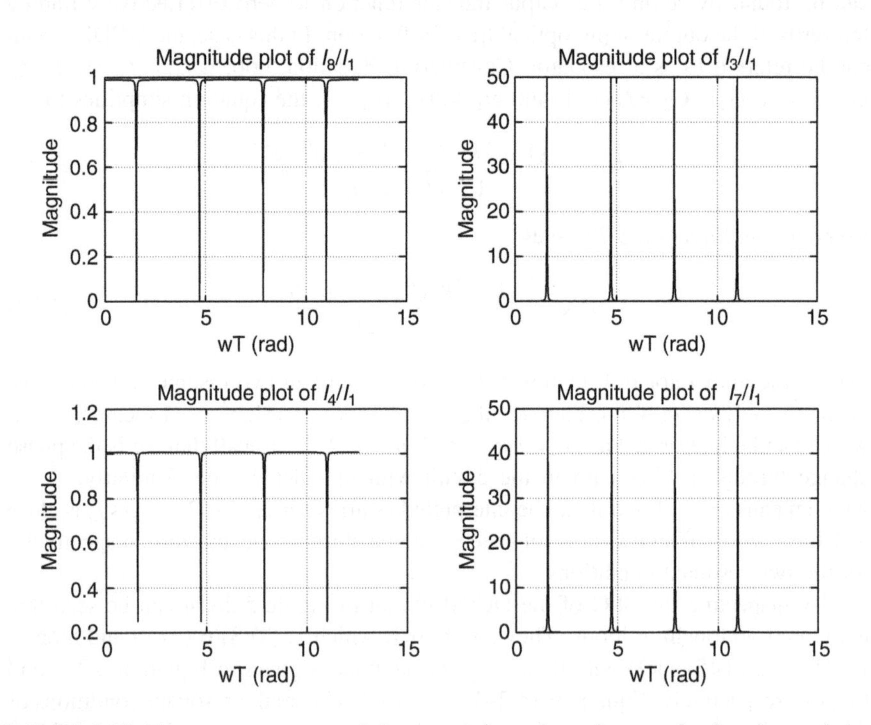

FIGURE 5.29 Frequency response of the output–input intensity transfer function and the circulating-input intensity transfer functions of the DCDR resonator under the resonant conditions of $t_a = 0.99$ and $k = 0.995$.

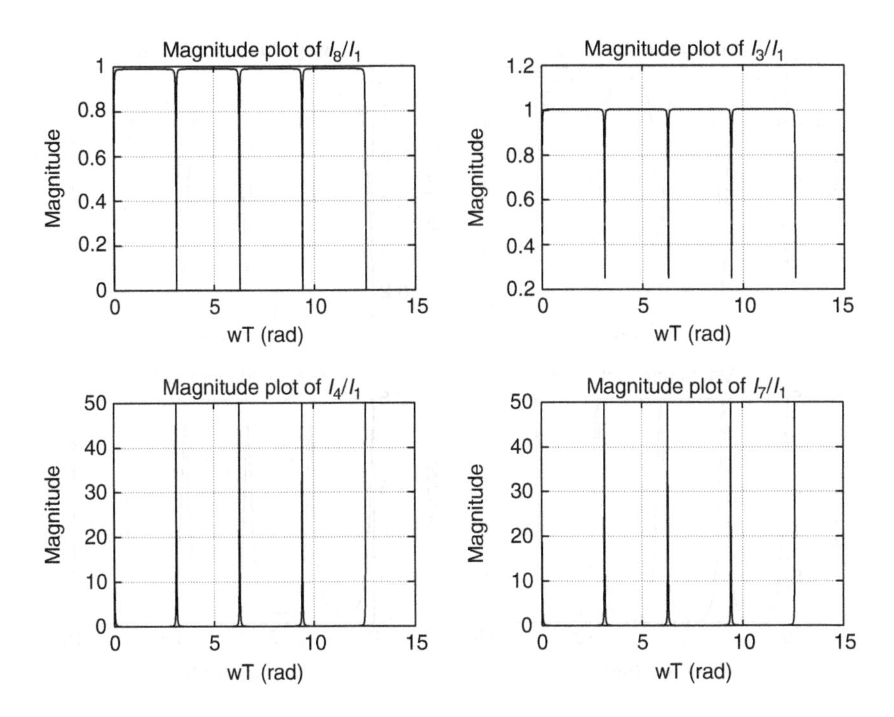

FIGURE 5.30 Frequency response of the output–input intensity transfer function and the circulating-input intensity transfer functions of the DCDR resonator under the resonant conditions of $t_a = 0.99$ and $k = 0.005$.

In Figure 5.29, there are maximums for the circulating intensity I_3 at $\omega\tau = \pi/2$, $3\pi/2$, $5\pi/2$, and $7\pi/2$ and correspondingly there are minimums for the output intensity I_8 and circulating intensity I_4 at these positions. This shows the resonance of loop 1 of the DCDR resonator. Indeed, the resonance criterion on the phase change for the signal in the circuit is that the round-loop phase change of the resonant loop needed to be $2n\pi$, where n is an integer. Looking at the criterion on $\omega\tau$ (the phase change per fiber path with unit delay) for resonance of loop 1 in Equation 4.2–16, it shows the above point. The round-loop phase change for loop 1 in this case is $2(n\pi - \pi/2)$ plus the phase change encountered across the couplers which is $\pi/2$ per coupler. It is obvious that this sum adds up to the value of $2n\pi$. This is another way of looking at the resonance condition; hence, it can be determined by examining the circuit's SFG alone. The resonance of loop 2 of the DCDR resonator is shown in Figure 5.30. There are maximums for the circulating intensity I_4 at $\omega\tau = \pi$, 2π, 3π, and 4π and correspondingly, there are minimums for the output intensity I_8 and circulating intensity I_4 at these positions.

It is also found that the sharpness of the resonance depends on the circuit parameters. In Figure 5.31, resonance behavior of the DCDR resonator is shown with $t_a = 0.8$ and $k = 0.1$, which corresponds to the resonance of loop 2 of the resonator. A smaller value of t_a is used here which indicates that the loss in the fiber path is larger. It is clearly seen that the maximum value of the circulating intensity I_4 is much smaller than the one given in Figure 5.30. Thus, in this case, it can be stated that low-loss fibers should be used in building up the resonator to have sharp resonant peaks.

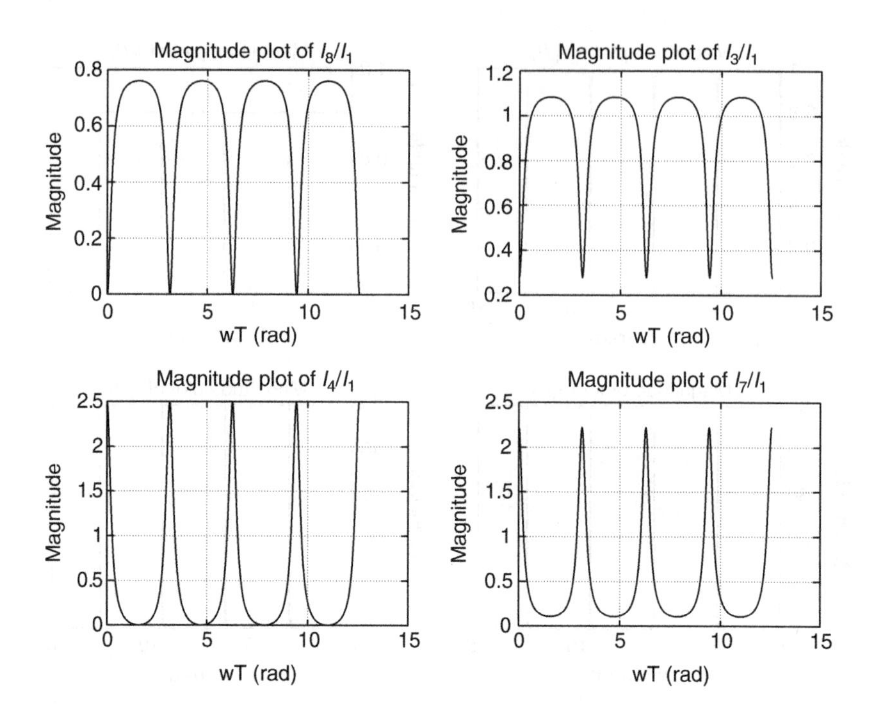

FIGURE 5.31 Frequency response of the output–input intensity transfer function and the circulating-input intensity transfer functions of the DCDR resonator under the resonant conditions of $t_a = 0.8$ and $k = 0.1$.

In the above results, it is observed that the energy can be stored in either of the two loops of the DCDR resonator depending on the circuit parameters. We see from Figures 5.29 and 5.30 that H_{17f} has maximums in both cases.

Next, we examine the stability of the circuit under the above resonant condition. Applying Jury's stability test to H_{18f} results in the stability condition of

$$|(1 - 2k)t_a| < 1 \tag{5.74}$$

From Equation 5.74, it can be seen that a stable operation of the passive DCDR resonator is always fulfilled provided that $0 < k < 1$ and $0 < t_a < 1$. If the DCDR resonator is operated in a different situation as mentioned above, for instance $G_1 > 1$ or $k_1 \neq k_2$, there would be other sets of resonant conditions for the resonator.

5.1.5.5 Transient Response of the DCDR Circuit

5.1.5.5.1 Effects of Source Coherence

In Section 5.1.4.3, we studied the response of the DCDR circuit to a monochromatic light source, i.e., single frequency, which can be called the steady-state response of the circuit. But, practically, a purely monochromatic source is not available. The best we can have is a single frequency source with finite linewidth. In the following analysis, the effect of the source coherence is taken into account in evaluating the transient response of the DCDR circuit.

We start the analysis by considering the electric field of the input light source which is expressed in the following form:

$$E_1(t) = E_s(t)\exp[\omega_0 t + \phi(t)] \tag{5.75}$$

where

$E_s(t)$ is the amplitude of the electric field at time t
ω_0 is the frequency of the source
$\phi(t)$ is the time-dependent phase which represents the phase fluctuation

This is often called the phase noise in optical fiber systems [3,29]. This phase fluctuation is the cause of broadening of the optical source spectrum, which results in finite linewidth of the source spectrum. Assuming that the source amplitude is time invariant and recalling some basic properties of the z-transform, the output–input field transfer function of the circuit $H_{18f} = H(z)$ can be stated in the form as given below:

$$\begin{aligned} H_{18f}(z) &= \sum_{n=-\infty}^{\infty} h[n]z^{-n} \\ &= \cdots + h[1]z^{1} + h[0] + h[1]z^{-1} + h[2]z^{-2} + h[3]z^{-3} + \cdots \end{aligned} \tag{5.76}$$

An inverse z-transform manipulation would convert H_{18f} from the z-domain to the time domain. For $m_1 = m_2 = m_3 = 1$, i.e., there is unity delay in each path of the DCDR circuit and $t_{a1} = t_{a2} = t_{a3} = 1$, the expansion of Equation 5.76 gives

$$\begin{aligned} h[n] &= 0 \quad \text{for all } n < 0 \\ h[0] &= 0 \\ h[1] &= \left(\sqrt{(1-k_1)(1-k_2)G_1} - \sqrt{k_1 k_2 G_2}\right) \\ h[2] &= 0 \\ h[3] &= -\left[\sqrt{k_1 k_2 (1-k_1)(1-k_2) G_3}\,(G_1 + G_2) + (k_1 + k_2 - 2k_1 k_2)\sqrt{G_1 G_2 G_3}\right] \end{aligned}$$

Hence, $h[n]$ represents the impulse response of the system at time index n. Although $h[n]$ is calculated for the restricted condition $m_1 = m_2 = m_3 = 1$ and $t_{a1} = t_{a2} = t_{a3} = 1$ for simplicity, the following analysis is applicable to the general situation. To compute the output of the resonator for an arbitrary input sequence (or pulse shape) $x[n]$, we perform the convolution in the time domain between $h[n]$ and $x[n]$, i.e., the output $y[n]$ is given by

$$y[n] = h[n] * x[n] \tag{5.77}$$

where $*$ represents the convolution operation. This is, in fact, the theory of operation to obtain the pulse response of the system.

Recall that the basic delay time of the circuit is denoted as τ and it is the sampling time as well. At time $t_n = n\tau$, the time averaged output intensity $I(t_n)$ of the DCDR circuit is

$$\langle I_8(t_n)\rangle = \langle E_8(t_n) E_8 * (t_n)\rangle \tag{5.78}$$

where $E_8*(t_n)$ is the complex conjugate of $E_8(t_n)$. The angular brackets denote ensemble average. From a different point of view, $E_8(t_n)$ which is the output at time $t_n = n\tau$ can be considered as the nth sampled output and similarly, the input $E_1(t_n)$ at time $t_n = n\tau$ can be regarded as the nth sampled input into the circuit. So the former term is, in fact, $y[n]$ in Equation 5.77. We can see the relationship between the discrete-time signal representation and the signals in the photonic circuit.

To include the source coherence contribution to our analysis, the phase fluctuation term $\exp[j\phi(t)]$ of the light source as given in Equation 5.75 needs to be examined. This time-varying phase fluctuates randomly and is statistical in nature. Traditionally, it is treated as random signal or process, and it is best described by its correlation function. This function has been determined by several works [1–6]. In general, we consider the autocorrelation function $R[(p-s)\tau]$ of the input optical wave field and it is given as [2]

$$R[(p-s)\tau] = \frac{\langle E_1(t-p\tau)E_1*(t-s\tau)\rangle}{[E_s(t-p\tau)E_s*(t-s\tau)]} \tag{5.79}$$

where p and s are integers. From Equation 5.79, it becomes

$$R[(p-s)\tau] = \exp\{j[(s-p)\omega_0\tau]\}\langle\exp[j\phi(t-p\tau)]\exp[-j\phi(t-s\tau)]\rangle \tag{5.80}$$

If the spectrum broadening of the laser due to the random phase fluctuation $\phi(t)$ is of the Lorentzian form, the phase function $\exp[j\phi(t)]$ would have the following correlation [1,2]:

$$\langle\exp[j\phi(t-p\tau)]\exp[-j\phi(t-s\tau)]\rangle = \exp(-|s-p|\Delta\omega\tau) \tag{5.81}$$

where $\Delta\omega/2\pi$ is the halfwidth at half-maximum of the Lorentzian spectrum. Also note that $\exp(-\Delta\omega\tau)$ is the Fourier transform of the Lorentzian spectrum. It is related to the autocorrelation function for the phase. Since the coherence time τ_c of the light source is equal to [3]

$$\tau_c = \frac{1}{2\Delta\omega} \tag{5.82}$$

Thus, this can be rewritten as

$$\langle\exp[j\phi(t-p\tau)]\exp[-j\phi(t-s\tau)]\rangle = \exp(-|s-p|\tau/2\tau_c) \tag{5.83}$$

that is the same expression as obtained in Refs. [2,5]. Hence, Equation 5.80 can be rewritten as

$$R[(p-s)\tau] = \exp\{j[(s-p)\omega_0\tau]\}\exp(-|s-p|\tau/2\tau_c)$$

or

$$R[(p-s)\tau] = \exp\{j[(s-p)\omega_0\tau]\}D^{|s-p|} \tag{5.84}$$

where $D = \exp(-\tau/2\tau_c)$. This gives an expression for the phase autocorrelation function in terms of the source coherence time. When computing $\langle I_8(t_n)\rangle$ in Equation 5.78, products of the input field, such as $\langle E_I(t_n)E_1*(t_n)\rangle$, $\langle E_1(t_n)E_1*(t_n-\tau)\rangle, \ldots, \langle E_1(t_n-p\tau)E_1*(t_n-s\tau)\rangle, \ldots$ etc., are involved as $E_8(t)$ is related to $E_1(t)$ by Equation 5.76.

The general algorithms of computing the transient response of the photonic circuit are as follows: (1) From the transfer function $H(z)$ of the optical circuit, obtain the impulse response or indeed the sequence $h[n]$ (Equation 5.77). (2) Convolute the impulse response obtained in step 1 with the input sequence to obtain the output of the system for the pulse input (Equation 5.76). (3) Obtain the expression for the phase fluctuation of the input, i.e., the correlation between values of phases at different times. (4) Compute the output intensity of the resonator (Equation 5.78) by using result obtained in step 2 in combination with expression in step 3. The results for the DCDR circuit are presented in the following section.

It can be seen from Equation 5.81 that the transient response of the circuit mainly depends on the ratio of the basic time delay to the source coherence time. For the two extreme cases, monochromatic source (very long coherence time, $\tau_c \gg \tau$) and temporal incoherent source (very short coherence time, $\tau_c \ll \tau$), the value of D is equal to 1 and 0, respectively. Hence, the range of D is [0, 1]. In general, the temporal response with monochromatic source is called the steady-state response. The transient response corresponds to the case where the source contains finite linewidth.

The transient responses are shown in Figures 5.32 and 5.33 for the passive DCDR circuit with unit delay in each path. In Figure 5.32, the circuit parameters

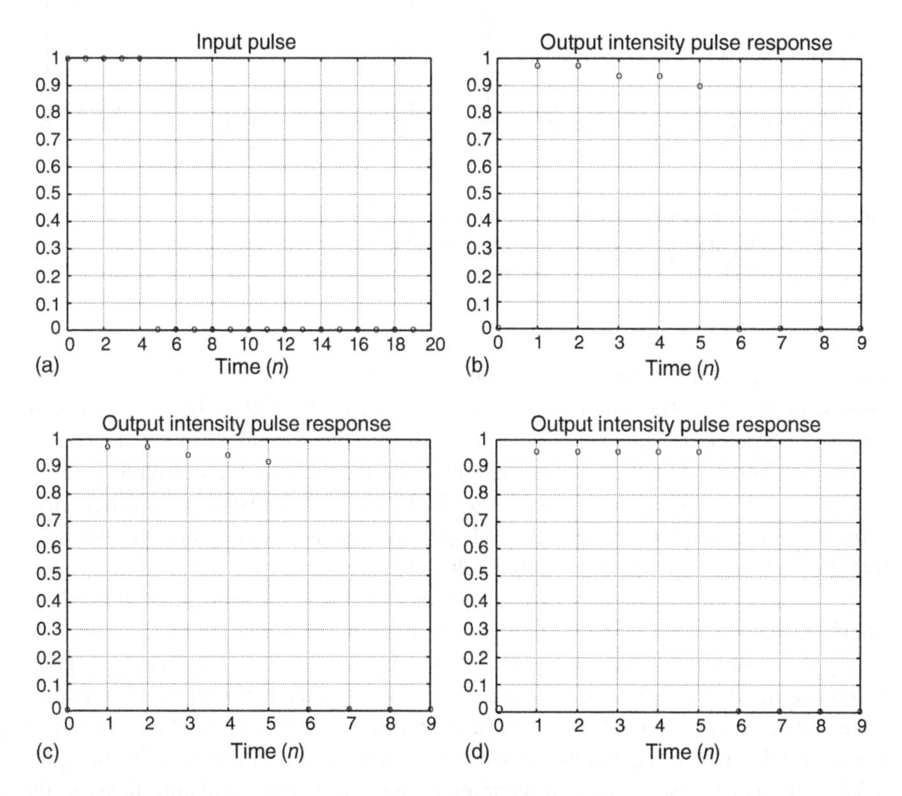

FIGURE 5.32 Transient response of the passive DCDR circuit with $t_a = 0.99$, $k = 0.995$ and $\omega_0\tau = \pi/2$. (a) Input pulse; (b) output intensity pulse with $\tau/\tau_c = 0.001$ (monochromatic); (c) output intensity pulse with $\tau/\tau_c = 0.2$; (d) output intensity pulse with $\tau/\tau_c = 500$ (temporal incoherent).

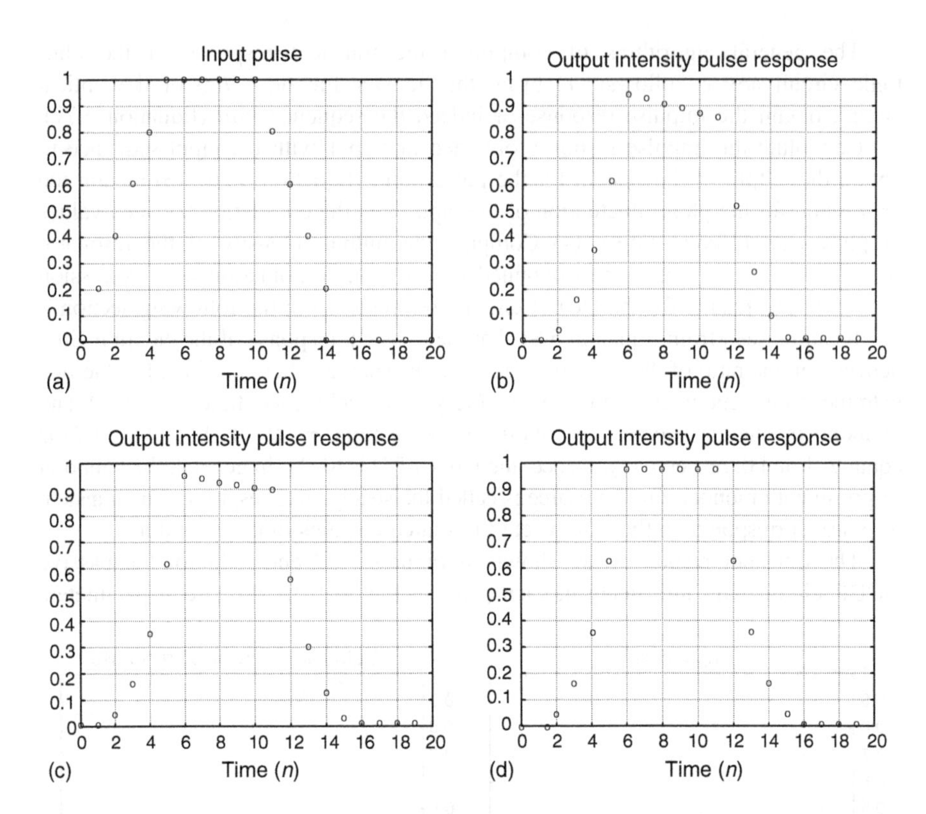

FIGURE 5.33 Transient response of the passive DCDR circuit with $t_a = 0.99$, $k = 0.995$, and $\omega_0\tau = \pi/2$. (a) Input pulse; (b) output intensity pulse with $\tau/\tau_c = 0.001$ (monochromatic); (c) output intensity pulse with $\tau/\tau_c = 0.2$; and (d) output intensity pulse with $\tau/\tau_c = 500$ (temporal incoherent).

satisfying the resonant condition can be used with $t_a = 0.99$, $k = 0.995$, and $\omega_0\tau$ is chosen to be $\pi/2$. The input pulse is shown in Figure 5.32a whereas the output intensity response is shown in Figures 5.32b through d with different degrees of source coherence. It is interesting to examine the incoherent case in Figure 5.32d in which the output is constant. In Figure 5.33, the circuit response to other input sequence (or in other words, the pulse shape) is examined. It is found in Figure 5.33 that the responses for different source coherence are similar.

In Figure 5.34, the circuit responses are computed with the same input pulse shape as in Figure 5.33a, but the circuit parameters here are different. The values of t_a and k still satisfy the resonant condition but nonresonant values of $\omega_0\tau$ have been used in Figures 5.34b and d. Also, effects of different degree of source coherence on the response are compared. It can be seen in Figures 5.34a and b that the responses oscillate. The output in Figure 5.34a which satisfies the resonant condition oscillates more than that in Figure 5.34b. This is because at resonant condition, there are many circulating fields in the circuit that interact with each other. Consider the geometry of our DCDR circuit, which has two forward paths and the signals in them add up at coupler 2 after each circulation.

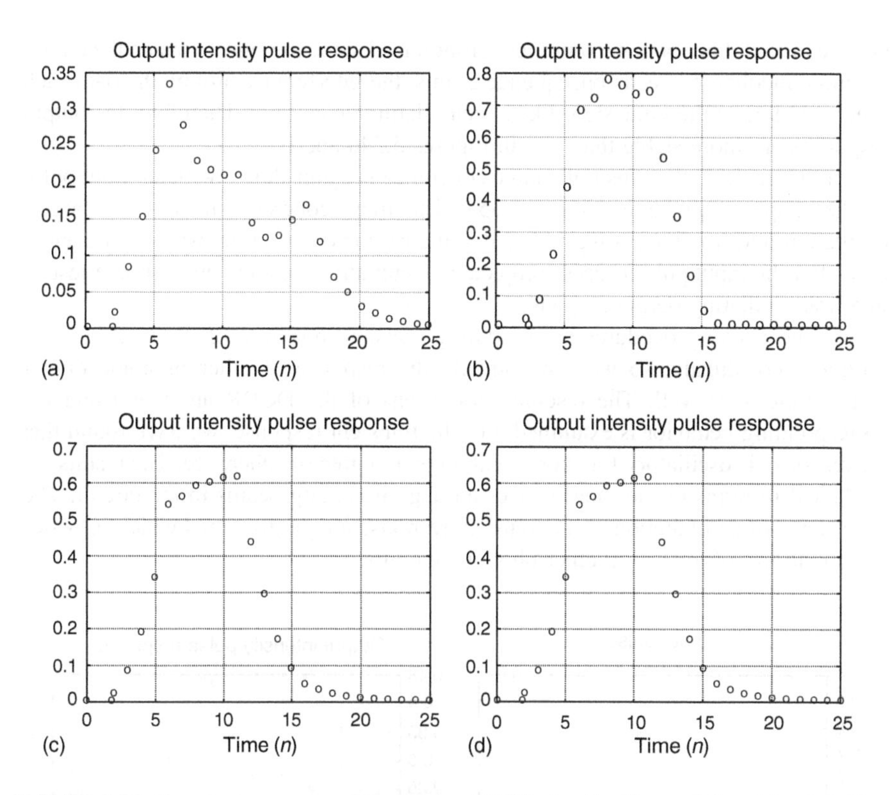

FIGURE 5.34 Transient response of the passive DCDR circuit with $t_a = 0.8$ and $k = 0.9$. The input pulse shape is the same as in Figure 5.33a. (a) Output intensity pulse with $\omega_0\tau = \pi/2$ and $\tau/\tau_c = 0.2$, (b) output intensity pulse with $\omega_0\tau = \pi$ and $\tau/\tau_c = 0.2$, (c) output intensity pulse with $\omega_0\tau = \pi/2$ and $\tau/\tau_c = 10$, and (d) output intensity pulse with $\omega_0\tau = \pi$ and $\tau/\tau_c = 10$.

They may not always have constructive interference occurred because of the relative magnitudes of the two beams and their "signs." The signs are determined by the phase change the signal experienced when traveling. For instance, after traveling along a fiber path whose $\omega_0\tau$ is denoted by $\pi/2$, this is equivalent to multiplying the complex magnitude of the signal by j, which represents the phase. In the case of Figure 5.34a, it may suggest that there are alternating constructive and destructive interferences at each circulation that results in the oscillations in the output response. In Figure 5.34b, there are less circulating fields taking part in the interaction as most of them travel only once inside the circuit and are out to the output directly. This explains why the average output intensity is larger in Figure 5.34b. The same reasoning can be applied to explain the difference between Figures 5.33c and 5.34a. In the former one, since the value of k is close to 1, most of the beams are coupled out of the circuit after they have traveled only one path in the circuit leaving less amount of circulating beams interacted with incoming beams. Thus, the former one does not have oscillation in the output response.

When we increase the value of τ/τ_c to 10 (i.e., a reduction in the source coherence) in Figures 5.34c and d, the oscillations cease. In this case, the phase change in each fiber path becomes less pronounced since the signals inside the circuit are more incoherent. As a consequence of this, the interference around the path each time would be of the same sign (like a recirculating delay line). Therefore, the output responses are more stable than that in Figures 5.34a and b.

In Figure 5.35, the effect of source coherence is again shown with an input pulse of the shape as given in Figure 5.35a. The most coherent one has the largest oscillation and it is interesting to see that the response has large oscillation even at time after the input pulse ceases completely. Generally, a less coherent source results in a less oscillating output response.

As the above procedures of computing transient response of the resonator are programmed already, so we can compute the response of other photonic circuit configurations as well. The resonant conditions of the DCDR are found and the corresponding behavior is examined. For the transient response, we have found that the response is oscillatory for source with higher degree of coherence. The results are explained in terms of the interferences among circulating beams in the circuit. The procedures of computation of the transient response are programmed which provides the possibility of analyzing other photonic circuits.

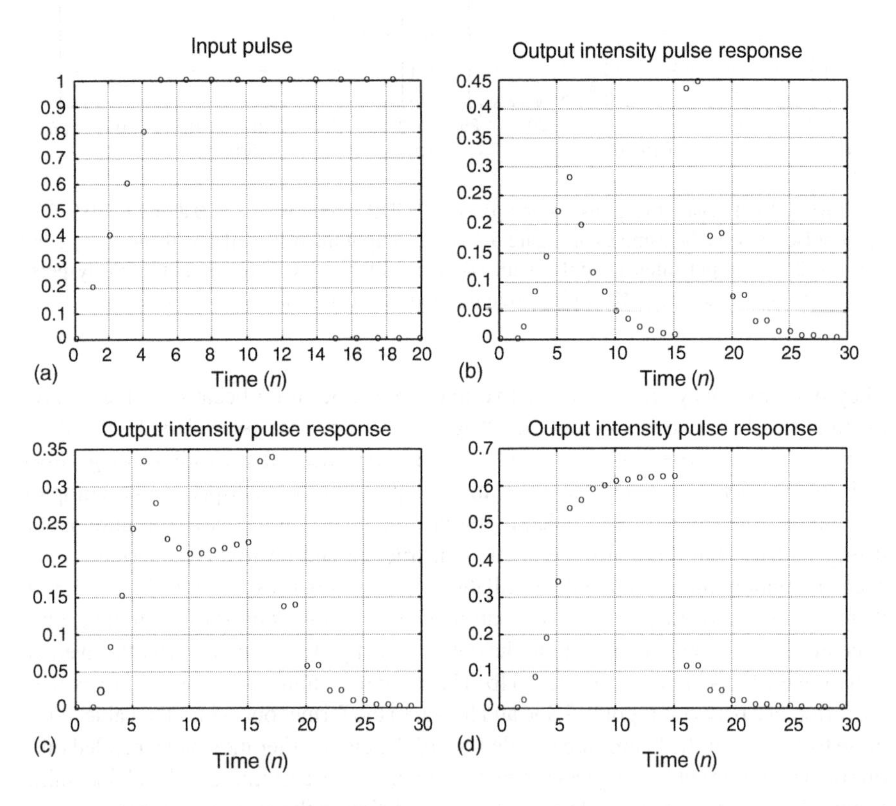

FIGURE 5.35 Transient response of the passive DCDR circuit with $t_a = 0.8$, $k = 0.9$, and $\omega_0\tau = \pi/2$. (a) Input pulse, (b) output intensity pulse with $\tau/\tau_c = 0.001$, (c) output intensity pulse with $\tau/\tau_c = 0.2$, and (d) output intensity pulse with $\tau/\tau_c = 10$.

5.1.6 Photonic Resonator as a Dispersion Equalizer

The performance of optical communication systems can be limited by several effects. One of these effects is the pulse distortion as a result of fiber chromatic dispersion.

The study of pulse propagation in dispersive media is important in many applications, including the transmission of optical pulses through the optical fibers [6] used in optical communication systems [5,9]. In the fiber-optic transmission system, compensators or dispersion equalizers [10–13,30–33] are required to compensate for the distortion resulted during the signal transmission in the fiber. This ensures that the signal distortion at the end of the transmission link is kept to the minimum so that the signal is received without error. In this chapter, the possibility of employing the DCDR resonator as a dispersion equalizer is investigated.

In the first part of this chapter, the equations governing the optical pulse transmission in single mode fiber are derived. These equations will include the effects of source coherence and the modulation signal bandwidth on the pulse-broadening behavior in the fiber. The application of DCDR circuit as the dispersion equalizer is studied in the second part of the chapter.

Since the optical fiber is a dispersive medium, an optical pulse broadens as it travels along a fiber. If several pulses are transmitted through the fiber, the broadening of the optical pulse will cause the overlapping among the pulses after traveling a long distance. This interference between adjacent pulses is called the intersymbol interference (ISI). The ISI will affect the signals received at the other end of the fiber that causes errors in the receiver. The fiber dispersion thus limits the information-carrying capacity of the fiber system as adjacent signal pulses cannot be transmitted too close to each other, hence lowering the rate of transmission.

There are two types of dispersion mechanisms of pulse spreading in fibers: intramodal dispersion and intermodal dispersion. Intramodal dispersion, sometimes called chromatic dispersion, is the pulse spreading that occurs within a single mode. The intermodal dispersion is the dispersion between the modes. Only chromatic dispersion will be considered in the following analysis, as pulse transmission in single mode fiber is concerned. Chromatic dispersion is mainly due to the wavelength dependence of the refractive index of the fiber's core material. Hence, the chromatic dispersion can be represented by the wavelength dependence of the propagation constant.

The main aim of this section is to obtain the signal at the output of the fiber link so that design of equalizer can be performed. As this chapter is mainly devoted to the study of equalization of fiber dispersion by the DCDR circuit, the discussion on pulse propagation will concentrate on typical cases rather than a general one. In this instance, only Gaussian pulse will be considered.

The expression for the signal at the output of a fiber link is derived. This output will be fed into the DCDR circuit at the end of the fiber link for dispersion equalization. We start with the signal entering the fiber, i.e., the light source. We assume that the light source emits an optical signal with Gaussian spectrum centered at $\omega = \omega_0$ which takes the form

$$S(\omega) = \frac{\sqrt{2\pi}}{W_s} \exp[5.5(\omega - \omega_0)^2/(2W_s^2)] \tag{5.85}$$

where

W_s is the spectral width at the 1/e points in units of angular frequency
ω is the angular optical frequency
ω_0 is the center angular frequency of the source

The corresponding temporal (time domain) function $s(t)$, by applying the Fourier transform on Equation 5.85, is

$$S(t) = \exp\left(\frac{-t^2 W_s^2}{2}\right)\exp(j\omega_0 t) \tag{5.86}$$

The signal needs to be modulated to get the light pulse. The modulation of the light source signal can be done either directly or externally. In direct modulation, one would modulate the driving current of the source to modulate the source output directly. In external modulation, the continuous optical signal from the source is modulated using an optical modulator following the source.

Suppose the source is amplitude modulated with a pulse of Gaussian shape of the form [34]

$$A_m(t) = \exp\left(\frac{-t^2}{2\tau_m^2}\right) \tag{5.87}$$

where τ_m is the 1/e half width. Thus, the light signal, or the electric field at the input of the fiber, $E_{\text{if}}(t)$ can be expressed as

$$E_{\text{if}}(t) = A_m(t)s(t) \tag{5.88}$$

where

$$E_{\text{if}}(t) = \exp\left[\frac{-t^2}{2}\left(\frac{1}{\tau_m^2} + W_s^2\right)\right]\exp(j\omega_0 t) \tag{5.89}$$

The electric field at the end of the fiber of length L is

$$E_{\text{of}}(t) = \int E_{\text{if}}(\omega)\exp[j(\omega t - \beta L)]\mathrm{d}\omega \tag{5.90}$$

where $E_{\text{if}}(\omega)$ is the Fourier transform of $E_{\text{if}}(t)$ and β is the propagation constant of the propagation mode in the fiber. Low-loss fiber is assumed to be used so that attenuation of the fiber can be neglected. The dispersion effect of the fiber can be represented by the fact that β is a function of frequency. By Taylor series expansion of β about $\omega = \omega_0$, β can be approximated by

$$\beta(\omega) = \beta(\omega_0) + \beta'(\omega_0)(\omega - \omega_0) + \frac{\beta''(\omega_0)}{2}(\omega - \omega_0)^2 + \cdots \tag{5.91}$$

For simplicity, we use β_0 to represent $\beta(\omega_0)$ thereafter. In the following analysis, only the terms up to the second derivative of the propagation constant are included. The higher-order derivative terms in the Taylor series expansion of the propagation constant are neglected. Moreover, the first two terms in Equation 5.91 represent the pure delay of the carrier and the envelope, respectively, and they do not contribute to the dispersion. Therefore, these two terms are ignored in the following analysis for clarity. After these simplifications, the effective propagation constant used in our analysis can be represented by

$$\beta(\omega) = \frac{\beta_0''}{2}(\omega - \omega_0)^2 \tag{5.92}$$

The value of the second derivative of the propagation constant is a measure of the so-called first-order dispersion [1,2]. For a silica fiber, the wavelength at which this first-order dispersion vanishes is about 1300 nm. At wavelength other than the zero-dispersion wavelength, the first-order dispersion is nonzero and it accounts for the broadening of optical pulse traveling in the fiber. Also, the values of higher-order derivatives which correspond to second- and higher-order dispersion are negligible at wavelength other than the zero-dispersion wavelength.

By using Equations 5.90 and 5.91, from Refs. [1,2], the output power at the fiber end is

$$I_{\text{of}}(t) = \frac{1}{[1 + 4\hat{D}^2(1 + V^2)]^{1/2}} \exp\left[\frac{-(t/\tau_m)^2}{1 + 4\hat{D}^2(1 + V^2)}\right] \tag{5.93}$$

with

$$V = \tau_m W_s \tag{5.94}$$

$$\hat{D} = \frac{\beta_0'' L}{2\tau_m^2} \tag{5.95}$$

Equation 5.93 shows that the Gaussian-shaped pulse remains Gaussian after traveling through the fiber; thus, the output power at the fiber end has a pulse width τ_0 given by the following:

$$\tau_0^2 = \tau_m^2[4\hat{D}^2(1 + V^2)] \tag{5.96}$$

It is clearly shown in Equation 5.96 that a larger value of V would produce a wider spreading pulse width τ_0 at the fiber output end. Equation 5.94 indicates that an incoherent source (large W_s) broadens the signals more than the case with a coherent source. Moreover, the broadening of pulses depends on the value of $\hat{D}$, hence the values of β_0'' and L.

Without loss of generality, the source is assumed to be monochromatic in the following analysis, i.e., $W_s = 0$. This makes $V = 0$. From Refs. [24,35,36], the electric field at the fiber output for a monochromatic light source is given by

$$E_{\text{of}}(t) = \frac{1}{(1+4\hat{D}^2)^{1/4}} \exp(j\omega_0 t) \exp\left[\frac{-(t/\tau_m^2)}{2(1+4\hat{D}^2)}\right] \exp\left[\frac{-j\hat{D}(t/\tau_m^2)}{(1+4\hat{D}^2)}\right] \tag{5.97}$$

Expressing Equation 5.97 in terms of the physical parameters, we obtain

$$E_{\text{of}}(t) = \frac{\tau_m}{\left[\tau_m^4 + (\beta_0''L)^2\right]^{1/4}} \exp(j\omega_0 t) \exp\left\{\frac{-t^2(\tau_m^2 + j\beta_0''L)}{2\left[\tau_m^4 + (\beta_0''L)^2\right]}\right\} \tag{5.98}$$

which is the same expression as obtained in Ref. [10] and is the field equation required for the subsequent analysis. This would be the field at the input of an optical equalizer.

We can now investigate the application of the DCDR circuit as an equalizer for single mode fiber dispersion. The principle of the fiber dispersion equalizer [31,37,38] lies in the fact that the equalizer has a group delay in the opposite sign to that introduced by the fiber. Thus, the group delay induced on the signal after the signal travels through the fiber would be partially or completely cancelled by that in the equalizer. Recall that the magnitude of the phase change induced by the fiber on the signal after the signal travels length L of fiber is βL. Thus, the group-delay time τ_f induced by the fiber can be expressed as

$$\tau_f = \beta_0''(\omega - \omega_0)L \tag{5.99}$$

which is obtained by differentiating βL with respect to ω making use of Equation 5.95. The chromatic dispersion induced by the fiber link of length L is thus equal to

$$\frac{d\tau_f}{d\omega} = \beta_0''L \tag{5.100}$$

Ideally, we need to use the linear portion (if available) of the equalizer group delay that has the suitable slope, both the magnitude and the sign, to counteract the fiber dispersion. In addition, that portion should be positioned to the center frequency of the signal. This point is called the operating point of the equalizer thereafter.

First, we consider the pulse at the fiber link output whose field equation is given in Equation 5.98. The following parameters are used in our analysis. The input pulse into the fiber is a Gaussian-shaped pulse with temporal half width at the $1/e$ points and τ_m equals to 50 ps. The working wavelength of the system, λ_0, is taken to be 1550 nm in the following analysis. As we concentrate only on the dispersive effect of the fiber, the attenuation effect of the fiber is ignored. Therefore, the fiber is assumed to be loseless. As the operating wavelength of the system is 1550 nm at which the loss of standard fiber is lowest, the above assumption is valid. Actually, the attenuation introduced by the fiber can be easily compensated by an amplifier or repeater at the

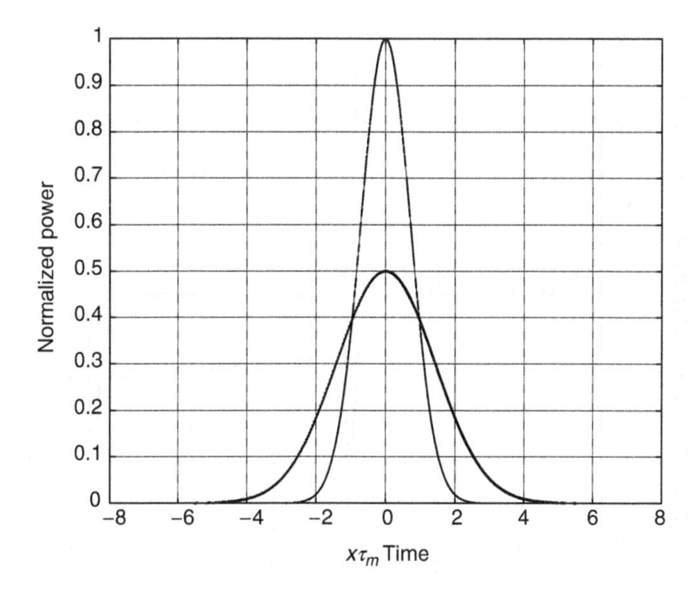

FIGURE 5.36 The intensities of the optical pulses at the input of the fiber (dotted line) and at the fiber output (solid line) with the above fiber parameters and operating conditions.

end of the fiber length but this is not the main point in our analysis. The fiber dispersion value is represented by a parameter D, which is 17 ps/nm/km for conventional silica fiber. The dispersion caused by fiber is given by $\beta_0'' L$ [36] which is

$$\beta_0'' L = \frac{D\lambda_0^2 L}{2\pi c} \tag{5.101}$$

where c is the speed of light in vacuum. The transmission fiber path length L is taken to be 200 km unless otherwise stated. Using the given parameters, the value of $\beta_0'' L$ is 4.3335e^{-21}. The intensities of the optical pulses at the input of the fiber and at the fiber output with the above fiber parameters and operating conditions are plotted in Figure 5.36. The dispersive properties of the fiber can be clearly observed when the difference between the pulse widths of the two pulses is compared. Obviously, the optical pulse broadens as it travels along the fiber.

5.1.6.1 Group Delay and Dispersion of the DCDR Resonator

The group delay and the dispersion of the DCDR resonator is evaluated numerically using programs written in MATLAB commands. Recall that τ is the basic delay time of the DCDR circuit and it is set to the value of 20 ps, which is equal to $\tau_m/2.5$. One result is shown in Figure 5.37 with $k_1 = k_2 = 0.1$, $t_{a1} = t_{a2} = t_{a3} = 1$, $G_1 = G_2 = G_3 = 1$, and $m_1 = m_2 = m_3 = 1$.

It is found in Figure 5.37 that the group delay of the equalizer is always negative whereas its dispersion has both positive and negative values. The negative portion of the equalizer dispersion can be employed to cancel the effect of fiber dispersion.

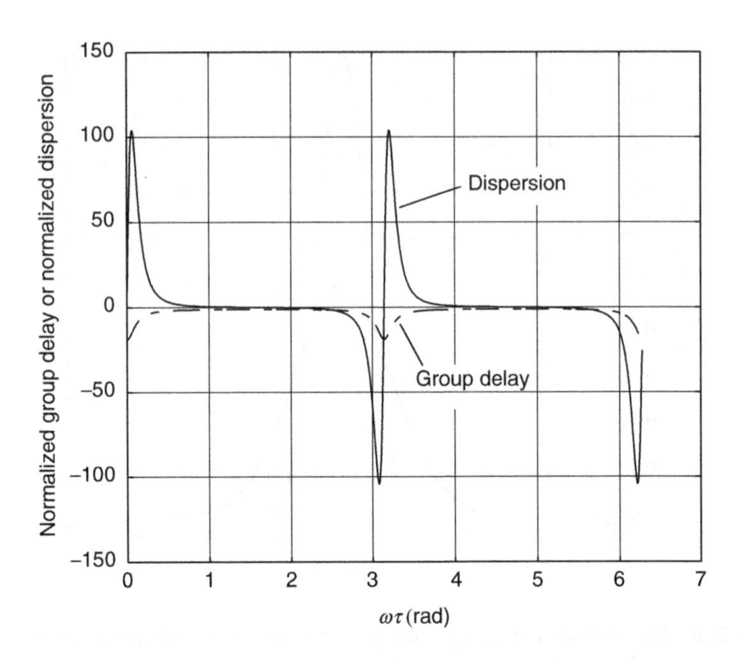

FIGURE 5.37 Group delay (– – –) and dispersion of the resonator with $k_1 = k_2 = 0.1$, $t_{a1} = t_{a2} = t_{a3} = 1$, $G_1 = G_2 = G_3 = 1$, and $m_1 = m_2 = m_3 = 1$. Since the group delay and dispersion values are normalized, the group-delay value read on the axis needed to be multiplied by τ to get the absolute value, it needs to be multiplied by τ^2 for the dispersion.

The task of the DCDR circuit is to compensate the dispersion created by the fiber path. Let the ratio of equalizer dispersion to the fiber dispersion be R, the ideal value of R would then be -1. We also need to choose the operating point of the equalizer so that we can get the desired equalizer dispersion to perform the equalization. The operating point of the equalizer, ϕ_o, is actually the operating $\omega\tau$ of the DCDR circuit. The operating point can be positioned by changing the length of the fiber path in the DCDR circuit using piezoelectric effect. To get better equalization, the equalizer dispersion should vary as small as possible with $\omega\tau$ in the vicinity of ϕ_o. This is because of the fact that we have assumed a uniform dispersion factor of the fiber in the vicinity of the source operating wavelength. The value of ϕ_o can be chosen from the resonator dispersion in Figure 5.37.

It is found that at $\omega\tau = 2.8225$ rad (161.7°), the magnitude of the equalizer dispersion is 4.3617e^{-21} which gives an R of -1.0065. The corresponding output intensity of the equalizer together with the input intensities of the fiber and equalizer is shown in Figure 5.38.

By the comparison between the resonator input intensity and the resonator output intensity in Figure 5.38, it can be seen that the broadening of the former pulse has been compensated to a great extent by the resonator. The resonator cannot give full equalization because its dispersion is not constant with frequency. For that reason, we attempt to design parameters giving the best rather than 100% equalization for a

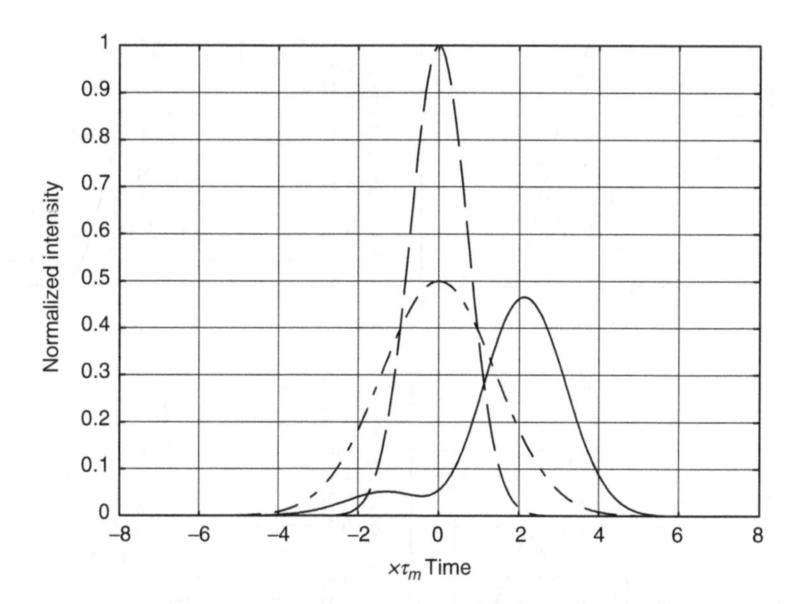

FIGURE 5.38 The intensities of the optical pulses at the input of the fiber (dashed – –), at the fiber output (i.e., resonator input) (dash dot –·–) and at the resonator output (solid line) with $k_1 = k_2 = 0.1$, $t_{a1} = t_{a2} = t_{a3} = 1$, $G_1 = G_2 = G_3 = 1$, and $m_1 = m_2 = m_3 = 1$, operating point of the resonator $\phi_o = 2.8225$ rad (161.7°).

given pulse. The loss in the magnitude of the equalized pulse can be compensated by an amplifier inserted after the resonator.

The effects of different operating points on the equalization are shown in Figure 5.39. The operating points used for the plots are given in Table 5.3.

The operating points for the plots in Figure 5.39 are all in the proximity of $\phi_o = 2.8225$ rad, i.e., the operating point used in Figure 5.38. In this range of ϕ_o, the resonator's dispersion is negative, which is appropriate for its use as an equalizer. It can be observed that when we move from the operating point $\phi_o = 2.5035$ to $\phi_o = 3.1355$, a trailing tail of the signal is building up. It becomes significant as compared to the original signal, thus causing other distortion effects. Therefore, ϕ_o in Figures 5.39c through e are not good operating points of the equalizer in this case.

To obtain a more visual presentation on the effect of broadened pulses, we consider a sequence of pulses. Figure 5.40b shows the overlapping between two adjacent pulses separated $6\tau_m$ in time at the end of the fiber link after undergoing dispersion. Figure 5.40a shows clearly that the two pulses are under noninterference at the transmitter end. If the sampling time of the detection system is $6\tau_m$ in this situation, the fiber dispersion would not cause any problem to the detection. But if the pulses are getting closer to each others, i.e., at a higher transmission rate, the ISI would limit the capacity of the system.

Eye pattern diagram is usually used to assess visually the system performance in a digital communication system. Firstly, random data signal patterns are generated.

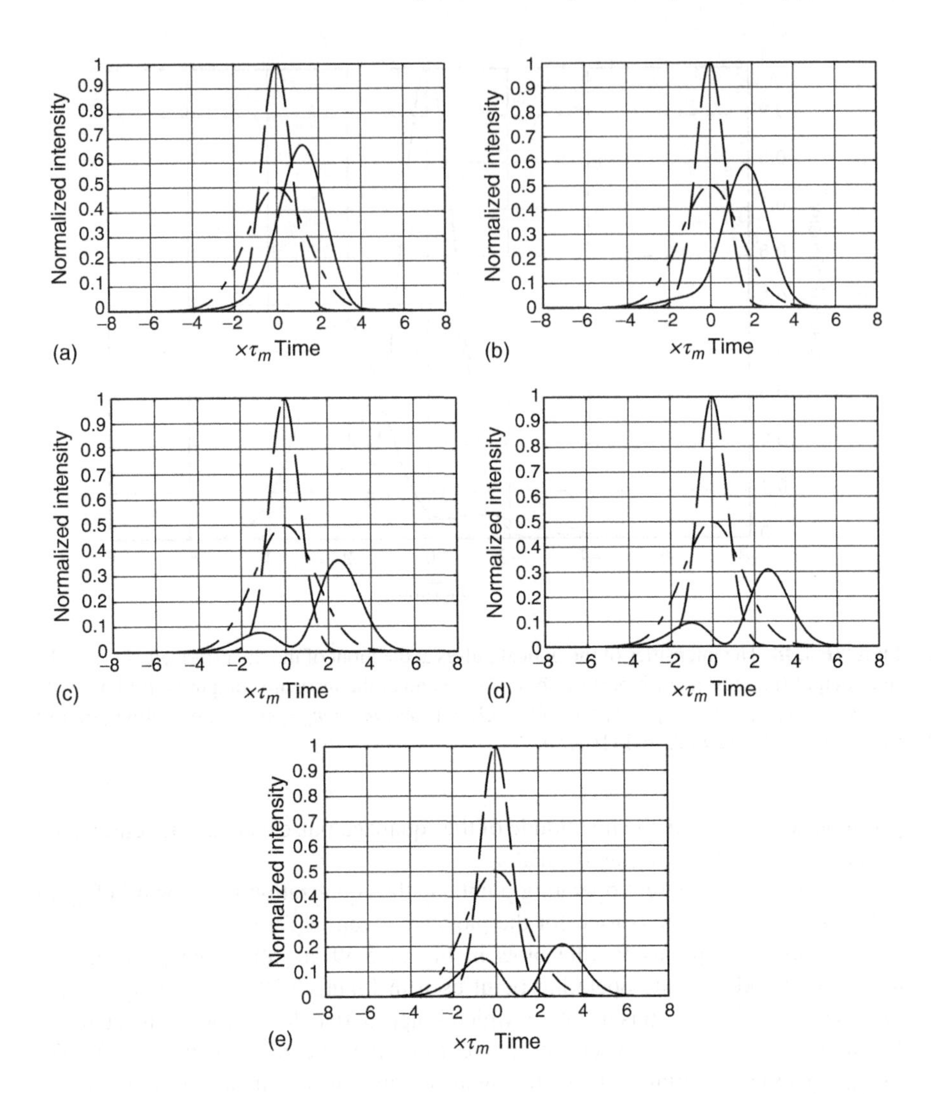

FIGURE 5.39 The intensities of the optical pulses at the input of the fiber (dashed – –), at the fiber output (i.e., resonator input) (dash dot –·–) and at the resonator output (solid line) with $k_1 = k_2 = 0.1$, $t_{a1} = t_{a2} = t_{a3} = 1$, $G_1 = G_2 = G_3 = 1$, and $m_1 = m_2 = m_3 = 1$ for different operating points of the resonator. (a) $\phi_o = 2.5035$ rad, (b) $\phi_o = 2.6814$ rad, (c) $\phi_o = 2.9391$ rad, (d) $\phi_o = 3.0005$ rad, and (e) $\phi_o = 3.1355$ rad.

For a 3-bit-long sequence, for instance, we could have "111, 011, 110," . . . , etc. When these combinations are superimposed simultaneously, an eye pattern diagram is formed. The eye pattern diagrams for the intensities before and after the equalizer corresponding to the above case are shown in Figures 5.41a and b, respectively. Recall that the operating point of the equalizer here is 2.8225 rad.

Figure 5.41b indicates that the eye-opening of the diagram is improved than that of Figure 5.41a; hence, the equalizer indeed enhances the quality of the received

TABLE 5.3
Operating Points of the Equalizer Together with the Value of *R* for Figure 5.4a through e

	ϕ_o (rad)	R
Figure 5.4a	2.5035	−0.1473
Figure 5.4b	2.6814	−0.3758
Figure 5.4c	2.9391	−2.9215
Figure 5.4d	3.0005	−5.5526
Figure 5.4e	3.1355	−1.6116

signals. In Figure 5.42, the eye pattern diagrams for two different transmission rates are shown. In Figure 5.42a, the pulses are pushed closer together, and the time between their adjacent peaks is $4\tau_m$. The corresponding time is $8\tau_m$ in Figure 5.42b. It is obvious that the eye closure in Figure 5.42a is greater than that in Figure 5.42b. This means that it is more difficult to detect signal correctly in the case of Figure 5.42a. By comparing among the eye pattern diagrams in Figures 5.41 and 5.42, we can state that the minimum spacing between transmitted pulses is about $6\tau_m$ in this instance. This, in turn, is a way of determining the maximum transmission rate of a transmission channel.

In Figure 5.43, we have shown that with $\phi_o = 2.5035$ rad and $R = -0.1473$ (first row in Table 5.3), the result is even better than that in Figure 5.6 with $\phi_o = 2.8225$ rad and $R = -1.0065$. These results show that the DCDR resonator can be used as an equalizer even in the passive operation.

Since the dispersion of equalizer is proportional to τ^2, it is expected that a larger value of τ would result in a greater magnitude of the equalizer dispersion. In Figure 5.44, two values of τ are used with the same operating point $\phi_o = 2.5035$ rad. It can be seen that a better compensation is obtained by the equalizer with a larger τ.

An active mode operation of the equalizer is shown in Figure 5.45. The parameters used are $k_1 = k_2 = 0.1$, $t_{a1} = t_{a2} = t_{a3} = 1$, $G_1 = 1.9$, $G_2 = G_3 = 1$, and $m_1 = m_2 =$

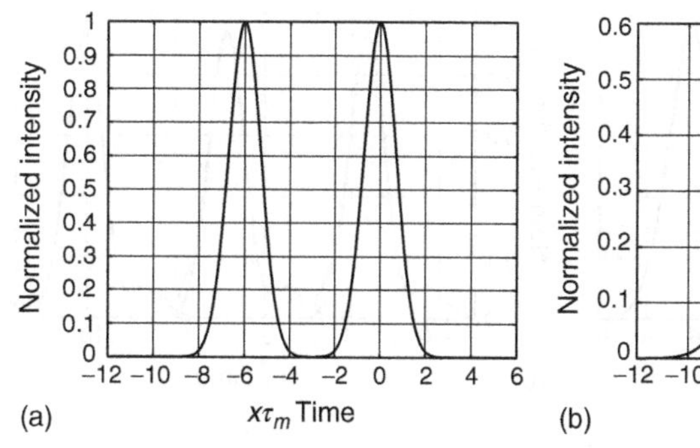

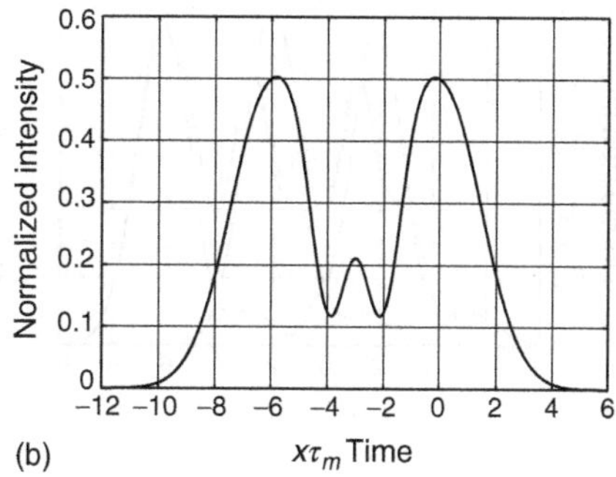

FIGURE 5.40 Two adjacent pulses (a) at the input of the fiber link and (b) at the end of the fiber link before the equalizer. The two pulses are separated $6\tau_m$ in time.

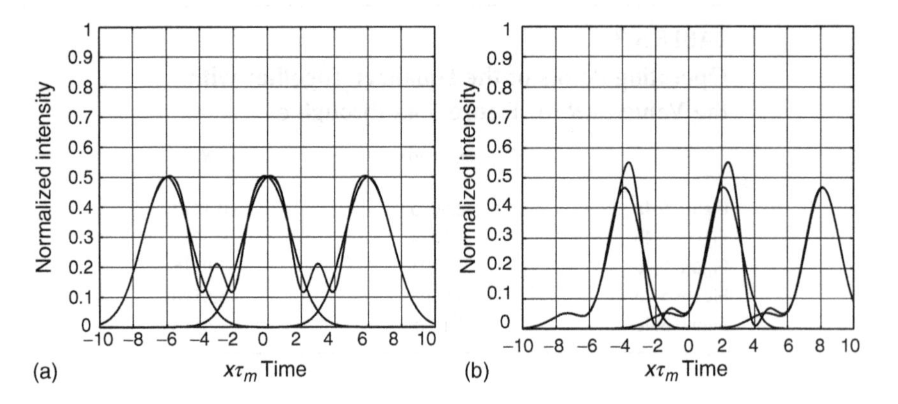

FIGURE 5.41 (a) The eye pattern diagram before the equalizer and (b) the eye pattern diagram after the equalizer. The parameters used in the equalizer are the same as those used in Figure 5.3. Pulses are separated by $6\tau_m$ in time.

$m_3 = 1$. The operating point of the resonator is located at 2.8225 rad which is the same as that in Figure 5.3. The value of R in this case is -1.0074. By comparing between Figures 5.42 and 5.43, it is obvious that the equalized pulse in the latter graph has a larger magnitude. Thus, the amplifier in the DCDR circuit compensates for the loss in magnitude as well. This is a useful feature in our amplified DCDR circuit as the dispersion compensation and the compensation for the pulse magnitude can be done with a single device. In other words, in addition to the dispersion equalization, the DCDR circuit can be tailored to compensate for the attenuation in the pulse amplitude. This would be an economical way of equalizer design.

The eye pattern diagram for the equalized pulse in Figures 5.46 and 5.38 is given in Figure 5.46. The eye pattern diagram before the equalizer is the same as that in Figure 5.41a. The eye pattern diagram in Figure 5.46 is to be compared with that in Figure 5.41b, where we have passive mode operation. In both cases, the equalizers have performed considerable amount of dispersion equalization.

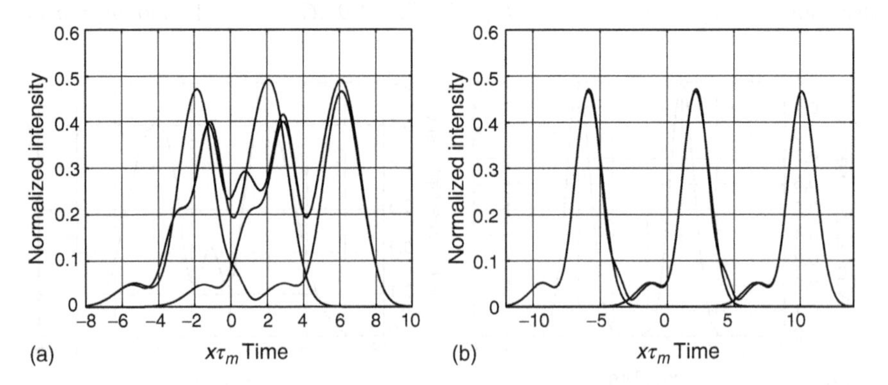

FIGURE 5.42 The eye pattern diagrams after the equalizer. The parameters used in the equalizer are the same as those used in Figure 5.3. (a) Pulses are separated by $4\tau_m$ in time and (b) pulses are separated by $8\tau_m$ in time.

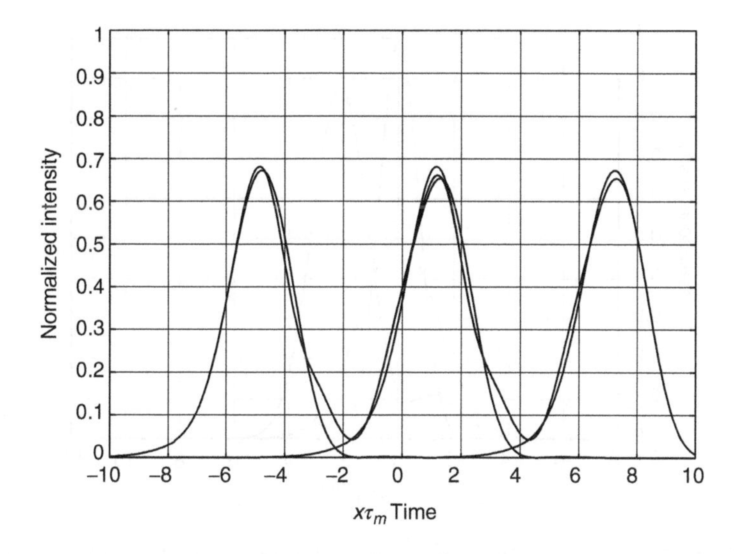

FIGURE 5.43 The eye pattern diagrams after the equalizer. The parameters of the equalizer are the same as those used in Figure 5.3, except that $\phi_o = 2.5035$ rad. Pulses are separated by $6\tau_m$ in time.

The programs have been tested for a single loop resonator using the same parameters as given in Ref. [10] after value-by-value comparison, our results are consistent with theirs. Thus, the programs can be used with great confidence. One advantage of our programs is that they are very generalized; they can handle other circuit configurations, provided that the input–output field transfer functions of the circuits are given.

The application of using the DCDR circuit as a dispersion equalizer is demonstrated. The effects of working the equalizer at several operating points are shown.

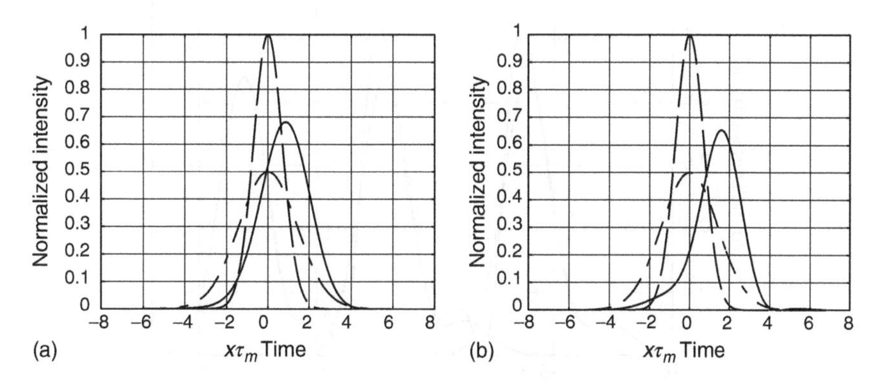

FIGURE 5.44 The intensities of the optical pulses at the input of the fiber (dashed – –), at the fiber output (i.e., resonator input) (dash dot –·–) and at the resonator output (solid line) with $k_1 = k_2 = 0.1$, $t_{a1} = t_{a2} = t_{a3} = 1$, $G_1 = G_2 = G_3 = 1$, and $m_1 = m_2 = m_3 = 1$, operating point of and the resonator $\phi_o = 2.5035$ rad. (a) $\tau = 15$ ps, (b) $\tau = 25$ ps.

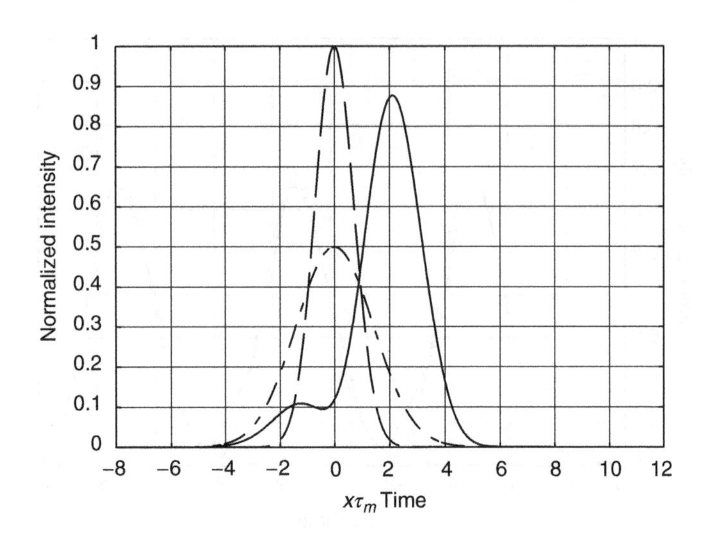

FIGURE 5.45 The intensities of the optical pulses at the input of the fiber (dashed – –), at the fiber output (i.e., resonator input) (dash dot –·–) and at the resonator output (solid line) with $k_1 = k_2 = 0.1$, $t_{a1} = t_{a2} = t_{a3} = 1$, $G_1 = 1.9$, $G_2 = G_3 = 1$, and $m_1 = m_2 = m_3 = 1$, operating point of the resonator $\phi_o = 2.8225$ rad.

It is found that the DCDR circuit can provide a reasonable amount of chromatic dispersion equalization. Active operation of the DCDR circuit can provide compensation for the pulse amplitude in addition to the dispersion compensation. Therefore, theoretically, the circuit can be employed as a dispersion equalizer in fiber-optical communication systems.

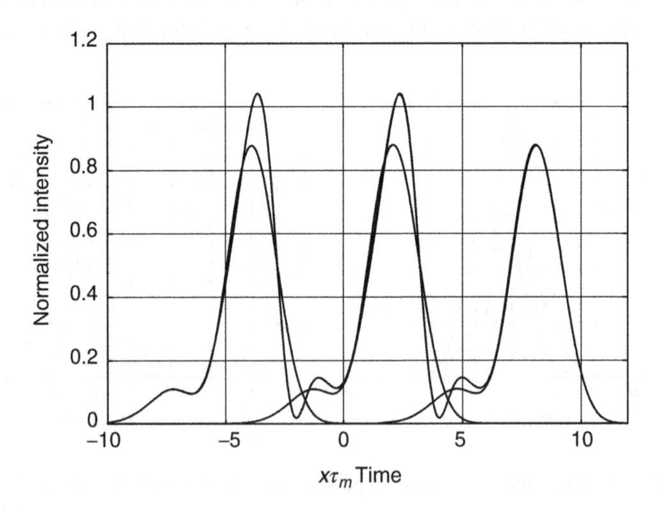

FIGURE 5.46 The eye pattern diagram after the equalizer. The parameters used in the equalizer are the same as those used in Figure 5.10. Pulses are separated by $6\tau_m$ in time.

The resonant behavior of the circuit operating as a resonator is observed. It is discovered that by proper choice of circuit parameters, the sharpness of the resonant peak can be adjusted. The transient responses of the DCDR circuit with different source coherences and different input pulse shapes are examined. The results show that responses of the circuit are oscillatory for highly coherent source. Moreover, the procedures of computation of transient response are generalized which could be applied to general photonic circuits.

It is shown to be possible to employ the DCDR circuit as a fiber dispersion equalizer. The circuit provides larger flexibility in equalizer design than simple configurations. One significant consequence is that the DCDR circuit could provide compensation for the pulse amplitude in addition to dispersion compensation.

In general, our novel graphical representation of optical circuits can tackle nonlinear optical elements. In other words, the transmittances in the SFG of the circuit can be nonlinear in nature. The use of our technique may lead to easier understanding of the performance of nonlinear optical circuits. Over the last decade, there were advancing research interests in the optical fiber communication systems. The use of an all-fiber photonic circuit as an optical equalizer for the fiber dispersion should be a considerable step in going towards an all-optical communication systems. Further research on this topic could be the investigations of possibilities of using other photonic circuits as the optical-based dispersion equalizer. Considerations may cover different signal modulation methods, various signal pulse shapes, and different detection techniques. We expect that the novel graphical representation and the associated manipulation rules of the photonic circuits would play an important role in future optical network analysis.

5.1.6.2 Optical Eigenfilter as Dispersion Compensators

In this section, a digital eigenfilter approach is employed to design linear optical dispersion eigencompensators (ODECs) for compensation of the combined effect of laser chirp and fiber dispersion at 1550 nm in high-speed long-haul IM/DD (intensity modulation/direct detection) lightwave systems [39,40]. Various techniques developed to mitigate this detrimental effect are reviewed in Section 5.2.11. Section 5.2.1 describes the formulation of the dispersive fiber channel which is used as a basis for designing ODECs, and also compares the performance of the proposed eigencompensating method with the Chebyshev technique in the frequency domain. A procedure for the synthesis of the ODECs using the PLC technology described in Section 5.2.5 is outlined in Section 5.3.3. Section 5.2.6 describes the transmission system model used to simulate the performance of the ODEC in an IM/DD optical communication system, and also compares the performance of the ODEC with that of the Chebyshev equalizer in the time domain. The reasons for comparing the eigenfilter approach with the Chebyshev technique are given in Section 5.2.7. The theory of coherent integrated-optic signal processing described in Section 2.3 is employed in this chapter where electric-field amplitude signals are considered.

5.1.6.3 Remarks

The advent of the erbium-doped fiber amplifier (EDFA) has created a need for optical dispersion compensators to be developed to solve the important problem of fiber chromatic dispersion in optical communication systems operating in the vicinity of the minimum-loss 1550 nm wavelength. The combined effect of laser chirp and fiber dispersion, which results in significant pulse broadening, is the critical factor limiting the maximum bit rate-distance in high-speed long-haul IM/DD systems. To exploit the potential benefits of the EDFA for long-distance transmission at 1550 nm, this detrimental effect must be minimized either by using dispersion-shifted fibers (DSFs) with zero-dispersion wavelength near 1550 nm or by using 1300 nm optimized standard fibers with some means of dispersion compensation. The use of DSFs to upgrade the already-embedded fiber network is clearly not an economical way to enhance network performance. The employment of DSFs in conjunction with EDFAs as repeaters is seen to be a flexible and cost-effective choice of technology for the generation of new ultralong-, ultrahigh-speed transoceanic systems. However, this technology still suffers from residual dispersion at the end of the link because each DSF span has its own zero-dispersion wavelength. In fact, it has been proposed that operating exactly at the zero-dispersion wavelength is undesirable because of the strong buildup of noise due to the four-wave mixing between the signal and the amplified noise [41]. The residual dispersion can be dealt with by placing an adjustable optical dispersion compensator at the receiver end. Thus, dispersion compensation will remain a key technological issue for both the upgrading of existing terrestrial fiber networks and the installation of new transoceanic systems for years to come.

Several nonlinear and linear optical dispersion-compensating techniques have been demonstrated to mitigate the primary effect of fiber dispersion on lightwave systems at 1550 nm. The nonlinear methods, such as bright-soliton transmission [42–45] or dark-soliton transmission as described in Chapter 4 and optical phase conjugation [41] using DSFs and EDFAs, have been demonstrated in ultralong transmission systems by means of nonlinear optical effects in fibers [46–54]. The linear techniques, the subject of this chapter, require the optical power level in the transmission fiber to be below that required for the onset of the nonlinear optical effects, and they are briefly described below.

Wedding et al. demonstrated a dispersion-supported transmission technique capable of increasing the bit rate-distance of a conventional direct-modulation system fourfold [55]. This method is based on the interferometric conversion of the optical frequency modulation generated by a directly modulated laser to an amplitude modulation by means of fiber dispersion. Despite the simplicity and low cost of this scheme, it still imposes a limit on the achievable bit rate-distance.

Henmi et al. reported prechirp and modified prechirp techniques for improving the dispersion limitation of standard external-modulation systems [41]. The prechirp scheme is based on a combination of direct frequency modulation of the laser and external amplitude modulation for extending the dispersion-limited distance to nearly double. The modified prechirp technique is based on time division multiplexing of two independent prechirp RZ (return-to-zero) signals by utilizing polarization multiplexing techniques for increasing the dispersion-limited distance to nearly four times. Although these techniques are simple and cheap to implement, they are not applicable to direct-modulation systems, and still set a limit on the achievable bit rate-distance.

Poole et al. developed a broadband fiber-based compensating technique capable of achieving negative chromatic dispersion as large as −770 ps/nm/km at 1555 nm [56]. The fiber compensator was designed to operate in a higher-order mode near its cutoff wavelength at which large negative dispersion could be obtained to counteract its positive counterpart in a standard fiber. Despite the broadband capability of this device, its insertion loss is high due to the high attenuation loss (~5 dB/km) and the long fiber length required. To achieve effective dispersion compensation, this device requires high-efficiency mode converters to excite the desired higher-order mode.

Onishi et al. designed a dispersion-compensating fiber with a negative chromatic dispersion of −80 ps/nm/km at 1550 nm for countering its positive counterpart in a conventional fiber [57]. This technique is based on optimization of the refractive index profile of the fiber core and special fluorine doping in the cladding. The low negative dispersion and relatively high loss (0.32 dB/km) of the compensating fiber reflect its high cost for system implementation.

Linearly chirped fiber Bragg grating has been demonstrated as an all-fiber dispersion compensator [58]. The spectral components of the dispersive input pulse are reflected at different locations along the grating length and hence experience different group velocities. It is this frequency dependence of the group velocity of the compensator which leads to dispersion compensation. Long grating length, which results in reduction of the filter bandwidth, is necessary for compensation of a long dispersive fiber. The grating chirp must also be carefully designed to obtain the desired frequency responses. Although this device is compact and has low insertion loss, it is limited to single-wavelength applications because of its aperiodic frequency characteristics. Also, this compensator has limited design flexibility resulting in bandwidth limitation.

Cimini et al. proposed a reflective fiber Fabry–Perot filter for dispersion compensation [59]. To avoid degradation of the filter performance, its mirror reflectivity must be accurately designed. Despite the structural simplicity of the filter, it is inherently bandwidth-limited because its group-delay response is only linear over a narrow optical bandwidth as a result of its limited design flexibility. Nevertheless, impressive performance of the Fabry–Perot equalizer has been experimentally demonstrated in an 8 Gb/s 130 km external-modulation system [60].

Ozeki proposed a non-recursive Chebyshev optical equalizer for dispersion compensation [61]. A dispersive waveform, which has higher and lower frequency components at the front and end of the pulse, is passed through the equalizer which slows down and speeds up the higher and lower frequency components of the pulse. In this way, the dispersive pulse is compressed by the equalizer resulting in dispersion compensation. The Chebyshev equalizers using the PLCs [61] and birefringent crystals [62] have been demonstrated with impressive results for dispersion compensation in external-modulation systems.

Most of the linear dispersion-compensating techniques [55–57], excluding [56,58,59] described above, were only applied to externally modulated systems for the purpose of increasing data rate and transmission distance. When these techniques are used in direct-modulation systems, the transmission speed and distance are expected to be reduced [63]. Thus, they are only effective in the absence of laser chirp.

In this chapter, a digital eigenfilter approach [64] is applied to the design of linear optical dispersion compensators for overcoming the combined effect of laser chirp and fiber dispersion at 1550 nm in high-speed long-haul IM/DD lightwave systems. This technique is called optical dispersion eigencompensating and hence the acronym ODECs, because it is adopted from the digital eigenfilter technique. Most of the work presented here has been described by Ngo et al. [64–66].

The fiber loss is not considered so that the underlying principle of the eigen-compensating technique can be demonstrated.

5.2 EIGENFILTER DESIGN FOR DISPERSION COMPENSATION

5.2.1 Formulation of Dispersive Optical Fiber Channel

Neglecting the nonlinearity and loss of the optical fiber, the frequency response of the standard fiber, containing only the quadratic phase response, is given by [67]

$$H_f(\omega) = \exp\left[-j\hat{D}(\omega T - \omega_c T)^2\right] \tag{5.102}$$

where the dimensionless first-order fiber chromatic dispersion parameter is defined as

$$\hat{D} = \frac{DL\lambda^2}{4\pi c T^2} \tag{5.103}$$

where

$j = \sqrt{-1}$
ω is the angular optical frequency
ω_c the central angular frequency of the optical source
D (−17 ps/nm/km) the chromatic dispersion at the operating wavelength λ (1550nm) of a 1300 nm optimized fiber
L is the fiber length
c is the speed of light in free space
T is the sampling period of the eigenfilter (or the Chebyshev filter)

In Equation 5.103, the constant and linear phase terms have been neglected since they do not contribute to pulse distortion, and the cubic and higher-order phase terms have also been ignored as insignificant at 1550 nm. Full equalization of the dispersive fiber channel can be achieved by means of an optical dispersion compensator, whose frequency response is opposite to that of the fiber is given by

$$F(\omega) = \exp\left[+j\hat{D}(\omega T - \omega_c T)^2\right] \tag{5.104}$$

5.2.2 Formulation of Optical Dispersion Eigencompensation

The eigenfilter technique [66] is based on the finite-impulse response (FIR) filter method which has been widely used for equalization of phase distortions in digital transmission channels [68]. This scheme is described for the design of the ODECs with approximately unity magnitude response and prescribed phase response that are complementary to that of the fiber. The eigenfilter technique is a least squares approach used to minimize some measure of the difference between the phase response of the ODEC and the desired fiber phase response. It is based on the formulation of minimizing a quadratic measure of the frequency band errors in the magnitude and phase responses. The coefficients of the FIR ODEC can be obtained from computation of the eigenvectors that correspond to the smallest eigenvalues of the real, symmetric, and positive-definite matrices. In this way, the magnitude and phase responses of the fiber are simultaneously equalized resulting in very effective dispersion compensation.

The frequency response $H_{eq}(\omega)$ of an N-tap FIR filter with tap coefficients $a(n)$, with $n=0, 1, \ldots, N-1$, is given by [69]

$$\exp(jM\omega T)H_{eq}(\omega) = \exp(jM\omega T)\sum_{n=0}^{N-1} a(n)\exp(-jn\omega T) \tag{5.105}$$

where $M=(N-1)/2$ with N being an odd integer. The linear phase factor $\exp(jM\omega T)$ has been included on both sides of Equation 5.105 to facilitate the eigenfilter technique. The eigenfilter method requires the real and imaginary parts of this equation to approximate the real and imaginary parts, respectively, of the desired eigenfilter frequency response

$$F(\omega) = \exp\left(j\hat{\phi}(\omega)\right) \tag{5.106}$$

where

$$\hat{\phi}(\omega) = \hat{D}(\omega T - \omega_c T)^2 \tag{5.107}$$

is the desired phase function of the eigenfilter. The eigenfilter algorithm requires Equation 5.107 to be an even function about $\omega T=\pi/2$; thus, $\omega_c T=\pi/2$ must be chosen in all designs. By separately minimizing the real and imaginary parts of the error functions, the elements of matrix Q_R are given by [67]

$$q_R(n,m;\,\omega_0) = \int_0^{\pi}\left[\frac{\cos 2n(\omega_0 T)}{\cos\hat{\phi}(\omega_0)}\cos\hat{\phi}(\omega) - \cos 2n(\omega T)\right]\cdot \left[\frac{\cos 2m(\omega_0 T)}{\cos\hat{\phi}(\omega_0)}\cos\hat{\phi}(\omega) - \cos 2m(\omega T)\right]d(\omega T) \quad \text{for } 0 \le n, m \le I \tag{5.108a}$$

and the elements of matrix Q_I are given by

$$q_I(n,m;\,\omega_0) = \int_0^{\pi}\left[\frac{\sin(2n-1)(\omega_0 T)}{\sin\hat{\phi}(\omega_0)}\sin\hat{\phi}(\omega) - \sin(2n-1)(\omega T)\right]\cdot \left[\frac{\sin(2m-1)(\omega_0 T)}{\sin\hat{\phi}(\omega_0)}\sin\hat{\phi}(\omega) - \sin(2m-1)(\omega T)\right]d(\omega T) \quad \text{with } 1 \le n, m \le J \tag{5.108b}$$

The reference frequency ω_0 is arbitrary except that $\cos\hat{\phi}(\omega_0) \neq 0$ and $\sin\hat{\phi}(\omega_0) \neq 0$, e.g., $\omega_0 T = 0.32\pi$. The eigenvectors of matrices Q_R and Q_I, that correspond to the smallest eigenvalues, are the coefficients $u(n)$ and $v(n)$, respectively, which relate to the tap coefficients according to

$$u(n) = \begin{cases} a(M), & n = 0 \\ 2a(M-2n), & n = 1, \ldots, I \end{cases} \tag{5.109a}$$

$$v(n) = 2a(M-2n+1), \quad n = 1, \ldots, J \tag{5.109b}$$

where $I = M/2$ and $J = M/2$ for even M or $I = (M-1)/2$ and $J = (M+1)/2$ for odd M. Thus, the eigenfilter technique requires the group-delay response of $\exp(jM\omega T)H_{eq}(\omega)$ to approximate the desired filter group-delay response

$$G(\omega) = -\frac{d\hat{\phi}(\omega)}{d\omega} = -2\hat{D}T(\omega T - \omega_c T) \tag{5.110}$$

Consequently, the group-delay response of the eigenfilter $H_{eq}(\omega)$ approximates the total desired group-delay response

$$G_{total}(\omega) = MT - 2\hat{D}T(\omega T - \omega_c T) \tag{5.111}$$

where the constant group-delay factor MT, which is the pure propagation delay of the eigenfilter, has no effect on channel equalization.

5.2.3 Design

To show the effectiveness of the eigenfilter technique, one particular design is considered. In this example, $\hat{D} = -8/\pi$ that corresponds to the typical values $L = 100$ km and $T = 20.63$ ps according to Equation 5.103, and $N = 21$ is chosen for reasons to be given shortly. Using the eigenfilter approach described above, Table 5.4 shows the computed tap coefficients of the ODEC which are bipolar and less than unity. Note that the coefficients have even or odd symmetry about the center point $a(10)$.

Figure 5.47 shows that the ODEC magnitude response lies within 1.5 dB over the Nyquist bandwidth (or maximum filter-bandwidth), given by $\Delta f_{max} = 1/(2T)$ (or 24.24 GHz in this case), whereas Figure 5.47 shows that the ODEC group-delay

TABLE 5.4
For $0 \leq n, m \leq I$, Computed Tap Coefficients of the ODEC for $\hat{D} = -8/\pi$ ($L = 100$ km, $T = 20.63$ ps), and $N = 21$

n	0	1	2	3	4	5	6
$a(n)$	−0.0423	+0.0708	+0.1320	+0.1917	+0.3058	+0.3310	+0.2641
n	7	8	9	10	11	12	13
$a(n)$	+0.0423	−0.3585	−0.2308	+0.2642	+0.2308	−0.3585	+0.0347
n	14	15	16	17	18	19	20
$a(n)$	+0.2461	−0.3311	+0.3508	−0.1912	+0.1320	−0.0718	+0.0423

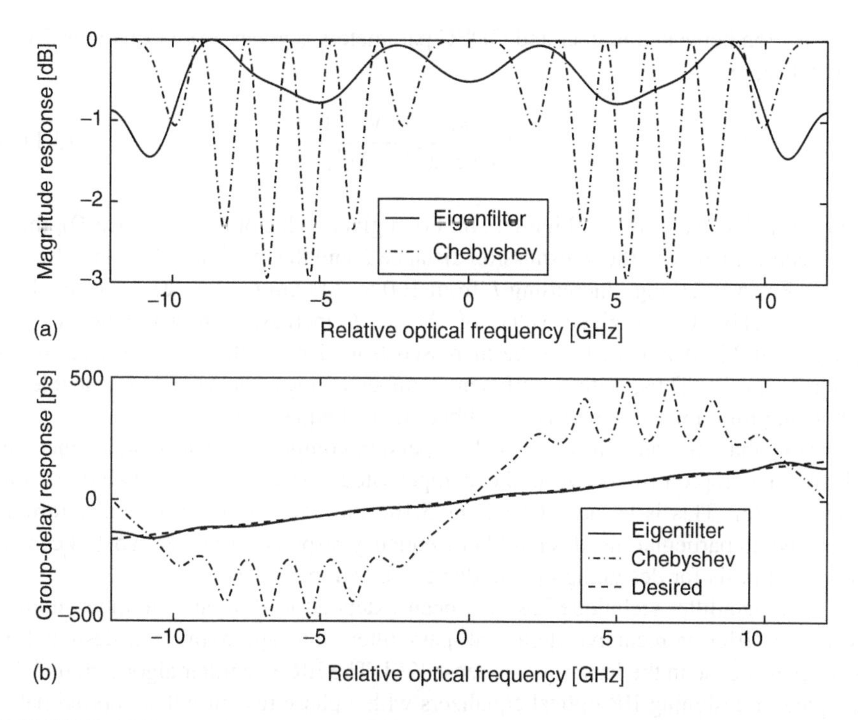

FIGURE 5.47 Frequency responses of the optical dispersion eigencompensator (ODEC) ($N=21$) with transfer function $\exp(jM\omega T)H_{eq}(\omega)$ and the Chebyshev equalizer ($N=12$) for $\hat{D}=-8/\pi$ ($L=100$ km, $T=20.63$ ps, $\Delta f_{max}=24.24$ GHz). (a) Magnitude response and (b) group-delay response over the maximum filter-bandwidth.

response approximates the desired group-delay response reasonably well over the Nyquist bandwidth. An ODEC dispersion of 1710 ps/nm is achieved over an operating bandwidth of $\Delta f_{op}=0.76\Delta f_{max}$, the frequency interval over which the ODEC group-delay response is linear to within 10%, which is the criteria used for choosing $N=21$. The maximum bandwidth of the ODEC can be obtained by increasing the number of taps, which also results in improving the accuracy of the frequency responses. For example, the maximum filter-bandwidth of $\Delta f_{op}=\Delta f_{max}=24.24$ GHz can be achieved with a 31-tap ODEC.

It is approximated from Table 5.5 that N increases linearly with $|\hat{D}|$ according to

$$N \cong 6.7|\hat{D}| + 4.3 \quad (13 \leq N \leq 47) \tag{5.112}$$

TABLE 5.5
Number of Taps N Required for Various Values of the Dispersion Parameter $\hat{D}$ for $\Delta f_{op} \geq 0.6\Delta f_{max}$

$\hat{D}$	$-4/\pi$	$-8/\pi$	-4	$-16/\pi$	$-20/\pi$
N	13	21	31	39	47

The relation between the fiber and ODEC parameters is given, from Equations 5.103 and 5.112, as

$$L = \left[\frac{\pi c}{6.7|D|\lambda^2}\right]\frac{(N-4.3)}{\Delta f_{\max}^2} \tag{5.113}$$

and is graphically shown in Figure 5.48. For a given value of N (and hence $\hat{D}$), there is a trade-off between the transmission distance L and the maximum filter-bandwidth $\Delta f_{\max}$. For $N=21$, e.g., increasing L from 100 to 200 km reduces $\Delta f_{\max}$ from 24.24 to 17.4 GHz. For a given value of $\Delta f_{\max}$, L increases linearly with N. For $\Delta f_{\max}=24.24$ GHz, e.g., L can be increased from 100 to 200 km by increasing N from 21 to 39. Thus, with a sufficient number of taps, the ODEC is capable of achieving high-bit rate long-distance fiber channel equalization.

Note that, for any linear optical dispersion-compensating technique, there is always a compromise between the compensated distance and bandwidth of the compensator. This is because Equations 5.102 through 5.104 are generally formulated and, in particular, the desired filter frequency response in Equation 5.104 can be used as a basis for designing optical dispersion compensators.

The eigenfilter technique has also been extended to designing infinite-impulse response (IIR) or recursive digital all-pass filters that approximate a desired frequency response in the least squares sense [70]. The IIR eigenfilter algorithm may be adapted in designing IIR optical equalizers with a phase response that approximates the desired phase function given in Equation 5.104. One possible optical synthesis of the IIR digital filters is to cascade the basic all-pole and all-zero optical filters to obtain the desired filtering characteristics as described in Chapter 6.

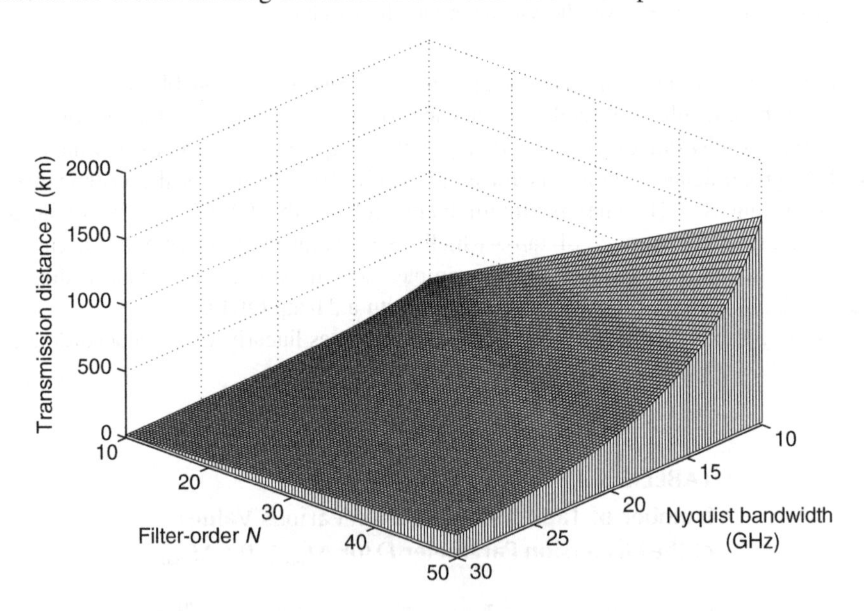

FIGURE 5.48 Plot showing the relation of the transmission distance L, filter-order N and maximum filter-bandwidth $\Delta f_{\max}$ of the ODEC.

5.2.4 Performance Comparison of Eigenfilter and Chebyshev Filter Techniques

The performance of the ODEC is compared with that of the Chebyshev optical equalizer in the frequency domain. The parameters of the Chebyshev equalizer using birefringent crystals are chosen as follows [62]: filter-order $N = 12$, optical axis rotation between adjacent birefringent crystals $\Theta = \pi/4$, and sampling period $T = 20.63$ ps. From Figure 5.47, the frequency responses of the ODEC compare favorably with those of the Chebyshev equalizer. For this particular Chebyshev filter design, optimum performances are achieved in both the frequency domain and the time domain. Increasing the order of the Chebyshev filter only deteriorates its performance instead of improving it. Unlike the eigenfilter technique, the Chebyshev technique is not systematic in the sense that its formulation does not approximate the desired magnitude and phase responses simultaneously. As a result, the Chebyshev equalizer is bandwidth-limited compared with the eigenfilter technique.

5.2.5 Synthesis of Optical Dispersion Eigencompensators

The ODECs can be synthesized using a P-tap FIR coherent optical filter (see Figure 5.49), which has been used in the synthesis of the higher-derivative FIR optical differentiators as described in Chapter 3. The FIR filter essentially consists of a $1 \times P$ optical splitter, a $P \times 1$ optical combiner, and P waveguide delay lines, into each of which a tunable coupler (TC) and a phase shifter (PS) are incorporated. A dispersive signal entering the splitter will be evenly distributed to P signals, which are then appropriately delayed by the delay lines and weighted by the TCs and PSs. These signals are then coherently collected by the combiner to generate the compensated signal.

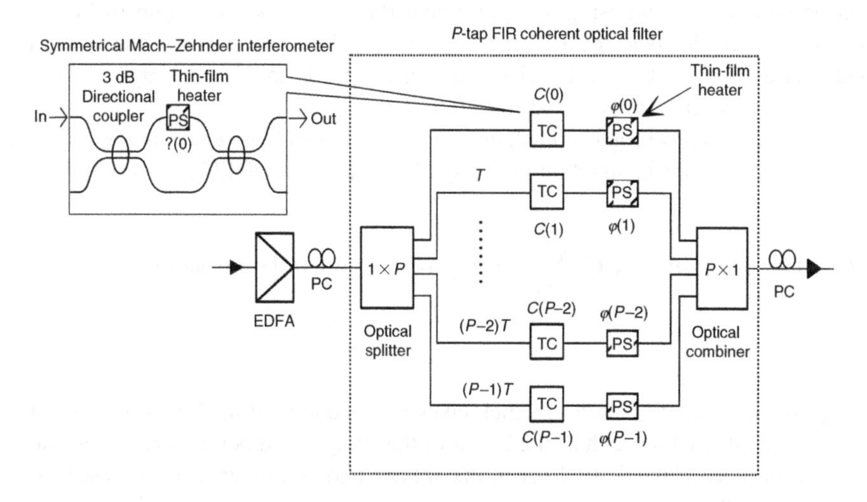

FIGURE 5.49 Schematic diagram of the P-tap FIR coherent optical filter used to synthesize the ODECs using the PLC technology. TC: tunable coupler, PS: phase shifter, and PC: polarization controller.

The ODECs can be constructed using the PLC technology, namely, silica-based waveguides embedded on a silicon substrate. In each kth delay line, the PS following the TC is a waveguide with a thin-film heater attached and uses the thermo-optic effect to induce a carrier phase change of $\phi(k)$. The PS can use the temperature dependence of the refractive index of the silica waveguide to compensate for any optical path-length differences of the waveguides. The TC is a symmetrical Mach–Zehnder interferometer (see the inset of Figure 5.49). It consists of two 3 dB directional couplers (DCs), two equal waveguide arms, and a thin-film heater, with a carrier phase change of $\varphi(k)$, attached to one of the arms to control the output amplitude.

Neglecting the insertion loss of the 3 dB DCs, the propagation delay and waveguide birefringence of the TC, the kth TC transfer function is given by

$$C(k) = |C(k)| \exp[j\angle C(k)] = 0.5\{\exp[j\varphi(k)] - 1\} \tag{5.114a}$$

where $\angle C(k)$ denotes the argument of $C(k)$,

$$|C(k)| = \sqrt{0.5 - 0.5\cos[\varphi(k)]} \tag{5.114b}$$

$$\varphi(k) = \cos^{-1}\left[1 - 2|C(k)|^2\right] \tag{5.114c}$$

and

$$\angle C(k) = \tan^{-1}\left(\frac{\sin[\varphi(k)]}{\cos[\varphi(k) - 1}\right) \tag{5.114d}$$

for $k = 0, 1, \ldots, P-1$. Equation 5.114b shows that the required amplitude of the TC can be obtained by choosing an appropriate PS phase $\varphi(k)$ according to Equation 5.114c, and this leads to the TC phase as given by Equation 5.114d. The amplitude and phase of the TC can be changed from 0 to 1 and from $-\pi/2$ to $+\pi/2$, respectively, when $\varphi(k)$ is varied from 0 to 2π.

Assuming an isotropic and neglecting the propagation delay and waveguide anisotropy, the transfer function of the synthesized P-tap ODEC is given by

$$H_{\text{eq}}(\omega) = \sqrt{l_{\text{path}}} \cdot P^{-1} \cdot \sqrt{G} \cdot \sum_{k=0}^{P-1} (-1)^k \cdot |C(k)| \exp[j\angle C(k)] \cdot \exp[j\phi(k)] \cdot \exp(-jk\omega T) \tag{5.115}$$

where

l_{path} is the intensity path loss which takes into account all the losses associated with each delay line such as the losses of the straight and bend waveguides, the insertion loss of the 3 dB DCs in the splitter and combiner, and the insertion loss of the TC

P^{-1} is the coupling loss factor as a result of a 3 dB coupling loss at each stage of the splitter and combiner

G is the intensity gain of the EDFA
$(-1)^k = \exp(jk\pi)$ is the phase shift factor due to the $\pi/2$ cross-coupled phase shift of the 3 dB DCs in the splitter and combiner
T, the sampling period of the ODEC, is the differential delay between neighboring delay lines

Optical synthesis of the digital coefficients requires the equality of Equations 5.111 and 5.112 such that

$$G = \frac{P^2}{l_{\text{path}}}, \quad |C(k)| = |a(k)| \quad \text{and} \quad \phi(k) = \angle a(k) - \angle(-1)^k - \angle C(k) \tag{5.116}$$

where $\angle a(k)$, $\angle(-1)^k \in (0,\pi)$. Thus, the insertion loss of the ODEC can be overcome by the EDFA with gain as given in Equation 5.113. In each kth delay line, Equation 5.113 shows that the amplitude of the digital coefficient can be optically implemented by the TC amplitude, and the PS must provide a phase shift according to Equation 5.113. Note that the TC amplitude can also incorporate, in addition to $|a(k)|$, the deviations of the coupling coefficients of the couplers from the desired 3 dB values as a result of fabrication errors.

5.2.6 IM/DD Transmission System Model

Figure 5.50 shows a block diagram of a 1550 nm IM/DD lightwave system with the ODEC. The optical source is assumed to be a single-longitudinal mode semiconductor laser, which is directly modulated by an 8 Gb/s non-return-to-zero (NRZ) (or RZ) with 50% duty cycle) current pulse with a pseudorandom 6-bit pattern of length 64.

$$i_c(t) = I_{\text{bias}} + \sum_{k=-\infty}^{\infty} A_k i_p(t - kT_b) \tag{5.117}$$

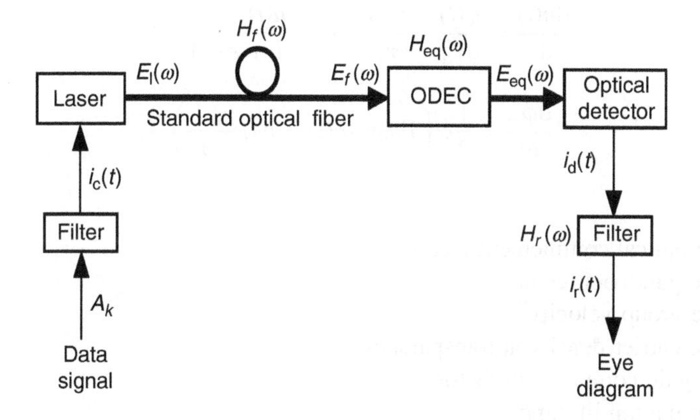

FIGURE 5.50 Block diagram of a 1550 nm IM/DD optical communication system with the ODEC. The injected laser current $i_c(t)$ is a pseudorandom digital pulse waveform.

where

- I_{bias} is the bias current
- $A_k \in (0,1)$ is the digital sequence
- $i_p(t)$ is the current pulse shape
- T_b is the bit period which is the inverse of the bit rate (or $1/T_b = 8$ Gb/s)

If a binary one (zero) is transmitted during the kth time interval, then $A_k = 1$ ($A_k = 0$). The current pulse shape is of the form

$$i_p(t) = \begin{cases} 0, & t < 0 \\ I_m[1 - \exp(-t/t_r)], & 0 \le t \le T_b' \\ I_m \exp\left[-(t - T_b')/t_r\right], & t > T_b' \end{cases} \tag{5.118}$$

where

- $T_b' = T_b$ for the NRZ encoding format and $T_b' = T_b/2$ for the RZ encoding format
- I_m is the peak modulation current
- t_r is the pulse rise and fall times (approximately 15%–85%)

Note that the injected laser current $i_c(t)$ is assumed to be the response of a first-order electrical filter when subject to an input signal A_k.

The optical field amplitude and phase of the semiconductor laser [34,36], in response to the injected current $i_c(t)$, are determined by numerically solving the large-signal laser rate equations describing the nonlinear modulation dynamics of the device. A fourth–fifth order Runge–Kutta algorithm is used to integrate the coupled set of first-order differential equations for the photon density $s(t)$, carrier density $n(t)$, and optical phase $\phi_m(t)$

$$\frac{ds(t)}{dt} = \left(\Gamma a_0 v_g \frac{n(t) - n_0}{1 + \varepsilon s(t)} - \frac{1}{\tau_p}\right) s(t) + \frac{\beta \Gamma n(t)}{\tau_n} \tag{5.119a}$$

$$\frac{dn(t)}{dt} = \frac{i_c(t)}{qV_a} - \frac{n(t)}{\tau_n} - a_0 v_g \frac{n(t) - n_0}{1 + \varepsilon s(t)} s(t) \tag{5.119b}$$

$$\frac{d\phi_m}{dt} = \frac{1}{2}\alpha\left(\Gamma a_0 v_g (n(t) - n_0) - \frac{1}{\tau_p}\right) \tag{5.119c}$$

where

- Γ is the optical confinement factor
- a_0 is the gain coefficient
- v_g is the group velocity
- n_0 is the carrier density at transparency
- ε is the gain compression factor
- τ_p is the photon lifetime
- β is the fraction of spontaneous emission coupled into the lasing mode
- τ_n is the carrier lifetime

q is the electron charge
V_a is the active layer volume
α is the linewidth enhancement factor or laser chirp parameter

The time variations of the optical power and laser chirp, respectively, are given by

$$m(t) = 0.5s(t)V_a\eta_0 hf_c/(\Gamma\tau_p) \tag{5.120}$$

$$\Delta v(t) = \frac{1}{2\pi}\frac{d\phi_m(t)}{dt} \tag{5.121}$$

where hf_c is the photon energy at the carrier frequency f_c and η_0 is the total differential quantum efficiency. The time-dependent electric-field amplitude at the laser output is given by

$$e_l(t) = \sqrt{m(t)}\exp[j\phi_m(t)] \tag{5.122}$$

which can be obtained by numerically solving Equations 5.114 through 5.116.

The frequency-dependent electric-field amplitude at the output of the dispersive fiber channel is given by

$$E_f(\omega) = E_l(\omega)H_f(\omega) \tag{5.123}$$

where $E_l(\omega) = \mathrm{FFT}\{e_l(t)\}$ with FFT denoting fast Fourier transform. The frequency-dependent electric-field amplitude at the output of the ODEC is given by

$$E_{eq}(\omega) = E_f(\omega)H_{eq}(\omega) \tag{5.124}$$

The time-dependent optical intensity at the output of the optical detector is given by

$$i_d(t) = \left|\mathrm{IFFT}\{E_{eq}(\omega)\}\right|^2 \tag{5.125}$$

where IFFT denotes the inverse FFT operation. The time-dependent filtered signal at the output of the baseband receiver filter is given by

$$i_r(t) = \mathrm{IFFT}\{\mathrm{FFT}\{i_d(t)\}H_r(\omega)\} \tag{5.126}$$

The receiver filter is a third-order Butterworth electrical lowpass filter, whose transfer function is given by

$$H_r(\omega) = \frac{1}{(S+1)(S+0.5-j0.866)(S+0.5+j0.866)} \tag{5.127}$$

where $S = j\omega/\omega_r$ is the S-transform parameter with ω_r being the 3 dB angular bandwidth. The 3 dB bandwidth of the receiver filter is $\omega_r/(2\pi) = 0.65/T_b = 5.2$ GHz for the NRZ signal or $\omega_r/(2\pi) = 0.55/T_b = 4.4$ GHz for the RZ signal.

The third derivative of the fiber propagation constant, typically $\beta_3 = 0.1\ \text{ps}^3/\text{km}$ [31], was also included in the simulation model although it has been neglected in Equation 5.102 for the purpose of facilitating the eigenfilter design. However, it was found to have negligible effect on the system performance. The parameter values used to obtain the numerical results are given in Table 5.6.

5.2.7 Performance Comparison of Optical Dispersion Eigencompensator and Chebyshev Optical Equalizer

The performances of the ODEC and Chebyshev equalizer are assessed in the time domain by means of the dispersion-induced optical power penalty P_{D} (in dB) at the baseband filter output, which is defined as [71]

$$P_{\text{D}} = 10 \log(y/x) \tag{5.128}$$

where x is the ideal eye-opening at the receiver baseband filter output with no fiber in place, and y is the eye-opening of a particular system under study.

Figure 5.51 shows the NRZ eye pattern diagrams of an 8 Gb/s 100 km 1550 nm uncompensated and compensated external-modulation system, in which the laser chirp factor is $\alpha = 0$ and the source spectrum is about 6 GHz. The ODEC and the Chebyshev equalizer improve the dispersion-induced optical power penalty from

TABLE 5.6
List of Parameters

Symbol	Parameter Value
I_{th} (threshold current)	29.8 mA
$I_{\text{m}} = I_{\text{th}}$	29.8 mA
$I_{\text{bias}} = 1.05 I_{\text{th}}$	31.3 mA
Γ	0.4
v_{g}	$7.5 \times 10^9\ \text{cm s}^{-1}$
a_0	$2.5 \times 10^{-16}\ \text{cm}^2$
n_0	$1.0 \times 10^{18}\ \text{cm}^{-3}$
ε	$6.0 \times 10^{-17}\ \text{cm}^3$
τ_{p}	0.9 ps
β	2.0×10^{-4}
τ_{n}	1.0 ns
V_{a}	$0.75 \times 10^{-10}\ \text{cm}^3$
η_0	0.13
q	1.6022×10^{-19} C
h	6.6262×10^{-34} Js
λ	1550 nm
D	−17 ps/nm/km
T_{b}	125 ps
$t_{\text{r}} = 0.15 T_{\text{b}}$	18.75 ps
α	$0 \leq \alpha \leq 6$

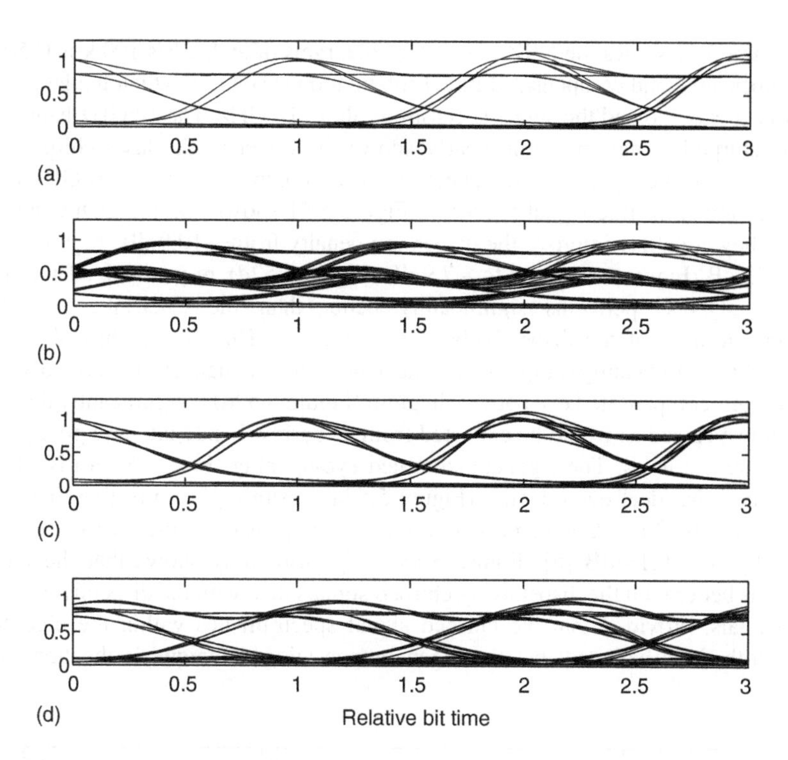

FIGURE 5.51 NRZ eye pattern diagrams of an 8 Gb/s 1550 nm IM/DD lightwave system for $\hat{D} = -8/\pi$ and $\alpha = 0$. (a) Uncompensated link ($L = 0$ km), (b) uncompensated link ($L = 100$ km), (c) compensated link ($L = 100$ km) using the ODEC with $T = 20.63$ ps and $N = 21$, and (d) compensated link ($L = 100$ km) using the Chebyshev equalizer with $T = 20.63$ ps, $\Theta = \pi/4$, and $N = 12$.

$P_D = -3.2$ dB to $P_D = -0.18$ dB and to $P_D = -0.77$ dB, respectively. This shows a 3.02 dB power improvement provided by the ODEC as compared with 2.43 dB by the Chebyshev equalizer. The superiority of the eigenfilter technique to the Chebyshev technique is as expected because the 18 GHz bandwidth provided by the ODEC is more than enough to cover the whole signal spectrum of 6 GHz, whereas the linear region of the group-delay response of the Chebyshev equalizer is less than 5 GHz. Furthermore, the Chebyshev equalizer can only provide partial dispersion compensation because its dispersion (or the gradient of the linear region of its group-delay response) is not closely matched to the desired filter dispersion. This explains why the Chebyshev-compensated eye pattern diagram in Figure 5.51d is slightly closed although detectable when compared with the eigencompensated eye pattern diagram in Figure 5.51c, which resembles the ideal eye pattern diagram in Figure 5.51a. The fact that the ODEC still experiences a penalty of -0.18 dB, even though its bandwidth covers the whole signal spectrum, may be due to a very small error in the group-delay response, resulting in a very small uncompensated residual dispersion.

This case study shows that, for an ideal chirp-free system, the ODEC outperforms the Chebyshev equalizer by 0.59 dB in the time domain.

Figure 5.51 shows the NRZ eye pattern diagrams of an 8 Gb/s 100 km 1550 nm uncompensated and compensated direct-modulation system, in which the laser chirp parameter is $\alpha = 6$ and the signal spectrum is about 18 GHz. The 18 GHz bandwidth of the chirped signal spectrum clearly shows the effect of the laser chirp on the broadening of the optical source spectrum when compared with the 6 GHz bandwidth of the chirp-free signal spectrum. Figure 5.51 shows that the ODEC and the Chebyshev equalizer improve the dispersion penalty from −10.9 dB (Figure 5.52b) to +0.72 dB (Figure 5.52c) and to −7.9 dB (Figure 5.52d), respectively. This shows that the ODEC performs significantly better than the Chebyshev equalizer in compensation of the dispersively chirped signals. Thus, for a chirped system, the ODEC significantly outperforms the Chebyshev equalizer by 8.62 dB. The Chebyshev-compensated eye pattern diagram (Figure 5.52d) is significantly distorted and almost closed as would be expected from its narrow bandwidth and poor group-delay characteristics. The eigencompensated eye-opening (Figure 5.53c) is slightly wider than the ideal eye-opening (Figure 5.53a), resulting in a positive dispersion penalty of +0.72 dB. Compared with the eigencompensated chirp-free system with a penalty of −0.18 dB [5] (Figure 5.53c, this case study shows that the ODEC performs better with the dispersively chirped signals than with the dispersively chirp-free signals, provided that the chirped signal spectrum lies within the eigenfilter bandwidth. It is believed that the eigencompensating technique is the first linear

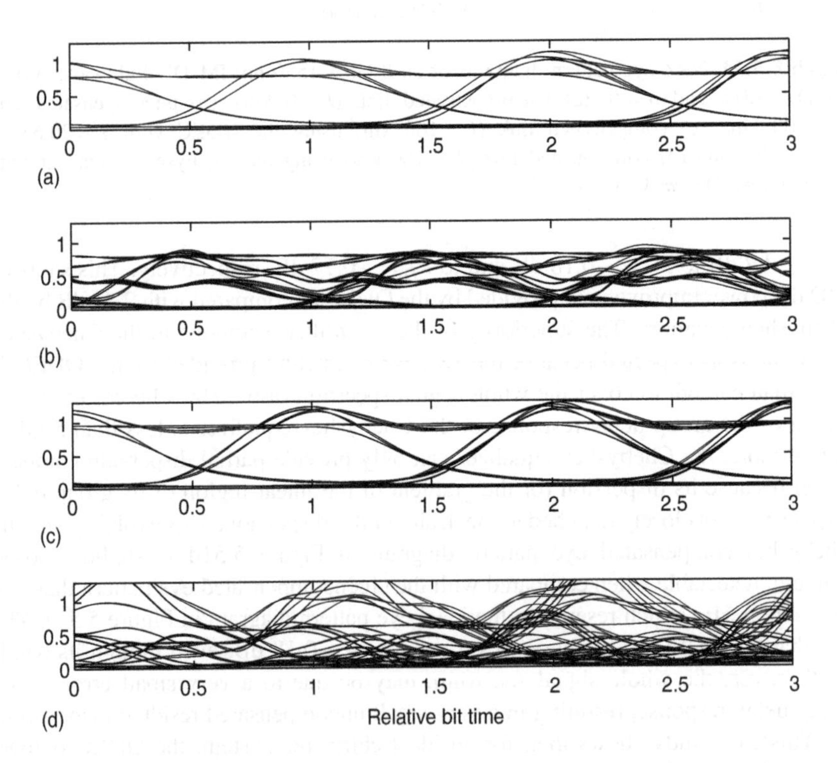

FIGURE 5.52 The conditions are exactly the same as those in Figure 5.51 except that $\alpha = 6$.

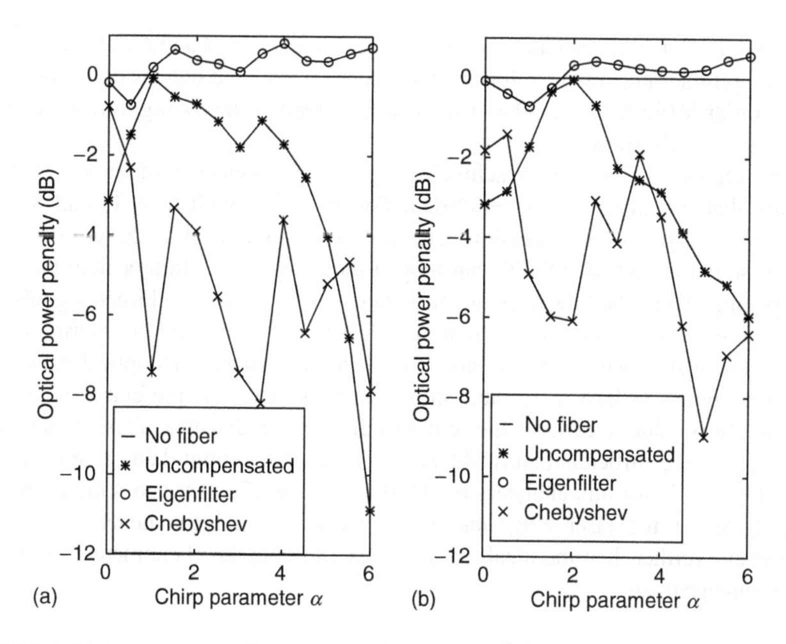

FIGURE 5.53 Dependence of the dispersion-induced ($\hat{D} = -8/\pi$) optical power penalty on the laser chirp parameter for the uncompensated link ($L = 0$ km), the uncompensated link ($L = 100$ km), the compensated link ($L = 100$ km) using the ODEC with $T = 20.63$ ps and $N = 21$, and the compensated link ($L = 100$ km) using the Chebyshev equalizer with $T = 20.63$ ps, $\Theta = \pi/4$, and $N = 12$. (a) An 8 Gb/s 1550 nm system with the NRZ signal, (b) an 8 Gb/s 1550 nm system with the RZ signal.

optical dispersion-compensating scheme that actually "likes" to work with chirped signals, instead of "hates" as in other techniques [56,57,72].

Figures 5.51a and b show the dispersion penalties of an 8 Gb/s 100 km 1550 nm uncompensated and compensated system using the NRZ and RZ signals, respectively. In these figures, α varies from 0 to 6 in step of 0.5, and this leads to varying the signal spectrum from 6 to 18 GHz for the NRZ signal and from 8 to 20 GHz for the RZ signal. There is an optimum value of the chirp parameter when the dispersion penalty of the uncompensated link is minimum. The uncompensated link experiences a minimum penalty of −0.08 dB for the NRZ signal with $\alpha = 1$ (which also agrees with the experimental result that the chirp parameter must be $\alpha < 1$ to achieve the longest transmission distance) and of −0.06 dB for the RZ signal with $\alpha = 2$. Some means of dispersion compensation are obviously required for large departures from the optimum chirp value.

For $\alpha = 0$, the Chebyshev equalizer improves the penalties from −3.2 to −0.77 dB for the NRZ signal and from −3.2 to −1.8 dB for the RZ signal. However, the performance of the Chebyshev equalizer deteriorates significantly for $\alpha \geq 1$ and, in most cases, becomes worse than that of the uncompensated link. This study shows that the interaction of the poor filter group-delay characteristics, laser chirp, and fiber dispersion further degrades the uncompensated eye pattern diagram, instead of improving it. Note that the Chebyshev equalizer with $\Theta = \pi/4$ and $N = 12$ was found to yield optimum performance for this particular system, and increasing (or decreasing) its order N (for the same $\Theta = \pi/4$) would, instead of improving, further degrade the eye pattern diagram.

The ODEC is capable of achieving significant reopening of an eye pattern diagram that would otherwise be closed. For the NRZ (with $\alpha \geq 1$) and the RZ (with $\alpha \geq 2$) signals, the dispersion penalties are positive for the eigencompensated system, showing that the ODEC can reopen the receiver eye further than the ideal eye-opening. Thus, the ODEC performs better with dispersively chirped signals and can hence be used in external-modulation high-speed systems without sacrificing for the data rate and transmission distance. This is because an external optical modulator can be designed to have a residual chirp, and, in fact, it is preferable to have a nonzero chirp value to achieve the longest transmission distance. This advantage is clearly due to the sufficient bandwidth and high-accuracy group-delay characteristics of the ODEC. From this analysis, the ODEC can significantly and simultaneously compensate for the laser chirp and fiber dispersion, and can hence reopen the receiver eye further than the ideal eye-opening, showing the phenomenon of optical power enhancement.

5.2.8 Eigencompensated System with Parameter Deviations of the Optical Dispersion Eigencompensator

The combined effect of the deviation of three ODEC parameters on the performance of the eigencompensated system can now be considered. Deviation of filter parameters will arise in practice due to fabrication error of device parameters. The deviation of the ODEC parameters to be considered are the TC amplitude (see Equation 5.114b), the PS phase following the TC (see Equation 5.114c), and the waveguide delay line.

Figures 5.54a and b show the dispersion penalties of an 8 Gb/s 100 km 1550 nm eigencompensated link ($\hat{D} = -8/\pi$, $\alpha = 6$) using the nonideal ODEC for the NRZ and RZ signals, respectively. The legend boxes represent, respectively, the percentage deviations of the TC amplitude and PS phase from below or above their nominal values. The TC amplitude and PS phase deviations are randomly chosen to be (0%, 0%), (±5%, ±5%), and (±10%, ±10%), and the delay-line deviation is randomly varied from 0% to ±5%.

The dispersion penalties of the nonideal ODEC vary from −2.5 to +0.72 dB for the NRZ signal and from −2 to +1.2 dB for the RZ signal. For the 0% delay-line deviation, Figure 5.54a indicates that the power enhancement is reduced from the ideal case of +0.72 to +0.43 dB for the (±5%, ±5%) case and to +0.39 dB for the (±10%, ±10%) case. For the ±5% delay-line deviation, the power penalty of the (0%, 0%) case, i.e., −2.5 dB, is surprisingly more severe than those of the (±5%, ±5%) case, i.e., −1.1 dB, and the (±10%, ±10%) case, i.e., −1.9 dB. The randomness of the parameter deviations makes it difficult to provide a reasonable explanation for this discrepancy. Similar explanations can be made for the case of the RZ signal as shown in Figure 5.54b. For a 1 dB penalty, the eigencompensated performance allows a ±2% delay-line deviation for the NRZ signal and a ±4%

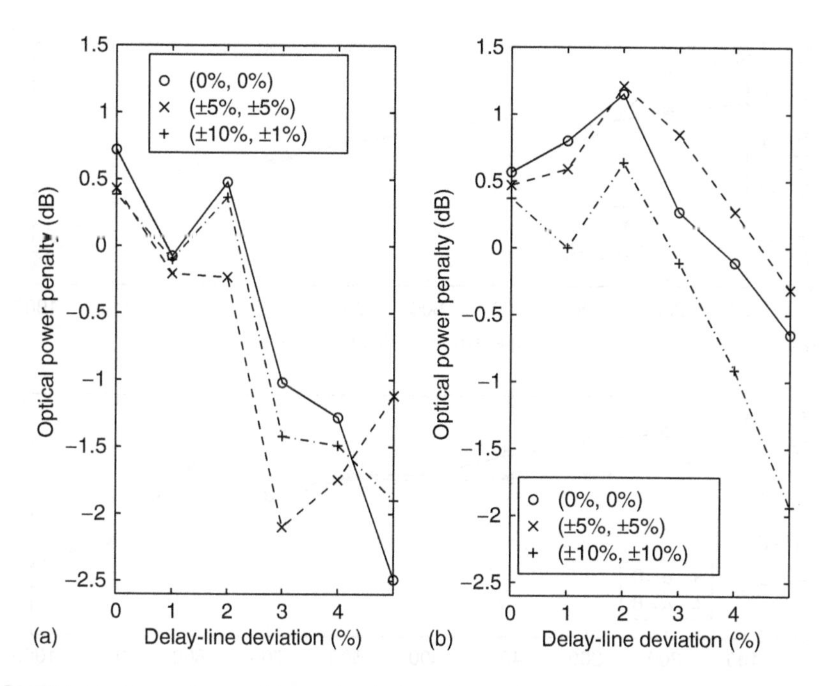

FIGURE 5.54 Dependence of dispersion-induced optical power penalty on delay-line deviation of the ODEC for an 8 Gb/s 100 km 1550 nm direct-modulation system ($\hat{D} = -8/\pi, \alpha = 6$). The curves correspond to the nonideal ODEC ($T = 20.63$ps, $N = 21$) with the TC amplitude and PS phase deviations of (0%, 0%), (±5%, ±5%), and (± 10%, ±10%), respectively. (a) NRZ signal and (b) RZ signal.

delay-line deviation for the RZ signal, and the TC amplitude and PS phase deviations of up to (±10%, ±10%) for both cases. It has been claimed that the waveguide delay line could be accurately fabricated to within 1% deviation. The effect of the delay-line deviation on the performance of the nonideal ODEC can thus be neglected, and the TC amplitude and PS phase deviations may be allowed to more than (±10%, ±10%) with less than 1 dB penalty.

5.2.9 Trade-Off between Transmission Distance and Eigenfilter Bandwidth

Section 5.2.3 has shown that there is a compromise between the eigencompensated distance and the eigenfilter bandwidth for a fixed number of taps N. Such an effect on the eigencompensated performance is now investigated.

Figure 5.55 shows the variation of the bandwidth-limited optical power penalty with the transmission distance for an 8 Gb/s 1550 nm eigencompensated system ($N = 21$, $\hat{D} = -8/\pi$) for both the NRZ and RZ signals with $\alpha = 0$ and $\alpha = 6$. As expected from Equation 5.113, increasing the distance from $L = 100$ to 1000 km increases the power penalty, since the eigenfilter bandwidth is reduced from $\Delta f_{max} = 24.24$ GHz ($T = 20.63$ ps) to $\Delta f_{max} = 7.67$ GHz ($T = 65.3$ ps). For $L > 100$ km, the vertical

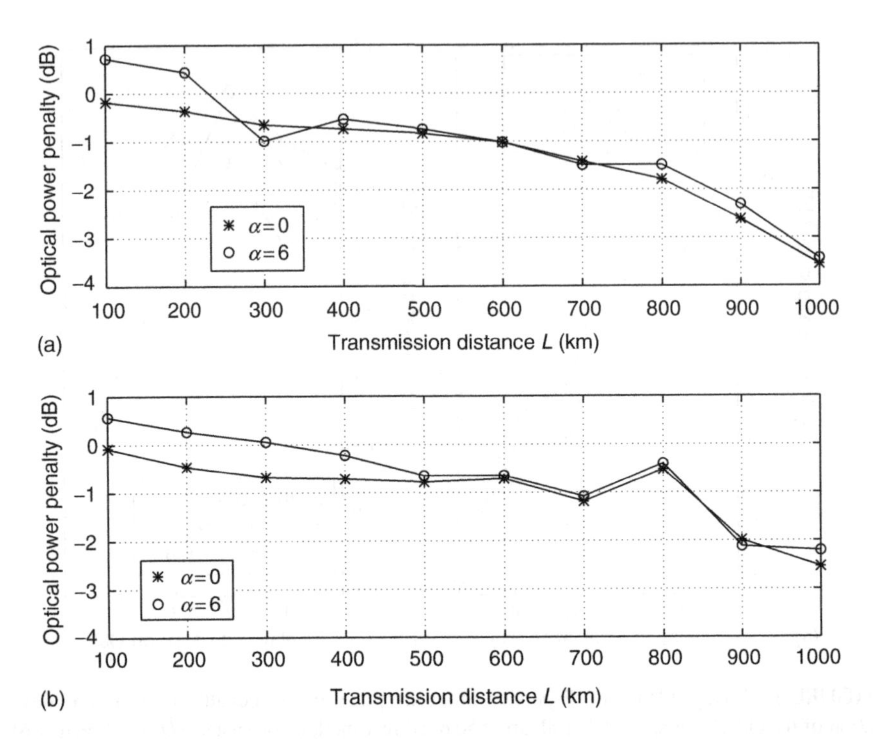

FIGURE 5.55 Dependence of bandwidth-limited optical power penalty on transmission distance of an 8 Gb/s 1550 nm IM/DD system with ODEC ($N = 21$, $\hat{D} = -8/\pi$) for $\alpha = 0$ and $\alpha = 6$. (a) NRZ signal and (b) RZ signal.

amplitudes of the ideal and eigencompensated eye pattern diagrams, which are still widely open, are smaller than those of the 100 km case. This can be explained in terms of either a reduction in bandwidth Δf_{max} or an increase in sampling period T of the whole system because of the interrelation between Δf_{max} and T. The sampling period must be chosen to be sufficiently small to capture most of the encoded information for transmission or the compensator bandwidth must be designed to be sufficiently large to cover the signal spectrum. Thus, the ideal eye pattern diagram of the 100 km case is used as a basis for computing the bandwidth-limited (but not dispersion-limited) optical power penalty for the $L > 100$ km cases.

For distances up to 200 km for the NRZ signal and up to 400 km for the RZ signal, the eigencompensated system performs significantly better with chirped than with chirp-free signals. For longer distances, the eigencompensated performance shows a little difference for both cases because of the reduced eigenfilter bandwidth. For a 1 dB bandwidth-limited penalty, error-free transmission may still be possible for this particular eigencompensated system with distances up to 600 km for both the NRZ and RZ chirped signals. The peaks and dips of these curves are difficult to explain because of the complicated effect of the interaction of the laser chirp, fiber dispersion, and eigenfilter group delay. However, they provide useful information about the optimum eigencompensated performance for a particular transmission distance, which can be greatly increased, without any reduction in bandwidth, by increasing the eigenfilter tap according to Equation 5.113.

5.2.10 Compensation Power of Eigencompensating Technique

The performance of the eigencompensated system is now characterized by means of the compensation power (CP) of the ODEC. Section 5.2.9 has shown that, for a 1 dB bandwidth-limited penalty, the maximum eigencompensated distance is about 600 km for both the NRZ and RZ chirped signals. In this case, $N = 21$, $\hat{D} = -8/\pi$, $L = 600$ km, $T = 50.56$ ps, and $\Delta f_{max} = 9.89$ GHz $= 1.24B$, where B denotes the bit rate. For a 1 dB bandwidth-limited penalty, the product of the square of the bit rate and the eigencompensated distance for both the NRZ and RZ chirped signals is thus given by

$$B^2L \leq 0.6(N - 4.3)\left(\frac{c}{2|D|\lambda^2}\right) \tag{5.129}$$

when $\Delta f_{max} \geq 1.24B$ has been substituted into Equation 5.113. For an uncompensated system, the B^2L value is given by [73]

$$B^2L \leq \frac{\pi}{\left[2(1+\alpha^2)^{1/2} + 2\alpha\right]}\left(\frac{c}{2|D|\lambda^2}\right) \tag{5.130}$$

which decreases with increasing chirp parameter α. For $\alpha = 0$, Equation 5.129 is almost the same as that defined in Ref. [70] and a factor of $\pi/2$ more than that given in Ref. [74], where a 1 dB penalty was used as a criterion in both references. Equations 5.129 and 5.130 are graphically illustrated in Figure 5.56, where the dispersion limits for both $\alpha = 0$ and $\alpha = 6$ are shown to be significantly improved by the ODEC. For example, for $B = 8$ Gb/s, the uncompensated distance is limited to only 7.5 km for $\alpha = 6$ but is extended to 90 km for $\alpha = 0$. However, the presence of the ODEC significantly extends the transmission distance to 575 km for $N = 21$ and $B = 8$ Gb/s.

The overall performance of the eigencompensated system can be characterized by the compensation power of the ODEC, which is a measure of the increase in the B^2L value and is defined as the ratio of Equations 5.129 and 5.130 as

$$\mathrm{CP} = 0.19(N - 4.3)\left[2(1+\alpha^2)^{1/2} + 2\alpha\right] \tag{5.131}$$

Figure 5.57 shows that the CP value increases with the eigenfilter order as well as with the laser chirp parameter. The above example gives $\mathrm{CP} = 77$ for $\alpha = 6$, showing that the B^2L value of the eigencompensated direct-modulation system is improved by a factor of 77.

The reasons for comparing the eigenfilter technique with the Chebyshev technique are twofold. First, the Chebyshev technique [62,69] has a greater design flexibility than those of the chirped fiber Bragg grating [63] and the Fabry–Perot equalizer [75,76] in modifying the filter frequency response. Second, the ODEC and

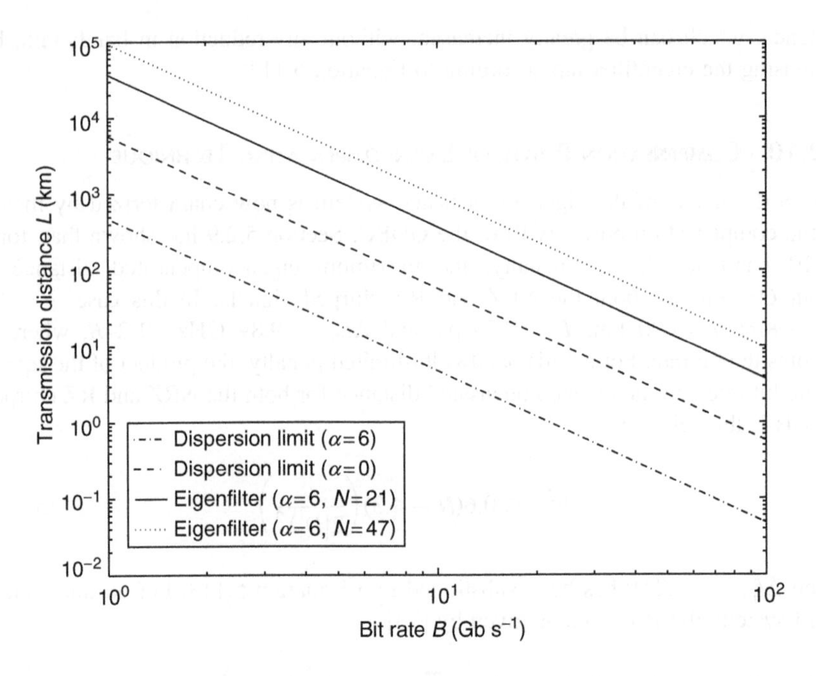

FIGURE 5.56 Dispersion limits for the uncompensated transmission using external modulation ($\alpha = 0$) and direct modulation ($\alpha = 6$), and dispersion improvements for the eigencompensated transmissions using direct modulation ($\alpha = 6$).

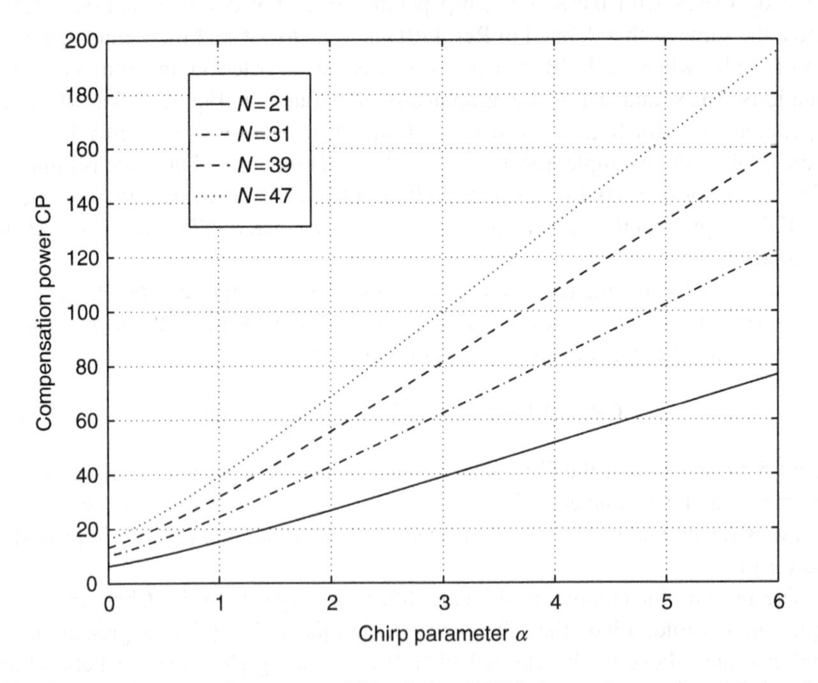

FIGURE 5.57 Compensation power (CP) of the ODEC for various values of the chirp parameter and filter order.

the Chebyshev equalizer have common features, such as linearity, nonrecursiveness, and periodic frequency response, and they both can be implemented using the same PLC technology. As briefly described in Section 5.2.7, there are advantages as well as disadvantages associated with various techniques, so that it is difficult to make a fair comparison between them. However, the unique feature of the ODEC is its ability to perform better with dispersively chirped signals than with dispersively chirp-free signals, an advantage which has not yet been claimed by other methods. In addition, the ODEC is compact, capable of operating stably and can perform high-speed signal compensation because of its integrated-optic form. Further progress in the PLC technology would make the ODEC even more attractive.

5.2.11 Remarks

- An effective digital eigenfilter approach has been employed to design linear ODECs for compensation of the combined effect of laser chirp and fiber dispersion at 1550 nm in high-bit rate long-distance IM/DD lightwave systems.
- The ODECs, which have been synthesized using an integrated-optic transversal filter, are very effective in equalization of a dispersively chirped optical communication channel.
- In an 8 Gb/s 100 km 1550 nm system, the performance of the ODEC is more impressive than that of the Chebyshev equalizer in both the frequency and time domains. The ODEC slightly outperforms the Chebyshev equalizer for external-modulation transmission but significantly outperforms the Chebyshev equalizer for direct-modulation transmission.
- For a 1 dB power penalty, a 21-tap ODEC may provide error-free transmission of an 8 Gb/s 1550 nm direct-modulation system for distances up to 600 km for both the NRZ and RZ chirped signals.
- The ODEC has a large value of compensation power, which increases with the eigenfilter order and also with the laser chirp parameter.
- The combined effect of the laser chirp, fiber dispersion, and ODEC group delay can reopen the receiver data eye further than that of the ideal eye-opening, resulting in the phenomenon of optical power enhancement, a feature which is not available with other compensation techniques.

REFERENCES

1. S.J. Mason, Feedback theory—some properties of signal flow graphs, *Proc. IRE*, 41, 1144–1156, September 1953.
2. S.J. Mason, Feedback theory—further properties of signal-flow graphs, *Proc. IRE*, 44, 920–926, July 1956.
3. A.V. Oppenheim and R.W. Schafer, *Discrete-Time Signal Processing*, Englewood Cliffs, NJ: Prentice-Hall, 1989.
4. B. Moslehi, J.W. Goodman, M. Tur, and H.J. Shaw, Fiber lattice optic signal processing, *Proc. IEEE*, 72(7), 909–930, 1984.
5. B. Moslehi, Fiber-optic filters employing optical amplifiers to provide design flexibility, *Electron. Lett.*, 28, 226–228, 1992.

6. B. Moslehi and J.W. Goodman, Novel amplified fiber-optic recirculating delay line processor, *J. Lightwave Technol.*, 10(8), 1142–1147, 1992.
7. B.C. Kuo, *Digital Control Systems*, Asia, Hong Kong: CBS Publications, 1987, Chapter 5, pp. 267–296.
8. Y.H. Ja, Single-mode optical fiber ring and loop resonators using degenerate two-wave mixing, *Appl. Opt.*, 30(18), 2424–2426, 1991.
9. L.N. Binh, N.Q. Ngo, and S.F. Luk, Graphical representation and analysis of the z-shaped double-coupler optical resonator, *J. Lightwave Technol.*, 11(11), 1782–1792, 1993.
10. Y.H. Ja, Generalized theory of optical fiber loop and ring resonators with multiple couplers: 1: Circulating and output fields and 2: General characteristics, *Appl. Opt.*, 29, 3517–3529, 1990.
11. Y.H. Ja, Optical fiber loop resonators with double couplers, *Opt. Commun.*, 75, 239–245, March 1990.
12. Y.H. Ja, A double-coupler optical fiber ring-loop resonator with degenerate two-wave mixing, *Opt. Commun.*, 81, 113–120, February 1991.
13. Y.H. Ja, On the configurations of double optical fiber loop or ring resonator with double couplers, *J. Opt. Commun.*, 12, 29–32, 1991.
14. B. Moslehi, J.W. Goodman, M. Tur, and H.J. Shaw, Fiber lattice optic signal processing, *Proc. IEEE*, 72(7), 909–930, 1984.
15. B. Moslehi and J.W. Goodman, Novel amplified fiber-optic recirculating delay line processor, *IEEE J. Lightwave Technol.*, 10(8), 1142–1147, 1992.
16. J.M. Vigoureux and F. Raba, A model for the optical transistor and optical switching, *J. Modern Opt.*, 38(12), 2521–2530, 1991.
17. B.E.A. Saleh and M.I. Irshid, Transmission of pulse sequences through monomode fibers, *Appl. Opt.*, 21(23), 4219–4222, 1982.
18. B.E.A. Saleh and M.I. Irshid, Coherence and intersymbol interference in digital fiber optic communication systems, *IEEE J. Quantum Electron.*, QE-18(6), 944–951, 1982.
19. K. Jürgensen, Transmission of Gaussian pulses through monomode dielectric optical waveguides, *Appl. Opt.*, 16(1): 22–23, 1977.
20. K. Jürgensen, Gaussian pulse transmission through monomode fibers, accounting for source linewidth, *Appl. Opt.*, 17(15), 2412–2415, 1978.
21. C. Lin and D. Marcuse, Optimum optical pulse width for high bandwidth single-mode fiber transmission, *Electron. Lett.*, 17(1), 54–55, 1981.
22. D. Marcuse and C. Lin, Low dispersion single-mode fiber transmission—the question of practical versus theoretical maximum transmission bandwidth, *IEEE J. Quantum Electron.*, QE-17(6), 869–878, 1981.
23. D. Marcuse, Pulse distortion in single-mode fibers, *Appl. Opt.*, 19, 1653–1660, 1980.
24. D. Marcuse, Selected topics in the theory of telecommunications fibers, *Optical Fiber Telecommunications II*, 2nd ed., Chapter 3, New York: Academic Press, Inc., 1988.
25. Y. Ohtsuka, Analysis of a fiber-optic passive loop-resonator gyroscope: Dependence on resonator parameters and light-source coherence, *J. Lightwave Technol.*, LT-3(2), 378–384, 1985.
26. G.F. Levy, Optical fiber ring resonators: Analysis of transient response, *IEE Proc. J.*, 139(5), 313–317, 1992.
27. B. Moslehi, Analysis of optical phase noise in fiber-optic systems employing a laser source with arbitrary coherence time, *J. Lightwave Technol.*, LT-4(9), 1334–1351, 1986.
28. B. Crosignani, A. Yariv, and P. Di Porto, Time-dependent analysis of a fiber-optic passive-loop resonator, *Opt. Lett.*, 11(4), 251–253, 1986.
29. Y. Ohtsuka, Optical coherence effects on a fiber-sensing Fabry–Perot interferometer, *Appl. Opt.*, 21(23), 4316–4320, 1982.

30. P.M. Shankar, Performance of LED/single mode fiberoptic systems, *J. Opt. Commun.*, 10(4), 132–137, 1989.
31. S. Dilwali and G. Soundra Pandian, Pulse response of a fiber dispersion equalizing scheme based on an optical resonator, *IEEE Photon. Technol. Lett.*, 4(8), 942–944, 1992.
32. S.D. Personick, Comparison of equalizing and nonequalizing repeaters for optical fiber systems, *Bell Syst. Tech. J.*, 55(7), 957–971, 1976.
33. S. Kitajima, N. Kikuchi, K. Kuboki, H. Tsushima, and S. Sasaki, Dispersion compensation with optical ring resonator in coherent optical transmission systems, International Conference on Integrated Optics and Optical Communications, San Francisco, pp. 353–356, 1993.
34. K. Petermann, Semiconductor laser noise in an interferometer system, *IEEE J. Quantum Electron.*, QE-17(7), 1251–1256, 1981.
35. S.D. Personick, *Optical Fiber Transmission Systems*, New York: Plenum Press, 1981.
36. G. Keiser, *Optical Fiber Communications*, New York: McGraw-Hill, Inc., 1989.
37. K. Iwashita and N. Takachio, Chromatic dispersion compensation in coherent optical communications, *J. Lightwave Technol.*, 8(3), 367–375, 1990.
38. L.J. Cimini, Jr., L.J. Greenstein, and A.A.M. Saleh, Optical equalization to combat the effects of laser chirp and fiber dispersion, *J. Lightwave Technol.*, 8(5), 649–659, 1990.
39. P.P. Iannone, A.H. Gnauck, and P.R. Prucnal, Dispersion-compensated 333-km 10-Gb/s transmission using mid-span spectral inversion in an injection-locked InGaAsP V-grove laser, *IEEE Photon. Technol. Lett.*, 6, 1046–1049, 1994.
40. N.Q. Ngo and L.N. Binh, Eigenfilter synthesis of optical dispersion eigencompensators, Proceedings of the IOOC'95, Hong Kong, vol. 2, pp. 173–174, 1995.
41. N. Henmi, T. Saito, and T. Ishida, Prechirp technique as a linear dispersion compensation for ultrahigh-speed long-span intensity modulation directed detection optical communication systems, *J. Lightwave Technol.*, 12, 1716–1729, 1994.
42. M. Nakazawa, Ultrahigh speed optical soliton communication and related technology, IOOC'95 Technical Digest Series, Hong Kong, vol. 4, pp. 96–98, 1995.
43. G. Aubin, E. Jeanney, T. Montalant, J. Moulu, F. Pirio, J.-B. Thomine, and F. Devaux, Electro absorption modulator for a 20 Gb/s soliton transmission experiment over 1 million km with a 140 km amplifier span, Post-deadline paper PD2–5, IOOC'95 Technical Digest Series, Hong Kong, vol. 5, pp. 29–30, 1995.
44. W. Zhao and E. Bourkoff, Propagation properties of dark solitons, *Opt. Lett.*, 14, 703–705, 1989.
45. W. Zhao and E. Bourkoff, Interactions between dark solitons, *Opt. Lett.*, 14, 1371–1373, 1989.
46. J.P. Hamaide, P. Emplit, and M. Haelterman, Dark-soliton jitter in amplified optical transmission systems, *Opt. Lett.*, 16, 1578–1580, 1991.
47. Y.S. Kivshar, M. Haelterman, P. Emplit, and J.P. Hamaide, Gordon–Haus effect on dark solitons, *Opt. Lett.*, 19, 19–21, 1994.
48. P. Emplit, J.P. Hamaide, F. Reynaud, C. Froehly, and A. Barthelemy, Picosecond steps and dark pulses through nonlinear single mode fibers, *Opt. Commun.*, 62, 374–379, 1987.
49. D. Krökel, N.J. Halas, G. Giuliani, and D. Grischkowsky, Dark-pulse propagation in optical fibers, *Phys. Rev. Lett.*, 60, 29–32, 1988.
50. A.M. Weiner, J.P. Heritage, R.J. Hawkins, R.N. Thurston, E.M. Kirschner, D.E. Leaird, and W.J. Tomlinson, Experimental observation of the fundamental dark soliton in optical fibers, *Phys. Rev. Lett.*, 61, 2445–2448, 1988.

51. D.J. Richardson, R.P. Chamberlin, L. Dong, and D.N. Payne, Experimental demonstration of 100 GHz dark soliton generation and propagation using a dispersion decreasing fibre, *Electron. Lett.*, 30, 1326–1327, 1994.
52. W. Zhao and E. Bourkoff, Generation of dark solitons under a cw background using waveguide electro-optic modulators, *Opt. Lett.*, 15, 405–407, 1990.
53. M. Nakazawa and K. Suzuki, Generation of a pseudorandom dark soliton data train and its coherent detection by one-bit-shifting with a Mach–Zehnder interferometer, *Electron. Lett.*, 31, 1084–1085, 1995.
54. M. Nakazawa and K. Suzuki, 10 Gbit/s pseudorandom dark soliton data transmission over 1200 km, *Electron. Lett.*, 31, 1076–1077, 1995.
55. B. Wedding, B. Franz, and B. Junginger, 10-Gb/s optical transmission up to 253 km via standard single-mode fiber using the method of dispersion-supported transmission, *J. Lightwave Technol.*, 12, 1730–1737, 1994.
56. C.D. Poole, J.M. Wiesenfeld, D.J. DiGiovanni, and A.M. Vengsarkar, Optical fiber-based dispersion compensation using higher order modes near cutoff, *J. Lightwave Technol.*, 12, 1756–1768, 1994.
57. M. Onishi, Y. Koyano, M. Shigematsu, H. Kanamori, and M. Nishimura, Dispersion compensating fiber with a high figure of merit of 250 ps/nm/dB, *Electron. Lett.*, 30, 161–164, 1994.
58. A.H. Gnauck, C.R. Giles, L.J. Cimini, Jr., J. Stone, L.W. Stulz, S.K. Korotky, and J.J. Veselka, 8-Gb/s-130 km transmission experiment using Er-doped fiber preamplifier and optical dispersion equalization, *IEEE Photon. Technol. Lett.*, 3, 1147–1149, 1991.
59. L.J. Cimini, Jr., L.J. Greenstein, and A.A.M. Saleh, Optical equalization to combat the effects of laser chirp and fiber dispersion, *J. Lightwave Technol.*, 8, 649–659, 1990.
60. M. Sharma, H. Ibe, and T. Ozeki, Optical circuits for equalizing group delay dispersion of optical fibers, *J. Lightwave Technol.*, 12, 1769–1775, 1994.
61. T. Ozeki, Optical equalizers, *Opt. Lett.*, 17, 375–377, 1992.
62. N.Q. Ngo, L.N. Binh, and X. Dai, Eigenfilter approach for designing FIR all-pass optical dispersion compensators for high-speed long-haul systems, Proceedings of the IREE, 19th Australian Conference on Optical Fibre Technology, Melbourne, pp. 355–358, 1994.
63. G.P. Agrawal, *Nonlinear Fiber Optics*, Boston, MA: Academic Press, 1989.
64. N.Q. Ngo, L.N. Binh, and X. Dai, Optical dispersion eigencompensators for high-speed long-haul IM/DD lightwave systems: Computer simulation, *J. Lightwave Technol.*, 14, 2097–2107, 1996.
65. D. Marcuse, Pulse distortion in single-mode fibers, *Appl. Opt.*, 19, 1653–1660, 1980.
66. A.P. Clark, *Equalizers for Digital Modems*. London: Pentech Press, 1985.
67. T.Q. Nguyen, T.I. Laakso, and R.D. Koilpillai, Eigenfilter approach for the design of allpass filters approximating a given phase response, *IEEE Trans. Signal Process.*, 42, 2257–2263, 1994.
68. J.C. Cartledge and A.F. Elrefaie, Effect of chirping-induced waveform distortion on the performance of direct detection receivers using traveling-wave semiconductor optical preamplifiers, *J. Lightwave Technol.*, 9, 209–219, 1991.
69. S.C. Pei and J.J. Shyu, Eigen-approach for designing FIR filters and all-pass phase equalizers with prescribed magnitude and phase response, *IEEE Trans. Circuits Syst.*, 39, 137–146, 1992.
70. A.F. Elrefaie, R.E. Wagner, D.A. Atlas, and D.G. Daut, Chromatic dispersion limitations in coherent lightwave transmission systems, *J. Lightwave Technol.*, 6, 704–709, 1988.
71. B.C. Kuo, *Digital Control Systems*, New York: Reinhart & Winston, Chapter 5, 1980.

72. P.A. Krug, T. Stephens, G. Yoffe, F. Ouellette, P. Hill, and G. Dhosi, Dispersion compensation over 270 km at 10 Gbit/s using an offset-core chirped fiber Bragg grating, *Electron. Lett.*, 31, 1091–1093, 1995.
73. F. Koyama and Y. Suematsu, Analysis of dynamic spectral width of dynamic-single-mode (DSM) lasers and related transmission bandwidth of single-mode fibers, *IEEE J. Quantum Electron.*, QE-21, 292–297, 1985.
74. P.S. Henry, Lightwave primer, *IEEE J. Quantum Electron.*, QE-21, 1862–1879, 1985.
75. N. Sugimoto, H. Terui, A. Tate, Y. Katoh, Y. Yamada, A. Sugita, A. Shibukawa, and Y. Inoue, A hybrid integrated waveguide isolator on a silica-based planar lightwave circuit, *J. Lightwave Technol.*, 14, 2537–2546, 1996.
76. D. Marcuse, Single-channel operation in very long nonlinear fibers with optical amplifiers at zero dispersion, *J. Lightwave Technol.*, 9, 356–361, 1991.

6 Tunable Optical Filters

Optical filters play important roles in photonic processing systems and optical communications as well as in many optical systems. Even more important and critical in the use of these filters are the types, the bandwidth, and center wavelength which can be tuned to desired regions and passband. In this chapter, a digital filter design method is employed to systematically design tunable optical filters with variable passband and center wavelength/frequency characteristics for lowpass, highpass, bandpass, and bandstop types. Potential applications of such filters and the existing design techniques in wavelength-division multiplexing (WDM) and dispersion equalization are also discussed in this chapter. Basic optical filter structures, namely, the first-order all-pole optical filter (FOAPOF) and the first-order all-zero optical filter (FOAZOF) are described, hence the design process of tunable optical filters and the design of the second-order Butterworth lowpass, highpass, bandpass, and bandstop tunable optical filters are described. An experimental development of the first-order Butterworth lowpass and highpass tunable fiber-optic filters is given.

6.1 INTRODUCTION

Tunable optical filters with variable bandwidth and center frequency characteristics are important in applications where dynamic changes in the bandwidth and center frequency of the filter are required. One such application is in frequency-division multiplexing (FDM) or multi-WDM optical systems, which utilizes the large bandwidth of optical fibers to increase the transmission capacity that is mainly limited by fiber dispersion. In WDM systems, tunable optical filters are used as optical demultiplexers at the receivers to select one or more desired channels at any wavelength.

There have been reports of several types of tunable optical filters such as the optical ring resonators [1], optical transversal filters [2], and Fabry–Perot interferometers [3,4]. These filters may be referred to as bandpass tunable optical filters because their magnitude responses have Gaussian-type characteristics. Another type of tunable optical filter is the cascaded-coupler Mach–Zehnder channel adding/dropping filter whose output ports constitute a power complementary pair; one port is used as a bandpass filter while the other port is used as a bandstop filter [5]. It is believed that there has been no previous report of a systematic filter design technique, which can be used to design tunable optical filters with the essential features of variable bandwidth and center frequency characteristics as well as general filtering characteristics such as lowpass, highpass, bandpass, and bandstop characteristics.

In this chapter, the design tunable optical filters with variable bandwidth and center frequency characteristics as well as lowpass, highpass, bandpass, and band-stop characteristics are discussed. Techniques for the design of digital filters are adopted and described in the Appendix as a reference. The filter design method employing a graphical approach described here is adopted from the method proposed by Ngo and Binh [6] where nontunable optical filters were designed. This chapter is organized as follows. Section 6.2 outlines the basic structures of tunable filters with much reference to the digital filter design techniques given in the Appendix. The composite design of optical filters whose filtered band and center optical frequency are tunable to a desired position is described in Section 6.3, illustrating the effectiveness of the systematic design procedures. Some experimental results are presented in Section 6.4. Concluding remarks of the design techniques are given thereafter.

6.2 BASIC STRUCTURES OF TUNABLE OPTICAL FILTERS

This section describes the basic filter structures, namely, FOAPOF and FOAZOF of tunable optical filters. These filters employ a fundamental approach of recursive filters whose characteristics are summarized in the Appendix.

6.2.1 First-Order All-Pole Optical Filter

Consider a basic optical filter structure shown in Figure 6.1. This figure represents the *k*th-stage of an FOAPOF* using the planar lightwave circuit (PLC) technology. The FOAPOF consists of an optical waveguide loop interconnected by two identical tunable couplers (TCs) and a thermo-optic phase shifter (PS). The TC, which is a symmetrical Mach–Zehnder interferometer, consists of, in this case, two 3 dB directional couplers interconnected by two waveguide arms of equal length.

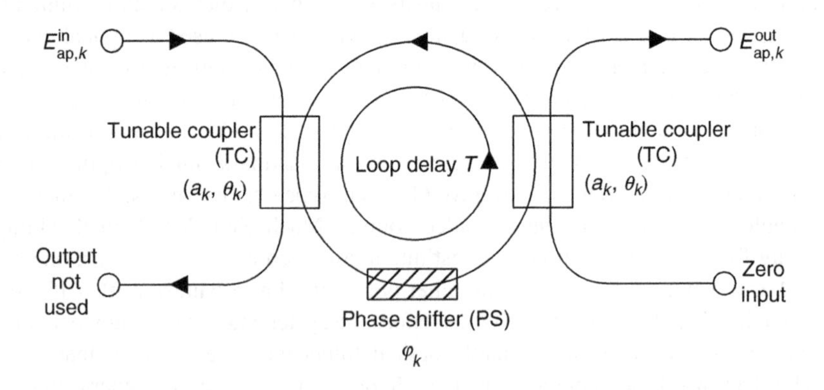

FIGURE 6.1 Schematic diagram of the *k*th-stage first-order all-pole optical filter (FOAPOF) using the PLC technology.

* Note that the FOAPOF has been used as an optical integrator for dark-soliton generation.

The TC has a

$$\text{Complex cross-coupled coefficient} = \sqrt{\gamma_{\text{w}} a_k}\ \exp(j\theta_k) \tag{6.1}$$

and

$$\text{Complex direct-coupled coefficient} = \sqrt{\gamma_{\text{w}}(1 - a_k)}\ \exp(j\theta_k) \tag{6.2}$$

where γ_{w} (typically $\gamma_{\text{w}} = 0.89$ for an insertion loss of 0.5 dB) is the intensity transmission coefficient of the TC. The cross-coupled intensity coefficient of the TC is given as

$$a_k = (1 + \cos\varphi_k)/2 \quad (0 \le a_k \le 1) \tag{6.3}$$

from which the phase shift of the PS on the upper arm of the TC is given by

$$\varphi_k = \cos^{-1}(2a_k - 1) \quad (0 \le \varphi_k \le 2\pi) \tag{6.4}$$

and the resulting phase shift of the TC is given as

$$\theta_k = \tan^{-1}\left[\frac{\sin\varphi_k}{\cos\varphi_k - 1}\right] \quad (-\pi/2 \le \theta_k \le \pi/2) \tag{6.5}$$

The optical transmittances of the upper (Λ_{1k}) and lower (Λ_{2k}) halves of the waveguide ring are defined as

$$\Lambda_{1k} = \exp(-\alpha_{\text{w}} L_{1k}) \exp(-j\omega T_{1k}) \tag{6.6}$$

$$\Lambda_{2k} = \exp(-\alpha_{\text{w}} L_{2k}) \exp(-j\omega T_{2k}) \exp(j\phi_k) \tag{6.7}$$

where the amplitude waveguide propagation loss is typically $\alpha_{\text{w}} = 0.01151\ \text{cm}^{-1}$ as determined from $\alpha_{\text{w}} = \alpha_{\text{w}}(\text{dB/cm})/8.686$ with $\alpha_{\text{w}}(\text{dB/cm}) = 0.1$ dB/cm, L_{1k} and L_{2k} the waveguide lengths of the upper and lower halves of the ring, T_{1k} and T_{2k} the corresponding time delays of the upper and lower halves of the ring, $j = \sqrt{-1}$, ω the angular optical frequency, and ϕ_k $(0 \le \phi_k \le 2\pi)$ the phase shift of the PS. Note that the lengths of the lower and upper halves of the ring do not necessarily need to be the same. Using the signal-flow graph method described in Ref. [6] together with Equations A.6.1 and A.6.2, the transfer function of the kth-stage FOAPOF is simply given, by inspection of Figure 6.1, as

$$H_{\text{ap},k}(\omega) = \frac{E_{\text{ap},k}^{\text{out}}}{E_{\text{ap},k}^{\text{in}}} = \frac{\gamma_{\text{w}} a_k \exp(j2\theta_k)\Lambda_{2k}}{1 - \gamma_{\text{w}}(1 - a_k)\exp(j2\theta_k)\Lambda_{1k}\Lambda_{2k}} \tag{6.8}$$

where $E_{\text{ap},k}^{\text{in}}$ and $E_{\text{ap},k}^{\text{out}}$ are the electric-field amplitudes at the input and output ports of the FOAPOF, respectively. It is useful to define the waveguide loop length L, the waveguide loop delay T, and the z-transform parameter as

$$L = L_{1k} + L_{2k} \tag{6.9}$$

$$T = T_{1k} + T_{2k} \tag{6.10}$$

$$z = \exp(j\omega T) \tag{6.11}$$

Substituting Equations 6.6, 6.7, and 6.9 through 6.11 into Equation 6.8, the z-transform transfer function of the kth-stage FOAPOF becomes

$$H_{\mathrm{ap},k}(z) = \frac{A_{\mathrm{ap},k} \exp[j(2\theta_k + \phi_k)] \exp(-j\omega T_{2k}) z}{z - p_k} \tag{6.12}$$

where the amplitude $A_{\mathrm{ap},k}$ and the pole location p_k in the z-plane are given by

$$A_{\mathrm{ap},k} = \gamma_{\mathrm{w}} a_k \exp(-\alpha_{\mathrm{w}} L_{2k}) \tag{6.13}$$

$$p_k = \gamma_{\mathrm{w}}(1 - a_k)\exp(-\alpha_{\mathrm{w}} L)\exp(j(2\theta_k + \phi_k)) \tag{6.14}$$

It is useful to express Equation 6.14 in the phasor form as

$$p_k = |p_k| \exp(j\arg(p_k)) \quad (0 \le |p_k| < 1) \tag{6.15}$$

where

$$|p_k| = \gamma_{\mathrm{w}}(1 - a_k)\exp(-\alpha_{\mathrm{w}} L) \tag{6.16}$$

$$\arg(p_k) = 2\theta_k + \phi_k \tag{6.17}$$

6.2.2 First-Order All-Zero Optical Filter

Figure 6.2 shows the schematic diagram of the kth-stage FOAZOF* using the PLC technology. The FOAZOF, which is an asymmetrical Mach–Zehnder interferometer,

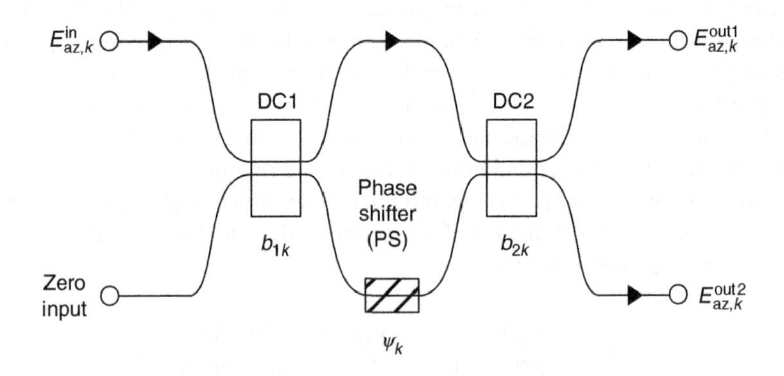

FIGURE 6.2 Schematic diagram of the kth-stage first-order all-zero optical filter (FOAZOF) using the PLC technology.

* Note that the FOAZOF has been used as an optical dark-soliton detector.

consists of two waveguide directional couplers DC1 and DC2, with cross-coupled intensity coefficients b_{1k} and b_{2k} ($0 \leq b_{1k}, b_{2k} \leq 1$), respectively, which are interconnected by two unequal waveguide arms with a differential time delay of T. The PS on the lower arm has a phase shift of ψ_k ($0 \leq \psi_k \leq 2\pi$).

The optical transmittances of the upper (Λ_{3k}) and lower (Λ_{4k}) waveguide arms are defined as

$$\Lambda_{3k} = \exp(-\alpha_{\mathrm{w}} L_{3k}) \exp(-j\omega T_{3k}) \tag{6.18}$$

$$\Lambda_{4k} = \exp(-\alpha_{\mathrm{w}} L_{4k}) \exp(-j\omega T_{4k}) \exp(j\psi_k) \tag{6.19}$$

where L_{3k} and L_{4k} are the waveguide lengths of the upper and lower arms, and T_{3k} and T_{4k} are the corresponding time delays of the upper and lower arms.

Using the signal-flow graph method as described in Ref. [6], the transfer function of the kth-stage FOAZOF (for the upper output port) is simply given, by inspection of Figure 6.2, as

$$H_{\mathrm{az},k}(\omega) = \frac{E_{\mathrm{az},k}^{\mathrm{out1}}}{E_{\mathrm{az},k}^{\mathrm{in}}} = \sqrt{\gamma_{\mathrm{az}}(1 - b_{1k})(1 - b_{2k})}\ \Lambda_{3k} - \sqrt{\gamma_{\mathrm{az}} b_{1k} b_{2k}}\ \Lambda_{4k} \tag{6.20}$$

where $E_{\mathrm{az},k}^{\mathrm{in}}$ and $E_{\mathrm{az},k}^{\mathrm{out1}}$ and $E_{\mathrm{az},k}^{\mathrm{out2}}$ describe the electric-field amplitudes at the input and output ports of the FOAZOF, respectively, and γ_{az} (which is assumed to be $\gamma_{\mathrm{az}} = \gamma_{\mathrm{w}} = 0.89$ [a typical value] for analytical simplicity) is the intensity transmission coefficient of the FOAZOF. It is useful to define the differential length* L and the differential time delay† T as

$$L = L_{4k} - L_{3k} \tag{6.21}$$

$$T = T_{4k} - T_{3k} \tag{6.22}$$

By substituting Equations 6.18 through 6.20, 6.22, and 6.11 into Equation 6.21, the z-transform transfer function of the kth-stage FOAZOF (for the upper output port) becomes

$$H_{\mathrm{az},k}(z) = A_{\mathrm{az},k} \exp(-j\omega T_{3k}) z^{-1} (z - z_k) \tag{6.23}$$

where the amplitude $A_{\mathrm{az},k}$ and the zero location z_k in the z-plane are given by

$$A_{\mathrm{az},k} = \sqrt{\gamma_{\mathrm{az}}(1 - b_{1k})(1 - b_{2k})}\ \exp(-\alpha_{\mathrm{w}} L_{3k}) \tag{6.24}$$

$$z_k = \sqrt{\frac{b_{1k} b_{2k}}{(1 - b_{1k})(1 - b_{2k})}}\ \exp(-\alpha_{\mathrm{w}} L) \exp(j\psi_k) \tag{6.25}$$

---

* L is the differential length of the kth-stage FOAZOF (see Equation 6.21) as well as the loop length of the kth-stage FOAPOF (see Equation 6.9).

† T is the differential time delay of the kth-stage FOAZOF (see Equation 6.22) as well as the loop delay of the kth-stage FOAPOF (see Equation 6.10).

It is useful to express Equation 6.25 in the phasor form as

$$z_k = |z_k| \exp(j\arg(z_k)) \tag{6.26}$$

where

$$|z_k| = \sqrt{\frac{b_{1k}b_{2k}}{(1-b_{1k})(1-b_{2k})}} \exp(-\alpha_w L) \tag{6.27}$$

$$\arg(z_k) = \psi_k \tag{6.28}$$

Similarly, the z-transform transfer function of the kth-stage FOAZOF (for the lower output port) is given by

$$H^*_{az,k}(z) = \frac{E^{out2}_{az,k}}{E^{in}_{az,k}} = A^*_{az,k} \exp(-j(\omega T_{3k} + \pi/2)) z^{-1}(z - z^*_k) \tag{6.29}$$

where the amplitude $A^*_{az,k}$ and the zero location z^*_k in the z-plane are given by

$$A^*_{az,k} = \sqrt{\gamma_{az} b_{2k}(1-b_{1k})} \exp(-\alpha_w L_{3k}) \tag{6.30}$$

$$z^*_k = \sqrt{\frac{b_{1k}(1-b_{2k})}{b_{2k}(1-b_{1k})}} \exp(-\alpha_w L) \exp(j(\psi_k + \pi)) \tag{6.31}$$

In the phasor form, Equation 6.31 becomes

$$z^*_k = |z^*_k| \exp(j\arg(z^*_k)) \tag{6.32}$$

where

$$|z^*_k| = \sqrt{\frac{b_{1k}(1-b_{2k})}{b_{2k}(1-b_{1k})}} \exp(-\alpha_w L) \tag{6.33}$$

$$\arg(z^*_k) = \psi_k + \pi \tag{6.34}$$

From Equations 6.28 and 6.34, the zero z_k (for the upper output port) is out of phase with the zero z^*_k (for the lower output port) by π. This means that the transfer functions $H_{az,k}(z)$ and $H^*_{az,k}(z)$ constitute a power complementary pair:

$$\left|H_{az,k}(z)\right|^2 + \left|H^*_{az,k}(z)\right|^2 = 1 \tag{6.35}$$

In other words, if $H_{az,k}(z)$ has a lowpass magnitude response then $H^*_{az,k}(z)$ has a highpass magnitude response, an important property useful in the design of tunable optical filters.

Note that only the transfer function $H_{az,k}(z)$ is used in the design of tunable optical filters as described in the following section.

6.2.3 Mth-Order Tunable Optical Filter

The transfer function of the Mth-order tunable all-pole optical filter, which is the transfer function of the cascade of M FOAPOFs, is defined as

$$H_{ap}(z) = \prod_{k=1}^{M} H_{ap,k}(z) \tag{6.36}$$

The transfer function of the Mth-order tunable all-zero optical filter, which is the transfer function of the cascade of M FOAZOFs, is defined as

$$H_{az}(z) = \prod_{k=1}^{M} H_{az,k}(z) \tag{6.37}$$

The transfer function of the Mth-order tunable optical filter can thus be written as

$$H(z) = G \cdot H_{ap}(z) \cdot H_{az}(z) \tag{6.38}$$

where G is the amplitude optical gain of the erbium-doped fiber amplifier (EDFA). Figure 6.3 is the block diagram representation of Equation 6.38, which describes the transfer function between the input port and the upper output port (i.e., output 1). The lower output port (output 2) is not used in filter design but its usefulness is investigated. Substituting Equations 6.12, 6.23, 6.36, and 6.37 into Equation 6.38, the transfer function of the Mth-order tunable optical filter is given by

$$H(z) = \exp\left[j\left(\sum_{k=1}^{M}(2\theta_k + \phi_k) - \omega\sum_{k=1}^{M}(T_{2k} + T_{3k})\right)\right] \cdot \left[A\prod_{k=1}^{M}\frac{(z - z_k)}{(z - p_k)}\right] \tag{6.39}$$

where the amplitude A is defined as

$$A = G\prod_{k=1}^{M} A_{ap,k}A_{az,k} \tag{6.40}$$

The proposed Mth-order tunable optical filter has the advantage that its poles and zeros can be adjusted independently of each other. Thus, a particular pole–zero pattern can easily be obtained to design filters with general characteristics.

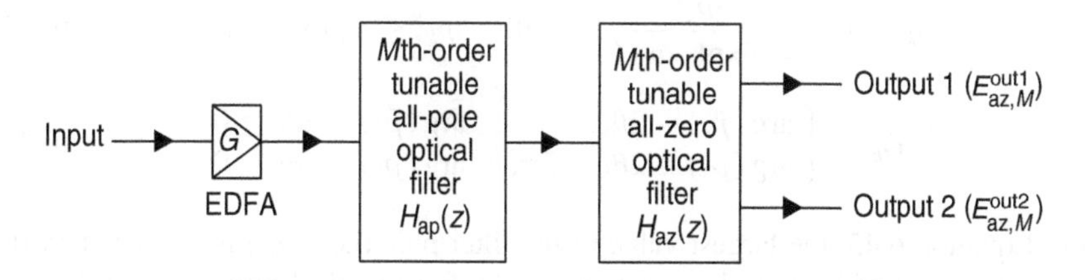

FIGURE 6.3 Block diagram representation of the Mth-order tunable optical filter.

6.3 TUNABLE OPTICAL FILTERS

This section describes the design of tunable optical filters and also presents a detailed design of the second-order Butterworth lowpass, highpass, bandpass, and bandstop tunable optical filters with variable bandwidth and center frequency characteristics.

In general, the design of a tunable optical filter from the characteristics of a digital filter involves the following stages: (1) the specification of the desired magnitude response of the optical filter in the optical domain, which is usually described by the spectrum of the optical signals to be processed by the filter, (2) the design of a digital filter whose magnitude response in the digital domain approximates the desired magnitude response of the optical filter in the optical domain, (3) the design of the optical filter structure whose transfer function is similar in form to the transfer function of the digital filter, (4) the design of the parameters of the optical filter structure using the pole–zero characteristics of the digital filter, and (5) the practical realization of the optical filter.

6.3.1 DESIGN EQUATIONS FOR TUNABLE OPTICAL FILTERS

For analytical simplicity, the exponential factor in Equation 6.47, which represents a linear phase term, is neglected because it has no effect on the magnitude response of the filter. The design of a tunable optical filter from the characteristics of a digital filter requires the second factor of Equation 6.47 to be equal to Equation 6.1 such that the following equations hold:

$$A = \hat{A} \tag{6.41}$$

$$p_k = \hat{p}_k \tag{6.42}$$

$$z_k = \hat{z}_k \tag{6.43}$$

Substituting Equations 6.13, 6.24, and 6.40 into Equation 6.41 results in

$$G = \frac{\left(\gamma_{\mathrm{w}}\gamma_{\mathrm{az}}^{1/2}\right)^{-M}\hat{A}}{\prod_{k=1}^{M} a_k[(1-b_{1k})(1-b_{2k})]^{1/2}\exp(-\alpha_{\mathrm{w}}(L_{2k}+L_{3k}))} \tag{6.44}$$

Using Equations A.6.2 and 6.15 through 6.18 and substituting these into Equation 6.42 results in

$$a_k = 1 - \frac{|\hat{p}_k|}{\gamma_{\mathrm{w}}\exp(-\alpha_{\mathrm{w}}L)} \quad \text{with} \quad |\hat{p}_k| \leq \gamma_{\mathrm{w}}\exp(-\alpha_{\mathrm{w}}L) \tag{6.45}$$

$$\phi_k = \begin{cases} \arg(\hat{p}_k) - 2\theta_k, & \arg(\hat{p}_k) - 2\theta_k \geq 0 \\ \arg(\hat{p}_k) - 2\theta_k + 2\pi, & \arg(\hat{p}_k) - 2\theta_k < 0 \end{cases} \tag{6.46}$$

From Equation 6.45, the largest value of the filter pole (i.e., $|\hat{p}_k|$) is restricted by the loss of the waveguide loop, i.e., $\gamma_{\mathrm{w}}\exp[-\alpha_{\mathrm{w}}L]$. For practical purposes, a full cycle phase shift of 2π has been added, without affecting the filter performance, to the second equation of Equation 6.46 so that ϕ_k takes a positive value (i.e., $0 \leq \phi_k \leq 2\pi$). Substituting Equations A.6.3 and 6.26 through 6.28 into Equation 6.43 and using the relation $|z_k| = |z_k^*| = |\hat{z}_k| = 1$, the zeros are located exactly on the unit circle, results in

$$b_{1k} = \frac{1}{1 + \exp(-2\alpha_w L)} \tag{6.47}$$

$$b_{2k} = 1/2 \tag{6.48}$$

$$\psi_k = \begin{cases} \arg(\hat{z}_k), & \arg(\hat{z}_k) \geq 0 \\ \arg(\hat{z}_k) + 2\pi, & \arg(\hat{z}_k) < 0 \end{cases} \tag{6.49}$$

To change the center frequency of the tunable optical filter without affecting its bandwidth, an additional phase shift of $\delta_0 (0 < \delta_0 < 2\pi)$ must be added to Equations 6.46 and 6.49, resulting in

$$\phi_k = \begin{cases} \arg(\hat{p}_k) - 2\theta_k + \delta_0, & \arg(\hat{p}_k) - 2\theta_k + \delta_0 \geq 0 \\ \arg(\hat{p}_k) - 2\theta_k + \delta_0 + 2\pi, & \arg(\hat{p}_k) - 2\theta_k + \delta_0 < 0 \end{cases} \tag{6.50}$$

$$\psi_k = \begin{cases} \arg(\hat{z}_k) + \delta_0, & \arg(\hat{z}_k) + \delta_0 \geq 0 \\ \arg(\hat{z}_k) + \delta_0 + 2\pi, & \arg(\hat{z}_k) + \delta_0 < 0 \end{cases} \tag{6.51}$$

In summary, the design equations for the design of tunable optical filters are Equations 6.4, 6.5, 6.44, 6.45, 6.47, 6.48, 6.50, and 6.51.

6.3.2 Design of Second-Order Butterworth Tunable Optical Filters

To demonstrate the effectiveness of the proposed filter design technique, this section describes the design of the second-order ($M = 2$) Butterworth lowpass, highpass, bandpass, and bandstop tunable optical filters with variable bandwidth and center frequency characteristics.

In this example, a $1/T = 5$ GHz filter is considered, resulting in $T = 200$ ps and $L = 40$ mm. Using $\gamma_w = \gamma_{az} = 0.89$, $\alpha_w = 0.01151$ cm^{-1}, and hence $\exp(-\alpha_w L) = 0.955$, Equations 6.45 and 6.47 become

$$a_k = 1 - 1.18|\hat{p}_k| \; |\hat{p}_k| \leq 0.85 \tag{6.52}$$

$$b_{1k} = 0.523 \tag{6.53}$$

The corresponding gain coefficient is

$$G = \frac{(0.392)^{-M}\hat{A}}{\prod_{k=1}^{M} a_k} \tag{6.54}$$

where $\exp[-\alpha_w(L_{2k} + L_{3k})] = \exp(-\alpha_w L) = 0.955$ has been assumed for numerical simplicity. In summary, Equations 6.4, 6.5, 6.48, and 6.50 through 6.54 are used in the design of the second-order Butterworth tunable optical filters.

The following definitions are used hereafter. The normalized optical frequency on the frequency axis of the squared magnitude response represents $\omega T/\pi$, the squared magnitude responses are plotted over the Nyquist interval, $0 \leq \omega T/\pi \leq 1$, ω_c is the 3 dB angular cutoff frequency of the lowpass and highpass filters, and ω_{c1} (ω_{c2}) is the 3 dB angular lower (upper) corner frequency of the bandpass and bandstop filters where $\omega_{c1} < \omega_{c2}$. For the lowpass and highpass filters, the normalized 3 dB bandwidth is defined as $\omega_c T/\pi$. For the bandpass and bandstop filters, the normalized 3 dB bandwidth is defined as $(\omega_{c2} - \omega_{c1})T/\pi$.

Using MATLAB, Tables 6.1 and 6.2 show the computed design parameters of the second-order Butterworth lowpass and highpass (Table 6.1) digital filters with various cutoff frequencies and bandpass and bandstop (Table 6.2) digital filters*

TABLE 6.1
Design Parameters of the Second-Order ($M=2$) Butterworth Lowpass and Highpass Digital Filters with Various Cutoff Frequencies

		Poles of $\hat{H}_{ap,k}(z)$, $(k=1,2)$		Zeros of $\hat{H}_{az,k}(z)$, $(k=1,2)$			
Filter Type	$\omega_c T/\pi$	$\hat{A}$	$\lvert\hat{p}_1\rvert = \lvert\hat{p}_2\rvert$	$\arg(\hat{p}_1)$	$\arg(\hat{p}_2) = -\arg(\hat{p}_1)$	$\lvert\hat{z}_1\rvert = \lvert\hat{z}_2\rvert$	$\arg(\hat{z}_1) = \arg(\hat{z}_2)$
Lowpass	0.1	0.0201	0.8008	−0.2258	0.2258	1	π
	0.2	0.0675	0.6425	−0.4746	0.4746	1	π
	0.3	0.1311	0.5217	−0.7718	0.7718	1	π
	0.4	0.2066	0.4425	−1.1401	1.1401	1	π
	0.5	0.2929	0.4142	−1.5708	1.5708	1	π
	0.6	0.3913	0.4425	−2.0015	2.0015	1	π
	0.7	0.5050	0.5217	−2.3698	2.3698	1	π
	0.8	0.6389	0.6425	−2.6670	2.6670	1	π
	0.9	0.8006	0.8008	−2.9158	2.9158	1	π
Highpass	0.1	0.8006	0.8008	−0.2258	0.2258	1	0
	0.2	0.6389	0.6425	−0.4746	0.4746	1	0
	0.3	0.5050	0.5217	−0.7718	0.7718	1	0
	0.4	0.3913	0.4425	−1.1401	1.1401	1	0
	0.5	0.2929	0.4142	−1.5708	1.5708	1	0
	0.6	0.2066	0.4425	−2.0015	2.0015	1	0
	0.7	0.1311	0.5217	−2.3698	2.3698	1	0
	0.8	0.0675	0.6425	−2.6670	2.6670	1	0
	0.9	0.0201	0.8008	−2.9158	2.9158	1	0

Note: For each bandwidth $\omega_c T/\pi$, both the lowpass and highpass filters have the same poles which occur in complex conjugate pairs. The zeros are located exactly on the unit circle in the z-plane but there is a phase difference of π between the zeros of the lowpass and highpass filters.

* These digital filters are described by the transfer function given in Equation 6.1.

TABLE 6.2
Design Parameters of the Second-Order ($M = 2$) Butterworth Bandpass and Bandstop Digital Filters with Various Sets of Lower ($\omega_{c1}T/\pi$) and Upper ($\omega_{c2}T/\pi$) Corner Frequencies

Filter				Poles of $\hat{H}_{ap,k}(z)$, ($k = 1, 2$)			Zeros of $\hat{H}_{az,k}(z)$, ($k = 1, 2$)		
Type	$\omega_{c1}T/\pi$	$\omega_{c2}T/\pi$	$\hat{A}$	$\|\hat{p}_1\| = \|\hat{p}_2\|$	$\arg(\hat{p}_1)$	$\arg(\hat{p}_2)$	$\|\hat{z}_1\| = \|\hat{z}_2\|$	$\arg(\hat{z}_1)$	$\arg(\hat{z}_2)$
Bandpass	0.40	0.60	0.2425	0.7138	$\pi/2$	$\pi/2$	1	0	π
	0.35	0.65	0.3375	0.5700	$-\pi/2$	$\pi/2$	1	0	π
	0.30	0.70	0.4208	0.3980	$-\pi/2$	$\pi/2$	1	0	π
	0.25	0.75	0.5000	0	0	0	1	0	π
	0.20	0.80	0.5792	0.3980	π	0	1	0	π
	0.15	0.85	0.6625	0.5700	π	0	1	0	π
	0.10	0.90	0.7548	0.7138	π	0	1	0	π
Bandstop	0.40	0.60	0.7548	0.7138	$-\pi/2$	$\pi/2$	1	$-\pi/2$	$\pi/2$
	0.35	0.65	0.6625	0.5700	$-\pi/2$	$\pi/2$	1	$-\pi/2$	$\pi/2$
	0.30	0.70	0.5792	0.3980	$-\pi/2$	$\pi/2$	1	$-\pi/2$	$\pi/2$
	0.25	0.75	0.5000	0	$-\pi/2$	0	1	$-\pi/2$	$\pi/2$
	0.20	0.80	0.4208	0.3980	0	0	1	$-\pi/2$	$\pi/2$
	0.15	0.85	0.3375	0.5700	π	0	1	$-\pi/2$	$\pi/2$
	0.10	0.90	0.2425	0.7138	π	0	1	$-\pi/2$	$\pi/2$

Note: For each set of corner frequencies, both the bandpass and bandstop filters have the same poles, which occur in complex conjugate pairs. The zeros are located exactly on the unit circle in the z-plane.

with various sets of lower and upper corner frequencies. Using Equations 6.4, 6.5, 6.48, and 6.50 through 6.54, Tables 6.3 and 6.4 show the computed design parameters of the second-order Butterworth lowpass and highpass (Table 6.3) and bandpass and bandstop (Table 6.4) tunable optical filters with variable bandwidth and fixed center frequency, $\delta_0 = 0$, characteristics.

6.3.3 Tuning Parameters of the Lowpass and Highpass Tunable Optical Filters

Figure 6.4 shows the characteristics of the tuning parameters versus the normalized bandwidth defined as $\omega_c T/\pi$, of the lowpass and highpass tunable optical filters with variable bandwidth, $0.1 \leq \omega_c T/\pi \leq 0.9$ and fixed center frequency $\delta_0 = 0$, characteristics. The values of these parameters are obtained from Table 6.3.

The gain curves in Figure 6.4a show that a large EDFA gain, which can be achieved by having two EDFAs in cascade, is required for a very small or

TABLE 6.3
Design Parameters of the Second-Order ($M=2$) Butterworth Lowpass and Highpass Tunable Optical Filters with Variable Bandwidth and Fixed Center Frequency (i.e., $\delta_0=0$) Characteristics

Filter	Parameters of $H_{ap,k}(z)$, $(k=1,2)$					Parameters of $H_{az,k}(z)$, $(k=1,2)$				
Type	$\omega_c T/\pi$	G	$a_1=a_2$	$\varphi_1=\varphi_2$	$\theta_1=\theta_2$	ϕ_1	ϕ_2	b_{1k}	b_{2k}	$\psi_1=\psi_2$
Lowpass	0.1	43.08	0.0551	2.6677	−0.2369	0.2480	0.6996	0.5230	0.5000	π
	0.2	7.507	0.2419	2.1132	−0.5142	0.5538	1.5030	0.5230	0.5000	π
	0.3	5.774	0.3844	1.8041	−0.6687	0.5656	2.1092	0.5230	0.5000	π
	0.4	5.887	0.4779	1.6150	−0.7633	0.3865	2.6667	0.5230	0.5000	π
	0.5	7.294	0.5112	1.5484	−0.7966	0.0224	3.1640	0.5230	0.5000	π
	0.6	11.15	0.4779	1.6150	−0.7633	5.8083	3.5281	0.5230	0.5000	π
	0.7	22.24	0.3844	1.8041	−0.6687	5.2508	3.7072	0.5230	0.5000	π
	0.8	71.05	0.2419	2.1132	−0.5142	4.6446	3.6954	0.5230	0.5000	π
	0.9	1716	0.0551	2.6677	−0.2369	3.8412	3.3896	0.5230	0.5000	π
Highpass	0.1	1716	0.0551	2.6677	−0.2369	0.2480	0.6996	0.5230	0.5000	0
	0.2	71.05	0.2419	2.1132	−0.5142	0.5538	1.5030	0.5230	0.5000	0
	0.3	22.24	0.3844	1.8041	−0.6687	0.5656	2.1092	0.5230	0.5000	0
	0.40	11.15	0.4779	1.6150	−0.7633	0.3865	2.6667	0.5230	0.5000	0
	0.5	7.294	0.5112	1.5484	−0.7966	0.0224	3.1640	0.5230	0.5000	0
	0.6	5.887	0.4779	1.6150	−0.7633	5.8083	3.5281	0.5230	0.5000	0
	0.7	5.774	0.3844	1.8041	−0.6687	5.2508	3.7072	0.5230	0.5000	0
	0.8	7.507	0.2419	2.1132	−0.5142	4.6446	3.6954	0.5230	0.5000	0
	0.9	43.08	0.0551	2.6677	−0.2369	3.8412	3.3896	0.5230	0.5000	0

Note: The values of these parameters are obtained from Equations 6.12, 6.13, 6.56, and 6.58 through 6.62 and Table 6.1. For each bandwidth $\omega_c T/\pi$, both the lowpass and highpass filters have the same poles and hence the same parameters of the FOAPOFs, i.e., the same coupling coefficients (i.e., a_1 and a_2) and the same phase shifts (i.e., φ_1, φ_2, ϕ_1, and ϕ_2). For each bandwidth $\omega_c T/\pi$, there is a phase difference of π between the FOAZOFs of the lowpass and highpass filters.

large filter bandwidth, e.g., $G=1716$ or 65 dB is required at $\omega_c T/\pi=0.1$ for the highpass filter and at $\omega_c T/\pi=0.9$ for the lowpass filter. These curves are symmetrical about the mid-band frequency, which means the required gain at $\omega_c T/\pi$ for the lowpass filter is the same as that at $1-\omega_c T/\pi$ for the highpass filter. Figure 6.4b shows the intensity coupling coefficients of the FOAPOFs, $a_1=a_2$, and FOAZOFs, when $b_{1k}=0.523$ and $b_{2k}=0.5$, for both the lowpass and highpass filters. The required coupling coefficients $a_1=a_2$ can be obtained by varying the phase shifts $\varphi_1=\varphi_2$ of the TCs. Note that both the curves of $a_1=a_2$ and $\varphi_1=\varphi_2$ are symmetrical about the mid-band frequency. The lowpass and highpass filters require the phase shifts $\psi_1=\psi_2=\pi$ and $\psi_1=\psi_2=0$, respectively, of the FOAZOFs.

TABLE 6.4

Design Parameters of the Second-Order ($M = 2$) Butterworth Bandpass and Bandstop Tunable Optical Filters with Variable Bandwidth and Fixed Center Frequency (i.e., $\delta_0 = 0$) Characteristics

	Parameters of $H_{ap,k}(z)$, $(k = 1, 2)$						Parameters of $H_{az,k}(z)$, $(k = 1, 2)$					
Filter Type	$\omega_{c1}T/\pi$	$\omega_{c2}T/\pi$	G	$a_1 = a_2$	$\varphi_1 = \varphi_2$	$\theta_1 = \theta_2$	ϕ_1	ϕ_2	b_{1k}	b_{2k}	ψ_1	ψ_2
Bandpass	0.40	0.60	63.46	0.1577	2.3249	−0.4083	5.5290	2.3874	0.5230	0.5000	0	π
	0.35	0.65	20.49	0.3274	1.9232	−0.6092	5.9308	2.7892	0.5230	0.5000	0	π
	0.30	0.70	9.734	0.5304	1.5100	−0.8158	0.0608	3.2024	0.5230	0.5000	0	π
	0.25	0.75	3.254	1.0000	0	−1.5708	3.1416	3.1416	0.5230	0.5000	0	π
	0.20	0.80	13.40	0.5304	1.5100	−0.8158	4.7732	1.6316	0.5230	0.5000	0	π
	0.15	0.85	40.22	0.3274	1.9232	−0.6092	4.3600	1.2184	0.5230	0.5000	0	π
	0.10	0.90	197.5	0.1577	2.3249	−0.4083	3.9582	0.8166	0.5230	0.5000	0	π
Bandstop	0.40	0.60	197.5	0.1577	2.3249	−0.4083	5.5290	2.3874	0.5230	0.5000	$3\pi/2$	$\pi/2$
	0.35	0.65	40.22	0.3274	1.9232	−0.6092	5.9308	2.7892	0.5230	0.5000	$3\pi/2$	$\pi/2$
	0.30	0.70	13.40	0.5304	1.5100	−0.8158	0.0608	3.2024	0.5230	0.5000	$3\pi/2$	$\pi/2$
	0.25	0.75	3.254	1.0000	0	−1.5708	3.1416	3.1416	0.5230	0.5000	$3\pi/2$	$\pi/2$
	0.20	0.80	9.734	0.5304	1.5100	−0.8158	4.7732	1.6316	0.5230	0.5000	$3\pi/2$	$\pi/2$
	0.15	0.85	20.49	0.3274	1.9232	−0.6092	4.3600	1.2184	0.5230	0.5000	$3\pi/2$	$\pi/2$
	0.10	0.90	63.46	0.1577	2.3249	−0.4083	3.9582	0.8166	0.5230	0.5000		

Note: The values of these parameters are obtained from Equations 6.12, 6.13, 6.56, and 6.58 through 6.62 and Table 6.2. For each bandwidth $(\omega_{c2} - \omega_{c1})T/\pi$, both the bandpass and bandstop filters have the same poles and hence the same parameters of the FOAPOFs, i.e., the same coupling coefficients (i.e., a_1 and a_2) and the same phase shifts (i.e., φ_1, φ_2, ϕ_1, and ϕ_2).

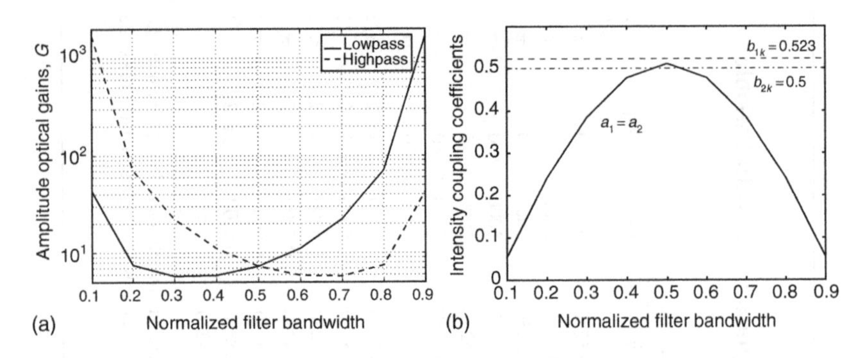

FIGURE 6.4 Characteristics of the tuning parameters versus the normalized bandwidth of the lowpass and highpass tunable optical filters with variable bandwidth and fixed center frequency, when $\delta_0=0$, characteristics, as obtained from Table 6.3. (a) EDFA amplitude gains, G and (b) intensity coupling coefficients of both the lowpass and highpass filters.

6.3.4 Tuning Parameters of Bandpass and Bandstop Tunable Optical Filters

Figure 6.5 shows the characteristics of the tuning parameters versus the normalized bandwidth defined as $(\omega_{c2}-\omega_{c1})T/\pi$ of the bandpass and bandstop tunable optical filters with variable bandwidth, $0.2 \le (\omega_{c2}-\omega_{c1})T/\pi \le 0.8$, and fixed center frequency, when $\delta_0=0$, characteristics. The values of these parameters are obtained from Table 6.4.

The gain curves in Figure 6.5a show that a large EDFA gain is required for a very small or large filter bandwidth (e.g., $G=197.5$ or 46 dB is required at $(\omega_{c2}-\omega_{c1})T/\pi=0.2$ for the bandstop filter and at $(\omega_{c2}-\omega_{c1})T/\pi=0.8$ for the bandpass filter). These curves are symmetrical about the mid-band frequency (i.e., the required gain at $(\omega_{c2}-\omega_{c1})T/\pi$ for the bandpass filter is the same as that at $1-(\omega_{c2}-\omega_{c1})T/\pi$ for the bandstop filter). Figure 6.5b shows the intensity coupling coefficients of the FOAPOFs (i.e., $a_1=a_2$) and FOAZOFs (i.e., $b_{1k}=0.523$, and $b_{2k}=0.5$) for both the bandpass and bandstop filters. The required coupling coefficients $a_1=a_2$ can be obtained by varying the phase shifts $\varphi_1=\varphi_2$ of the TCs (see the dotted–dotted curve in Figure 6.5c). Note that both the curves of $a_1=a_2$ and $\varphi_1=\varphi_2$ are symmetrical about the mid-band frequency (i.e., the required values at $(\omega_{c2}-\omega_{c1})T/\pi$ are the same as those at $1-(\omega_{c2}-\omega_{c1})T/\pi$). The dotted–dashed curves in Figure 6.5c show the required phase shifts ϕ_1 and ϕ_2 of the FOAPOFs for both the bandpass and bandstop filters. The bandpass and bandstop filters require the phase shifts ($\psi_1=0$, $\psi_2=\pi$) (see the solid curves) and ($\psi_1=3\pi/2$, $\psi_2=\pi/2$) (see the dashed–dashed curves), respectively, of the FOAZOFs.

6.3.5 Summary of Tuning Parameters of Tunable Optical Filters

The above discussion of the tunable optical filters with lowpass, highpass, bandpass, and bandstop characteristics is summarized below.

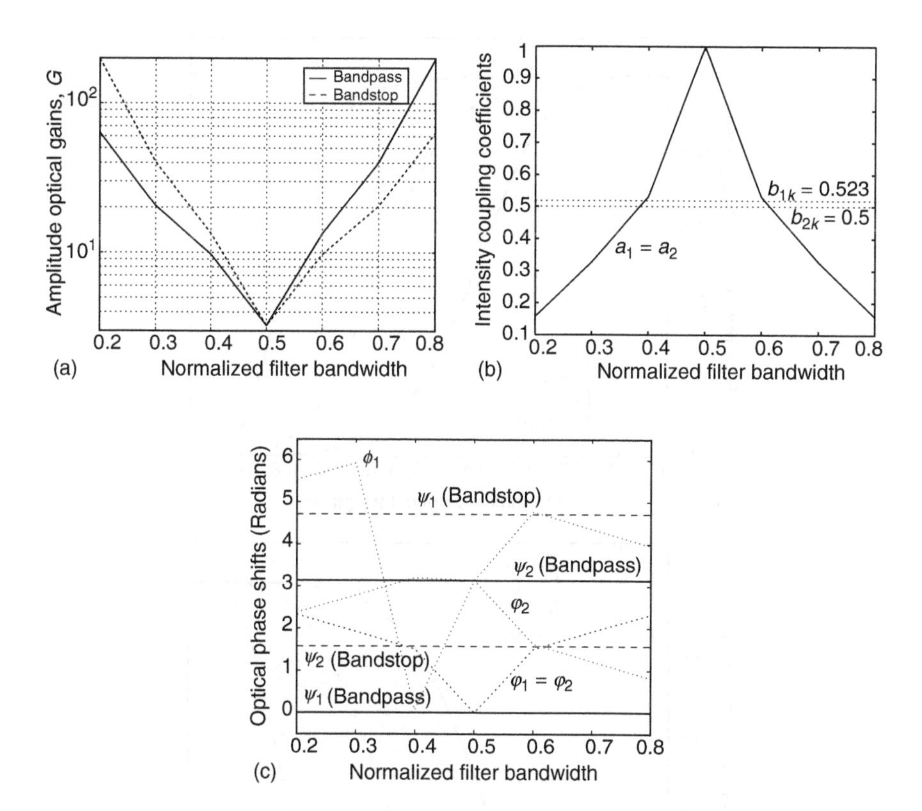

FIGURE 6.5 Characteristics of the tuning parameters versus the normalized bandwidth of the bandpass and bandstop tunable optical filters with variable bandwidth and fixed center frequency (i.e., $\delta_0 = 0$) characteristics, as obtained from Table 6.4. (a) EDFA amplitude gains, G, (b) intensity coupling coefficients of both the bandpass and bandstop filters, and (c) optical phase shifts, where ϕ_1, ϕ_2 and $\varphi_1 = \varphi_2$ are the phase shifts of both the bandpass and bandstop filters.

- For a particular filter bandwidth, the phase shifts (ψ_1, ψ_2) of the FOAZOFs determine the complementary characteristics of the filter. That is, a lowpass filter can be transformed into a highpass filter or vice versa, and a bandpass filter can be transformed into a bandstop filter or vice versa.
- For a particular set of phase shifts (ψ_1, ψ_2) of the FOAZOFs, the tuning parameters (i.e., the phase shifts $\varphi_1 = \varphi_2$, ϕ_1, and ϕ_2) of the FOAPOFs determine the bandwidth characteristics of a particular filter type (i.e., lowpass, highpass, bandpass, or bandstop).

6.3.6 Magnitude Responses of Tunable Optical Filters with Variable Bandwidth and Fixed Center Frequency Characteristics

Figures 6.6 and 6.7 show the squared magnitude responses of the lowpass (Figure 6.6a) and highpass (Figure 6.6b) and bandpass (Figure 6.7a) and bandstop (Figure 6.7b) tunable optical filters with variable bandwidth and fixed center

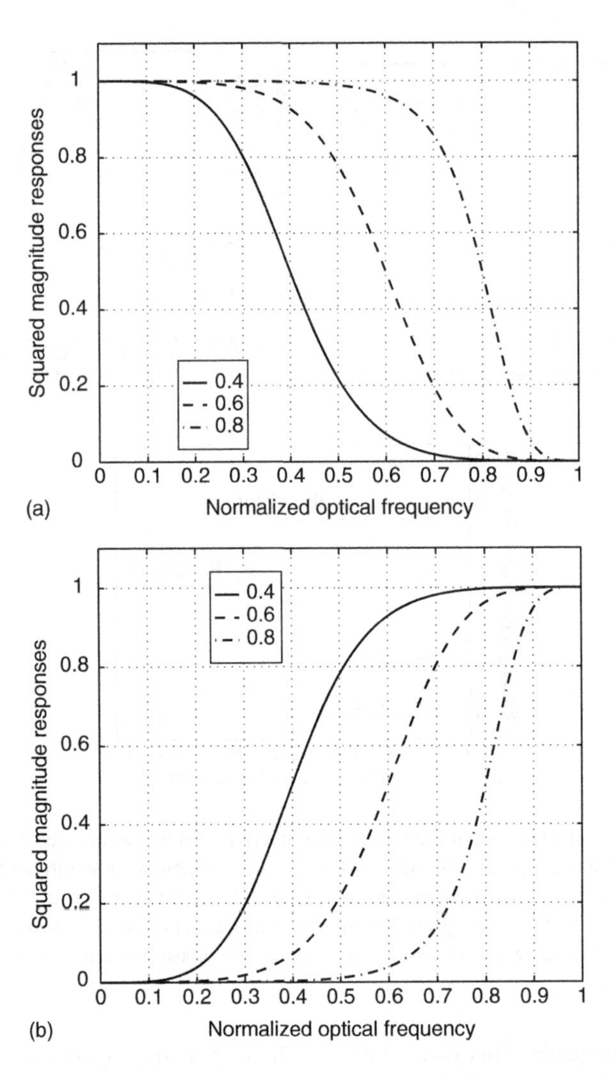

FIGURE 6.6 Squared magnitude responses of the lowpass and highpass tunable optical filters with variable bandwidth and fixed center frequency (i.e., $\delta_0 = 0$) characteristics. (a) Lowpass and (b) highpass. The numbers inside the legend box represent the normalized 3 dB cutoff frequencies (i.e., $\omega_c T/\pi$), which also correspond to the normalized filter bandwidths. Note that the normalized center frequency is designed at $\omega T/\pi = 0$. The filter parameters are given in Table 6.3 and shown in Figure 6.4.

frequency (i.e., $\delta_0 = 0$) characteristics. Figure 6.6 shows that the bandwidth of each filter type (lowpass or highpass) can be varied from $\omega_c T = 0.4$ to 0.8 by varying the phase shifts of the FOAPOFs (i.e., φ, ϕ_1 and ϕ_2) and by keeping the phase shifts of the FOAZOFs unchanged (i.e., $\varphi_1 = \varphi_2 = \pi$ for lowpass and $\varphi_1 = \varphi_2 = 0$ for highpass), see Figure 6.4. Similarly, Figure 6.7 shows that the bandwidth of each filter type (bandpass or bandstop) can be varied from $(\omega_{c1} - \omega_{c2})T/\pi = 0.2$ to 0.6 by varying the phase shifts of the FOAPOFs (i.e., $\varphi_1 = \varphi_2$, ϕ_1 and ϕ_2) and by keeping

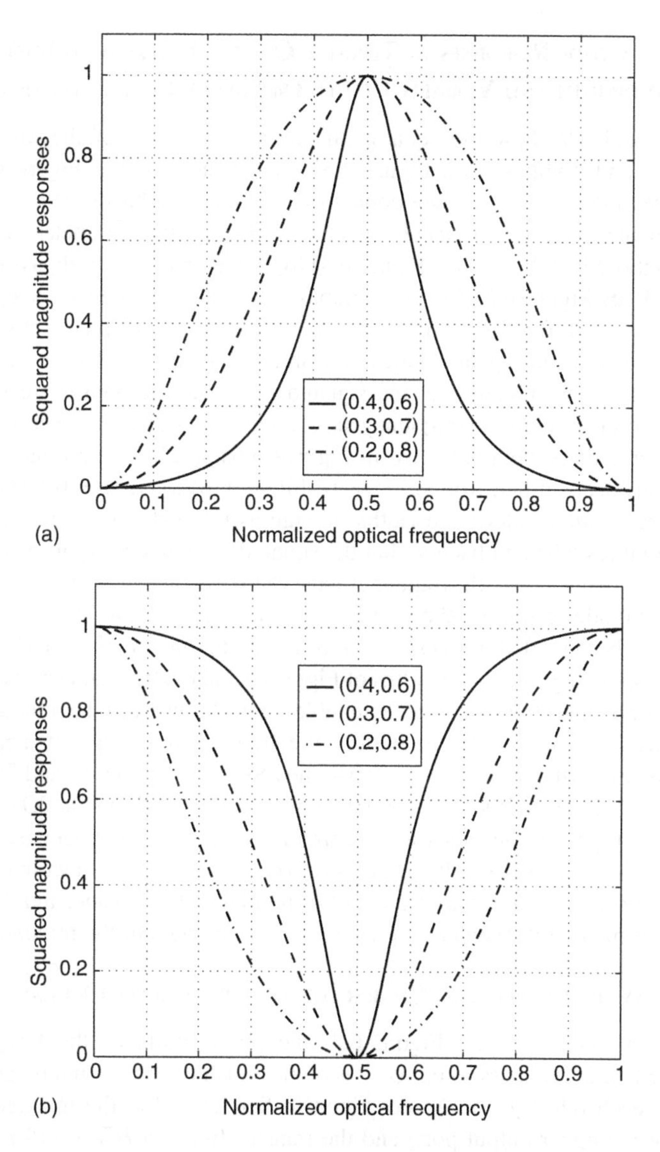

FIGURE 6.7 Squared magnitude responses of the bandpass and bandstop tunable optical filters with variable bandwidth and fixed center frequency (i.e., $\delta_0 = 0$) characteristics. (a) Bandpass and (b) bandstop. The numbers inside the legend box represent the normalized 3 dB lower and upper corner frequencies ($\omega_{c1}T/\pi, \omega_{c2}T/\pi$). The normalized filter bandwidth is given by ($\omega_{c2} - \omega_{c1}T/\pi$). Note that the normalized center frequency is designed at $\omega T/\pi = 0.5$. The filter parameters are given in Table 6.4 and shown in Figure 6.5.

the phase shifts of the FOAZOFs unchanged (i.e., $\varphi_1 = 0$ and $\varphi_2 = \pi$ for bandpass and $\varphi_1 = 3\pi/2$ and $\varphi_2 = \pi/2$ for bandstop), see Figure 6.5. Note that the normalized center frequencies are designed at $\omega T/\pi = 0$ for the lowpass and highpass filters and at $\omega T/\pi = 0.5$ for the bandpass and bandstop filters.

6.3.7 Magnitude Responses of Tunable Optical Filters with Fixed Bandwidth and Variable Center Frequency Characteristics

Figures 6.8 and 6.9 show the squared magnitude responses of the lowpass (see Figure 6.8a) and highpass (see Figure 6.8b) and bandpass (see Figure 6.9a) and bandstop (see Figure 6.9b) tunable optical filters with fixed bandwidth and variable center frequency (i.e., $\delta_0 = 0.1\pi$) characteristics. The design parameters are exactly the same as those in Table 6.3 (or Figure 6.4) for the lowpass and highpass filters and in Table 6.4 (or Figure 6.5) for the bandpass and bandstop filters, except that an additional phase shift of $\delta_0 = 0.1\pi$ has been added to the phase shifts of the FOAPOFs (i.e., ϕ_1, and ϕ_2) and to the phase shifts of the FOAZOFs (i.e., ψ_1 and ψ_2).

Figures 6.8a and b show that the squared magnitude responses are shifted by $\omega T/\pi = 0.1$ to the right of the frequency axis when compared with the corresponding squared magnitude responses shown in Figures 6.6a and b. The normalized center frequency has been shifted from $\omega T/\pi = 0$ (Figure 6.6) to $\omega T/\pi = 0.1$ (Figure 6.8) but the corresponding filter bandwidths of Figures 6.6 and 6.8 remain unchanged. Similarly, Figures 6.9a and b show that the squared magnitude responses are shifted by $\omega T/\pi = 0.1$ to the right of the frequency axis when compared with the corresponding squared magnitude responses shown in Figures 6.7a and b. The normalized center frequency has been shifted from $\omega T/\pi = 0.5$ (Figure 6.7) to $\omega T/\pi = 0.6$ (Figure 6.9), but the corresponding filter bandwidths of Figures 6.7 and 6.9 remain unchanged.

Thus, the center frequency of a tunable optical filter can be tuned, without affecting the filter bandwidth, to within one free spectral range by applying an additional phase shift of δ_0 ($0 < \delta_0 < 2\pi$) to the PSs of the FOAPOFs and FOAZOFs. From the point of view of the pole–zero pattern, the effect of δ_0 on the FOAPOFs and FOAZOFs is to rotate the poles and zeros in the angular anticlockwise direction relative to the z-plane. As a result, the pole–zero pattern of the resulting tunable optical filter rotates by some angular movement of δ_0 relative to the z-plane, and this has the effect of shifting the filter center frequency by δ_0 to the right of the frequency axis.

6.3.8 Summary of Filtering Characteristics of Tunable Optical Filters

As shown in Table 6.3 and Figure 6.4, the phase shifts of the lowpass (i.e., $\psi_1 = \psi_2 = \pi$) and highpass (i.e., $\psi_1 = \psi_2 = 0$) tunable optical filters are out of phase with each other by π. As described in Section 6.3.2, the transfer function $H_{az,k}(z)$ (for the upper output port) and the transfer function $H^*_{az,k}(z)$ (for the lower output port) are out of phase with each other by π. Thus, if the upper output port of the tunable optical filter has a lowpass magnitude response, then its lower output port has a highpass magnitude response. As a result, the tunable optical filter can be used as a channel adding/dropping filter that passes certain wavelength channels in one output port, while leaving the other channels undisturbed in the other output port. As shown in Table 6.4 and Figure 6.5, the phase shifts ψ_1 and ψ_2 of a particular filter type (bandpass or bandstop[†]) are out of phase with each other by π. As a result, both

[†] For the bandstop filter, it has been found from MATLAB that the phase shifts ψ_1 and ψ_2 are only out of phase with each other by π if the normalized center frequency is at the mid-band frequency (i.e., $\omega T/\pi = 0.5$), which is the case being considered here.

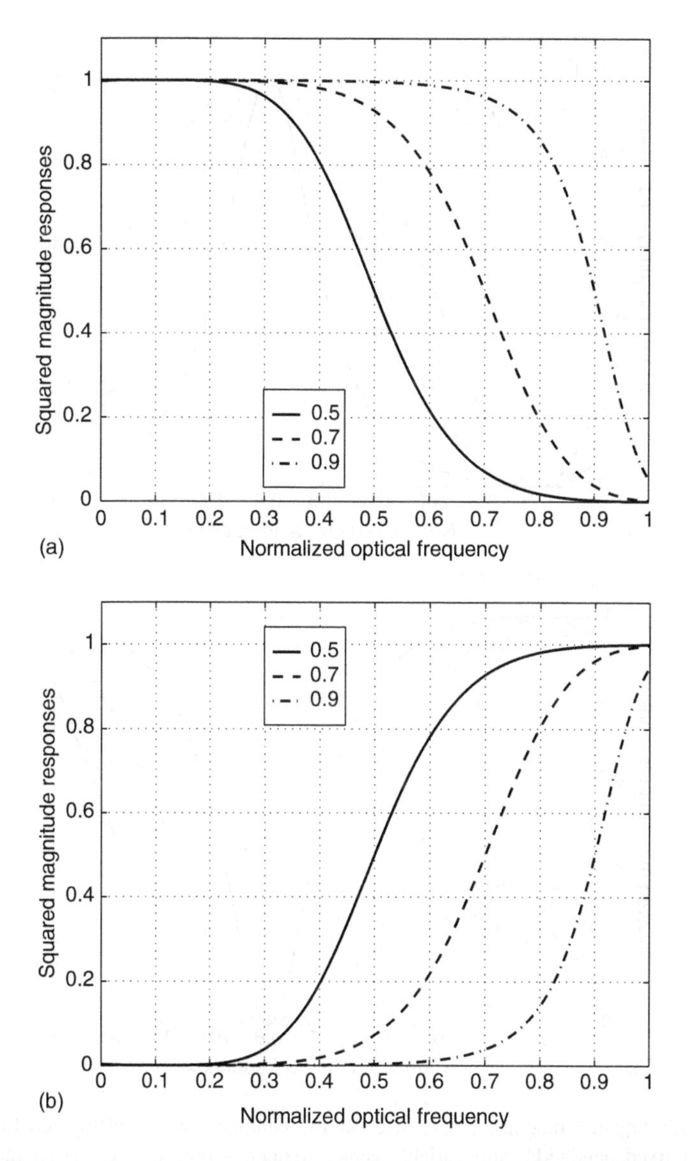

FIGURE 6.8 Squared magnitude responses of the lowpass and highpass tunable optical filters with fixed bandwidth and variable center frequency (i.e., $\delta_0 = 0.1\pi$) characteristics. (a) Lowpass and (b) highpass. The numbers inside the legend box represent the new normalized 3 dB cutoff frequencies (i.e., $\omega_c' T/\pi = \omega_c T/\pi + \delta_0/\pi$). The normalized filter bandwidths are still the same as those in Figure 6.6 (i.e., $\omega_c T/\pi = \omega_c' T/\pi - \delta_0/\pi$). Note that the new normalized center frequency is at $\omega T/\pi = 0.1$.

the output ports of the tunable optical filter have the same filtering characteristics (bandpass or bandstop). In summary, the filtering characteristics of the tunable optical filter are shown in Table 6.5.

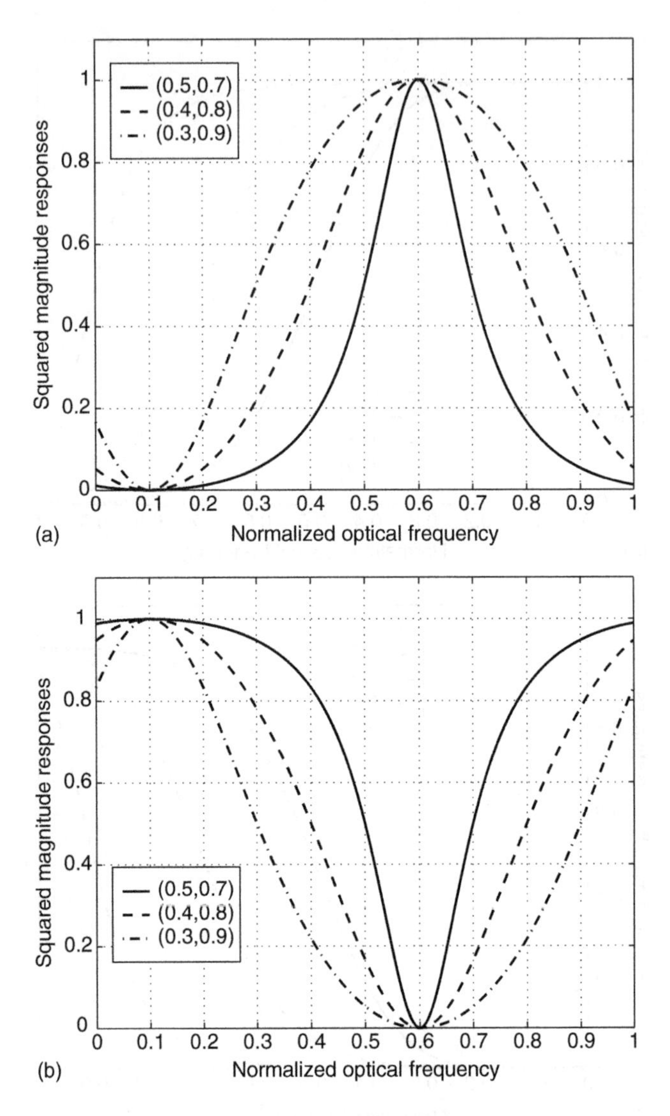

FIGURE 6.9 Squared magnitude responses of the bandpass and bandstop tunable optical filters with fixed bandwidth and variable center frequency (i.e., $\delta_0 = 0.1\pi$) characteristics. (a) Bandpass and (b) bandstop. The numbers inside the legend box represent the new normalized 3 dB lower and upper corner frequencies $\omega'_{c1}T/\pi, \omega'_{c2}T/\pi$, where $\omega'_{c1}T/\pi = \omega_{c1}T/\pi + \delta_0/\pi$ and $\omega'_{c2}T/\pi = \omega_{c2}T/\pi + \delta_0/\pi$. The normalized filter bandwidths are still the same as those in Figure 6.7, i.e., $(\omega_{c2} - \omega_{c1})T/\pi = (\omega'_{c2} - \omega'_{c1})T/\pi)$. Note that the new normalized center frequency is at $\omega T/\pi = 0.6$.

6.3.9 Discussions

The largest value of the filter pole, which is limited by the loss of the waveguide loop (see Equation 6.48), is restricted to be $|\hat{p}_k| \leq 0.85$ (see Equation 6.52). As a result, a tunable optical filter cannot be designed to have a very narrow or broad bandwidth,

TABLE 6.5
Filtering Characteristics at the Output Ports of the Second-Order Butterworth Tunable Optical Filter

Output Ports	Filtering Characteristics			
Upper output port (output 1)	Lowpass	Highpass	Bandpass	Bandstop
Lower output port (output 2)	Highpass	Lowpass	Bandpass	Bandstop

which requires $|\hat{p}_k| > 0.85$. Thus, the allowable normalized bandwidths of the tunable optical filter are in the range of $0.1 \leq \omega_c T/\pi \leq 0.9$ for the lowpass and highpass filters and $0.2 \leq (\omega_{c2} - \omega_{c1})T/\pi \leq 0.8$ for the bandpass and bandstop filters. These ranges of filter bandwidths are adequate for many filtering applications.

The normalized filter bandwidth can be extended to its full range (i.e., between 0 and 1) by incorporating an erbium-doped waveguide amplifier (EDWA) into the waveguide loop of the FOAPOF to compensate for the loop loss. However, this has the drawback of increasing the cost as well as the complexity of the filter structure, where the latter may degrade the filter performance unless undesirable effects associated with the EDWA are minimized.

Note that the presented design method is applicable to higher-order filters. The roll-off steepness of the magnitude responses of the tunable optical filter can be increased by increasing the filter order and hence the number of the FOAPOFs and FOAZOFs. However, the lowest filter order should be used to meet a prescribed set of filter specifications to keep the cost and complexity of the filter structure to a minimum.

The filter design technique is presented in a general manner and thus applicable in the design of other types of tunable optical filters such as the Chebyshev I and II and elliptic filters, whose properties have been summarized in the Appendix. Obviously, the choice of a particular filter type would depend on the specific application.

6.4 AN EXPERIMENTAL FIRST-ORDER BUTTERWORTH LOWPASS AND HIGHPASS TUNABLE FILTERS

This section describes an experimental development of the first-order Butterworth lowpass and highpass tunable fiber-optic filters. Although optical, waveguide components are required in the proposed filter design technique; optical fiber components were used in the experiment to demonstrate the effectiveness of the proposed method.

Figure 6.10 shows the experimental setup of the asymmetrical Mach–Zehnder interferometer used to implement the first-order Butterworth lowpass and highpass tunable fiber-optic filters. A 25 mW diode-pumped solid-state Nd:YAG laser with a linewidth of 5 kHz at an operating wavelength of 1319 nm was used to excite the interferometer. A built-in isolator (with a high isolation of >60 dB) was also incorporated inside the laser module. The lengths of the upper and lower fiber

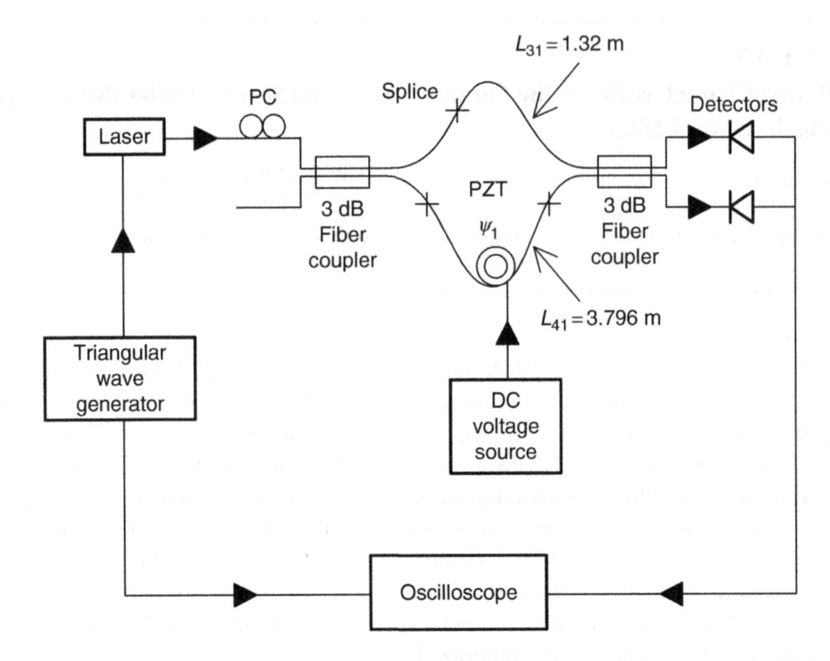

FIGURE 6.10 Experimental setup of the asymmetrical Mach–Zehnder interferometer used to implement the first-order Butterworth lowpass and highpass tunable fiber-optic filters. PC: polarization controller; PZT: piezoelectric phase modulator.

arms are $L_{31} = 1.32$ m and $L_{41} = 3.796$ m, respectively. The requirement for coherent operation of the interferometer is thus satisfied because the coherence length (about 60 km) of the optical source is much greater than the differential length of the interferometer ($L = L_{41} - L_{31} = 2.476$ m). The corresponding free spectral range (or sampling frequency) of the interferometer is 81 MHz. The polarization controller (PC), which was made of two rotatable fiber loops, was used to counter any birefringence induced in the fiber arms. The piezoelectric phase modulator (PZT) with a phase shift of ψ_1, which was constructed by winding several turns of fiber around a piezoelectric cylinder, expanded radially to modulate the phase when a DC voltage was applied to it.

In theory, for $\psi_1 = 0$, the response of the upper (lower) output port corresponds to the response of the first-order Butterworth highpass (lowpass) fiber-optic filter with a normalized 3 dB cutoff frequency of $\omega T/\pi = 0.5$. While, for $\psi_1 = \pi$, the upper (lower) output port has a lowpass (highpass) response.

To observe the filter output responses, the frequency of the laser was externally modulated by a low-frequency triangular waveform. When the PZT phase modulator was not applied (i.e., $\psi_1 = 0$), the measured output responses of the highpass and lowpass tunable fiber-optic filters are shown in Figure 6.11a and b, respectively. The triangular waveforms are shown on the upper traces of these figures. When the PZT phase modulator was applied (i.e., $\psi_1 = \pi$), the measured output responses of the lowpass (instead of highpass as in Figure 6.11a) and highpass (instead of lowpass as in

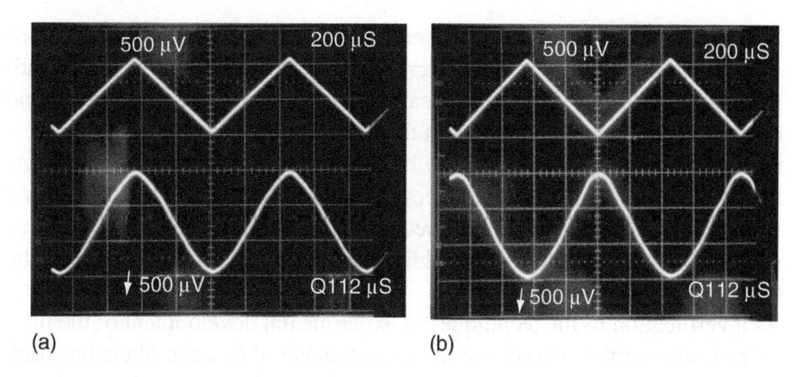

(a) (b)

FIGURE 6.11 Measured output responses of the first-order Butterworth highpass and lowpass tunable fiber-optic filters ($\psi_1 = 0$): (a) Upper port: highpass and (b) lower port: lowpass.

Figure 6.11b) tunable fiber-optic filters are shown in Figures 6.12a and b, respectively. Thus, a lowpass filter can be transformed into a highpass filter or vice versa by applying a phase shift of $\psi_1 = \pi$ to the PZT phase modulator. The interferometer was shielded in a container when taking measurements to minimize its instability due to environmental fluctuations (such as temperature and pressure changes) and frequency fluctuations of the laser. However, as described in Chapter 2, the filter stability can be greatly improved by using optical waveguide components.

6.5 REMARKS

1. A digital filter design technique has been employed to systematically design tunable optical filters with variable bandwidth and center frequency characteristics as well as lowpass, highpass, bandpass, and bandstop characteristics. An Mth-order tunable optical filter, which has been designed using integrated-optic structures, consists of a cascade of M FOAPOFs with a cascade of M FOAZOFs.

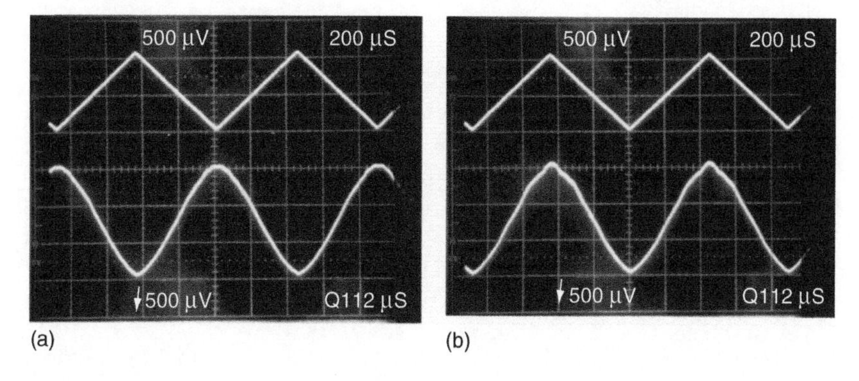

(a) (b)

FIGURE 6.12 Measured output responses of the first-order Butterworth lowpass and highpass tunable fiber-optic filters ($\psi_1 = \pi$). (a) Upper port: lowpass and (b) lower port: highpass.

2. The effectiveness of the optical filter design method has been demonstrated with the design of the second-order Butterworth lowpass, highpass, bandpass, and bandstop tunable optical filters with variable bandwidth and center frequency characteristics. In this design, for a fixed center frequency, the filter bandwidth can be varied by varying the parameters of the FOAPOFs and by keeping the parameters of the FOAZOFs unchanged; and for a fixed bandwidth, the filter center frequency can be varied, to within one free spectral range, by adding an additional phase shift to the PSs of the FOAPOFs and FOAZOFs.
3. As a verification of the technique, an experimental development of the first-order Butterworth lowpass and highpass tunable fiber-optic filters has been carried out. In addition to the Butterworth filters, the proposed filter design technique is applicable to the design of other types of tunable optical filters such as the Chebyshev I and II and elliptic filters, depending on the specific application.

Appendix: Fundamental Characteristics of Recursive Digital Filters

It is essential that the fundamental properties of various design techniques and general characteristics of recursive (or IIR) digital filters are summarized for readers who are unfamiliar with these filters. Further details can be found in Refs. [7–9]. This basic knowledge is required for the design of tunable optical filters as described in this chapter.

The common approach to the design of recursive digital filters involves the transformation of a recursive analog* (or continuous-time) filter into a recursive digital filter for a given set of prescribed specifications. This is because of the availability of well-developed techniques for analog filters, which are often described by simple closed-form design formulas. In such transformations, the essential properties of the frequency response of the analog filter are preserved in the frequency response of the resulting digital filter. The two well-known techniques used for converting Butterworth, Chebyshev I and II, and elliptic analog filters to their corresponding digital filters are the impulse invariance and bilinear transformation methods.

In the impulse invariance method, the impulse response of a digital filter is determined by sampling the impulse response of an analog filter. This technique requires the analog filter to be band-limited to avoid the aliasing (or interference) effect, and thus is only effective for lowpass and bandpass analog filters. If it is to be used for highpass and bandstop analog filters, then additional band-limiting is required on these filters to avoid severe aliasing distortion.

The bilinear transformation method involves an algebraic transformation between the variables s and z that maps the entire imaginary axis in the s-plane to one revolution of the unit circle in the z-plane. Thus, a stable analog filter (with poles in the left half-side of the s-plane) can be transformed into a stable digital filter (with poles inside the unit circle in the z-plane). As a result, unlike the impulse invariance method, this method does not suffer from the effect of aliasing distortion. However, it suffers from the effect of nonlinear compression of the frequency axis, and thus, is only useful if this undesirable effect can be tolerated or compensated for.

* Analog filters are commonly designed using standard approximation methods, namely, the Taylor series approximations and the Chebyshev approximations in various combinations. Specifically, these approximation methods are used to approximate the desired frequency responses of four different types of analog filters: the Butterworth, Chebyshev I and II, and elliptic filters.

An alternative approach, which is valid for both the impulse invariance and bilinear transformation methods, is to design a digital prototype lowpass filter and then perform a frequency transformation on it to obtain the desired lowpass, high-pass, bandpass, and bandstop digital filters. However, the use of the frequency transformation technique in the design of a digital filter is not so straightforward because the analog prototype lowpass filter is generally not known to the designer. Thus, it is necessary to "find" an analog prototype lowpass filter such that, after transformation, the resulting digital filter would meet a given set of specifications. Rapid advances in the field of digital signal processing in recent years led to digital filter design techniques with standard functions in MATLAB*. The design of the digital filters and hence tuneable optical filters is described in this chapter.

The four common types of recursive digital filters are the Butterworth, Chebyshev I and II, and elliptic filters, and their properties are summarized as follows:

- Butterworth digital filters: They are characterized by a magnitude response that is maximally flat in the passband and monotonic overall. They sacrifice roll-off steepness for monotonicity in the passband and stopband. If the Butterworth filter smoothness is not required, a Chebyshev or an elliptic filter can generally provide steeper roll-off characteristics with a lower filter order.
- Chebyshev I and II digital filters: Chebyshev I filters are equiripple in the passband and monotonic in the stopband, while Chebyshev II filters are monotonic in the passband and equiripple in the stopband. Chebyshev I filters roll off faster than Chebyshev II filters but at the expense of passband ripple. Chebyshev II filters have stopbands that do not approach zero like Chebyshev I filters but are free of passband ripple. Both the Chebyshev I and II filters have the same filter order for a given set of filter specifications.
- Elliptic digital filters: They are equiripple in both passband and stopband. They offer steeper roll-off characteristics than the Butterworth and Chebyshev filters but suffer from passband and stopband ripples. In general, elliptic filters, although the most complicated for computation, would satisfy a given set of filter specifications with the lowest filter order.

These types of digital filters have zeros located on the unit circle in the z-plane (i.e., $|z| = 1$), which greatly simplify the design of tunable optical filters.

For analytical clarity, the variables with a cap (e.g., $\hat{H}(z)$) are associated with digital filters while the corresponding variables without a cap (e.g., $H(z)$) are associated with optical filters. The transfer function of the Mth-order recursive digital filter can be expressed in a rational form as

* MATLAB, which is a registered trademark of the MathWorks, employs the frequency transformation technique together with the bilinear transformation method with frequency prewarping in the design of the Butterworth, Chebyshev I and II, and elliptic digital filters with lowpass, highpass, bandpass and bandstop characteristics.

$$\hat{H}(z) = \hat{A}\prod_{k=1}^{M}\frac{(z-\hat{z}_k)}{(z-\hat{p}_k)}$$
$$= \hat{A}\frac{(z-\hat{z}_1)(z-\hat{z}_2)\ldots(z-\hat{z}_M)}{(z-\hat{p}_1)(z-\hat{p}_2)\ldots(z-\hat{p}_M)} \quad \text{(A.6.1)}$$

where $\hat{A}$ is a constant and z is the z transform parameter [10,11]. Furthermore, $\hat{p}_k$ and $\hat{z}_k$ are the kth pole and zero in the z-plane, which can be expressed in the phasor forms as

$$\hat{p}_k = |\hat{p}_k|\exp(j\arg(\hat{p}_k)) \quad 0 \le |\hat{p}_k| < 1 \quad \text{(A.6.2)}$$

$$\hat{z}_k = |\hat{z}_k|\exp(j\arg(\hat{z}_k)) \quad |\hat{z}_k| = 1 \quad \text{(A.6.3)}$$

where arg denotes the argument. Note that the system stability requires the poles to be located inside the unit circle in the z-plane as described by the condition given in Equation 6.2.

Let the transfer function of the kth-stage first-order all-pole digital filter be defined as

$$\hat{H}_{\text{ap},k}(z) = \tfrac{1}{(z-\hat{p}_k)} \quad \text{(A.6.4)}$$

and the transfer function of the kth-stage first-order all-zero digital filter be defined as

$$\hat{H}_{\text{az},k}(z) = (z-\hat{z}_k) \quad \text{(A.6.5)}$$

where the subscripts *ap* and *az* denote all-pole and all-zero, respectively. The transfer function of the Mth-order all-pole digital filter, which is the transfer function of the cascade of M first-order all-pole digital filters, is given by

$$\hat{H}_{\text{ap}}(z) = \prod_{k=1}^{M}\hat{H}_{\text{ap},k}(z) \quad \text{(A.6.6)}$$

The transfer function of the Mth-order all-zero digital filter, which is the transfer function of the cascade of M first-order all-zero digital filters, is given by

$$\hat{H}_{\text{az}}(z) = \prod_{k=1}^{M}\hat{H}_{\text{az},k}(z) \quad \text{(A.6.7)}$$

The transfer function of the Mth-order recursive digital filter, as given in Equation A.6.1, can be written alternatively as

$$\hat{H}(z) = \hat{A}\cdot\hat{H}_{\text{ap}}(z)\cdot\hat{H}_{\text{az}}(z) \quad \text{(A.6.8)}$$

REFERENCES

1. S. Suzuki, K. Oda, and Y. Hibino, Integrated-optic double-ring resonators with a wide free spectral range of 100 GHz, *J. Lightwave Technol.*, 13, 1766–1771, 1995.
2. E. Pawlowski, K. Takiguchi, M. Okuno, K. Sasayama, A. Himeno, K. Okamato, and Y. Ohmori, Variable bandwidth and tunable centre frequency filter using transversal-form programmable optical filter, *Electron. Lett.*, 32, 113–114, 1996.
3. I.P. Kaminow, P.P. Iannone, J. Stone, and L.W. Stulz, FDMA–FSK star network with a tunable optical filter demultiplexer, *J. Lightwave Technol.*, 6, 1406–1414, 1988.
4. A.A.M. Saleh and J. Stone, Two-stage Fabry–Perot filters as demultiplexers in optical FDMA LAN's, *J. Lightwave Technol.*, 7, 323–330, 1989.
5. M. Kuznetsov, Cascaded coupler Mach–Zehnder channel dropping filters for wavelength-division-multiplexed optical systems, *J. Lightwave Technol.*, 12, 226–230, 1994.
6. N.Q. Ngo and L.N. Binh, Novel realization of monotonic Butterworth-type lowpass, highpass and bandpass optical filters using phase-modulated fiber-optic interferometers and ring resonators, *J. Lightwave Technol.*, 12, 827–841, 1994.
7. A.V. Oppenheim and R.W. Schafer, *Discrete-Time Signal Processing*, Englewood Cliffs, NJ: Prentice-Hall, 1989.
8. T.W. Parks and C.S. Burrus, *Digital Filter Design*, New York: John Wiley & Sons, 1987.
9. R.E. Bogner and A.G. Constantinides, *Introduction to Digital Filtering*, London: John Wiley & Sons, 1975.
10. N.Q. Ngo, X. Dai, and L.N. Binh, Realization of first-order monotonic Butterworth-type lowpass and highpass optical filters: Experimental verification, *Microwave Opt. Technol. Lett.*, 8, 306–309, 1995.
11. R.E. Bogner and A.G. Constantinides, *Introduction to Digital Filtering*, London: John Wiley & Sons, 1975.

Index

E

F

G

H

I

J

L

M

N

R

S

T

U

W